T0399376

REGENERATED ORGANS

REGENERATED ORGANS

Future Perspectives

Edited by

CHANDRA P. SHARMA FBSE

Department of Pharmaceutical Biotechnology, Manipal College of Pharmaceutical Sciences,
Manipal University, Manipal, Karnataka
College of Biomedical Engineering & Applied Sciences, Purbanchal University, Kathmandu, Nepal
Biomedical Technology Wing, Sree Chitra Tirunal Institute for Medical Sciences & Technology (SCTIMST),
Thiruvananthapuram, India

ELSEVIER

ACADEMIC PRESS

An imprint of Elsevier

Academic Press is an imprint of Elsevier
125 London Wall, London EC2Y 5AS, United Kingdom
525 B Street, Suite 1650, San Diego, CA 92101, United States
50 Hampshire Street, 5th Floor, Cambridge, MA 02139, United States
The Boulevard, Langford Lane, Kidlington, Oxford OX5 1GB, United Kingdom

Notices

Knowledge and best practice in this field are constantly changing. As new research and experience broaden our understanding, changes in research methods, professional practices, or medical treatment may become necessary.

Practitioners and researchers must always rely on their own experience and knowledge in evaluating and using any information, methods, compounds, or experiments described herein. In using such information or methods they should be mindful of their own safety and the safety of others, including parties for whom they have a professional responsibility.

To the fullest extent of the law, neither the Publisher nor the authors, contributors, or editors, assume any liability for any injury and/or damage to persons or property as a matter of products liability, negligence or otherwise, or from any use or operation of any methods, products, instructions, or ideas contained in the material herein.

British Library Cataloguing-in-Publication Data
A catalogue record for this book is available from the British Library

Library of Congress Cataloging-in-Publication Data
A catalog record for this book is available from the Library of Congress

ISBN: 978-0-12-821085-7

For Information on all Academic Press publications
visit our website at https://www.elsevier.com/books-and-journals

Publisher: Stacy Masucci
Acquisitions Editor: Elizabeth Brown
Editorial Project Manager: Billie Jean Fernandez
Production Project Manager: Selvaraj Raviraj
Cover Designer: Mark Rogers

Typeset by MPS Limited, Chennai, India

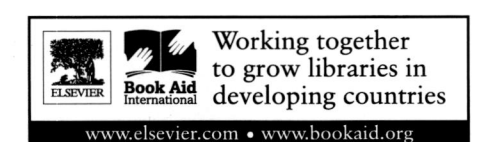

Working together
to grow libraries in
developing countries

www.elsevier.com • www.bookaid.org

Contents

2

Cardiovascular System

3

Musculoskeletal Regeneration of Tissues

10. Regenerative technologies for oral structures

Prachi Hanwatkar and Ajay Kashi

11. State-of-the-art strategies and future interventions in bone and cartilage repair for personalized regenerative therapy

Yogendra Pratap Singh, Joseph Christakiran Moses, Ashutosh Bandyopadhyay, Bibrita Bhar, Bhaskar Birru, Nandana Bhardwaj and Biman B. Mandal

12. Muscle tissue engineering — A materials perspective

John P. Bradford, Gerardo Hernandez-Moreno and Vinoy Thomas

4
Regenerative Neuroscience

13. Recent developments and new potentials for neuroregeneration

Sreekanth Sreekumaran, Anitha Radhakrishnan and Sanju P. Joy

5
Respiratory Research

14. Lung disease and repair — Is regeneration the answer?

S.S. Pradeep Kumar and A. Maya Nandkumar

6

Key Enabling Technologies for Regenerative Medicine Future Outlook and Conclusions

List of contributors

Devendra K. Agrawal Department of Translational Research, Western University of Health Sciences, Pomona, CA, United States

Swati Agrawal Department of Surgery, Creighton University School of Medicine, Omaha, NE, United States

Nidhi Arora SS Infotech, Indore, MP, India

Ashutosh Bandyopadhyay Biomaterial and Tissue Engineering Laboratory, Department of Biosciences and Bioengineering, Indian Institute of Technology Guwahati, Guwahati, India

Bibrita Bhar Biomaterial and Tissue Engineering Laboratory, Department of Biosciences and Bioengineering, Indian Institute of Technology Guwahati, Guwahati, India

Nandana Bhardwaj Department of Biotechnology, National Institute of Pharmaceutical Education and Research Guwahati, Guwahati, India

Bhaskar Birru Biomaterial and Tissue Engineering Laboratory, Department of Biosciences and Bioengineering, Indian Institute of Technology Guwahati, Guwahati, India

Hema Bora Biomaterials and Tissue Engineering Laboratory, School of Medical Science and Technology (SMST), Indian Institute of Technology – Kharagpur, Kharagpur, India

John P. Bradford Polymers & Healthcare Materials/Devices, Department of Materials Science & Engineering University of Alabama at Birmingham (UAB), Birmingham, AL, United States

Soumya K. Chandrasekhar Department of Zoology, KKTM Government College, Pullut, Calicut University, Kerala, India; Department of Zoology, Christ College, Irinjalakuda, Calicut University, Kerala, India

Thomas Chandy Phillips Medisize LLC, Hudson, WI, United States

Subrata K. Das Biomaterials and Tissue Engineering Laboratory, School of Medical Science and Technology (SMST), Indian Institute of Technology – Kharagpur, Kharagpur, India

Santanu Dhara Biomaterials and Tissue Engineering Laboratory, School of Medical Science and Technology (SMST), Indian Institute of Technology – Kharagpur, Kharagpur, India

Abir Dutta Biomaterials and Tissue Engineering Laboratory, School of Medical Science and Technology (SMST), Indian Institute of Technology – Kharagpur, Kharagpur, India

Prachi Hanwatkar Rochester General Hospital, Rochester, NY, United States

Santanu Hati Department of Biomedical Science, Creighton University School of Medicine, Omaha, NE, United States

Gerardo Hernandez-Moreno Polymers & Healthcare Materials/Devices, Department of Materials Science & Engineering University of Alabama at Birmingham (UAB), Birmingham, AL, United States

Maji Jose Department of Oral Pathology & Microbiology, Yenepoya Dental College, Yenepoya (Deemed to be University), Mangalore, India

Sanju P. Joy Neurology, National Institute of Mental Health and Neuro-Sciences, Bangalore, India

Ajay Kashi Private Practice, Rochester, NY, United States

Oormila Kovilam St Joseph's Hospital, Indiana, United States

Merlin Rajesh Lal LifeCell International (Pvt) Ltd, Chennai, India

Biman B. Mandal Biomaterial and Tissue Engineering Laboratory, Department of Biosciences and Bioengineering, Indian Institute of Technology Guwahati, Guwahati, India; Centre for Nanotechnology, Indian Institute of Technology Guwahati, Guwahati, India

A. Maya Nandkumar Division of Microbial Technology, Sree Chitra Tirunal Institute for Medical Sciences & Technology Thiruvananthapuram, India

Joseph Christakiran Moses Biomaterial and Tissue Engineering Laboratory, Department of Biosciences and Bioengineering, Indian Institute of Technology Guwahati, Guwahati, India

Raghav A. Murthy Department of Cardiovascular Surgery, Icahn School of Medicine at Mount Sinai, New York, NY, United States

Prabha D. Nair Division of Tissue Engineering and Regeneration Technologies, Biomedical Technology Wing, Sree Chitra Tirunal Institute for Medical Sciences and Technology (SCTIMST), Poojapura, Thiruvananthapuram, India

Joshi C. Ouseph Department of Zoology, Christ College, Irinjalakuda, Calicut University, Kerala, India

Willi Paul Biomedical Technology Wing, Sree Chitra Tirunal Institute for Medical Sciences & Technology, Thiruvananthapuram, India

S.S. Pradeep Kumar Division of Microbial Technology, Sree Chitra Tirunal Institute for Medical Sciences & Technology Thiruvananthapuram, India

Anitha Radhakrishnan Research and Development Department, Pharmaceutical Corporation (Indian Medicine), Thrissur, India

D.P. Rahul Department of Orthodontics and Dentofacial Orthopedics, School of Dentistry, Amrita Vishwa Vidyapeetham, Amrita Institute of Medical Sciences, Kochi, India

Vikrant Rai Department of Biomedical Science, Creighton University School of Medicine, Omaha, NE, United States

Vrinda Rajagopal Department of Biochemistry, University of Kerala, Karyavattom, India

Ragavi Rajasekaran Biomaterials and Tissue Engineering Laboratory, School of Medical Science and Technology (SMST), Indian Institute of Technology — Kharagpur, Kharagpur, India

Greeshma Ratheesh Institute of Health and Biomedical Innovation, Queensland University of Technology, Brisbane, QLD, Australia

Preetam Guha Ray Biomaterials and Tissue Engineering Laboratory, School of Medical Science and Technology (SMST), Indian Institute of Technology — Kharagpur, Kharagpur, India

Trina Roy Biomaterials and Tissue Engineering Laboratory, School of Medical Science and Technology (SMST), Indian Institute of Technology — Kharagpur, Kharagpur, India

Baisakhee Saha Biomaterials and Tissue Engineering Laboratory, School of Medical Science and Technology (SMST), Indian Institute of Technology — Kharagpur, Kharagpur, India

Aditya Sengupta Department of Cardiovascular Surgery, Icahn School of Medicine at Mount Sinai, New York, NY, United States

Chandra P. Sharma Biomedical Technology Wing, Sree Chitra Tirunal Institute for Medical Sciences & Technology, Thiruvananthapuram, India

Yogendra Pratap Singh Biomaterial and Tissue Engineering Laboratory, Department of Biosciences and Bioengineering, Indian Institute of Technology Guwahati, Guwahati, India

V.P. Sivadas Division of Tissue Engineering and Regeneration Technologies, Biomedical Technology Wing, Sree Chitra Tirunal Institute for Medical Sciences and Technology (SCTIMST), Poojapura, Thiruvananthapuram, India

Prashant Sonar School of Chemistry and Physics, Queensland University of Technology, Brisbane, QLD, Australia

Sreekanth Sreekumaran Department of Biochemistry, University of Kerala, Thiruvananthapuram, India

Finosh G. Thankam Department of Translational Research, Western University of Health Sciences, Pomona, CA, United States

Vinoy Thomas Polymers & Healthcare Materials/Devices, Department of Materials Science & Engineering University of Alabama at Birmingham (UAB), Birmingham, AL, United States

Aynur Unal Digital Monzoukuri, Palo Alto, CA, United States

Cedryck Vaquette School of Dentistry, The University of Queensland, Brisbane, QLD, Australia

Yin Xiao Institute of Health and Biomedical Innovation, Queensland University of Technology, Brisbane, QLD, Australia; Australia-China Centre for Tissue Engineering and Regenerative Medicine, Centre for Biomedical Technologies, Queensland University of Technology, Brisbane, QLD, Australia

Preface

Regenerated organs is an emerging area to encourage the patient specific organ development a reality in future. Therefore, the objective of this book is to bring all the related interdisciplinary concepts together and discuss the comprehensive developments possible of this field currently with future directions. The book contains six sections — Section 1: Engineering approaches: from scaffolding to bioprinting applications, Section 2: Cardiovascular system, Section 3: Musculoskeletal regeneration of tissues, Section 4: Regenerative Neuroscience, Section 5: Respiratory research, Section 6: Key enabling technologies for regenerative medicine future outlook and conclusions.

Each chapter has been written by experts in their specialized area.

This book is expected to be an essential reference resource for young graduate students, academic faculty and collaborating industrial partners who are interested in advancing the knowledge and translational research in the area of Regenerated Organs.

I thank all the authors for their efforts of preparing excellent contributions and Ms. Billie Jean Fernandez for her effective coordination of this project.

I also appreciate very much and thank my wife Aruna Sharma for her sustained support during the course of this project.

Chandra P. Sharma

Engineering Approaches: From Scaffolding to Bioprinting Applications

1

Tissue and organ regeneration: An introduction

Willi Paul and Chandra P. Sharma

Biomedical Technology Wing, Sree Chitra Tirunal Institute for Medical Sciences & Technology, Thiruvananthapuram, India

1.1 Introduction

Tissue repair or healing is a natural, complicated and continuous process in any living organism, i.e. restoration of tissue function and architecture after the injury. It comprises of two essential components; Regeneration and Repair. Repair after injury can occur by regeneration of cells or tissues that restores normal tissue structure, or by healing, which leads to the formation of a scar. In case of regeneration, the damaged or lost tissue is replaced by the proliferation of surrounding undamaged cells and tissue. The normal structure of the tissue is restored or complete regeneration occurs in epidermis, GI tract epithelium and hematopoietic system where the cells have high proliferative capacity. In the case of stable tissues like liver and kidney, compensatory growth occurs rather than true regeneration. However, repair predominantly is the deposition of collagen to form a scar. But will certainly depend upon the ability of the tissue to regenerate and the extent of the injury. Wounds on superficial skin heal through regeneration of surface epithelium (regeneration); however, restoration of original ECM damaged by severe injury involves collagen deposition and scar formation (repair) thereby the normal structure of the tissue is permanently altered. Chronic inflammation may cause massive fibrosis.

Different cell types have different capacity for regeneration. Labile cells which acts as physical barriers have unlimited regenerative capacity. They are cells found on skin, GI tract, respiratory tract and urinary tract that are characterized by continuous regeneration. Quiescent cell types found in most of the internal organs like liver, kidney, endocrine and mesenchymal cells (fibroblasts, smooth muscle, vascular) have limited regenerative capacity and are in response to stimuli. It requires an intact basement membrane for organized regeneration.

Regenerated Organs
DOI: https://doi.org/10.1016/B978-0-12-821085-7.00001-4

3

Permanent cell types like CNS neurons, skeletal and cardiac muscle cells have very little regenerative capacity and its repair forms scar.

Mammals and humans are generally considered as a poor example for regeneration when compared with most vertebrate species due to the differences in genetics, development, immune systems and tissue complexity [1]. In case of mammals, scar-free healing and regeneration normally occurs during the early stages of life. The ability to regenerate is lost during adulthood, but many non-mammalian vertebrates retain the capacity to regenerate organs and limbs after injury as depicted in Fig. 1.1 [1]. Physiological regeneration in mammals is limited to tissues with high proliferative capacity. The epithelia of the skin and gastrointestinal tract, the hematopoietic system (red blood cell replacement), hair cycling and antler regeneration are examples. This forms the basis of guided tissue regeneration which may be necessary for efficient restoration of damaged tissues. Thus the original function and form seems to be mimicked as closely as possible by the regenerated tissues. Regeneration thus requires an intact connective tissue scaffold.

1.2 Guided tissue regeneration

Guided tissue regeneration is a procedure where a biodegradable conduit provides contact guidance for enhancing the opportunity for one cell type to populate an area for regenerating tissue. The conduit or a biomaterial construct should be biocompatible and should not make any damage or be rejected by the host tissue. Regeneration is classified into guided tissue regeneration (GTR) which refers to the regeneration of periodontal attachment and guided bone regeneration (GBR) that refers to ridge augmentation and focused on development of hard tissues in addition to the soft tissue regeneration.

Ridge augmentation technique is required for successful implant placement in the right prosthodontic positions. Guided tissue regeneration is one technique used for ridge augmentation in rehabilitation of atrophic jaws with dental implants [2]. It uses barrier membranes with or without bone grafts or substitutes for osseous regeneration for exclusion of cells impeding bone formation. Epithelium and connective tissue are excluded from the root surface in the belief that they interfere with regeneration. This was based on the assumption that only periodontal ligament cells have the potential for regeneration. Theoretically guided tissue regeneration was developed by Melcher in 1976 [3]. Primarily, there are four stages for a successful bone or tissue regeneration, which are generally abbreviated as PASS. (1) Primary closure of the wound to promote undisturbed and uninterrupted healing; (2) angiogenesis to provide necessary blood supply and undifferentiated mesenchymal cells; (3) space creation and maintenance to facilitate space for bone ingrowth, and (4) stability of the wound to induce blood clot formation and allow uneventful healing.

Advantages of GTR membranes are that other tissues that interfere with the osteogenesis and bone formation can be prevented by using a barrier. This barrier also acts as a dressing for the wound coverage and anchorage for the blood clot. Prevent bacterial invasion and inflammation and provide suitable micro environment for regeneration. There are several GTR membranes used clinically which ranges from acellular dermal allograft to polymeric membranes both resorbable as well as non resorbable.

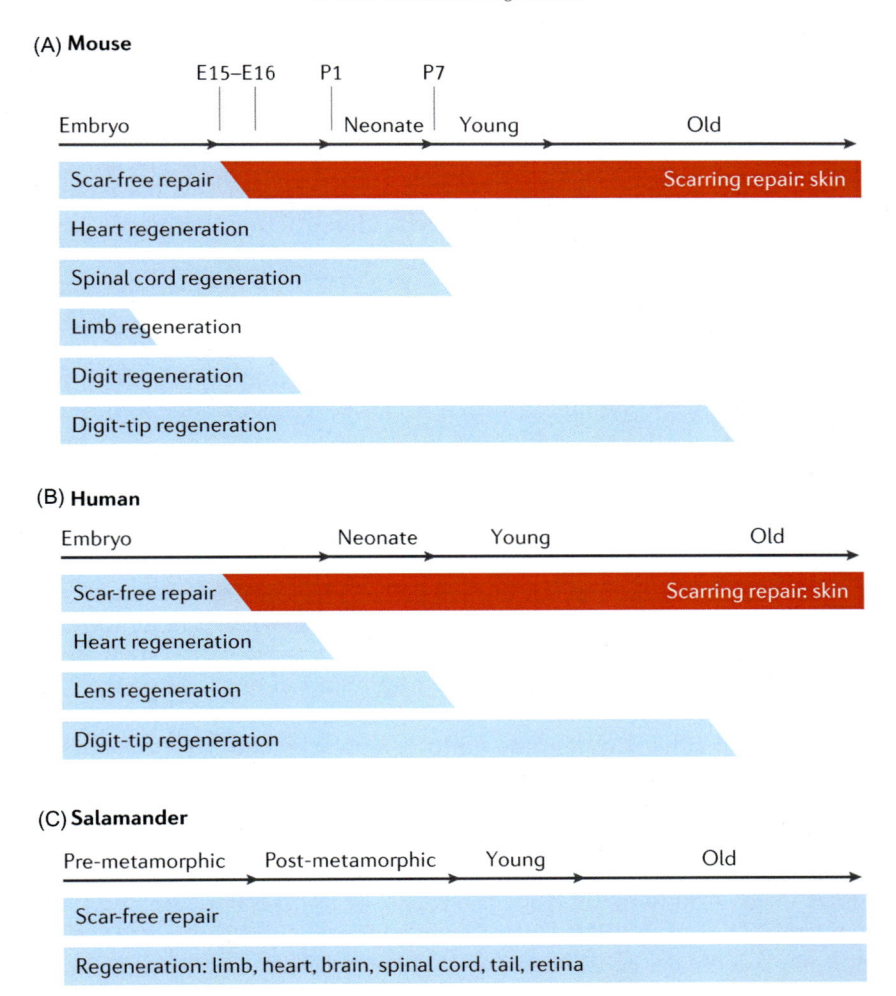

FIGURE 1.1   Tissue regeneration in different species. The capacity of tissue and organ regeneration varies in different animal species. (A) In mice, the capacity for scar-free repair decreases between embryonic day 15 (E15) and E16. The capacity for heart and spinal cord regeneration is lost in early postnatal life between postnatal day 1 (P1) and P7. Limb regeneration is lost early in development (before E15). Whole digits can be regenerated until E16, and digit tip regeneration is maintained throughout the development stages. (B) In humans, the regenerative potential is limited to early developmental stages, similar to mice. (C) In salamanders, the ability for scar-free repair and the regeneration of limbs, heart, brain, spinal cord, tail and retina are maintained throughout their life. *Adapted from Xia HM, et al. Tissue repair and regeneration with endogenous stem cells. Nat Rev Mater, 2018;3(7):174−93, under license from Springer Nature.*

1.3 Stem cells in tissue regeneration

Stem cells are the core of the modern regenerative medicine. Stem cells have prolonged self-renewal capacity and ability to asymmetric replication and are found in specialized niches within each tissue. During normal homeostasis the dead cells will be replenished by the stem cells, and also repair damaged tissue. The extrinsic signals interact with the proteins expressed by the stem cells in a dynamic manner in the niche microenvironment that influence the ability of stem cells to self-renew. In asymmetric replication, in every cell division, one cell will be identical to the original stem cell whereas the other one terminally differentiates. In stochastic differentiation, one stem cell develops into two differentiated daughter cells. Whereas a second one produces two stem cells identical to the original.

There are different sources for stem cells. Some come from embryos that are 3–5 days old called embryonic stem cells. They are pluripotent cells, can devide into many stem cells and can differentiate into any type of cell in a body. Thus these cells are versatile and can be used to regenerate and repair of any diseased tissue or organ. Adult stem cells are found in small numbers in adult tissue such as bone marrow or fat and have limited differentiation potential. By genetic reprogramming, adult cells can be transformed into embryonic stem cells. Stem cells are also found in amniotic fluid as well as umbilical cord blood, and are called perinatal stem cells, which also have the ability to change into specialized cells.

It has been reported that stem cells exist in two distinct states depending upon their relative activity and wound-induced regeneration. The amount of tissue generated is affected by the timing and length of stem cell activity. A recent study on hair follicle has shown that signals emanating from both heterologous niche cells and from lineage progeny influence the timing and length of stem cell activity [4]. The capacity of bone to regenerate and repair itself depends on the size of the wound and the presence of certain diseases. Large bone defects may require surgical intervention. Implantation of the bone stem and progenitor cells with tissue engineered scaffolds has immense potential in fracture bone healing [5]. Mesenchymal stem cells differentiates into osteoblasts, chondrocytes, and adipocytes and are critically important for musculoskeletal tissue regeneration and repair [6]. Stem cells have been explored for its regenerative ability widely in bone regeneration studies. Both adipose derived mesenchymal stem cells (ASC) and bone marrow stem cells (BMSC) have showed almost similar potential in bone regeneration, although BMSC has shown better results in vitro. A new method for the repair of injured bone or periodontal disease using bone marrow stem cells (BMSC) has been reported [7]. Proliferation and osteogenic differentiation to osteoblast cells has been achieved using red-light absorbing carbon nitride sheets used along with BMSC. It has been shown that the material absorbs red-light and emits fluorescence that speeds up bone regeneration. BMSC therapy has been shown as a promising choice in bone regeneration and repair particularly for critical-sized defects. However, study on the cellular and local interaction in the process of bone regeneration is required for the approval of Food and Drug Administration.

Traumatic muscle injuries are challenging to treat. Cell based approach have shown promising results in many pre-clinical studies. Myogenic stem cells as well as non myogenic cells are studied in muscle regeneration. Satellite cells (SC) give rise to large number of progeny which forms myofibers and repopulate the SC niche in host muscles. Mesenchymal stem cells (MSC) can modulate the function of myoblasts such as their fusion into myotubes, and their migration and proliferation kinetics. Bone marrow derived MSC has been shown to improve contractile muscle function after intramuscular implantation [8]. A clinical study reports the implantation of an acellular biological scaffold at the muscle injury site and providing the patient with aggressive physical therapy has shown significant functional

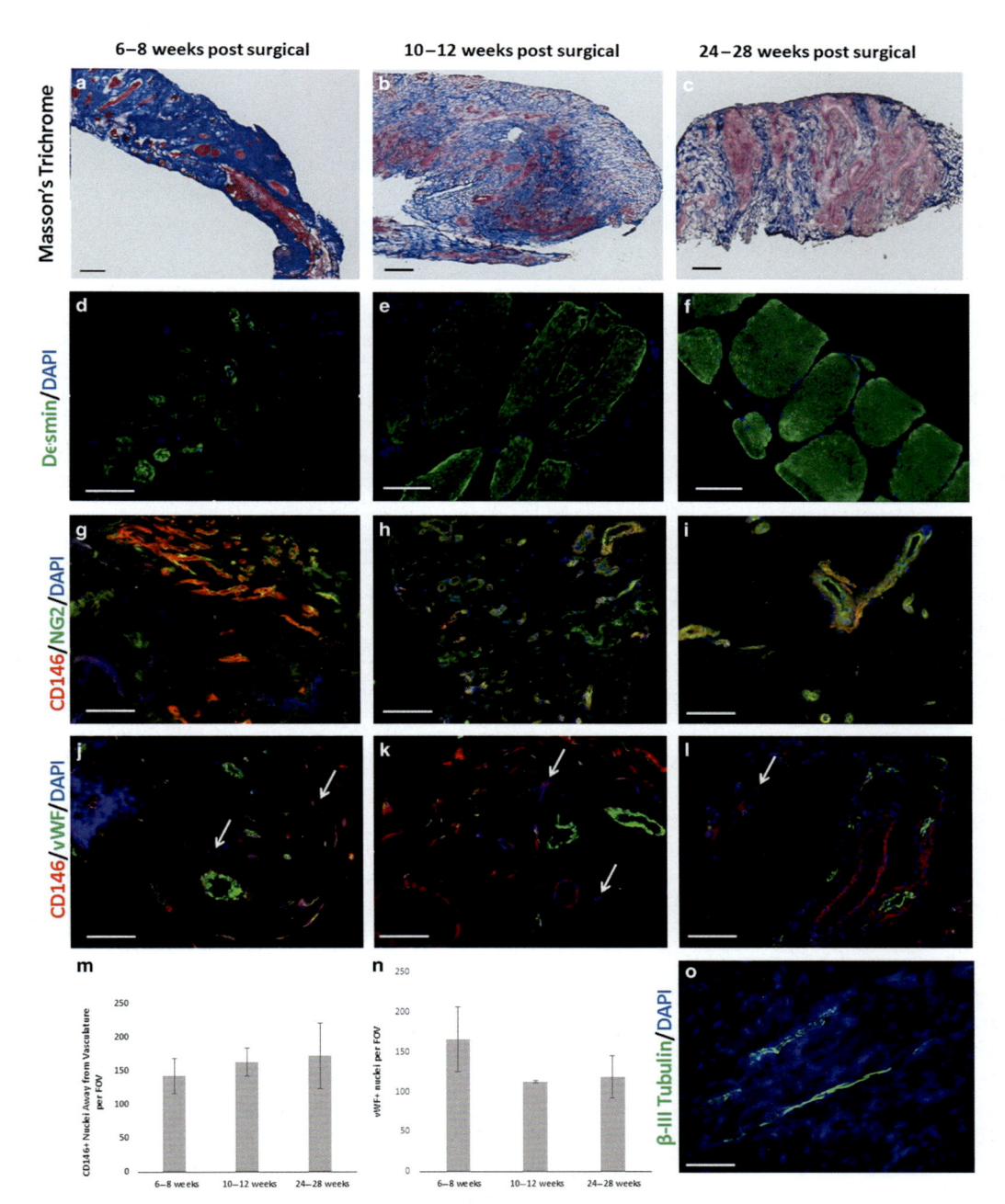

FIGURE 1.2 Site-appropriate tissue remodeling by ECM bioscaffolds. (A–C) Massons trichrome staining of human muscle biopsies shows islands of skeletal muscle present at 6–8 weeks, 10–12 weeks and 24–28 weeks post surgery, respectively. (D–F) Human muscle biopsies are characterized by desmin expression at all time points, indicating new muscle formation within the site of implantation. (G–I) ECM bioscaffold implantation is associated with the presence of CD146 + NG2 + perivascular stem cells. (J–L) PVSCs were shown to migrate away from their normal vessel-associated anatomic location at all time points. Arrows indicate CD146 + PVSCs migrating away from vessels. (M, N) Migrating PVSCs and vascularity was quantified using Cell Profiler image analysis software. (O) At 24–28 weeks post surgery, ECM bioscaffold implantation was associated with the presence of β-III tubulin + cells, implicating innervated skeletal muscle. (Scale bars = 50 μm). *Adapted from Dziki J, et al. An acellular biologic scaffold treatment for volumetric muscle loss: results of a 13-patient cohort study. Npj Regenerative Med 2016;1, under Creative Commons License.*

improvement in thirteen patients with volumetric muscle loss. As the scaffolds started degrading the stem cells migrate to the area and get differentiated into muscle cells [9]. New muscle formation and presence of neurogenic cells at the remodeling site is evident in Fig. 1.2. Although various studies have provided a positive outlook, an innovative cell-based therapy is yet to be standardized for traumatic muscle injuries.

1.4 Conclusion and future perspective

Tissue engineering concepts have been widely experimented for cartilage, skin, bone, vascular and nerve tissue regeneration. The 3D structure and its physical properties are equally important like its combination of materials, the cell-cell and the cell-matrix interactions. Traditional scaffold fabrication technique has its limitation that the complex structure of the real organs cannot be duplicated. 3D bio printing technique has been studied now a day as a strategy to improve regeneration of organs. The invention of stereolithography in 1983 later led to the development of 3D bio printing method for printing artificial human organs. It's a versatile 3D printing method utilizing bio-ink for printing artificial organs like blood vessel, skeleton and skin. 3D bio printing technology is highly precise and fast, and has the benefit of individualized medical treatment. It has been demonstrated with tricalcium phosphate that 3D printed scaffolds can have precise and controllable pore structure with optimal mechanical strength comparable with human cancellous bone [10]. This scaffold was biocompatible, and had adherence and rapid proliferation of bone mesenchymal stem cells (hMSC) for its application in load-bearing bone. A new bio-ink with precise control over printability, mechanical and degradation properties has demonstrated endochondral differentiation of encapsulated hMSCs [11]. This could 3D print patient specific bone tissue for regeneration of diseased bone. Similarly 3D printing could also be used in bio printing of heart valves and heart muscles for the treatment of cardiac patients [12]. A critical review by Deo et al. [13] discusses various design criteria and processing parameters of bio ink to help fabrication of complex structures for bioartificial organ manufacturing. There is a shortage of donor organs worldwide which projects the urgency of development of biocompatible 3D printed artificial organs. The strategies and process parameters for bio printing of organs like skin, cardiac tissue, bone, cartilage, liver, lung, neural tissues, pancreas etc. are reviewed in detail by Matai et al. [14]. The progress made in organ bio printing in regeneration has made considerable progress; however, still various challenges like structural stability in vivo and degradation, biocompatibility, maintenance of sterility etc. need to be optimized before clinical translation. The versatility of the bio printing could improve with the latest innovation like 5D printing of additive manufacturing (where the printing can achieve curved paths making the artificial organs more realistic). The advent of 4D printing where there is a fourth dimension added seems more dynamic which makes a smart material that responds to a stimulus. These seem to be more suitable for bioartificial organ regeneration.

References

[1] Xia HM, et al. Tissue repair and regeneration with endogenous stem cells. Nat Rev Mater 2018;3(7):174—93.
[2] Liu J, Kerns DG. Mechanisms of guided bone regeneration: a review. Open Dent J 2014;8:56—65.
[3] Melcher AH. On the repair potential of periodontal tissues. J Periodontol 1976;47(5):256—60.

[4] Blanpain C, Fuchs E. Stem cell plasticity, plasticity of epithelial stem cells in tissue regeneration. Science 2014;344(6189):1243 - + .

[5] Walmsley GG, et al. Stem cells in bone regeneration. Stem Cell Rev Rep 2016;12(5):524−9.

[6] Chen Y, et al. Mesenchymal stem cells: a promising candidate in regenerative medicine. Int J Biochem Cell Biol 2008;40(5):815−20.

[7] Tiwari JN, et al. Accelerated bone regeneration by two-photon photoactivated carbon nitride nanosheets. Acs Nano 2017;11(1):742−51.

[8] Qazi TH, et al. Cell therapy to improve regeneration of skeletal muscle injuries. J Cachexia Sarcopenia Muscle 2019;10(3):501−16.

[9] Dziki J, et al. An acellular biologic scaffold treatment for volumetric muscle loss: results of a 13-patient cohort study. Npj Regenerative Med 2016;1.

[10] Man X, et al. Research on sintering process of tricalcium phosphate bone tissue engineering scaffold based on three-dimensional printing. Sheng Wu Yi Xue Gong Cheng Xue Za Zhi 2020;37(1):112−18.

[11] Chimene D, et al. Nanoengineered osteoinductive bioink for 3D bioprinting bone tissue. ACS Appl Mater Interfaces 2020.

[12] Birla RK, Williams SK. 3D bioprinting and its potential impact on cardiac failure treatment: an industry perspective. APL Bioeng 2020;4(1):010903.

[13] Deo K, et al. Bioprinting 101: design, fabrication and evaluation of cell-laden 3D bioprinted scaffolds. Tissue Eng Part A 2020.

[14] Matai I, et al. Progress in 3D bioprinting technology for tissue/organ regenerative engineering. Biomaterials 2020;226:119536.

Tissue repair with natural extracellular matrix (ECM) scaffolds

Thomas Chandy

Phillips Medisize LLC, Hudson, WI, United States

2.1 Summary

Extracellular matrix (ECM) scaffolds that provide a conducive environment for normal cellular growth, differentiation and angiogenesis are important components of tissue engineered grafts for long term viability. ECM has shown to be an effective scaffold for the repair and reconstitution of several tissues, including blood vessels, skin graft, dural repair, soft tissue grafts, hernia repair, myocardial repair, urinary tract structures, ophthalmic reconstruction and nerve tissue regeneration. These ECM scaffolds are completely degraded in vivo and induce a host cellular response that supports constructive remodeling rather than scar tissue formation. Several naturally occurring scaffold materials have been investigated, including small intestinal submucosa (SIS), acellular dermis (Allo Derm) bladder acellular matrix graft (UBM), amniotic membrane tissue (anthromatrix, ambiodry, amniograft), cadaveric fascia (Tutoplast) and porcine pericardium (IO patch). Common features of ECM-associated tissue remodeling include extensive angiogenesis, recruitment of circulating progenitor cells, rapid scaffold degradation and constructive remodeling of damaged or missing tissues. The sources, the methods of procurement and processing, and the effects of these naturally occurring, materials on angiogenesis and tissue deposition are reviewed. SIS has found its application in a variety of tissue interfaces for tissue repair and reconstruction. The bladder acellular matrix is very similar to SIS structurally. However, extensive processing methods are needed to separate the attached muscular bladder wall from the submucosal membrane. These harsh chemical and enzymatic treatments on bladder matrix causes deleterious results on long term implants on tissue reconstruction and repair. Acellular human amniotic membrane shows promising in tissue repair and neovascularization due to their improved strength, flexibility, suturability, antibacterial effects and low immunogenicity. It seems human amnion, which is processed to yield a uniform, acellular biofabric, is a superior material for a variety of product applications.

11

Crosslinking is an effective means of controlling the biodegradation rate of collagen-based biomaterials. Crosslinked collagen or collagen based materials has a greater modulus of elasticity (Young's modulus), greater resistance to proteases, and a lower degree of swelling than uncrosslinked collagen. Glutaraldehyde fixation of bio-prosthetic tissue has been used successfully for almost 40 years. However, it is generally recognized that glutaraldehyde fixation of bio-prostheses is associated with the occurrence of calcification. Accordingly, many efforts have been undertaken to develop techniques for the fixation of bio-prostheses, which will not lead to calcification. Several alternative crosslinking techniques have been explored in different applications, including physical methods such as UV irradiation, dehydrothermal, freeze drying, etc. and the use of chemical reagents, such as diepoxides, diisocyanates, carbodiimides, diisothiocyanates, and glycidylethers.

The crosslinking of tissues reduces immunogenicity of the material and increase resistance to degradation by host and bacterial enzymes. It is suggested that the functionality of the amniotic membrane may be improved via selecting a suitable crosslinking technique to suit a specific application. However, the successful utilization of mammalian ECM as a therapeutic device will depend in large part upon our ability to understand and take advantage of the native structure/function relationships of the biological scaffold material.

This review also presents the recent advances in the 3D bioprinting and their relative components, including the bioinks, the cells, and applications for organ regeneration. Although challenges still remain in this research field, further multidisciplinary research to advance printing techniques, printable bioink materials and engineering designs can address the current challenges and realize the emerging potential of 3D organ bioprinting. We conclude this chapter by highlighting ongoing challenges and opportunities associated with growth factor (GF) delivery and address the biomaterials selection criteria for the fabrication of traditional and modern nano delivery systems that accomplish the spatiotemporal release of single/multiple GFs for functional regeneration of complex tissues.

2.2 Background

Effective repair and regeneration of injured tissues and organs depends on early reestablishment of the blood flow needed for cellular infiltration and metabolic support. Implantable biomaterials designed to replace damaged or diseased tissues must act as supports (i.e., scaffolds) into which cells can migrate and establish this needed blood supply [1–4]. One approach to treating damaged or diseased tissues relies upon synthetically derived biocompatible polymer scaffolds to serve as backbones for tissue repair and regeneration. Although many synthetic biopolymers have been used to replace damaged vascular structures, and while long-term patency rates have risen over the years, the ideal vascular graft scaffold remains elusive. For example, no synthetic biopolymer currently available for clinical use can restore normal structure and function to injured vascular tissues while avoiding severe complications such as thrombosis, neointimal hyperplasia, accelerated atherosclerosis, and/or approach to repair and regeneration of damaged tissues uses intact extracellular matrix obtained from animal tissues as the growth support for host cells. The extracellular matrix (ECM) is a complex mixture of structural functional proteins, proteoglycans and glycoproteins arranged in a unique, tissue specific three-dimensional ultrastructure. These proteins provide structural support and tensile strength for the organs and they deliver diverse host processes as angiogenesis and vasculogenesis, cell migration, cell proliferation and orientation, inflammation, immune responsiveness and wound healing. Implantable biomaterials designed to replace damaged or diseased tissues must act as supports (i.e., scaffolds) into which cells can migrate and establish this needed blood supply [1,2]. Similarly, this ECM must be strong enough to withstand the physiologic demands placed upon them when implanted into a site- specific organ system and must retain their mechanical properties over time.

The most common constituent of the ECM is the structural protein, collagen. When harvested from the tissue source and fabricated into a graft prosthesis, these ECM materials may be referred to as naturally occurring polymeric scaffolds, bio-scaffolds, biomatrices, ECM Scaffolds, or naturally occurring biopolymers [3,5–7]. These materials are harvested from several different body systems, but they share similarities when processed into a graft material. Specifically, since they are subjected to minimal processing after they are removed from the source animal, they retain a structure and composition nearly identical to their native state. The host cells are removed and the scaffolds are implanted acellularly to replace diseased or damaged tissues (Table 2.1).

Naturally occurring biopolymers include small intestinal submucosa, acellular dermis, cadaveric fascia, porcine pericardia, the bladder acellular matrix graft and

TABLE 2.1 ECM Scaffolds and selected investigational uses.

Scaffold material/ commercial name	Tissue source	Primary uses (Ref.)
Small intestinal submucosa (Surgisis® soft tissue graft, Cook)	Porcine small intestine	Vascular conduits [8], Skin graft [9] Dural repair, Soft tissue graft [10] Hernia repair [11], Ligament reconstruction [12] Myocardial repair [5], Urinary reconstruction [13]
Acellular dermis (AlloDerm®)	Pig skin	Skin graft [14], Dural repair [15], Urinary reconstruction Plastic and cosmetic surgery [16]
Bladder acellular matrix (Acellvet V1000-LY)	Porcine bladder	Vascular conduits [17], Bladder reconstruction [18] Esophagus reconstruction, Cardiac tissue repair f13J,
Amniotic membrane (Ambiodry, Amniograft, Anthromatrix)	Human amnion	Skin graft [19], Urinary tract reconstruction, Ophthalmology [20], Abdominal hernia repair, Closure of pericardium [21], Vascular repair [22] Nerve tissue regeneration [23]
Fascia Lata (Tutoplast™)	Human cadaver Fascia lata	Ligament reconstruction, Dural repair [24], Craniofacial reconstruction, Heart valves [25]
Pericardium (IO patch™)	Mammalian Pericardium (porcine, calf)	Heart valves [26], Corneal repair, Skin graft

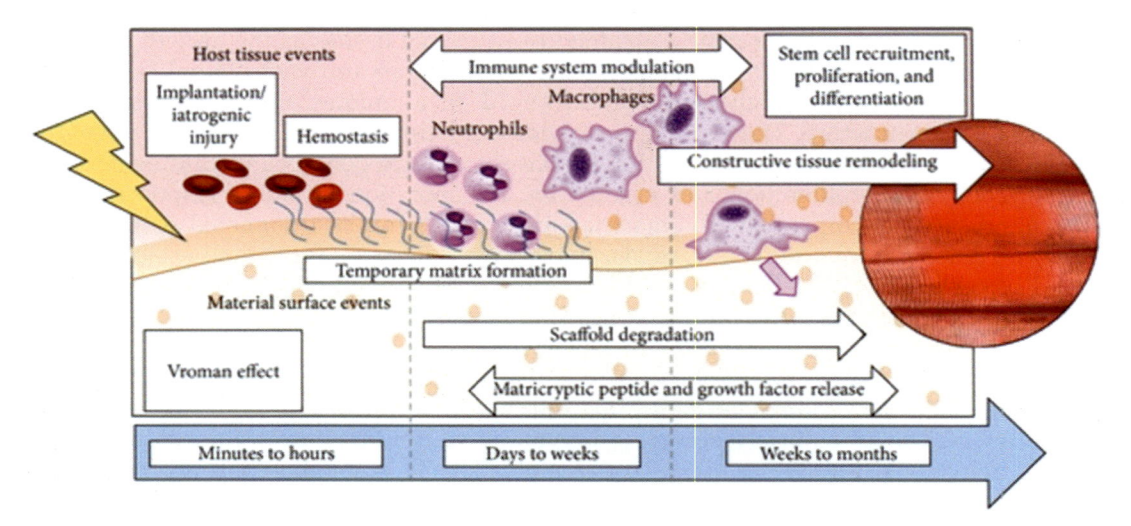

FIGURE 2.1   Remodeled tissue.

amniotic membrane [5,27]. These naturally occurring materials offer promising alternatives to synthetically engineered polymeric scaffolds for tissue repair and regeneration [28–30]. These naturally occurring scaffolds can be processed in such a way as to retain growth factors, such as basic fibroblast growth factor (FGF-2), transforming growth factor-β, vascular endothelial cell growth factor (VEGF), and epidermal growth factor (EGF) [31–33], glycosaminoglycans, such as heparin, hyaluronic acid, dermatan sulfate, chondroitin sulfate A and C [15,34], and structural elements such as fibronectin, elastin and collagen [27,34]. All ECMs share the common features of providing structural support and serving as a reservoir of growth factors and cytokines [1,5]. These materials prevent many of the complications associated with foreign material implants because they provide a natural environment onto which cells can attach and migrate, within which they can proliferate and differentiate. These naturally occurring biopolymers have been shown to interact quickly with the host's tissues, induce the deposition of cells and additional ECM, and promote rapid angiogenesis-functions that are essential to the restoration of functional soft tissue. In this manner, the ECM affects local concentrations and biologic activity of growth factors and cytokines and makes the ECM an ideal scaffold for tissue repair and reconstruction.

The ideal biomaterial must allow tissue incorporation and result in remodeled, functional tissue (Fig. 2.1) without leading to encapsulation, breakdown of the material, tissue erosion, or adhesion formation. The purpose of this literature analysis is to present an overview detailing the use of naturally occurring polymers as acellular bio-scaffolds and review the current knowledge about the biochemical composition of these materials that contribute to their ability to elicit an appropriate angiogenic response. It is assumed that the ECM scaffolds that retain essentially unchanged from native ECM elicit a host response that promote cell infiltration and rapid scaffold degradation, deposition of host derived neo-matrix and eventually constructive tissue remodeling with minimum of scar tissue [1,35]. Several of these materials and their primary uses are listed in Table 2.1 their known biochemical composition is summarized in Table 2.2.

TABLE 2.2 Biochemical composition of ECM scaffolds.

Scaffold material	Components identified°
Small intestinal Submucosa (30 μm) Size: 2 × 3 cm upto 7 × 10 cm	Collagen Types: I, III, IV, V, VI Other Proteins: fibronectin Proteoglycan, laminin. Glycosaminoglycans: hyaluronic acid, Heparin, heparin sulfate, chondroitin Sulfate A, dermatan sulfate, Chondroitin sulfate C Growth factors: FGF − 2 TGF − 8, VEGF
Acellular dermis (40 μm)	Collagen Types: I, IV, VII Other Proteins: elastin Glycosaminoglycans: Not detected (ND) Growth factors: ND
Bladder acellular matrix (60 μm) Size: 10 × 7 cm	Collagen Types: I, III, IV Other Proteins: elastin, fibronectin. Glycosaminoglycans: hyaluronic acid, Heparin, heparin sulfate, chondroitin Sulfate A, dermatan sulfate, Growth factors: FGF-2. TGF-0. VEGF
Amniotic membrane (20−30 μm) Sizes: 1 × 2 cm, 2 × 3 cm 4 × 4 cm	Collagen Types: I, III, IV, V, VII Other Proteins: laminin, fibronectin, decorin. Glycosaminoglycans: hyaluronic acid, heparin sulfate, Growth factors: EGF. FGF-2. TGF-8. TGF-o, KGF
Fascia lata (400−650 pm) Size: 0.3 × 15 cm	Collagen Type: I Other Proteins: ND. Glycosaminoglycans: ND Growth factors: ND
Pericardium (400−1000 μm) Size: 1 × 3 cm 1.5 × 1.5 cm	Collagen: Type: I Other Proteins: ND. Glycosaminoglycans: ND Growth factors: ND

2.3 Small intestinal submucosa

Small intestinal submucosa (SIS) is a resorbable, acellular bio-scaffold composed of extracellular matrix (ECM) proteins derived from the jejunum of pigs. SIS has characteristic of an ideal tissue engineered biomaterial and can act as a bioscaffold for remodeling of many body tissues including skin, body wall, musculoskeletal structure, urinary bladder, blood vessels, and supports new blood vessel growth [8,11−13]. SIS consists of three distinct layers of the mammalian small intestine: the lamina propria and muscularis mucosae of the intestinal mucosa, and the tunica submucosa (Fig. 2.2) [1]. The tunica submucosa is the layer of connective tissue arranged immediately under the mucosa layer of the intestine and is a 100−200 μm thick interstitial ECM: it makes up the bulk of the SIS biopolymer scaffold. SIS induces site-specific remodeling of both organs and the tissue depending on the site of implantation [27]. SIS stimulates host cells to proliferate and differentiate into site-specific connective tissue structures, and this replaces the SIS material within 90 days [36]. SIS's ability to induce tissue remodeling is associated with angiogenesis, cell migration and differentiation and deposition of ECM [36].

Bovine type I collagen (i.e., reconstituted collagen) is perhaps the most widely used biological scaffold for therapeutic applications due to its abundant source and its history of

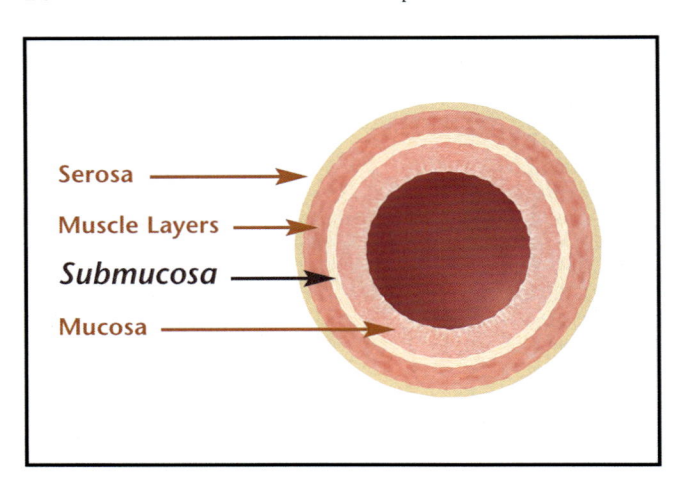

FIGURE 2.2   Cross-section diagram of small intestine.

successful use. Scaffolds for tissue reconstruction and replacement must have both appropriate structural and functional properties. Collagen types other than type I exist naturally occurring ECM like SIS [8−10]. These alternative collagen types each provide distinct mechanical and physical properties to the ECM and contribute to the utility of the intact ECM (as opposed to the isolated components of ECM) as a scaffold for tissue repair. Structurally, SIS consists of type I, III, IV, V and VI collagen [9−11] in addition to other components as shown in Table 2.2. This diversity of collagens and their structural arrangement within a single scaffold material is particularly responsible for the distinctive biological activity of SIS scaffold when compared to single reconstituted collagen matrix.

SIS is prepared from porcine jejunum [10] immediately after harvesting the intestine. The superficial layers of the tunica mucosa are removed by mechanical delamination. The tissue is then turned to the opposite side and the tunica muscularis externa and tunica serosa layers are mechanically removed. The remaining tissue represented the SIS and consisted of the tunica submucosa and basilar layers of the tunica mucosa. The biopolymer is thoroughly rinsed in water, treated with an aqueous solution of 0.1% peracetic acid, and rinsed in sequential exchanges of water and phosphate buffered saline. It is then stored in antibiotic solution containing 0.05% Gentamycin sulfate [10,34].

SURGISIS ES sheets have a thickness and mechanical strength that is several times that of a single-layer SURGISIS sheet [1]. Nominal properties for SURGISIS ES and single-layer SURGISIS sheets are listed in Table 2.3.

The mechanical properties and complement activation of SIS is indicated in Tables 2.3 and 2.4. respectively. The material has good mechanical and suture retention strength and has no complement activation. SIS has found its application in a variety of tissue interfaces for tissue repair and reconstruction. SIS has significant potential as a vascular graft material and was experimentally evaluated to repair large diameter (−10 mm ID) vascular graft [37], small diameter arteries and veins, vena cava, carotid arteries and heart valves [37,38]. In addition to vascular applications, the SIS biomaterial has been used extensively in the genitourinary system to repair congenital abnormalities of the bladder patch [13] has shown rapid and aggressive regeneration of bladder tissue within 2−4 weeks. SIS has been used to treat abdominal hernias and repair body wall, to treat chronic dermal wounds, to repair dura mater, and to replace tendon and ligament in orthopedic applications [1,5,9,27]. In all of these cases, SIS supported angiogenesis and caused replacement of damaged structures leading to the restoration of functional tissues. However, mild inflammation and anti-SIS antibody production have been reported following implantation, the immune response elicited by SIS have not lead to a rejection immune response [39].

Table 2.5 provides the in vivo (Dog implantation body wall repair model) degradation of SIS and tissue repair profile. There is a rapid decrease in strength of the repair device at

TABLE 2.3 Mechanical properties of SIS.

Property	SURGISIS[a] single layer	SURGISIS[b] enhanced strength
Nominal Thickness (mm)	0.20	0.42
Suture Retention Strength* (lb)	0.67 ± 0.1	1.70 ± 0.4
Burst Force** (lb)	5.20 ± 0.4	28 ± 6.8

[a]*single-layer SURGISIS sheets are designed to tolerate the mechanical stresses associated with low-stress body systems.*
[b]*SURGISIS ES sheets (2 sheets) are designed to tolerate the mechanical stresses associated with higher-stress body systems.*
*5-0 suture with 2 mm bite depth.
**9.5 mm diameter sphere.

TABLE 2.4 Complement activation with SIS.

Material	C3 complement activation (ng/mL)
Negative control	148 ± 42
SIS material	115 ± 24
Positive control	2449 ± 930

TABLE 2.5 Mechanical properties of explanted SISHRD using the ball burst test[a].

Survival time (days)	Burst load (lb)	
	Mean	SD
1	66.91	5.15
4	51.95*	7.49
7	42.77*	19.66
10	39.97*	18.03
30	72.65	38.12
90	109.63	63.06
180	120.72	39.47
720	157.20*	26.03

[a]*Implant ready SIS devices (n = 40) were tested for preimplant strength values. The mean burst load was 73.37 $\pm$ 11.45 lb. *$P < 0.05$.*

the surgical site during the first 10 days postsurgery to a value of 40.0 pounds. All subsequent time points of evaluation ranging from 1 month to 2 years show a progressive increase in strength of the surgical site [37,39]. It appears that the naturally occurring SIS show rapid degradation with associated and subsequent remodeling to a tissue with strength that exceeds that of the native tissue when used as a body wall repair device. Thus, the SIS consists of a complex mixture of structural and functional proteins and serves an important role in tissue and organ morphogenesis, maintenance of cell and tissue structure and function, and in the host response to injury.

2.4 Acellular dermis

Skin is comprised of two primary layers that differ in function, thickness and strength: the epidermis and the dermis. The dermis underlies the epidermis and is the thicker layer of the skin. The dermis is composed of fibroblasts, which produce a collagen-containing ECM that provides elasticity and support to skin. Acellular dermis is mainly the normal dermal tissue structures that remain after the cells are removed. Like other naturally occurring biopolymers, acellular dermis is rich in collagen Type 1. Collagen type IV and type VII are also retained during processing in the dermal skin [15] along with elastin.

Acellular dermis is harvested from either pig skin or human cadaver skin. The epidermis is removed by soaking the skin in 1 M NaCI for 8 h. Dermal fibroblasts and epithelial cells are removed by incubation of the material in 2% deoxycholic acid containing 10 mM ethylene diamine tetra acetate (EDTA). When implanted as an acellular tissue graft, acellular dermis supported reendothelialization of repaired vascular structures, inhibited excessive wound contraction and supported host cell incorporation and capillary in growth into the grafted site [15,40,41].

Allogenic acellular dermis is used clinically to manage full-thickness burns. The skin healed with minimal scar formation in comparison to an untreated site on patient's [14]. This study demonstrated that cellular dermis supported fibroblast infiltration, neovascularization, and epithelialization in the absence of an inflammatory response [14,16]. Acellular dermis has been used in several other tissue repair applications, including dermal replacement, internal soft-tissue repair, and small-diameter vascular replacement [16,42]. However, the allograft use for long-term performance of the tissues remains questions regarding their source tissue, in terms of viral and other disease transfer.

2.5 Bladder acellular matrix

Bladder acellular matrix graft (BAMG) was first described in 1975 [43] and was derived from a layer of the urinary bladder that is analogous to the submucosal tissue comprising the bulk of the SIS biomaterial. Structurally, the BAMG is determined to be composed of type I, type III and type IV collagen and elastin, but also contains other ECM components, including fibronectin, Glycosamino-glycans (hyaluronic acid, heparin sulfate, chondroitin sulfate A, dermatan sulfate) and several growth factors (FGF-2, TGF-β, VEGF) (Fig. 2.3).

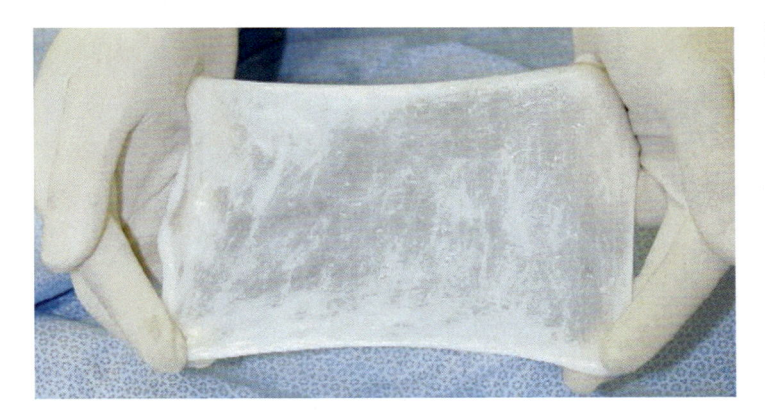

FIGURE 2.3 ECM harvested from porcine urinary bladder.

The submucosa of the urinary bladder is intimately attached to the muscular bladder wall and the complete mechanical separation of the layers has proven tedious and difficult [1]. The bladder submucosa is prepared by various chemical and or enzymatic treatments such as sodium hydroxide, sodium desoxycholate, sodium dodecyl sulfate or deoxyribonuclease [1,17]. In one processing method, whole bladders were soaked in an EDTA solution for 48 hours followed by additional soaking in Tris-potassium chloride-EDTA solution containing Triton X. Bladders were then rinsed in phosphate buffer solution, incubated overnight with deoxyribonuclease and ribonuclease to remove cytoplasmic and nuclear materials, and further extracted in a solution containing Tris and sodium dodecyl sulfate (SDS). The extracted bladders were then submerged in 70% ethanol for 24 hours to remove any residual SDS, washed in phosphate buffer and stored in refrigerated saline until use [18].

Porcine derived BAMG have been extensively used for the repair of lower urinary tract and for restoration of bladder function [18,44–47]. It is also investigated for replacing the damaged cardiac tissues, heart valves, portions of GIT and vascular grafts [2,27,48–52]. There are few drawbacks in using the BAMG in the present form due to their extensive processing methods and possible stone formation in long term applications. Unlike the intestinal submucosa, the submucosa of the urinary bladder is intimately attached to the muscular bladder wall. The complete removal of the muscle cells from the mucosa is tedious and difficult and strong treatments with proteolytic enzymes and detergents are needed. These chemical and enzymatic treatments can damage the structural integrity of the ECM components of the BAMG and subsequent deleterious results. This drawback limits the use of this tissue material for cardiovascular applications. Only through investigational work with this material in other body systems will the full effect of this harsh processing-whether deleterious or beneficial-be elucidated.

2.6 Amniotic membrane

The amniotic membrane is the innermost of the three layers forming the fetal membranes (Figs. 2.4 and 2.5). It is a translucent membrane composed of an inner layer of epithelial cells, planted on a basement membrane that in turn is connected to a thin

FIGURE 2.4 The avascular amniotic membrane and the vascularized chorion are only loosely attached to each other.

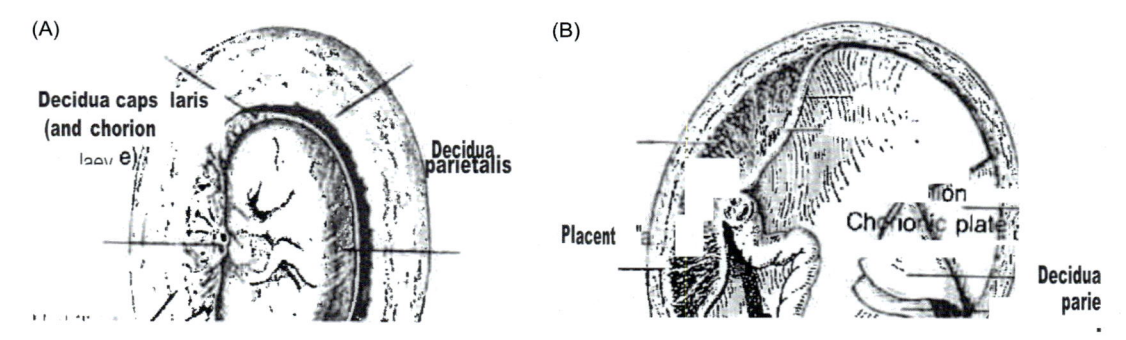

FIGURE 2.5 The amniotic membrane seen in placenta. The amniotic membrane is the innermost layer enwrapping the fetus in the amniotic cavity, which contains the amniotic fluid (A). The amniotic membrane is fused with the chorion to form the fetal membrane, when the fetus enlarges at the end of the first trimester (B). These two membranes are contiguous with and a part of the placenta.

connective tissue membrane by filamentous strand [53]. Outside the amnion is the chorion laevae comprising of connective tissue containing the fetal vessels.

The outermost layer of the fetal membranes, the decidua capsularis, is the only component of the maternal origin and is composed of modified endothelium [32,53]. The amnion varies in thickness from 0.02mm to 0.5 mm in thickness. It contains no blood vessels and has no direct blood supply [20]. Bourne described the amnion as consisting of five layers from within outward: (a) epithelium, (b) basement membrane, (c) compact layer,

(d) fibroblast layer, and (e) spongy layer [20]. The epithelial layer consists of a single layer of amniotic membrane epithelial cells. The basement membrane is a thin layer composed of reticular fibers. The compact layer is a dense layer almost totally devoid of cells and consists mainly of a complex reticular network. The fibroblastic layer is the thickest layer of the amnion and consists of fibroblasts embedded in a loose network of reticulum. The outermost spongy layer forms the interfaced between the amnion and chorion and consists of wavy bundles of reticulum bathed in mucin.

Amnion structure contains large amount of collagen, hyaluronan (hyaluronic acid, heparin sulfate), small proteoglycans (biglycan, decorin) and several growth factors (EGF, FGF-2, TGF-β, TGF-α, KGF). Collagen types I, III, IV, and VII, laminin and fibronectin have been identified in amniotic basement membrane and stromal amnion. Processing and preparation of the membrane is carried out under sterile conditions. Amniotic membrane is obtained at parturition and cleaned of blood with saline containing penicillin, streptomycin, amphotericin B and clindamycin [20]. It is separated from the chorion by blunt dissection, Washed in sterile water, and treated by socking for 3 h in a 10% solution of trypsin to lyse the cells. The membrane is then sterilized with gamma irradiation and frozen until clinical use [54].

Fakuda et al. [55] demonstrated similarities between the laminin-1, laminin-5, fibronectin and type VII collagen components of the basement membranes of conjunctiva, cornea and amniotic membrane. Table 2.6 provides the structural similarities of Amniograft, cornea and conjunctiva. The current popularity of amniotic membrane as an occular graft is often due to their structural similarities, which is thought to inhibit conjunctival overgrowth and provide a good substrate for normal epithelial migration [56]. When implanted as a graft, amniotic membrane promoted reepithelialization of the corneal surface with cells that expressed corneal type keratin. It has also been reported to prevent fibrosis, adhesion and infection and act as physical barrier to protect wounds [53–55]. Because it lacks most of the major histocompatibility antigens, it is unable to initiate a strong immune response, thereby limiting the amount of inflammation observed following implantation. Table 2.7 provides a summary of the biological properties of amniotic membrane to that of other ECM.

Human amnion, which is processed to yield a uniform, acellular biofabric, is a superior material for a variety of product applications [19]. The report provides a method for preparing a collagen biofabric from human placental membrane having a chorionic and amnionic membrane, by completely decellularizing the amniotic membrane (Fig. 2.6). This

TABLE 2.6   Structures of aminograft, cornea and conjunctiva.

Natural cornea	Natural conjunctiva	Aminograft
Collagen IV (α3, α4 and α5 chains)	Collagen IV (α1 and α2 chains)	Collagen IV (α3, α4 and α5 chains)
Collagen VII	Collagen VII	Collagen VII
Laminin (α1, α3, β1, β3, Υ1 and Υ2 chains)	Laminin (α1, α3, β1, β3, Υ1 and Υ2 chains)	Laminin (α1, α3, β1, β3, Υ1 and Υ2 chains)
Heparin sulfate	Heparin sulfate	Heparin sulfate
Proteoglycans	Proteoglycans	Proteoglycans

TABLE 2.7   Properties of matrix membranes.

Membrane	Biological properties
SIS	Angiogenic effect Anti-inflammatory Non-immunogenic Biocompatible, endothelialization No calcification in-subcutaneous implantation.
Amniotic membrane	Anti-bacterial effect Non-immunogenic Anti-inflammatory Angiogenic on acellular membrane No calcification on subcutaneous Implantation, biocompatible.
BAMG	Angiogenic effect Non-immunogenic Biocompatible, endothelialization Calcification? Cells are not removed completely

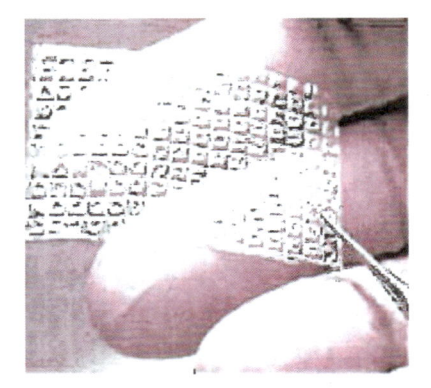

FIGURE 2.6   Dehydrated human amniotic membrane allograft (ambiodry).

collagen biofabric of the invention has utilities in the medical and surgical field including for example, blood vessels, tendon and ligament replacement, wound dressing, surgical grafts, ophthalmic uses, sutures and cardiac tissue repair. The benefits of the biofabric are, in part, due to its physical properties such as biomechanical strength, flexibility, suturability and low immunogenicity [19,23].

The amniotic membrane is separated from chorionic membrane and washed extensively with sterile 0.9% NaCI solution, at 4 °C. The amnion is then decellularized completely by removing all cellular material and cellular debris to get 10−40/m thick smooth and clear material. The decellularization method consists of selective removal of cellular components from the membrane such as cells and nuclear membranes and the DNA proteins. The applied procedure involves the use of two different detergents Triton X-100 and sodium dodecyl sulfate (SDS). Triton X-100, a nonionic surfactant was used initially to treat the amnion at a concentration of 0.1% in saline, with the addition of 1 mM phenyl methane sulfonyl fluoride (PMSF) as a proteinase inhibitor to prevent degradation of the extracellular matrix. The samples were immersed in the surfactant solution and soaked overnight at room temperature. This treatment removes the cellular membrane and proteins by disrupting the lipids.

The second detergent treatment was performed with 0.5% SDS with1mM PMSF in saline for 24 hours at room temperature in a mechanical shaker at 100 rpm. The SDS dissolves the nuclear envelope and subsequently removes them. The membrane is also extracted with 1% ethylene diamine tetra acetic acid (EDTA) to inhibit metallic enzymes and finally treated with DNase and RNAase to remove all nucleic acids from the membrane. The membranes were washed thoroughly with sterile saline and dried at 35 °C under vacuum [19]. This collagen biofabric has numerous utility in the medical field due, in part to its physical properties, such as biomechanical strength, flexibility, suturability, low immunogenicity in comparison to the traditional membranes used earlier [19]. This collagen biofabric may find a variety of medical applications as a natural material for tissue regeneration and repair in the future to come. The biofabric may be surface coated with appropriate biomolecule (prostaglandins, heparin, hirudin), growth factors (insulin like growth factor, VEGF etc.), antibiotics, antibacterial agents, tissue engineering molecules for specific applications.

The Table 2.8 provides the ECM components of various cardiovascular tissues and amnion. The structural components vary with different tissues and may relate to their functions. However, most of these ECM contain collagen, laminin and glycosaminoglycan as common structural components that found in amnion. The mechanical properties and degradation of amnion is shown in Table 2.9. Amniotic membrane degraded completely in

TABLE 2.8 ECM of blood vessels, myocardium, heart valve and amnion.

Blood vessels [57]	Myocardium [58]	Heart valve [59]	Amnion [20]
Collagen IV	Collagen I, III	Collagen	Collagen IV
Collagen XV	Collagen IV, V	Elastin	Collagen VII
Collagen XVIII	Fibronectin	Fibronectin	Laminin
Laminin	Laminin	Chondroitin sulfate	Heparin sulfate
Nidogen/entactin Agrin Heparin sulfate	Tenascin		Proteoglycans
Proteoglycans			

TABLE 2.9 The mechanical properties and degradation of amniotic membrane.[a]

Material	Degradation (in 0.1% of collagenase)	Elongation (%)	Tensile strength (kg/mm^2)
Amniotic memb	7–21 d	76.8	3.9
Crosslinked Amniotic memb. (0.1%o GA, 30 min)	90 d	175	4.35

[a]Amniotic membrane exposed to 0.1% glutaraldehyde (GA) for 30 min.

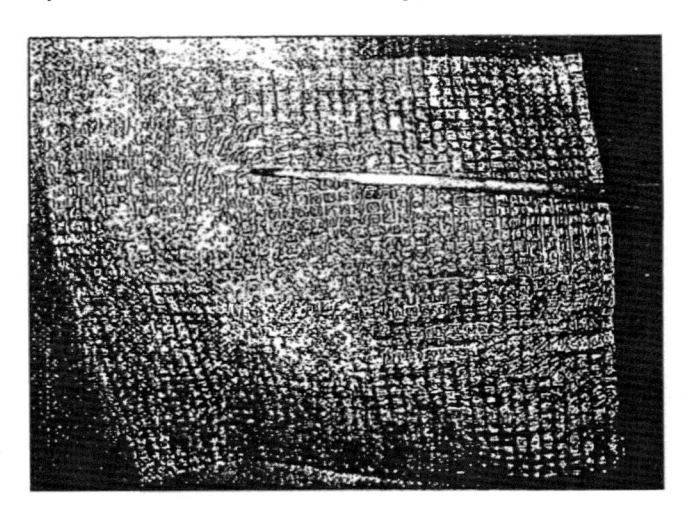

FIGURE 2.7 Acellular collagen fabric (anthromatrix).

0.1% collagenase solution within 7—21 days, however the glutaraldehyde fixed membrane retained its strength up to 90 days. The percent elongation for glutaraldehyde crosslinked amnion was significant (175%), as compared to fresh amnion (76.8%). A mild glutaraldehyde crosslinking does not alter the tensile strength as evident in Table 2.9.

Amniotic membrane is available in different trade names for medical applications and they vary in their properties due to variation in membrane preparation.

1. Ambiodry, Dehydrated human amniotic membrane allograft, packed for room temperature storage. Translucent, lightweight dry membrane. Ambiodry has basement membrane and stromal matrix surfaces. Basement membrane can be identified by its concave grid pattern and the stromal surface can be identified by its convex grid pattern. The stromal side should be in contact with surgical site, when using as allograft. May contain metallo enzymes and groMh factors and some cellular structures Used for ophthalmic applications mainly. (Ambiodry, 3151 Airway Ave, Suite 1-1, Costa Mesa, CA 92626).

2. Anthromatrix, acellular biofabric, packed for room temperature storage (Fig. 2.7). Translucent, optically clear. They claim to be completely acellular and free of metallo enzymes and cell components and so better strength, elasticity and slow degradation rate compared to normal membranes. May be used for a variety of tissue repair applications. (Anthrogenesis Corporation, Hanover Technical Center, Suite 110, 45 Horsehill Road, Cedar Knolls, NJ 07927).

3. Amnio Graft, Freeze-dried, opaque or cloudy. Available in frozen forms and mainly claimed for repairing damaged ocular surface. (Bio-Tissue Inc., 7000 SW 97 Ave, Suite 211, Miami, FL 33173).

2.7 Pericardium and fascia lata

Pericardial tissue and fascia lata are investigated for a variety of biomedical applications such as fabricating tissue valves, dermal patches, ophthalmic applications. Bovine or

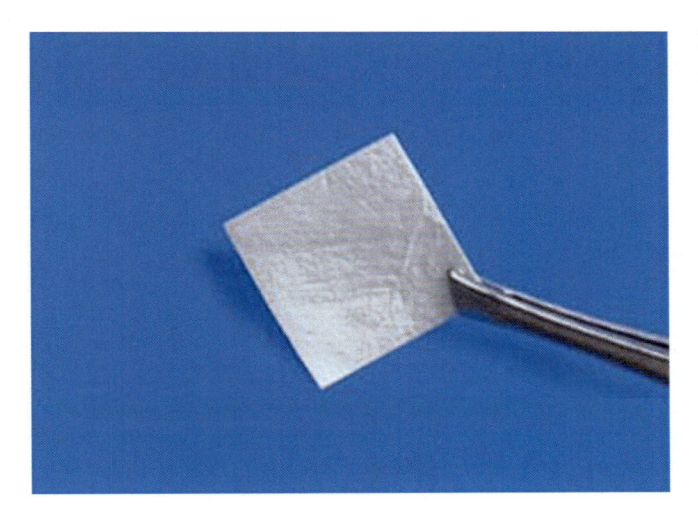

FIGURE 2.8 IO patch pericardium patch graft.

porcine pericardial tissues are collected and extracted with detergents and decellularized as that of dermal or BAMG tissues. Pericardium consists of type I collagen fibrils and are packed in bundles to get thick (400−600 μm) membranes [60]. The main problem in using pericardial tissue for long term implants is their degradation and hence crosslinking of the membrane is needed. The crosslinking of the membrane causes calcification and degeneration of the graft material (i.e., tissue valves) and hence, various surface treatments (amino oleic acid, heparin, ferric irons etc.) are proposed [26,61]. However, IO Patch, Ophthalmic patch grafts (Tutoplast®), is extensively employed for treating a variety of ocular conditions (Fig. 2.8).

The fascia lata is the thick band of connective tissue attaching the pelvis to the knee on the lateral side of the leg. Its muscular components at the hip join to thick connective tissues that help stabilize and steady the hip and the distal section of the fascia lata is harvested for graft prosthesis. In its native state, the fascia lata is composed of heavy, parallel bundles of type I collagenous fibers that are held together by extracellular matrix tissue. Processing of fascia involves ethanol extraction followed by high-pressure washing with antibiotics. The extracted tissue is then lyophilized and terminally sterilized with gamma irradiation. Evaluation of the graft material revealed the absence of cells and an almost pure collagen tissue structure [24]. The fascia lata has been used successfully to repair damaged ligament and tendon structures following orthopedic injury. Other uses of cadaveric fascia have included repair of the orbital wall following traumatic facial fractures, as a dura graft following trauma and for urinary tract repair [24,25]. Graft breakdown and tissue erosion into urethra [25] has been reported. The strength of this biopolymer for tissue regeneration needs to be improved for using them in tissue engineering applications.

Decellularized extracellular matrix (dECM) derived from myocardium has been widely explored as a nature scaffold for cardiac tissue engineering applications [48,62−64]. Cardiac dECM offers many unique advantages such as preservation of organ-specific ECM microstructure and composition, demonstration of tissue-mimetic mechanical properties and retention of biochemical cues in favor of subsequent recellularization. However, current processes of dECM decellularization and recellularization still face many challenges including the need for balance between cell removal and extracellular matrix preservation, efficient recellularization of dECM for obtaining homogenous cell distribution, tailoring material properties of dECM for enhancing bioactivity and prevascularization of thick dECM [64−66].

2.8 ECM for repair of damaged muscle

Effective clinical treatments for volumetric muscle loss resulting from traumatic injury or resection of a large amount of muscle mass are not available to date. Tissue engineering may represent an alternative treatment approach [67−69]. Decellularization of tissues and whole organs is a recently introduced platform technology for creating scaffolding materials for tissue engineering and regenerative medicine. The muscle stem cell niche is composed of a three-dimensional architecture of fibrous proteins, proteoglycans, and glycosaminoglycans, synthesized by the resident cells that form an intricate extracellular matrix (ECM) network in equilibrium with the surrounding cells and growth factors. A consistent body of evidence indicates that ECM proteins regulate stem cell differentiation and renewal and are highly relevant to tissue engineering applications. The ECM also provides a supportive medium for blood or lymphatic vessels and for nerves. Thus, the ECM is the nature's ideal biological scaffold material. ECM-based bioscaffolds can be recellularized to create potentially functional constructs as a regenerative medicine strategy for organ replacement or tissue repopulation. The bioscaffolds obtained from animal ECM by decellularization of small intestinal submucosa (SIS), urinary bladder mucosa (UB), and skeletal muscle, are some innovative ECM repair approaches for the application of such strategies in the clinical setting.

The ECM in skeletal muscle is critical for tissue development, structural support, and force transmission [70]. The main components of ECM are largely conserved across animal species [17,70−72]. The ECM of skeletal muscle is organized into three layers; the endomysium layer surrounds the individual muscle fibers, the perimysium surrounds the bundles of muscle fibers known as fascicles, and the epimysium surrounds the entire muscle. Laminin and collagen Type IV form the principal ECM components of the basal lamina layer [73,74]. The reticular lamina, located below the basement membrane, is composed mainly of fibrils of collagens (type I, III, and VI) and fibronectin in a proteoglycan-rich gel [28,73,74].

Stem cells in skeletal muscle are sensitive to biochemical and biophysical cues provided by the ECM [75]. For instance, loss of regenerative capacity in laminin-deficient (dy/dy) mice, as well as enhanced satellite cell activity with laminin-111 supplementation in vivo, suggest a pivotal role for this ECM protein in the regulation of stem cell function post-injury [75−77]. Similarly, the absence of collagen VI in Col6a − / − mice can impair satellite cell self-renewal and repair following injury [77,78]. Conversely, stem cells can also influence ECM composition. While fibroblasts are considered the leading contributor of ECM production in the skeletal [78]. The major extracellular and intracellular components responsible for regulating skeletal muscle function [79]. These decellularized scaffolds preserve the ultrastructure and composition of the ECM [71,72] and are known to contain basement membrane structural proteins, growth factors, and glycosaminoglycans (GAG). Therefore, these scaffolds possess the potential to recruit endogenous host cells while evading the problems associated with the delivery of exogenous cells such as cellular apoptosis, immunogenicity and ineffective delivery [78—80].

2.9 Stem cells for skeletal muscle regeneration

The skeletal muscle microenvironment is heterogeneous, with diverse cell populations that can be influenced by local structural and biochemical cues. Skeletal muscle is endowed with a remarkable capacity for regeneration, primarily due to the reserve pool of muscle-resident satellite cells. The anatomic location of satellite cells is in proximity to vasculature where they interact with other muscle-resident stem/stromal cells such as mesenchymal stem cells (MSCs) and pericytes through paracrine mechanisms [81,82]. A variety of other stem cell populations have also been identified in skeletal muscle including, side population cells [83,84], fibro/adipogenic progenitors [85,86] and interstitial stem cells [87]. These cell types share many features with MSCs such as multipotency and cell surface marker expression and are known to undergo proliferation in response to muscle injury [88]. Additionally, muscle-resident or marrow-derived hematopoietic stem cells (HSCs) are known to rapidly colonize skeletal muscle post-injury [81].

Muscle regeneration and recovery is a complex process that involves several different stem cell populations and ECM components. While the delivery of stem cells in injured or dystrophic muscles has been associated with improvements in muscle repair and function, the exact mechanism by which these cells contribute to muscle repair is unclear. Future studies should focus on identification of the intrinsic and extrinsic regulatory mechanisms that govern satellite and non-satellite cell differentiation and trophic factor secretion during the muscle regeneration process. Additionally, the mechanical and biochemical cues provided by key ECM components that can promote or dysregulate stem cell activity should be examined. This information will be crucial in the discovery of biomaterial-based strategies to augment the stem cell mediated muscle repair.

2.10 3D bioprinting for organ regeneration

Human organs are very complex structures formed by the combined, functional organization of multiple tissue types. The cells in these organs are highly specialized and group together to perform distinctive functions [89]. Organ dysfunction or failure is drastically increasing due to traumatic injury and disease [90]. Often, clinical treatments are limited by a paucity of available donors and immune rejection of donated tissue [91]. In the search for alternatives to conventional treatment strategies for the repair or replacement of missing or malfunctioning human tissues and organs, tissue engineering approaches are being explored as a promising solution [90—93]. Presently, tissue engineering approaches have been widely studied in cartilage, bone, skin, vascular tissue and nerve regeneration, among others [92—96]. When designing a tissue engineered scaffold, the combination of material, biological and engineering requirements must be considered in an application-specific manner [93—97]. Biomimetic design of the scaffolds, including 3D structural characteristics and physical properties, can substantially enhance the physiological performance through appropriate cell—cell and cell—matrix interactions, further enhancing biological functions [96—99].

The conventional methodology makes use of scaffolds as matrices to load cells [100]. These scaffolds can be fabricated from either naturally derived polymers such as gelatin [101—103], collagen [103—105], hyaluronic acid [103,106], and alginate [101,107], or synthetic polymers such as poly(ε-caprolactone) (PCL), poly(lactic acid) (PLA), poly(glycolic acid) (PGA), and poly(lactic-co-glycolic acid) (PLGA) [108—112]. The scaffolds serve as three-dimensional (3D) templates that support cells to attach, proliferate, and expand throughout the entire structure before they develop their own extracellular matrix (ECM), which eventually leads to the generation of mature cell-laden grafts with comparable properties to their native counterparts. Studies have shown that the phenotypes of seeded cells can be regulated in the scaffolds by applying a combination of different biological and physical stimuli, including growth factors [112,113] shear stress [114,115], as well as electrical [116—118] and mechanical cues [119,120]. However, there are limitations for these conventional scaffold-based approaches, including the intrinsic inability to mimic the complex microstructures of biological tissues [100].

Three-dimensional (3D) bioprinting is evolving into an unparalleled bio-manufacturing technology due to its high-integration potential for patient-specific designs, precise and rapid manufacturing capabilities with high resolution, and unprecedented versatility. It enables precise control over multiple compositions, spatial distributions, and architectural accuracy/complexity, therefore achieving effective recapitulation of microstructure, architecture, mechanical properties, and biological functions of target tissues and organs. 3D printing is a rapid prototyping and additive manufacturing technique used to fabricate complex architecture with high precision through a layer-by-layer building process [121]. This automated, additive process facilitates the manufacturing of 3D products having precisely controlled architecture (external shape, internal pore geometry, and interconnectivity) with highly reproducibility and repeatability [95,122]. Bioprinting utilizes biomaterials, cells or cell factors as a "bioink" to fabricate prospective tissue structures.

Biomaterial parameters such as biocompatibility, cell viability and the cellular microenvironment strongly influence the printed product. Various printing technologies have been investigated, and great progress has been made in printing various types of tissue, including vasculature, heart, bone, cartilage, skin and liver. Therefore, in the regeneration field, it can provide an excellent alternative for biomimetic scaffold fabrication by accurately positioning multiple cell types and biofactors simultaneously into complex multiscale architectures that better represent the structural and biochemical complexity of living tissues or organs [121,123,124]. In the past three decades, 3D bioprinting has been widely developed to directly or indirectly fabricate 3D cell scaffolds or medical implants for the [121,125—127] field of regenerative medicine. It offers very precise spatiotemporal control on placement of cells, proteins, DNA, drugs, growth factors, and other bioactive substances to better guide tissue formation for patient-specific therapy.

The 3D imaged tissue or organ model is divided into 2D horizontal slices that are imported into a 3D bioprinter system for the layer-by-layer deposition. Considering the available 3D bioprinting techniques, the cell types (differentiated or undifferentiated), biomaterials (synthetic or natural), and supporting biochemical factors are then selected, and the configuration of these printing components drives the construction of the 3D tissues and organs [123]. This integrated technique (imaging-design-fabrication) can recreate more complex 3D organ level structures and incorporate mechanical as well as biochemical cues that are crucial elements of the whole organ architecture. [123,128]. In addition, this technique has the capacity to build a 3D tissue-or organ-specific microenvironment by mimicking the natural, highly dynamic yet variable 3D structures, mechanical properties, and biochemical microenvironments [129]. In this manner, 3D bioprinting for organ regeneration involves additional strategies for printing multiple living cells, including vasculature and neural network integration, and eventually developing the specific functions of 3D bioprinted organ analogues.

A unique aspect of this technology is its ability to achieve a personalized therapeutic schedule to address individual patient needs [130]. Moreover, advanced materials engineering approaches featuring biologically dynamic variations will further allow temporal evolution of bioprinted tissue constructs that potentially meet the requirements of dynamic tissue remodeling during developmental processes. Furthermore, 3D bioprinting techniques have shown the potential to facilitate the development of realistic tissue/organ models, therefore this technology is also expected to translate advancing the needs of other specific applications such as models for pharmaceutical/toxicological screening [131].

Over the past few years, researchers not only have demonstrated proof-of-concept examples of different bioprinting technologies, but also have showed possibilities how 3D bioprinting may change the future of tissue engineering, ranging from fabrication of organ and tissue constructs for functional regeneration to relevant models for pharmacological investigations [132]. The 3D cell-embedding volumes of biomaterials generated by bioprinting could serve as biomimetic constructs with desired composition, structure, and architecture to ensure better cell viability and more importantly support the functionality of the tissues, as demonstrated by numerous studies where tissues such as vasculature, heart, liver, cartilage, bladder [132–134] and skin [132] have been bioprinted. Each of these tissues/organs is highly complex and may require a combination of several bioprinting techniques along with specifically designed bioinks to introduce structural heterogeneity and functionality. For example the sacrificial bioprinting strategy may be integrated into other deposition methods to produce hierarchically vascularized tissues; and bioinks derived from tissue-specific dECM may be fitted on a multi-material bioprinter to enable spatially defined deposition of bioinks that matches the architecture of the target organs to be printed. Although challenges still present, with new niches for technological developments on the instrumentation with improved spatial and temporal resolutions as well as optimized bioinks and cell sources for specific organs, it is expected that 3D bioprinting will eventually become one of the most efficient, reliable, and convenient methods to biofabricate tissue constructs in the near future. Combination with the stem cell technologies [133,135] and advanced materials engineering approaches featuring stimuli-responsiveness [135–137] will further allow temporal evolution of bioprinted tissue constructs that potentially meet the requirements of dynamic tissue remodeling during developmental processes.

2.11 Nanosystem delivery of cellular mediators

Tissue-regeneration strategies are often broken down into three categories: (i) direct injection of bolus cells into the tissue of interest or the systematic circulation, (ii) implantation of cells after they have been combined to form a three-dimensional tissue structure, often within a bioreactor, and (iii) scaffold-based delivery of signaling molecules such as low-molecular-weight drugs, proteins and oligonucleotides that stimulate cell migration, growth and differentiation [138]. These signaling molecules, are broadly grouped into the overlapping categories of mitogens (stimulate cell division), growth factors (originally identified by their proliferation-inducing effects, but have multiple functions) and morphogens (control generation of tissue form). Precise control over the signaling of these factors in a local area may potentially allow control over a regenerative process. Growth factors are soluble-secreted signaling polypeptides capable of instructing specific cellular responses in a biological environment [139] The specific cellular response trigged by growth factor signaling can result in a very wide range of cell actions, including cell survival, and control over migration, differentiation or proliferation of a specific subset of cells.

Two distinct strategies for biomaterial presentation of growth factors in tissue engineering have been pursued: (i) chemical immobilization of the growth factor into or onto the matrix and (ii) physical encapsulation of growth factors in the delivery systems. The former approach typically involves chemical binding or affinity interaction between the growth factor-containing polymer substrate and a cell or a tissue [138]. The latter approach is achieved by the encapsulation, diffusion and pre-programmed release of growth factor from substrate into the surrounding tissue. The selection of ideal delivery systems is as important as the properties of the GFs and biomaterials used to achieve the successful regeneration of complex tissues. The rate controlling delivery systems should be simple to fabricate with high loading efficiency, cost effectiveness, and stability for in vivo applications. These delivery systems should also enhance bioavailability, minimize immune responses, and overcome presystemic metabolism, clearance, nonspecific cellular uptake, and secondary damage to cells [140].

Several properties of these delivery systems are important to achieve increased loading efficiency and optimal GF release kinetics. Traditional delivery systems are classified by shape (particle, spheres, and capsules), by size (nano: 1−100 nm; submicrometer: 100 nm−1 μm; micro: 1−100 μm; macroscale >100 μm), and by surface properties (porous vs nonporous) [140,141]. Various other delivery systems have been developed and specialized to improve the clinical translational efficacy of GFs by exercising control over their spatiotemporal release. GFs are often characterized by their pH and MW, which each affect their ability to bind with their delivery-system-affiliated biomaterials [142]. There are also biomaterials-side factors that affect these interactions, mainly MW and surface charge of the selected biomaterial. These characteristics create many complex interactions that may prevent or facilitate GF loading and release, yet the rationale behind the specific combinations of various GFs and delivery systems is often unclear or poorly defined in many instances.

The most cogent strategy often involves designing and engineering GF delivery systems that mimic the natural ECM microenvironment. In the pursuit of the recreation of this highly complex environment, scientists have developed delivery systems to optimize and further control the loading efficiency and release kinetics of GFs. These delivery systems are often developed to optimize an exact GF/biomaterial combination for precise release kinetics and loading efficiency to address a specific defect or injury model. They include traditional systems such as simple hydrogel loading strategies that provide a burst release followed by a sustained release [140−142]. In recent years, modern systems, such as complex architectures that enable sequential or on-demand release, have also become increasingly representative of physiological GF expression profiles. However, the scientific community has many obstacles to overcome before delivery systems reach the regenerative potential of their physiological counterparts. Realizing the full potential of therapeutic GF delivery will require engineered delivery systems that provide the right GF at the right time to optimally promote tissue regeneration. As tissue engineering advances, this further control and GF profile optimization will enable scientists to direct endogenous cells' regenerative mechanisms, ultimately improving tissue healing when translated into the clinical setting.

Major advances have been made over recent years in the construction of polymer-based growth factor-delivery systems that allow the controlled release of growth factors, but there remain a number of challenges that will need to be addressed in the future. These include enhancing the stability of encapsulated proteins to allow release for extended times (e.g. weeks to months), the difficulty in scaling up certain approaches, determining the appropriate structure/compartmentalization of delivery materials to allow multiple factors to be released with distinct kinetics and the release of certain factors owing to their hydrophobic nature or strong charge—charge interactions between the polymers and the growth factors. Also, a major challenge is the adaptation of these approaches to clinical use. The majority of work carried out to date has been done with animal models and it is unclear how well much of this work will translate to the use of human recombinant growth factors in human patients [139—142]. In a broader perspective, further advances in the field will rely on multidisciplinary approaches that combine medicine, chemistry, engineering and pathology to develop effective strategies to treat complex wounds and pathologies.

2.12 Perspective

Tissue engineering methods designed to restore diseased and damaged tissues depend on the presence of a matrix structure that is amenable to cell growth and proliferation. In order to support the development of complex biologic structures, such biomaterials must effectively interact with the surrounding tissue and incite the host to populate the graft with new tissue. However, due to the highly specialized composite structure of native ECM, generation of biomaterials to replicate the complex ECM/cell interactions remains a significant challenge.

Understanding the embryonic genesis and patterning of the ECM and the mechanisms of bidirectional signaling during morphogenesis will provide critical information to direct regenerative medicine strategies that attempt to recapitulate the signaling mechanisms used by ECM during healing and regeneration. The ECM graft acts as the scaffold into which host cells grow, proliferate, and differentiate. It seems that the acellular amniotic membrane may be a good ECM scaffold, which need to be explored for tissue regeneration and repair.

Native ECMs provide a complex in vivo architecture and native physical and mechanical properties that support high biocompatibility. However, the applications of native ECMs are limited due to their tissue-specificity and chemical complexity. Artificial ECMs have been fabricated in an attempt to create a broadly applicable scaffold by using controllable components and a uniform formulation. On the other hands, artificial ECMs fail to mimic the properties of a native ECM; consequently, their applications in tissues are also limited. Understanding the structures and functions of ECMs in each tissue, controlling the fibrillogenesis mechanism using individual ECM components, and the recent development of new tools for bio-printing and imaging will greatly help to implement such a hybrid ECM in tissue engineering in the near future. It appears, although challenges still remain in this research field, further multidisciplinary research to advance printing techniques, printable bioink materials, engineering designs, stem cell technology and nanosystem delivery of growth factors can address the current challenges and realize the emerging potential of 3D organ bioprinting.

References

[1] Badylak SF. The extracellular matrix as a scaffold for tissue reconstruction. SemCell Developmental Biol 2002;13:377–83.

[2] Han ZC, Liu Y. Angiogenesis, state of the art. Int J Hematol 1999;70:68–82.

[3] Swinehart IT, Badylak SF. Extracellular matrix bioscaffolds in tissue remodeling and morphogenesis. Dev Dyn 2016;245:351–60.

[4] Reing JE, Brown BN, Daly KA, Freund JM, Gilbert TW, Hsiong SX, et al. The effects of processing methods upon mechanical and biologic properties of porcine dermal extracellular matrix scaffolds. Biomaterials 2010;31:8626–33.

[5] Badylak SF. Extracellular matrix for myocardial repair. Heart Surg Forum 2003;6:E20–6.

[6] Badylak SF, Freytes DO, Gilbert TW. Extracellular matrix as a biological scaffold material: Structure and function. Acta Biomater 2009;5:1–13.

[7] Grauss RW, Hazekamp MG, Oppenhuizen F, van Munsteren CJ, Gittenberger-de-Groot AC, DeRuiter MC. Histological evaluation of decellularised porcine aortic valves: matrix changes due to different decellularisation methods. Eur J Cardiothorac Surg 2005;27:566–71.

[8] Badylak SF, Lantz GC, Coffey A, Geddes LA. Small intestinal submucosa as a large diameter vascular graft in the dog. J Surg Res 1989;47:74–80.

[9] Dolla vecchia L, Engum S, Kogon B, et al. Evaluation of small intestine submucosa and acellular dermis as diaphragmatic prostheses. J Pediatr Surg 1999;34:167–71.

[10] Lantz GC, Badylak SF, Coffey A, et al. Small intestinal submucosa as a small diameter arterial autograft in the dog. J Invest Surg 1990;3:217–27.

[11] Clarke KM, Lantz GC, Salisbury SK, Badylak SF, Hiles MC, Voytik SL. Intestine submucosa and poly propylene mesh for abdominal wall repair in dogs. J Surg Res 1996;60:107–14.

[12] Aiken SW, Badylak SF, Toombs JP, Shelbourne KD, Niles MC, Lantz GC, et al. Small intestinal submucosa as an intra-articular ligamentous repair material: a pilot study in dogs. Vet Comp Ortho Traumatol 1994;7:124–8.

[13] Kropp BP, Eppley BL, Prevel CD, Rippy MK, Harruff RC, Badylak SF, et al. Experimental assessment of small intestine submucosa as a bladder wall substitute. Urology 1995;46:396–400.

[14] Wainwright DJ. Use of an acellular allograft dermal matrix (AIIoDerm) in the management of full-thickness burns. Burns 1995;21:243–8.

[15] Chaplin JM, Costantino PD, Wolpoe ME, Bederson JB, Griffey ES, Zhang WZ. Use of an acellular dermal allograft for dural replacement: an experimental study. Neurosurgery 1999;45:320–6.

[16] Isch JA, Engum SA, Ruble CA, Davis MM, Grosfeld JL. Patch esophagoplasty using AlloDerm as a tissue scaffold. J Pediatr Surg 2001;36:266–8.

[17] Wefer J, Sievert KD, Schlote N, Wefer AE, Nune SL, Dahiya R, et al. Time-dependent smooth muscle regeneration and maturation in a bladder acellular matrix graft: histological studies and in vivo functional evaluation. J Urol 2001;165:1755–9.

[18] Reddy PP, Barrieras DJ, Wilson G, Bagli DJ, McLorie GA, Khoury AE, et al. Regeneration of functional bladder substitute using large segment acellular matrix allografts in a porcine model. J Urol 2000;164:936–41.

[19] Hariri RJ, Kaplunovsky AM, Murphy PA. Collagen biofabric and methods of preparation and use therefor, US Patent, 2004/0048796A1, March 2004.

[20] Dua HS, Gomes JAP, King AJ, Maharajan VS. The amniotic membrane in ophthalmology. Surv Ophthalmol 2004;49:51−77.

[21] Lafrance H, Bergeron F, Roberge C, Germain L, Auger F. Tissue engineered heart valve, US Patent, US 2003/00 27332A1, February 6, 2003.

[22] Eugene B. Method and construct for producing graft tissue from an extracellular matrix, US Patent, 5800537, September 1, 1998.

[23] Mligiliche N, Endo K, Okamoto K, Fujimoto E, Ide C. Extracellular matrix of human amnion manufactured into tubes as conduits for peripheral nerve regeneration. J Biomed Mater Res 2002;63:591−600.

[24] Blander DS, Zimmern PE. Cadaveric fascia lata sling: analysis of five recent adverse outcomes. Urology 2000;56:596−9.

[25] Golomb J, Groutz A, Mor Y, Leibovitch I, Ramon J. Management of urethral erosion caused by a pubovaginal fascia sling. Urology 2001;57:159−60.

[26] Schoen FJ, Tsao JW, Levy RJ. Calcification of bovine pericardium used in cardiac valve bioprostheses, Implications for the mechanism of bioprosthetic tissue mineralization. Am J Pathol 1986;123:134−45.

[27] Hodde J. Naturally occurring scaffolds for soft tissue repair and regeneration. Tissue Eng 2002;8:295−308.

[28] Yue B. Biology of the extracellular matrix: an overview. J Glaucoma 2014;23:S20−3.

[29] Halper J, Kjaer M. Basic components of connective tissues and extracellular matrix: elastin, fibrillin, fibulins, fibrinogen, fibronectin, laminin, tenascins and thrombospondins. Adv Exp Med Biol 2014;802:31−47.

[30] Mouw JK, Ou G, Weaver VM. Extracellular matrix assembly: a multiscale deconstruction. Nat Rev Mol Cell Biol 2014;15:771−85.

[31] Hodde JP, Badylak SF, Brightman AO, Vortik-Harbin SL. Glycosaminoglycan content of small intestinal mucosa: a bioscaffold for tissue replacement. Tissue Eng 1996;2:209−17.

[32] Meinert M, Eriksen GV, Petersen AC, Helmig RB, Laurent C, Uldbjerg N. Proteoglycans and hyaluronan in human fetal membranes. Am J Obster Gynecol 2001;184:679−85.

[33] McPherson TB, Badylak SF. Characterization of fibronectin derived from porcine small intestinal submucosa. Tissue Eng 1998;4:75−83.

[34] Badylak S, Meurling S, Chen M, Spievack A, Simmrns-Byrd A. Resorbable bioscaffolds for esophageal repair in a dog model. J Pediatr Surg 2000;35:1097−103.

[35] Schwarzbauer JE. Basement membranes putting up the barriers. Curr Biol 1999;9:R242−4.

[36] Badylak SF, Tullius R, Kokin K, Shelbourne KD, Klootwyk T, Voytik SL, et al. The use of xenogenic small intestinal submucosa as a biomaterial for Archilles tendon repairs in a dog model. J Biomed Mater Res 1995;29:977−85.

[37] Lantz GC, Badylak SF, Coffey AC, Geddes LA, Sandusky GE. Small intestinal submucosa as a superior vena cava graft in the dog. J Surg Res 1992;53:175−81.

[38] Matheny RG, Hutchison ML, Dryden PE, Hiles MD, Shaar CJ. Porcine small intestine submucosa as a pulmonary valve leaflet substitute. J Heart Valve Dis 2000;9:769−74.

[39] Altman AJ, Mcpherson TB, Badylak SF, Merrill LC, Kallakury B, Sheehan C, et al. Xenogenic extracellular matrix grafts elicit a Th2 restricted immune response. Transplantation 2001;71:1631−40.

[40] Walden JL, Garcia H, Hawkins H, Crouchet JR, Traber L, Gore DC. Both dermal matrix and epidermis contribute to an inhibition of wound contraction. Ann Plast Surg 2000;4:162−6.

[41] Dalla vevecchia L, Engum S, kogon B, Jensen E, Davis M, Grosfeld J. Evaluation of small intestine submucosa and cellular dermis as diaphragmatic prosthesis. J Pediatr Surg 1999;34:167−71.

[42] Inoue Y, Anthony JP, Lieon P, Young DM. Acellular human dermal matrix as a small vessel substitute. J Reconstr Microsurgery 1996;12:307−11.

[43] Meezan E, Hjelle JT, Brendel K. A simple, versatile nondisruptive method for the isolation of morphologically and chemically pure basement membranes from several tissues. Life Sci 1975;17:1721−32.

[44] Piechota HJ, Dahms SE, Probst M, Gleason CA, Nunes LS, Dahiya R, et al. Functional rat bladder generation through xenotransplantation of the bladder acellular matrix graft. Br J Urol 1998;81:548−59.

[45] Da LC, Huang YZ, Xie HQ. Progress in development of bioderived materials for dermal wound healing. Regen Biomater 2017;4:325−34.

[46] Rosario DJ, Reilly GC, Ali Salah E, Glover M, Bullock AJ, Macneil S. Decellularization and sterilization of porcine urinary bladder matrix for tissue engineering in the lower urinary tract. Regen Med 2008;3:145−56.

[47] Simoes IN, Vale P, Soker S, Atala A, Keller D, Noiva R, et al. Acellular urethra bioscaffold: decellularization of whole urethras for tissue engineering applications. Sci Rep 2017;7:41934.

[48] Sreejit P, Verma RS. Natural ECM as biomaterial for scaffold based cardiac regeneration using adult bone marrow derived stem cells. Stem Cell Rev 2013;9:158—71.

[49] Aamodt JM, Grainger DW. Extracellular matrix-based biomaterial scaffolds and the host response. Biomaterials 2016;86:68—82.

[50] Roy DC, Wilke-Mounts SJ, Hocking DC. Chimeric fibronectin matrix mimetic as a functional growth- and migration-promoting adhesive substrate. Biomaterials 2011;32:2077—87.

[51] Roy DC, Hocking DC. Recombinant fibronectin matrix mimetics specify integrin adhesion and extracellular matrix assembly. Tissue Eng Part A 2013;19:558—70.

[52] Dubey G, Mequanint K. Conjugation of fibronectin onto three-dimensional porous scaffolds for vascular tissue engineering applications. Acta Biomater 2011;7:1114—25.

[53] Duanforth DN, Hull RW. The microscopic anatomy of the fetal membranes with particular reference to the detailed structure of amnion. Am J Obstet Gynecol 1958;75:536—47.

[54] Young RL, Cota J, Zund G. The use of an amniotic membrane graft to prevent postoperative adhesions. Fertil Steril 1991;55:624—8.

[55] Fukuda K, Chikama T, Nakamura M, Nishida T. Differential distribution of subchains of the basement membrane components type IV collagen and laminin among the amniotic membrane, cornea and conjunctiva. Cornea 1999;18:73—9.

[56] Kim JC, Tseng SCG. Transplantation of preserved human amniotic membrane for surface reconstruction in severely damaged rabbit corneas. Cornea 1995;14:473—84.

[57] Kalluri R. Basement membranes: structure, assembly and role in tumor angiogenesis. Nat Rev Cancer 2003;3:422—33.

[58] Schacherer CH, Koops D, Wiemer J, Hartmann A, Weis M, Klepzig H, et al. Extracellular matrix structure after heart transplantation. Internatl J Cardiology 1999;68:115—20.

[59] Grauss RW, Hazekamp MG, van Vliet S, Gittenberger-de Groot AC, DeRuiter MC. Decellularization of rat aortic valve allografts reduces matrix remodeling. J Thorac Cardiovasc Surg 2003;126:2003—10.

[60] Sporel E, Wollensak G, Reber F, Pillunat L. Crosslinking of human amniotic membrane by glutaraldehyde. Ophthalmic Res 2004;36:71—7.

[61] Sindhu CV, Chandy T. Effect of alternative crosslinking techniques on the enzymatic degradation of bovine pericardium and their calcification. J Biomed Mater Res 1997;35:357—69.

[62] Crapo PM, Gilbert TW, Badylak SF. An overview of tissue and whole organ decellularization processes. Biomaterials 2011;32:3233—43.

[63] Badylak SF, Taylor D, Uygun K. Whole-organ tissue engineering: decellularization and recellularization of three-dimensional matrix scaffolds. Annu Rev Biomed Eng 2011;13:27—53.

[64] Nakamura N, Kimura T, Kishida A. Overview of the development, applications, and future perspectives of decellularized tissues and organs. ACS Biomater Sci Eng 2017;3:1236—44.

[65] Silva AC, Rodrigues SC, Caldeira J, Nunes AM. Three-dimensional scaffolds of fetal decellularized hearts exhibit enhanced potential to support cardiac cells in comparison to the adult. Biomaterials 2016;104:52—64.

[66] Lee KM, Kim H, Nemeno JG, Yang W, Yoon J. Natural cardiac extracellular matrix sheet as a biomaterial for cardiomyocyte transplantation. Transpl Proc 2015;47:751—6.

[67] LauraTeodori S, Costa A, Marzio R, Perniconi B, Coletti D, Adamo S, et al. Native extracellular matrix: a new scaffolding platform for repair of damaged muscle. Front Physiol 2014;5:1—7.

[68] Agrawal V, Brown BN, Beattie AJ, Gilbert TW, Badylak SF. Evidence of innervation following extracellular matrix scaffold- mediated remodeling of muscular tissues. J Tissue Eng Regen Med 2009;3:590—600.

[69] Dai X, Xu Q. Nanostructured substrate fabricated by sectioning tendon using a microtome for tissue engineering. Nanotechnology 2011;22:494008.

[70] Thomas K, Engler AJ, Meyer GA. Extracellular matrix regulation in the muscle satellite cell niche. Connect Tissue Res 2015;56:1—8.

[71] Londono R, Badylak SF. Biologic scaffolds for regenerative medicine: mechanisms of in vivo remodeling. Ann Biomed Eng 2015;43:577—92.

[72] Wolf MT, Dearth CL, Sonnenberg SB, Loboa EG. Naturally derived and synthetic scaffolds for skeletal muscle reconstruction. Adv Drug Delivery Rev 2015;84:208—21.

[73] Kjaer M. Role of extracellular matrix in adaptation of tendon and skeletal muscle to mechanical loading. Physiol Rev 2004;4:649—98.

[74] Sanes JR. The basement membrane/basal lamina of skeletal muscle. J Biol Chem 2003;278:12601—4.

[75] Zou K, De Lisio M, Huntsman HD, Pincu Y, Mahmassani Z, Miller M, et al. Laminin-111 improves skeletal muscle stem cell quantity and function following eccentric exercise. Stem Cell Transl Med 2014;3:1013—22.

[76] Rooney JE, Gurpur PB, Burkin DJ. Laminin-111 protein therapy prevents muscle disease in the mdx mouse model for Duchenne muscular dystrophy. Proc Natl Acad Sci USA 2009;106:7991—6.

[77] Bönnemann CG, Laing NG. Myopathies resulting from mutations in sarcomeric proteins. Curr OpNeurol 2004;17:529—37.

[78] Sicari BM, Rubin JP, Dearth CL, Wolf MT, Ambrosio F, Boninger M, et al. An acellular biologic scaffold promotes skeletal muscle formation in mice and humans with volumetric muscle loss. Sci Transl Med 2014;6 234ra58.

[79] Elmashhady H, Kraemer BA, Patel KH, Sell SA, Garg K. Decellularized extracellular matrices for tissue engineering applications. Electrospinnning 2017;1:87—99.

[80] Grounds MD. Obstacles and challenges for tissue engineering and regenerative medicine: Australian nuances. Clin Exp Pharmacol Physiol 2018;45:390—400.

[81] Kuang S, Gillespie MA, Rudnicki MA. Niche regulation of muscle satellite cell self-renewal and differentiation. Cell Stem Cell 2008;2:22—31.

[82] Thomas K, Engler AJ, Meyer GA. Extracellular matrix regulation in the muscle satellite cell niche. Connect Tissue Res 2015;56:1—8.

[83] Challen GA, Little MH. A side order of stem cells: the SP phenotype. Stem Cell 2006;24:3—12.

[84] Motohashi N, Uezumi A, Yada E, Fukuda S, Fukushima K, Imaizumi K, et al. Muscle CD31(−) CD45(−) side population cells promote muscle regeneration by stimulating proliferation and migration of myoblasts. Am J Pathol 2008;173:781—91.

[85] Uezumi A, Fukada S, Yamamoto N, Takeda S, Tsuchida K. Mesenchymal progenitors distinct from satellite cells contribute to ectopic fat cell formation in skeletal muscle. Nat Cell Biol 2010;12:143—52.

[86] Joe AW, Yi L, Natarajan A, Le Grand F, So L, Wang J, et al. Muscle injury activates resident fibro/adipogenic progenitors that facilitate myogenesis. Nat Cell Biol 2010;12:153—63.

[87] Cottle BJ, Lewis FC, Shone V, Ellison-Hughes GM. Skeletal muscle-derived interstitial progenitor cells (PICs) display stem cell properties, being clonogenic, self-renewing, and multi-potent in vitro and in vivo. Stem Cell Res Ther 2017;8:158.

[88] Boppart MD, De Lisio M, Zou K, Huntsman HD. Defining a role for non-satellite stem cells in the regulation of muscle repair following exercise. Front Physiol 2013;4:310—16.

[89] Badylak SF, Weiss DJ, Caplan A, Macchiarini P. Engineered whole organs and complex tissues. Lancet 2012;379:943—52.

[90] Rustad KC, Sorkin M, Levi B, Longaker MT, Gurtner GC. Strategies for organ level tissue engineering. Organogenesis 2010;6:151—7.

[91] Zhang LG, Khademhosseini A, Webster TJ. Tissue and organ regeneration: advances in micro and nanotechnology. Pan Stanf Publishing; Stanf 2014.

[92] Cui H, Wang Y, Cui L, Zhang P, Wang X, Wei Y, et al. In vitro studies on regulation of osteogenic activities by electrical stimulus on biodegradable electroactive polyelectrolyte multilayers. Biomacromolecules 2014;15:3146—57.

[93] Place ES, Evans ND, Stevens MM. Complexity in biomaterials for tissue engineering. Nat Mater 2009;8:457—70.

[94] Webster TJ, Zhang LJ. Nanotechnology and nanomaterials: promises for improved tissue regeneration. Nano Today 2009;4:66—80.

[95] Derby B. Printing and prototyping of tissues and scaffolds. Science 2012;338:921—6.

[96] Griffith LG, Swartz MA. Capturing complex 3D tissue physiology in vitro. Nat Rev Mol Cell Biol 2006;7:211—24.

[97] Dvir T, Timko BP, Kohane DS, Langer R. Nanotechnological strategies for engineering complex tissues. Nat Nanotechnol 2011;6:13—22.

[98] Zan G, Wu Q. Biomimetic and bioinspired synthesis of nanomaterials/nanostructures. Adv Mater 2016;28:2099—147.

[99] Billiet T, Vandenhaute M, Schelfhout J, Van Vlierberghe S, Dubruel P. A review of trends and limitations in hydrogel-rapid prototyping for tissue engineering. Biomaterials 2012;33:6020–41.

[100] Leijten J, Rouwkema J, Zhang YS, Nasajpour A, Dokmeci MR, Khademhosseini A. Advancing tissue engineering: a tale of nano, micro and macro scale integration. Small 2015.

[101] Drury JL, Mooney DJ. Hydrogels for tissue engineering: scaffold design variables and applications. Biomaterials 2003;24:4337–51.

[102] Nichol JW, Koshy ST, Bae H, Hwang CM, Yamanlar S, Khademhosseini A. Cell-laden microengineered gelatin methacrylate hydrogels. Biomaterials 2010;31:5536–44.

[103] Hoffman AS. Hydrogels for biomedical applications. Adv Drug Del Rev 2012;64:18–23.

[104] Cen L, Liu W, Cui L, Zhang W, Cao Y. Collagen tissue engineering: development of novel biomaterials and applications. Pediatr Res 2008;63:492–6.

[105] Glowacki J, Mizuno S. Collagen scaffolds for tissue engineering. Biopolymers 2008;89:338–44.

[106] Burdick JA, Prestwich GD. Hyaluronic acid hydrogels for biomedical applications. Adv Mater 2011;23: H41–56.

[107] Augst AD, Kong HJ, Mooney DJ. Alginate hydrogels as biomaterials. Macromol Biosci 2006;6:623–33.

[108] Zhang YS, Xia Y. Inverse opal scaffolds for applications in regenerative medicine. Soft Matter 2013;9:9747–54.

[109] Hubbell JA. Biomaterials in tissue engineering. Biotechnology 1995;13:565–76.

[110] Hutmacher DW. Scaffold design and fabrication technologies for engineering tissues — state of the art and future perspectives. J Biomater Sci Polym Ed 2001;12:107–24.

[111] Shin H, Jo S, Mikos AG. Biomimetic materials for tissue engineering. Biomaterials 2003;24:4353–64.

[112] Yu SZ, Kan Y, Julio A, Kamyar MM, Syeda MB, Jingzhou Y, et al. 3D bioprinting for tissue and organ fabrication. Ann Biomed Eng 2017;45:148–63.

[113] Tayalia P, Mooney DJ. Controlled growth factor delivery for tissue engineering. Adv Mater 2009;21:3269–85.

[114] Niklason LE, Gao J, Abbott WM, Hirschi KK, Houser S, Marini R, et al. Functional arteries grown in vitro. Science 1999;284:489–93.

[115] Ratcliffe A. Tissue engineering of vascular grafts. Matrix Biol 2000;19:353–7.

[116] Tandon N, Cannizzaro C, Chao PH, Maidhof R, Marsano A, Au HT, et al. Electrical stimulation systems for cardiac tissue engineering. Nat Protoc 2009;4:155–73.

[117] Nunes SS, Miklas JW, Liu J, Aschar-Sobbi R, Xiao Y, Zhang B, et al. Biowire: a platform for maturation of human pluripotent stem cell-derived cardiomyocytes. Nat Methods 2013;10:781–7.

[118] Zhang YS, Aleman J, Arneri A, Bersini S, Piraino F, Shin SR, et al. From cardiac tissue engineering to heart-on-a-chip: Beating challenges. Biomed Mater 2015;10:034006.

[119] Huebsch N, Arany PR, Mao AS, Shvartsman D, Ali OA, Bencherif SA, et al. Harnessing traction-mediated manipulation of the cell/matrix interface to control stem-cell fate. Nat Mater 2010;9:518–26.

[120] Khetan S, Guvendiren M, Legant WR, Cohen DM, Chen CS, Burdick JA. Degradation-mediated cellular traction directs stem cell fate in covalently crosslinked three-dimensional hydrogels. Nat Mater 2013;12:458–65.

[121] O'Brien CM, Holmes B, Faucett S, Zhang LG. Three-dimensional printing of nanomaterial scaffolds for complex tissue regeneration. Tissue Eng, Part B 2015;21:103–14.

[122] Farahani RD, Dube M, Therriault D. Three-dimensional printing of multifunctional nanocomposites: manufacturing techniques and applications. Adv Mater 2016;28:5794–821.

[123] Murphy SV, Atala A. 3D bioprinting of tissues and organs. Nat Biotechnol 2014;32:773–85.

[124] Do AV, Khorsand B, Geary SM, Salem AK. 3D Printing of Scaffolds for Tissue Regeneration Applications. Adv Healthc Mater 2015;4:1742–62.

[125] Park JH, Jang J, Lee JS, Cho DW. Three-dimensional printing of tissue/organ analogues containing living cells. Ann Biomed Eng 2017;45:180–94.

[126] Mironov V, Richard VK, Visconti P, Forgacs G, Drake CJ, Markwald RR. Organ printing: tissue spheroids as building blocks. Biomater Sci 2009;30:2164–74.

[127] Zorlutuna P, Annabi N, Camci-Unal G, Nikkhah M, Cha JM, Nichol JW, et al. Microfabricated biomaterials for engineering 3D tissue. Adv Mater 2012;24:1782–804.

[128] Ozbolat IT. Bioprinting scale-up tissue and organ constructs for transplantation. Trends Biotechnol 2015;33:395–400.

[129] Gao B, Yang Q, Zhao X, Jin G, Ma Y, Xu F. 4D bioprinting for biomedical applications. Trends Biotechnol 2016;34:746−56.

[130] Jonathan G, Karim A. 3D printing in pharmaceutics: a new tool for designing customized drug delivery systems. Int J Pharm 2016;499:376−94.

[131] Peng W, Unutmaz D, Ozbolat IT. Bioprinting towards physiologically relevant tissue models for pharmaceutics. Trends Biotechnol 2016;34:722−32.

[132] Zhang YS, Yue K, Aleman J, Moghaddam KM, Bakht SM, Yang J, et al. 3D bioprinting for tissue and organ fabrication. Ann Biomed Eng 2017;45:148−63.

[133] Henmi C, Nakamura M, Nishiyama Y, Yamaguchi K, Mochizuki S, Takiura K, et al. Development of an effective three dimensional fabrication technique using inkjet technology for tissue model samples. AATEX 2007;14:689−92.

[134] Fullhase C, Soler R, Atala A, Andersson K-E, Yoo JJ. A novel hybrid printing system for the generation of organized bladder tissue. J Urol 2009;181:282−3.

[135] Lee V, Singh G, Trasatti JP, Bjornsson C, Xu X, Tran TN, et al. Design and fabrication of human skin by three-dimensional bioprinting. Tissue Eng Part C Methods 2014;20:473−84.

[136] Faulkner-Jones A, Fyfe C, Cornelissen D-J, Gardner J, King J, Courtney A, et al. Bioprinting of human pluripotent stem cells and their directed differentiation into hepatocyte-like cells for the generation of mini-livers in 3d. Biofabrication 2015;7:044102.

[137] Mehrban N, Teoh GZ, Birchall MA. 3d bioprinting for tissue engineering: stem cells in hydrogels. Int J Bioprinting 2016;2:1−14.

[138] Lee K, Silva EA, Mooney DJ. Growth factor delivery-based tissue engineering: general approaches and a review of recent developments. J R Soc Interface 2011;8:153−70.

[139] Cao L, Arany PR, Wang YS, Mooney DJ. Promoting angiogenesis via manipulation of VEGFresponsiveness with notch signaling. Biomaterials 2009;30:4085−93.

[140] Badeau BA, Comerford MP, Arakawa CK, Shadish JA, DeForest CA. Engineered modular biomaterial logic gates for environmentally triggered therapeutic delivery. Nat Chem 2018;10:251−8.

[141] Shi J, Votruba AR, Farokhzad OC, Langer R. Nanotechnology in drug delivery and tissue engineering: from discovery to applications. Nano Lett 2010;10:3223−30.

[142] Subbiah R, Guldberg RE. Materials science and design principles of growth factor delivery systems in tissue engineering and regenerative medicine. Adv Healthc Mater 2019;8:1−24.

3

Engineered surfaces: A plausible alternative in overviewing critical barriers for reconstructing modern therapeutics or biomimetic scaffolds

Preetam Guha Ray, Ragavi Rajasekaran, Trina Roy, Abir Dutta, Baisakhee Saha, Hema Bora, Subrata K. Das and Santanu Dhara

Biomaterials and Tissue Engineering Laboratory, School of Medical Science and Technology (SMST), Indian Institute of Technology — Kharagpur, Kharagpur, India

3.1 Introduction

In the realm of diagnostics and emerging therapeutics, healthcare sectors like therapeutics, point-of-care diagnostics (POCT) and theranostics have been witnessing a paradigm shift. The complexity of life-style in present scenario demands sophisticated and dedicated medical devices that can be deployed for obtaining desired outcomes both in quick succession and also impart long-term benefits. The advent of such emerging healthcare devices will immensely impact patient health especially when the turn-around time is critical, chronic injuries, burn patients, cardiac arrest or brain injury being some of the examples for such conditions. It is of paramount importance to highlight that a well-engineered platform is the core component of any therapeutic system. Designing platforms with multidimensional functionality provides a stable microenvironment, capable of mimicking the dynamic *in vivo* conditions thus securing the success aspect of any medical device. The interaction of cells, tissues and biomolecules with its microenvironment involve sequential coherence of complex biochemical events leading to occurrence of desired bio-affinitive reactions. For a heterogeneous system, bio-affinitive reactions demands precise control over the microenvironment for specific therapeutic applications as fate of biomolecules depends on their surroundings and may face steric hindrance and easily lose their activity when brought in close proximity to such platforms.

The advent of nanotechnology has revolutionized areas like tissue engineering, drug delivery, *in vivo* imaging, *in vitro* diagnostics or molecular therapeutics [1]. It is significant to compare the field of nanomedicine in the current scenario wherein the usage of nanoengineered components are deployed as drug carrier for targeted drug delivery or has been utilized as direct therapeutics for physical destruction in malignant tissues [2,3]. However, it is important to note that after decades of continuous research and development worth billions of dollars, only few successful products could actually make it to the patient's bed side. Products including dressing materials, implants or grafts for tissue-engineering, point-of-care diagnostic devices for quick detection of blood glucose, lactate, cholesterol, triglycerides, urea or creatinine, and other liposome or protein based delivery system as nanocarriers for doxorubicin or paclitaxel respectively are being given approval for clinical use [4–7]. It is clear that transformation of innovations, promised in the field of theranostics by self-guided precise nanotherapeutics is still a prevalent challenge that needs to be addressed in order to surface its idea for real time applications. Overall it was clear that therapeutic application of theranostic products is only possible when we can develop and also have a better understanding of the surface chemistry, interaction between interparticle and also with biological elements, for example serum proteins with its bioenvironment [8–11].

The uprising of therapeutics has also witnessed the upscale growth of global market for tissue engineering with organ specific tissue engineered grafts gathering much attention in treating chronic injuries. As of 1998, the global market size for tissue engineered products was recorded at an estimated US$ 1.7 billion [12]. In 2016, it was estimated that the global tissue engineering market had a market size of US$ 5 billion and was expected to potentially growth at an exponential rate in near future. Tissue engineering based medical devices can render solution which was believed to replace the existing methods of tissue repair which may include and not limited to mechanical devices, transplants and surgical reconstruction. Additionally, it has become apparent that with increasing occurrence of lifestyle disorders like obesity and diabetes or the surge in trauma cases with rapidly aging population, the need for next generation tissue engineering devices with advance health benefits is a prerequisite.

3.2 Current status of medical devices and relative complications involved

The development of medical devices, capable of catering to the above dimensions' demands precision in order to make it compact and relatively inexpensive. The core component of any medical device consist of a well-engineered platform that could act as substrate to promote bio-affinitive reactions which exposed to in vitro and in vivo conditions. The rendition of such platforms eventually require a well-controlled surface chemistry that could provide a stable matrix for interaction of biomolecules to generate optimum results. The perfect amalgamation of a suitable fabrication technique alongside a suitable procedure for controlled surface functionalization is a prerequisite for developing medical devices for tissue engineering or point-of-care diagnostic technology.

Tissue engineering is evolving as an interdisciplinary subject which integrates dimensionally different subjects like biology, material science, engineering, chemistry, and pharmaceuticals in order to find perfect amalgamation of the same to develop a biological substitute capable of restoring, replacing or maintaining a particular organ function. For the past few decades, substantial progress has been made in the field of tissue engineering, of which some are already been used in humans for clinical regeneration activities (e.g., skin, bone and cartilage) while some are in clinical trials (e.g., bladder and blood vessels). Nevertheless, all tissue engineering strategies are based on the principle that under optimum bioreactor conditions, cell loaded three dimensional biocompatible scaffolds can potentially grow and differentiate into functional structures that closely resemble their native counterfeit. In the present scenario, it is an accepted fact that in order to replicate cell behavior or function as found in *in vivo* conditions, it is important to recapitulate a tissue specific microenvironment. Cells are always surrounded by extracellular matrix, which is nothing but a hierarchical arrangement of fibrillar structure. In order to mimic the microenvironment, it was essential to develop nanoscale structures which can closely resemble the extracellular matrix structure, thus providing a suitable environment for the cell to grow and proliferate. It was further elucidated that immobilization or incorporation of growth promoters or growth factors like collagen, GAGs, keratin or other proteins and glycoproteins will ensure faster regeneration kinetics. Moreover, with the advancement of nanotechnology, engineers and scientist are able to maneuver biomaterials into hierarchical nanostructures that can closely mimic the complexity and functionality of the targeted tissue. Notably, these nanostructures also offer the privilege to functionalize, encapsulate, embed, immobilize or decorated the scaffold with growth promoting factors responsible for the regeneration process.

3.3 Substrates deployed in biomedical applications

The choice of substrate is of paramount importance as the subsequent steps of surface functionalization and its associated application in biomedical research, depends on the kind of reactive functional groups present on the surface. Substrate is the core component of any biomedical device therefore its selection is a primary step.

3.3.1 Polymeric membrane based substrates

Polymers have substantial role in the field of tissue engineering owing to their inherent properties like mechanical strength, ease of processability, inertness in biological environment, abrasive resistance, cellular adhesion, hemocompatibility and permeability to oxygen to mention a few among others, especially in the development of tissue engineered instructive scaffolds. There are several types of polymeric membranes as listed in subsequent paragraphs.

3.3.1.1 Natural polymers

Natural polymers are the most promising biodegradable scaffold material as they show better matrix- cell interactions owing to the presence of natural peptide, glycosaminoglycans, glycosidic bonds in the main chain of biopolymer endows them inherent quality of interacting positively with wide range of primary and secondary cells, coupled with their bond strength. Certain examples of the same are chitosan, collagen, hyaluronic acid and silk fibroin. Chitosan, a polycationic polysaccharide derived from chitin, comprises of functional groups β-(1—4)-linked D-glucosamine and N-acetyl-D-glucosamine groups which serves as a cross-linking site. Chitosan was chosen owing to its desirable characteristics like excellent biocompatibility, non-immunogenicity, nontoxicity, biodegradability and its antimicrobial activity [13]. As of 2011, about 28 different varieties of collagen have been identified but among them the most common ones are Type I to Type v. Collagen I itself forms almost 90% of the total collagen content in human body [14]. The capacity of collagen to support in the growth and development of organs has made it find applications in a wide variety of hard and soft tissue engineering scaffolds. Apart from chitosan and collagen, another natural polymer which has notable contribution in tissue engineering is silk fibroin. Silk fibroin, an insoluble fibrous protein, is extruded by glands of silkworms (Bombyx mori, Antherea mylitta and other domestic variants) and by the larvae of spiders. The main structural component of silk fibroin consist of repeating amino acid sequences (Gly-Ser-Gly-Ala-Gly-Ala)n and arranged in antiparallel β-sheets. Silk finds wide application in every domain of tissue engineering owing to its exceptional biocompatibility, high stiffness and minimum immunogenicity [15]. The above natural polymers are explored extensively in the form of gels, sponges, films, nanofibers in hard and soft tissue engineering applications. Hyaluronic acid (HA) is a nonadhesive glycosaminoglycan which is mostly found in connective, epithelial and neural tissues, applicable in both hard and soft tissue regenerations [16]. Eggshell Membrane (ESM) is a microfibrous, thin, protein based membrane found in the region between mineralized eggshell and egg yolk and which provides protection against bacterial invasion. Moreover, ESM matrix is believed to closely resemble the extra cellular matrix and its individual components thus making it a perfect choice of skin tissue engineering applications.

3.3.1.2 Synthetic biodegradable polymers

Synthetic biodegradable polymers, inclusive of various polyesters, are most widely deployable material for scaffold fabrication in tissue engineering. Synthetic polymers can be tailor-made with complete precise and control owing to the reason that these polymers are developed by reacting monomers with subsequent bond formation through primary covalent bonding. Few notable examples which are extensively used in tissue engineering industry are Poly (lactic acid) (PLA) and its copolymers, Poly (glycolic acid) (PGA) and its copolymers, polycaprolactone (PCL), polyorthoesters and Polyanhydrides. PLA is synthesized commercially from lactic acid through ring opening polymerization procedure in presence of metallic catalysts. PLA generally exists in isomers 'L' and 'D' and in racemic form (D,L), among which the 'L' isomer is most commonly used in major tissue applications, owing to its structural property. PLA finds its utilization in cardiovascular, muscular, cartilage tissue engineering and drug delivery. PGA is another biodegradable polymer formed from glycolic acid either through polycondensation or ring opening polymerization. PGA has high crystallinity (46—50%) due to which it has high glass transition and melting temperature. It is generally used as a suture material, implants and drug delivery applications. PLGA is the combination of the above two polymers PLA and PGA in different weight ratios and finds application in drug delivery, tissue engineering and customization implants. All the above polyesters are approved by Food and Drug Administration (FDA) for clinical use. The degradation products of above biodegradable polyesters (lactic acid and glycolic acid) are absolutely non-toxic without posing harmful effects on the physiological system.

Polycaprolactone (PCL), produced from ε-caprolactone by ring opening polymerization, is a slow biodegrading polymer with low glass transition ($-60\,°C$) and melting ($60\,°C$) temperature. PCL is flexible and elastic in nature and usually exhibits better blending with other polymers. However, the cellular activity of PCL is insufficient because of its high hydrophobic nature, as such it needs some form of functionalization/surface modification or biomolecule immobilization to enhance its activity. Polyorthoesters are class of biodegradable polymers which are generally suitable for orthopedic applications but have limited mechanical properties and induces certain inflammatory responses. Polyanhydrides are biodegradable polymers with biocompatibility and excellent controlled release characteristics. Langer and coworkers have synthesized polyanhydrides for drug delivery applications, and used in delivering anti-cancer drug at the tumor site [16].

3.3.1.3 *Non degradable polymers*

Non degradable polymers like ultrahigh molecular weight polyethylene (UHMWPE), polypropylene, ethylene-covinyl acetate (EVA) (Elvax®), polytetrafluoroethylene (PTFE) (Teflon®), poly(dimethylsiloxane) (PDMS) (Silastic®), poly urethanes (PU) (Tecoflex®, Tecothane®, BioSpan®) are commonly used as nondegradable synthetic polymers in tissue engineering. Due to the high flexural and compressive modulus, high density polyethylene (HDPE) and UHMWPE is extensively used in load bearing applications. However, the PE particulate debris due to frictional wear and tear might result in long term health risks.

3.3.2 Metallic substrates

Metallic implants are used today in majority of the surgically invasive cases for diseases or injuries because of its multifaceted advantages over ceramic or polymer implants. Primary stability is one of the major reasons, where the micromotion between the host and the implant is restricted to few microns, providing excellent stability for the host-implant integrity towards appropriate facilitation of tissue ingrowth. However, the secondary stability, provided by the host and implant osseointegration is not at par for the solid or bulk implant, which led the focus shifted towards porous and surface engineered implant. However, porous implants have also limitations in terms of the strength, but the enhanced osseointegration is always an antidote for the bane-like situations like poor fixation, aseptic loosening as well as stress shielding appeared in the cases of excellently strong bulk implants. Titanium (Ti) and its alloys are now globally accepted material for orthopedic and maxillofacial applications [17]. Lightweight, high compressive strength [18], resistance to corrosion, and excellent biocompatibility [19] are amongst those many reasons to make it as a standalone choice for both the surgeons as well as the engineers.

3.3.3 Organoids

Organoids are defined as *in vitro* synthesized, self organized 3D structure derived from stem cells and recapitulate some functions of the respective native organ [20,21]. The supplementation of 3D scaffold with biochemical factors, differentiated cells from stem cells will self-organize to form tissue-specific organoids e.g., optic cup [22,23], brain [20], intestine [24], liver [25], and kidney [26,27]. Organoids do mimic majority of the biological parameters e.g., tissue specific heterogeneous cells nicely compartmentalized, cell—cell and cell—matrix interactions capable of performing physiological function. It acts as a model system to study the tissue—organ biology, development, regeneration, disease modeling, improvements in organ transplantation, drug discovery/response studies.

3.4 Types of surface modification techniques towards ligand specific activation

As discussed in the previous sections that surface modification of substrates and gaps are one of the major steps in preparation of medical device for optimum performance. Especially in the field of biosensors, homogeneous distribution of such modified layer facilitates detection of the target analyte and also maintains perfect orientation of the capture probes. There are primarily three types of prevalent designs which are used as physical structures: self-assembled monolayers (SAMs), layer-by-layer(LbL) multilayers, and polymer brushes (Fig. 3.1).

3.4.1 Self-assembled monolayers (SAMs)

SAMs are well organized structures comprising of single layered organic molecules which are chemically grafted on the surface of the substrate [28]. SAMs are typically robust conformal coatings possessing a uniform thickness of around 1—4 nm, having dense crystalline packing in order to act as ultrathin coatings. The monolayers generally constitute of predetermined surface chemistries which can substantially protect the biomolecules from undergoing conformational changes. Stringent conditions prevail while preparing self-assembled monolayers due to the presence of specific surface sites, thus placing constraints in choice of molecules too. However,

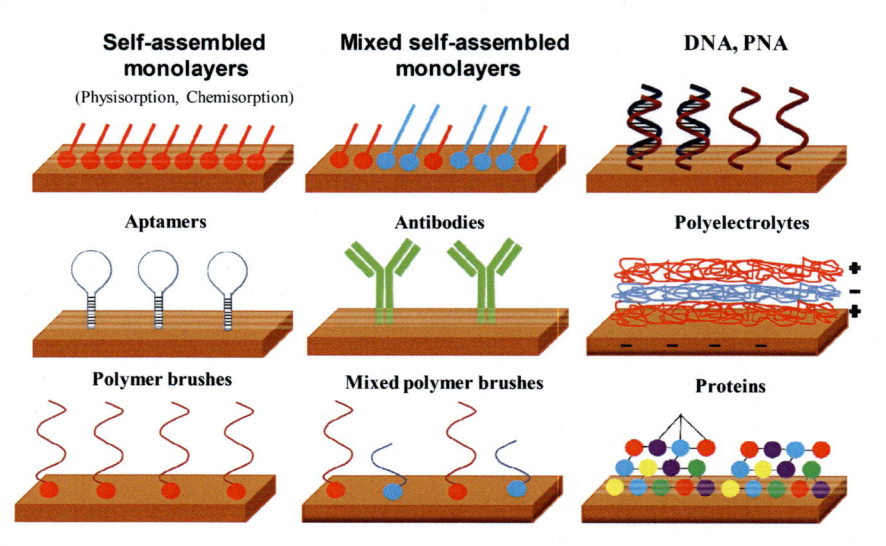

FIGURE 3.1 Different schemes of surface modifications used for varying substrates.

the monolayers can be further modified to attain the desired choice of surface functional groups as required for a particular set of bio-affinitive reactions. The most prevalent and extensively used chemical moieties for creating SAMs are alkylsilanes like APTES or MPTMS for oxidized surface or alkanethiols for gold surfaces [28]. The robust ultrathin SAMs are one of the preferred surface modification used in developing micro-devices as it is believe to impart excellent anti-fouling properties, oxidation resistance and lubrication [29−31].

3.4.2 Polymer brushes

Polymer brushes are generally long-chain chemical moieties grafted on the surface of a substrate and are multiple nanometer thick with polymeric chains extending in the vertical direction [32−36]. Further a uniform layer of polymer brushes can be tuned to achieve precise thickness following appropriate polymerization techniques. Additionally, various copolymers can also be deployed to form brush layers thus providing an opportunity for large degree of customization to meet the requirement of a particular application. Polymer brushes can be grafted using two procedures, (a) Grafting-to and (b) Grafting-from. Grafting-to is a relatively easy procedure wherein an existing chain of polymer can be covalently substituted to the surface of the substrate possessing a layer of linker group. Grafting-from on the other hand is a procedure wherein an entire chain of polymer brush is prepared from its monomer units by utilizing strong surface initiated chemistries, which also confirms attachment of the same in the growing chain. The long chain polymeric layer formed on the surface using above techniques encourages good packing density and thickness of the formed layer.

3.4.3 Layer-by-layer (LBL) multilayers

The formation of LBL structures are based on assembly technique wherein alternating layers comprising of oppositely charged materials are stacked one upon another and all the layers of a LBL structure is held together by complementary interactions. This approach of surface modification is distinctively different from the other two, SAMs and polymer brushes as no chemical bonding is required during the process. This approach is mainly adopted for polyelectrolytes while fabricating free-standing films, hollow microcapsules or coatings [37]. The process involves stepwise deposition of multilayers on sacrificial templates, substrates or cores [38]. LBL assembly can also be performed on the surface of solid substrates or particles [39,40]. Thickness and composition of multilayers in a LBL structure can be controlled by repeating the deposition of individual layers over again. In case of sacrificial templates, necessary crosslinking steps or surface treatments are performed in order to selectively dissolve the same, only leaving hollow microcapsule or films of LBL structures for further use. The LBL deposition can be significantly used in surface modification of optical fibers for application in biosensor or chemical sensor technology.

3.5 Surface engineering of polymeric substrates

The modification of polymers is not limited up to its bulk properties but it extends to its surface dynamics (surface charge density, roughness, wettability, protein adsorption kinetics)

TABLE 3.1 Different techniques of surface modification.

Process	Description of approach	Surface modification	References
Addition of functional groups onto the surface			
Oxygen plasma etching	The substrate is etched with oxygen plasma to introduce functional groups	Surface gets modified by oxygen containing functional groups	[14]
UV/ozone treatment	Oxidation of substrate material with UV/ozone	Surface gets active by introduction of oxygen and nitrogen functional groups	[15]
Covalent grafting of peptides	Surface is activated by addition of functional groups from outside or reacting with carbodiimide and subsequently peptides are grafted onto the substrate	Grafting of peptides on surfaces	[16]
Coating of surface with thin layer of polymer/chemical moieties			
Plasma polymerization	Surface is coated with monomer vapors under moderate temperatures	Presentation of functional groups of plasma polymers	[17,18]
Physical adsorption	Scaffolds are immersed in protein or peptide grafted polymer solutions	The peptides are physically adsorbed on the surface	[19,20]
Layer by layer assembly	The initial surface is coated with a charged layer following which another layer of opposite charge is applied on it	Repeated process of this yields self-assembled bilayers	[2,21]
In situ apatite formation	Scaffolds are soaked in simulated body fluids (SBF) for apatite formation	In situ deposition of apatite layers on scaffold	[22]

which are subtle yet have profound implications in initial cell adhesion/biomolecule capture and establishing contact guidance between cell–cell and cell–material interactions. The better this crosstalk between polymer surfaces and bioentities present in surrounding physiological environment occurs, the more efficiency and effectiveness in the performance of implants/scaffolds is observed. Of late, surface engineering has emerged as an exceptional domain wherein material surfaces are designed and decorated (chemically or physically) to impart biomimetic touch resembling close to its native structure/hierarchy. The biomimetic aspect enhances focal adhesion contact point among cells, anchorage to substrates avoiding undesirable surface adsorption and biofouling. Several approaches are being explored in giving biomimetic property in the form of physical, chemical and biological, physiological and topographical cues (Table 3.1).

3.5.1 Physical cues

The different techniques of scaffolds fabrication (film casting, freeze drying, electrospinning, 3D printing, lithography, chemical and physical modification of surfaces) in tissue engineering, imparts the scaffold matrix with its inherent properties of surface roughness, stiffness, wettability, which in turn instructs the biological activity of the cells on the matrix.

3.5.2 Surface roughness and wettability

Surface texture of the scaffolds particularly roughness affects the wettability of the surfaces to a significant extent. Cellular response and protein adsorption kinetics are greatly affected by changes in the surface roughness. Cells favor to adhere and proliferate preferentially on moderately hydrophilic surfaces with the optimum range lying between $40°$ and $60°$ [23]. Increasing surface roughness renders the scaffold's surface hydrophobic (contact angle $\geq 120°$) and vice-versa. With extreme changes in roughness, it bestows the surface either with superhydrophobicity (contact angle $>150°$) or superhydrophilicity (contact angle $<5°$) [24]. Depending on the requirements of specific tissue engineering application, the surface wettability could be tuned by combining block copolymers having repeating units with different hydrophilic and hydrophobic compositions, thus achieving scaffold structures with controlled hydrophilicities.

Another approach to tune the wettability of rough substrates is through subjecting the polymer surface to plasma treatment. It has been observed that plasma treatments, particularly oxygen and argon plasma, reduces the wettability of surfaces as such, cell adhesion is improved considerably and over a wide range of surface hydrophilicities.

3.5.3 Surface stiffness

The initial anchorage of cells on the substrate is formed through certain adhesion links between the integrin receptors unique to specific cell and the substrate matrix. The cytoskeleton of cells generates forces to this adhesive bond which further helps in spreading of cells across the matrix. The magnitude of the generated forces along with extent of cell spreading is determined by the surface/substrate stiffness. The extent of anchorage and proliferation, differentiation of cells is dependant on the response of each specific cell lines to the magnitude of surface stiffness [41]. It has been observed, growth and proliferation of neurons is generally governed by soft substrates [42], whereas differentiation of myocites is influenced by substrates with intermediate stiffness, rather than very high or low stiffness [43].

Directional migration could be observed in vascular smooth muscle cells where cells proliferate preferentially from softer to considerably stiffer regions [44]. Normal cells are observed to attain polarization and maintained their polarized state in larger surface area on softer substrates irrespective of other topographies. In contrast, cancer cells' polarization state and retaining of cell shape appears to be insensitive to any changes in surface stiffness [45]. This behavior of normal and cancer cells in response to substrate stiffness could open future avenues in biomimetic design of scaffolds, inhibiting metastatic transition of healthy normal cells.

3.5.4 Chemical and biological cues

The conjugation of different chemical and biological cues enhances cell-selectivity of surfaces to specified target cells repelling unwanted cells or molecules. The polymer surfaces are grafted with ligand molecules (peptides, antibodies, and other chemotactic cytokines) which promote recognition of particular specific cells followed by establishing linkage with the target cells.

3.5.5 Peptides

Peptides are one of the most important biological molecules which are formed by essential α-amino acids joined together through amide bonds. They have specialized selective affinity to certain cells while no effects on other types. The immobilization of peptides on the biomaterial surface occurs through formation of amide bond between the peptide and the carboxylic group of surfaces [46]. Peptide immobilization is performed in three consecutive steps: (1) activation of surface functional groups, (2) surface modification and (3) grafting or immobilization of peptides. Activation of surface functional groups are achieved through N-hydroxysuccinimide (NHS), plasma treatment, electrochemical etching and aminolysis [47−49]. Further surface modification is performed by SI-ATRP (Surface-initiated Atom Transfer Radical Polymerization), γ-irradiation, azido-alkyne click chemistry, layer by layer (LbL) self-assembly, thiol functionalization of peptides [50,51]. The availability of functional groups on the polymer surfaces forms the most vital part of peptide immobilization process. The lack of sufficient functional groups on the polymer surfaces makes it essential to graft/ incorporate biopolymers on the surface to improve their peptide grafting ability.

Arg-Gly-Asp (RGD) tripeptides find extensive application in peptide conjugation for promoting cellular adhesion and proliferation but it has affinity for platelet adhesion as it recognizes platelet adhesion receptors αIIbβ3, as such it cannot be used effectively as a peptide on vascular grafts or cardiovascular implants [52]. cRRE peptide exhibits lower affinity for $\alpha_{IIb}\beta_3$ but higher affinity for adhesion receptors $\alpha_5\beta_1$ of endothelial cells, through interaction between respective Trp residue and acceptor of cRRE and α_5 [53]. cRRE peptide exhibits favorable response for adhesion receptors α5β1 of endothelial cells, through interaction between respective Trp residue and acceptor of cRRE and α5 [54]. Cys-Ala-Gly (CAG), a tripeptide, is found to be extremely effective in attracting endothelial cells (EC) and inhibiting smooth muscle cells (SMC) when blended with PCL electrospun fibers [54]. The adhesion of EC improved twice compared with SMC, with the former having wide spreading morphology and the latter a shrunken, round morphology. In another study, polycarbonate urethane (PCU) surface was grafted with hydrophilic poly (ethylene glycol) methyl methacrylate (PEGMA) and pentafluorophenyl methacrylate copolymers through SI-ATRP process, and CAG peptide was immobilized enhancing EC adhesion [51].

Another important peptide is a fibronectin derived tetrapeptide, Arg-Glu-Asp-Val (REDV) possessing the ability of binding specifically to $\alpha_4\beta_1$ integrin found on ECs abundantly but scarcely on SMCs [55]. REDV is incorporated in hydrophilic copolymer brushes of N-(2-hydroxypropyl) methacrylamide (HPMA) and eugenyl methacrylate (EgMA) by thiol-ene click reaction [50], which not only improves affinity of EC adhesion but also inhibits platelet adhesion. REDV is also immobilized on heparin (HEP) and chitosan containing multilayers to improve and oppose the cell-resistant property of HEP/ chitosan towards ECs [56]. Tyr-Ile-Gly-Ser-Arg (YIGSR) is a pentapeptide which binds specifically to laminin receptor. Laminin is an important noncollagenous glycoprotein which forms the basal lamina of extracellular matrix (ECM). YIGSR peptide has been found to stimulate epidermal proliferation and basement membrane formation in skin equivalent by reducing TGF-β1 secretion level [57]. Ile-Lys-Val-Ala-Val (IKVAV) peptide which binds to laminin-1 sequence is found to essentially promote and sustain neuronal attachment and growth [58].

3.5.6 Antibodies

Cells' surface has specific antigens which bind specially with the respective antibodies having particular cell selectivity. CD 34 antigen is expressed in bone marrow-derived circulating endothelial progenitor cells (EPC) in human [59]. Vascular grafts immobilized with anti-CD34 antibody exhibit minimal non-specific adsorption, enhancing formation of mature endothelium and repressing restenosis [60]. Hydrophobins are amphiphilic molecules produced by some filamentous fungi (7–15 kDa molecular weights). They self-assemble themselves to convert a surface from hydrophilic to hydrophobic and vice-versa. Anti-CD31 antibody has been immobilized to vascular grafts through HFBI (class II hydrophobins) which binds to platelet endothelial cell adhesion molecule-1 (PECAM1/CD31), a special cell marker expressed by ECs [61]. Anti-CD 133 antibody immobilized on ePTFE vascular grafts *in vivo* binds to CD 133 (found on EPC surfaces) depicting long term patency of grafts with intact lumen [62].

3.5.7 Antifouling surfaces

In conjugation with other biomolecules, antifouling surfaces have occupied an important position in surface modification and surface mimicking to native biological structures. The concept of 'antifouling surface' dwells in preparing surfaces which repels or prevents unwanted and undesirable molecular adhesion. Hydration layers is the common method of antifouling, it is formed by hydrogen bonding or electrostatic forces generated by ion solvation [63]. The interaction between water molecules and terminal end of proteins increases the system entropy, which further prevents protein adsorption [63]. Steric hindrance is another way to apply antifouling property to surfaces. When proteins approach long chain polymers, they tend to squeeze the polymers giving rise to system entropy which resists the adherence of proteins to the polymer chains. Long chain poly (ethylene glycol) (PEG) is the gold standard used as antifouling material [64]. Apart from this, zwitterionic polymers such as poly(2-methacryloyloxyethyl phosphorylcholine) (PMPC), poly(3-dimethyl(methacryloyloxyethyl) ammonium propanesulfonate) (poly(DMAPS)), hydroxypropyl methacrylate (HPMA) are used as antifouling material either as a coating or responsive polymer brushes [65].

3.5.8 Physiological cues

Target cells are sometimes attracted towards substrate by the influence of certain physiological environment, wherein the desired cells are drawn by in situ secreted chemokines and cytokines. Tissue regeneration is a complex process where healing occurs through various synergistic effects of secreted cytokines, inflammatory cells and signaling molecules. Inflammations have a direct relation with healing process. Immune cells attracted to the wound/implant produces certain cytokines and chemokines which cause chemotaxis of epithelial and endothelial cells at the wound site promoting regeneration/vascularization. Generally, monocytes in circulation polarize into two main phenotypes: M1 and M2, which have distinct effect on recruiting and performance of cells [66,67]. Directed migration and oriented differentiation of stem cells could be achieved by cultivating immune cells on the material surface. M1 macrophages promote long distance rostral migration of neural stem cells in CXCR4 dependent pathway [68]. Macrophages are found to capture mesenchymal stem cells (MSCs) and direct their own phenotype transition from M1 to M2 as determined by their M2/ M1 ratio marker (CD163/CCR7) positive cells [69].

3.5.9 Topographical cues

Topographical patterns are imprinted on the substrate through micro and nanofabrication techniques like soft-, photo- and electron-beam lithography. The effects of topological features (microgrooves, microridges, nanochannels, etc.) have been evaluated for cell proliferation and migration. Contact guidance is established between cells, where elongation and alignment occurs along the direction of the grooves. The degree of alignment is responsive on the groove depth and width and varies according to different cells. In general it has been observed that between groove depths 1−25 μm, cell orientation increases significantly [70]. Osteoblasts have been found to align and orient better along micropatterned films compared to flat or macropatterned substrates [71]. Fibroblasts generally exhibits contact guidance on surfaces above 35 nm or ridge width greater than 100 nm [72]. Cell proliferation has been increased with decreasing pit diameter [73] but cell adherence has been more profound on microscale pillars [74]. Endothelial cell adhesion and proliferation is most pronounced on 1 μm groove, which inhibits smooth muscle cells and prevents platelet adhesion and activation [75] (Tables 3.2 and 3.3).

3.6 Surface activation of metallic substrates

The mechanical surface modification methods are utilized when the focus is primarily on micro-roughness of the surface. The increment of surface area as well as inducing of a distinct topographical change in the surfaces (in the form of pits, microarrays of valleys, or

TABLE 3.2 Effect on cell adhesion following physical/topographical modification.

Cell line	Substrate	Physical modification	Outcome	References
bPASMCs (bovine pulmonary artery smooth muscle cells)	PDMS	Microposts, 2−10 μm diameter, 3−50 μm height	Cellular adhesion and spreading with deflected multiple posts	[60]
hCECs (human corneal epithelial cells)	Silicon	Nanogrooves, 70 nm width and 600 nm depth	Elongation of cells along micro/ nanometer sized grooves and ridges	[61,62]
hECs (human endothelial cells)	PDMS	Nanogrooves, 1200 nm width and 600 nm depth	Elongation along ridges with formation of capillary tubes on Matrigel	[63]
NIH 3T3 fibroblast	PUA (poly (urethane acrylate))	Gradient microgrooves, 1−9.1 μm	Improved cell adherence, alignment and elongation along the direction of ridges	[64]

TABLE 3.3 Effect on cell adhesion following chemical/biological modification.

Cell line	Substrate	Chemical modification	Outcome	References
Human keratinocytes	TCPS (tissue culture polystyrene)	Plasma copolymerization of acrylic acid/1,7-octadiene and allyl amine/1,7-octadiene	Better keratinocytes adhesion on acrylic acid/1,7-octadiene with low concentration of carboxylic acid groups	[65]
Human osteoprogenitor cells	PDLLA & PLLA	Physical adsorption of RGD-PLL and fibronectin to PDLLA substrate	Osteogenic differentiation to mature osteogenic phenotype could be observed for both modifications	[20]
bAECs (bovine aortic endothelial cells)	PDLLA	Adsorption of PLL-GRGDS to substrate	Improved cell adhesion but inhibition of spreading at high concentration of PLL-GRGDS	[19]
Human fibroblasts	Glass or silicon	Silanization of glass or silicon surfaces, surface ends terminated with CH_3, Br, $CH=CH_2$ or PEG	Strong adhesion, spreading, enhanced activity of integrins on $-COOH$ and $-NH_2$ terminated surface, weak interaction with $-CH_3$, -PEG and $-OH$	[66]
MC3T3-E1 osteoblasts	PLA	Surface entrapment of gelatin	Increase in hydrophilicity following gelatin incorporation, significant cell adhesion	[67]
3T3 fibroblast	Glass	Plasma polymerization of allyl amine and hexane	Increased cell density	[18]
SaOS − 2	Gelatin/bioglass	In situ apatite formation	Increased secretion of ECM	[68]

pillars) [76–78] facilitate the mineralization from the host live tissue environment, more faster compared to the bare implant surfaces without any modification. In this context, the protein adsorption is also increased due to an increase in the gross surface area of the exposed implant surface to the tissue [79]. Another wider approach of tailoring the microroughness of the implant surface is blasting [80]. Grit blasting [81] or sand blasting [82] on the Ti surfaces may push an increase in the surface reactivity. Another mechanical surface modification technique is micro-arrays of specific surface topography on the implant surface, which has gained popularity in recent days. Arrays of microspheres, pits or valleys of different shapes like hexagon, circular, or square, are embedded onto the surface of the implant surfaces in order to promote excellent adhesion of the cells and tissue ongrowth [78]. However, there is a distinct limitation of this process worth to be mentioned. These microarrays of specific topographical structures are only applicable for solid or bulk implants. The embossing or mechanical embedding required for the microarrays could prematurely disturb the structure of the porous implants, leading to failure of the sole purpose. Now, the solid implants are reported to be prone to failure due to aseptic loosening and stress shielding [83]. In this regard, whether in shape of grafts or implants, porous structures have reasonably become the state-of-the-art approach. However, porous structures too suffer from low endurance limit compared to the solid ones. So, the surface morphology and biochemistry involved within that have been explored in order to provide an enhanced tissue-structure induction and integration towards establishing a stable bond between the host and the implant. This would compensate the structural insufficiency towards a successful tissue integration.

The surface morphology and the inherent chemistry have been rigorously investigated to achieve intended modifications on the surface [84]. The primary objective for tailoring the surface chemistry of the Ti surfaces is to enhance biocompatibility, tissue inductivity, resistance to corrosion as well as to make the surface friendly to the targeted tissue environment. Presence of oxidative species or a thin layer of oxide films is always an advantage if surface modification techniques are to be undertaken on the structure [85]. There are chemical treatments reported to precisely immobilize a thin stable oxide layer on the Ti surfaces. Notable amongst those are acid and alkali treatment [86,87], peroxide treatment [88,89], anodization [90,91], chemical vapor deposition (CVD) [92,93]. Chemical acid treatments of Ti surfaces are performed popularly using any of these following acids, notably sulfuric acid (H_2SO_4) [94], hydrochloric acid (HCl) [95], and hydrogen peroxide (H_2O_2) [96]. It is primarily performed to clean the contamination and expose the pure Ti layer. In many of the cases, instead of using single type of acid, a combination of different acids is used to obtain intended results [97,98]. However, acid treatments generally yield very thin layer of oxides (<10 nm), which may be further affected by the environmental oxidation, resulting in deposition of layers up to few nanometers [92,99]. Hydrogen peroxide treatment is undertaken mainly aiming towards further apatite formation on the Ti surfaces in order to enhance the neo-bone generation on the surfaces [100]. It is well observed that Ti-oxide (TiO2) layer deposited as a result of Ti-peroxide reaction, is an excellent host for apatite formation after treatment using simulated body fluid (SBF) [101]. The TiO_2 layers have three distinct types of structures, rutile, anatase, and brookite [102]. Further heat treatment in the low temperature ranges from 300 to 600 °C may lead to crystalline state of the anatase layer. However, rutile phase with decreased bioactivity is observed when the temperature is more than 700 °C. Investigation towards betterment in achieving bioactivity leads to exploring alkali treatment on Ti surfaces [103,104]. The methodology for alkali treatment is pretty straightforward. The Ti surfaces are dipped into alkali solutions (NaOH or KOH) for 16—24 hours. The surfaces are duly cleaned by sonication in distilled water, followed by drying in room temperature in a vacuum oven. Further, the treated Ti surfaces are heat-treated, similar to the temperature range as in acid treatment. However, the thickness and stability of the oxide layer are optimized according to the required bioactivity and further immobilization of other moieties on the surface. Furthermore, the alkali-treated Ti surfaces also provide stable apatite formation while treated with SBF [104]. The duration for SBF treatment is subject to variation depending on the thickness and morphological characteristics of the apatite layer. Another potential surface modification strategy, tailoring the micro-nano topography of Ti surfaces, is anodization technique [105,106]. The metal surface (Ti) is used as anode in an aqueous or organic electrolyte solution to obtain the oxide layer by applying potential difference between anode and cathode. Enhanced adherence of the oxide layer to the bare Ti surface and its stable bonding are the primary characteristics of this process. The thickness of the oxide layers is dependent on the applied voltage. Hydrothermal treatment of Ti surfaces in Ca or apatite enriched mediums also provide excellent nanotopography on the modified surfaces with deposition of Ca and phosphates, which enhances the osteoinductivity [107]. Another effective but, costly process for Ti surface modification is chemical vapor deposition (CVD) [108]. In this process chemicals in their gaseous phase is reacted with the substrate surface, which yields a deposition of a chemical compound on the surface. Improved wear, chemical corrosion resistance and enhanced biocompatibility are major advantages of this chemical modification process.

However, the majority of these chemical modification processes are based on either synthesized or naturally available chemicals, which may lead to foreign body reaction from the host site of the patients in a questionable state. This led to opening another genre of modification methodology as biochemical surface modification, which involves biological moieties as well to make the host friendlier to the implant surface [109]. The extracellular matrix (ECM) human bone tissue is mainly composed of 90% of collagen-based and 10% noncollagen (osteocalcin, osteopontin, fibronectin, growth factors etc.) based proteins [110]. The inorganic part of the bone is mainly structured by hydroxyapatite. So, there will be directional or target-based bone apposition if the implant surface contains proteins or organic biomolecules which are present in the native bone tissue of the host. This would also induce signaling pathways towards specific type of bone tissue formation. The surface modification of Ti surfaces with Hydroxyapatite (or apatite or CaP), the inorganic component of the bone, has been widely explored [111,112]. These CaP coatings are reported to be advantageous for inducing enhanced osteoconductivity and bone ongrowth on the surface. However, poor control on the phases of CaP deposition, particle leeching, delamination of the CaP layers, and brittleness are few drawbacks of this [109]. On the other hand, collagen, proteins, peptide sequences (specific as well as nonspecific), cell signaling moieties, DNA chains are amongst purely organic biomolecules which are potentially capable of stimulating neo-bone apposition in the peri-implant area establishing a stable interlocking between the host bone and the implant.

Amongst many types of methods, two major types are reported extensively to graft these organic biomolecules on the Ti surfaces, (1) adsorption and (2) covalent immobilization [113,114]. In the case of adsorption, the Ti surfaces are dipped into protein or biomolecule enriched mediums, with an optimized pH and temperature. The duration of the dipping is also optimized as per the desired surface activity. However, the surface loading achieved by this procedure is low as compared to the covalent immobilization technique. Moreover, a mild amount of proteins or biomolecules are also lost from the modified Ti surfaces while processing for further experiments or biological assays. In contrast, covalent immobilization technique is capable of yielding high surface loading of biomolecules, as well as the attachment is more stable leading to very small amount of loss. Bare Ti surfaces are grafted primarily with reactive groups notably amino groups or aldehyde groups [115]. Biomolecules are then grafted on those groups present on the Ti surfaces. Covalent immobilization of organic biomolecules is popularly performed on the bare surfaces using silane chemistry. ECM protein grafting on Ti surfaces are widely investigated owing to its capability to enhanced adhesion of the osteoblasts on the modified surfaces [116,117]. Amongst many other proteins present in the ECM of bone, amino acid sequence Arginine-Glycine-Aspartic (RGD) are reported as responsible for the enhanced cell adhesion [118]. The cell adhesion of the biofunctional Ti surfaces is promoted by employing these ECM proteins on the surface. Moreover, these proteins also stimulate migration, proliferation, differentiation to site-specific bone cells induces morphological change, which further decides the neo-bone structure. However, these ECM proteins are long-chain biomolecules, which allows less control over its immobilization on the surface. Random placement of these proteins is also possible due to its long-chain nature. In this context, short-chain peptide sequences have attracted the researchers due to precise control over its grafting on the surface, as well as possessing similar osteoinductive properties of the ECM proteins [119]. Moreover, these short fragment peptide sequences can lead to development of artificial ECM on the Ti surfaces allowing more ground for enhanced tissue integration. Peptide sequences are covalently conjugated to the Ti surfaces with the help of functional groups (e.g., hydroxyl, or amino groups) present on the Ti surface prior to the peptide grafting. Neo bone generation has been reported to be accelerated by grafting the growth factors on the Ti surfaces. The main advantage of these growth factors is that can induce cell signaling pathways for accelerated functions of osteoblasts leading to neo bone generation. Transforming growth factor-β and Bone Morphogenetic Protein 2 and 7 (BMP-2 and BMP-7) are notable amongst several other growth factors (such as fibroblast growth factor, platelet-derived growth factor, insulin derived growth factor, etc.), which are grafted to the Ti surfaces by adsorption mechanism as well as covalent immobilization. However, a limitation regarding the utilization of BMPs is its brittle nature of the neo bone tissue, which is obviously not an intended outcome [120].

Deoxyribonucleic acids (DNA) are also reported to be used for surface modification of Ti implants and grafts. The advantage of using DNA as a potential surface modification moiety are primarily high phosphate contents, antibacterial activities of the DNA−lipid conjugates and its ability to activation and conjugation of other molecules utilizing groove binding and intercalation [121]. In this regard, DNA loaded surfaces are also advantageous to utilize as targeted drug delivery substances which would enhance the bone growth. Electrostatic self-assembly technique is mostly utilized to graft DNA onto the Ti surfaces [122]. This technique is based on the layerwise deposition of the biomolecule to the surface, which takes care of the coating to be successful in the areas of highly complicated geometrical contours of the implants. Multi-layered coatings of DNA are performed using this process by employing a stable electrostatic interaction between positively charged phosphatase groups of DNA and the negatively charged polyelectrolytes. Stable and enriched deposition of CaP on the DNA grafted Ti surfaces are reported. The surface modification techniques have still been evolving as the focus is shifted towards the patient-specific requirements. In that regard, along with tailored implant and its surface structures, these inorganic and biomolecule based techniques can be employed to achieve successful osseointegration, thus reducing the failure of implants.

3.7 Engineering organoids

Cells grown in 2D layers have homogenous growth and proliferation which is ideal to conduct experiments with same cell population [123]. However, this property of 2D cell culture has become the major disadvantage for the advanced research where scientists are curious about the interaction of different cells in the organ. The 2D cell cultures lack cell−cell, cell−extracellular matrix interactions which are the natural phenomenon in functioning of any organ. This lacuna has been filled by the introduction of three dimensional (3D) cultures. The 3D cell culturing allows the cells to expose adhesion molecules distributed across the cell surface allowing interactions of cell to cell, extracellular matrix and surrounding microenvironment influencing the cellular functions e.g., morphology, gene and protein expression, cell proliferation, differentiation. The cell responses in 3D cultures are much similar to in vivo behavior [124]. Three dimensional culture systems have been extensively used in the tissue engineering, cancer research and drug discovery. The 3D cultures can be done in either scaffold free (aggregate cultures and spheroids, hydrogels) or scaffold based manner (scaffold are prepared from synthetic or natural materials). The continuous efforts to mimic the biological system and latest advancements in the field of stem cell technology has conjointly driven the 3D culture technology to path of tissue engineering applications by generating "organ mimicking *in vitro* models" called organoids.

Although various organoids have been generated still there are various limitations associated with organoids developments e.g., most organoids lack some components of tissue or certain cell types ultimately hampering the cell—cell or cell—matrix interactions. Intestinal organoids showed the absence of villous structure when cultured in matrigel; may be due to lack of cellular or extracellular components of *in vivo* e.g., laminin-511, which plays important role in villous formation and some nonepithelial cell types [125,126]. In another report, transplanted human intestinal organoids could maximally achieve the maturation status of fetal human intestine, which limits its use in transplantation studies [127]. Another major challenge in organoids development is reproducibility owing to lack of information on the fact about how do the different types of cells behave as a unit in organoids which is quite different from the single cell culture studies [128]. In addition, the matrix which is used, do play a crucial role in development of organoids. Hence, to use the organoids technology in various applications, it is quite important to overcome this inconsistency factor.

Bioengineering approaches have been utilized to address these limitations; various factors such as bioactive cues, mechanical and physical modulations of matrix can help in improvement of cell—cell interactions, cell—matrix interactions which are crucial for functioning of organoids. Primarily, every organ has its unique requirement; hence modulation of factors can be beneficial for development of different kinds of organoids and their applications. Organoids maturation, reproducibility and scalability can be improved by engineering the culture environment, using instrumental tools for micro-manipulation and computational models and data interpretation. Thereafter, these mature organoids can be used in various translational applications. Various organs have been recapitulated as organoids model, the focus of the chapter will be on the intestinal organoids as they are vastly studied.

3.7.1 Engineering local matrix properties

Different stages of organoids formation require different mechanical environment and ECM constituents. Mouse intestinal stem cells (ISCs) when grown in PEG based soft hydrogels did not survive, however fibronectin or Arg-Gly-Asp (RGD) functionalized PEG hydrogels demonstrated survival and proliferation of ISCs. Matrix stiffness do play a role in ISCs expansion and differentiation. High stiffness significantly enhanced ISC expansion whereas for differentiation and organoid formation, soft matrix and laminin-based adhesion is prerequisite [129]. The four-armed, maleimide-terminated PEG hydrogels (PEG-4MAL) functionalised with the RGD adhesive peptide and crosslinked with the degradable peptide supported the growth of intestinal organoids from hPSC-derived spheroids. Upon injection of PEG hydrogels and HIOs into the mucosal wound, *in situ* polymerized hydrogel assisted the organoids engrafting at the site and enhanced the wound repairing process [130]. This delivery system can be used to develop HIO-based therapies to treat intestinal disorders in humans. A 3D matrix was designed using fibrin/laminin hydrogel which can act as a substitute for Matrigel to culture small intestine derived human epithelial organoids. Fibrin provided the physical support and have Arg-Gly-Asp (RGD) adhesion domains on the scaffold and upon accessorized with laminin served as a matrix for intestinal organoids development [131]. In another instance, collagen I matrix was appended with components of Wnt signaling to culture repairing epithelium-like phenotype *in vitro* and could be successful in treatment for ulcerative colitis [132].

Density of the hydrogels also plays a crucial role in the growth of organoids. Human intestinal organoids were grown in native or unmodified alginate and found to differentiate and mature *in vivo* similarly to HIOs grown in Matrigel; 1−2% alginate concentration was found to best for the HIOs and being nondegradable in nature, it maintained the culture upto 30 days without passaging, thus reducing the cost [133]. PEG-4MAL was tested to study the influence of hydrogel polymer density on the growth of HIOs and 4.0% PEG-4MAL showed high viability for 7 days in culture which is comparable to Matrigel. However, PEG is costly due to degradable nature as the PEG grown HIOs need passage every week [134]. Further, in case of intestine, villi formation occurs when the radial layers of smooth muscles of the gut differentiates and generates a stress which creates the folds in intestine whereas crypt formation is independent of this mechanism [126]. Human intestinal organoids (HIOs), manipulated with Nickle-Titanium (nitinol) spring *in vivo* resulted in the significant increase in villus height, crypt depth in comparison to HIOs devoid of spring [135]. Strain on cells can be applied through various techniques like atomic force microscopy, laser severing or magnetic-driven deformation [136,137]. The intestinal crypt budding can be biomechanically initiated by stem cells whose stiffness is less than the surrounding Paneth cells and thus require less stress force to deform. As stem cells divide, compression forces are generated, leading to bulging of cell material out of the crypt wall and bud formation [138].

3.7.2 Genetic engineering of organoids

With the introduction of powerful genome editing technologies, such as CRISPR/Cas9 with organoids culture systems, the organoids can be genetically transformed into a resourceful culture system. The primary organoid culture system can be manipulated to overexpress and knock down the expression of genes to study mammalian gene function *ex vivo*. Alterations in genes involved in intestinal epithelial functions play major role in the pathogenesis of inflammatory bowel disease. In addition, CRISPR Cas9 technology can aid in the introduction of disease associated mutations into the healthy organoid cultures and then engineered organoids can be used in disease modeling and drug screening. APC and P53 genes were mutated in intestinal organoids and found to be involved in intestinal cancer [139]. The patient-specific iPSCs derived intestinal organoids, (isogenic and gene-corrected organoids), revealed a feedback loop between the Wnt pathway and telomere function [140]. CRISPR-Cas9 technology successfully restored the CFTR function in cystic fibrosis patient-derived intestinal organoids [141]. Gastrointestinal epithelial organoids have also been used to study microvillus inclusion disease (MVID) and multiple intestinal atresia (MIA) [142,143]. Although genetic technologies have shown promising results but their full potential is yet to discover keeping in mind the epigenetic factors contributing in the pathology of disease especially in cases where genetic and epigenetic variants do not translate to a direct effect on the epithelial behavior or phenotype. Epigenome editing in organoids can definitely provide exciting opportunities in such cases.

Bioengineered organoids can be used to map the events in course of transformation, growth, and development of cancers. Patient's derived engineered organoids can act as powerful tools to regenerate intestinal tissues bypassing problems associated with immune rejection. Engineered organoids from individual patient can serve as models to screen the therapeutic agents leading to advancement in personalized medicine.

3.8 The concept of engineered *in vivo* system (organ-on-chip devices)

Along with surface engineered substrates, it is also important to develop Micro or nano-scale engineered platforms to incorporate the same in order to exactly mimic the *in vivo* system so as to provide a modern advancement in the field of tissue engineering. Lab-on-a-chip (LoC) is a recently emerging microfluidic concept in multidisciplinary approach with biology, engineering and chemistry allows for preclinical testing such as biochemical detection [144]. For diagnostic LOC are typically used for rapid analyzing with a very low volume and highly précised biochemical detection are studied by creating a micro channel with electrode, microelectronics and electrical field [145]. Over last few decades the research progression on nanomaterials combined with microfluidic platform presented a humanoid system and incipient due to its inherent with very low volume, cost effectiveness and highly efficient characterization. Organ-on-a-chip (OoC) mimics the tissue or organ functions in 3 dimension along with fluidic physics with full functionality depending on the detailed condition such as pH, osmatic pressure, nutrient content, flow rate and pressure, etc. [144].

Microfluidics are fabricated using various material, design and technique; among the well documented was soft lithography which was familiarized by Whitesides [146]. Recently, variety of cost effective, less time consuming and using cheaper materials are well explored using fabrication techniques such as mold making, laminates, 3D printed and nanoscale lithography [147]. In late 1990s poly(dimethylsiloxane) (PDMS) based microfluidics channels was detonated for biological applications [146,148]. In 2004 notable paper was published on mimicking human physiology of lung and liver interaction in a microfluidic chip using cells [149]. Further, microelectromechanical systems (MEMS) whereas, electrodes and sensors integrated in the microfluidic channel to develop a microchip was gradually evolved and advanced in bioengineering and therapeutic aspects [150].

In drug delivery industry such 3D miniaturized physiological functions or system engineered has proved its potential by reducing the divergence in clinical or preclinical trials [144]. PDMS based microfluidics is widely used due to its optical transparency, oxygen permeability and standard fabrication technology [151]. To design a OoC device contains several components and as well as engineering aspects as briefly shown in Fig. 3.2. Through synergistic engineering such as (1) patterning and geometrical confinement allows for the cell with guided spatial captivate with multi cellular coculture; (2) presence of fluid flow from inlet to outlet with cell to liquid or media ratio; (3) environment control

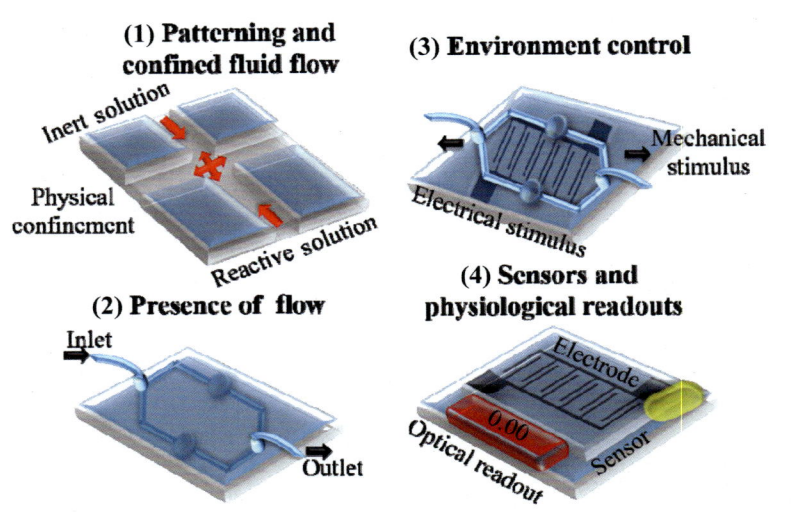

(1) Patterning and confined fluid flow

(2) Presence of flow

(3) Environment control

(4) Sensors and physiological readouts

FIGURE 3.2 Organ-on-a-chip system.

that provides simulation such as dynamic mechanical, electrical and electromagnetic force along with control of required biochemical microenvironment such as nutrient, growth factor or O_2 and CO_2 supply and (4) sensor and physiological readouts where microfluidics built-in with electrodes and sensors for biochemical measurements [152].

However, further surface modification would be required to enhance the surface wettability and to support the specific protein absorbance for supporting cell and material interaction [145,153,154]. This section highlights the recent progress in OoC and some of the surface modified microfluidic based *in vitro* models of lung-on-a-chip, heart-on-a-chip, liver-on-a-chip, kidney-on-a-chip and brain-on-a-chip.

3.8.1 Lung on a chip

Lung based investigation could improve the Lung tissue engineering whereas, OoC models which mimics the physiological function to understand the efficient therapy or repair could replace the animal test and rapid analysis with précised evaluation for pulmonary diseases [155]. PDMS based micro fabricated microfluidics has an adverse effect with physiological samples and affects the efficacy although, surface chemical modification could overcome the drawback and support the normal live responds [156]. Huh et al., have fabricated and validated the PDMS based lung-on-a-chip with hostile effect of physio mechanical force for providing new opportunity to screen and study the nanotoxicology of silica nanoparticle. The blood-air barrier was mimicked as shown in Fig. 3.3 whereas, the tissue—tissue interface was produced using a coculture system of single layer of human alveolar epithelium and closely against the monolayer of pulmonary microvascular endothelial cells to reconstitute organ function [157].

Further, Huh et al., demonstrated the prospective of bioengineered surface by coating with extracellular matrix (ECM) were it supported cell adhesion and chamber adjacent to microchannel for providing mechanical stimulus to reconstruct the micro physiological

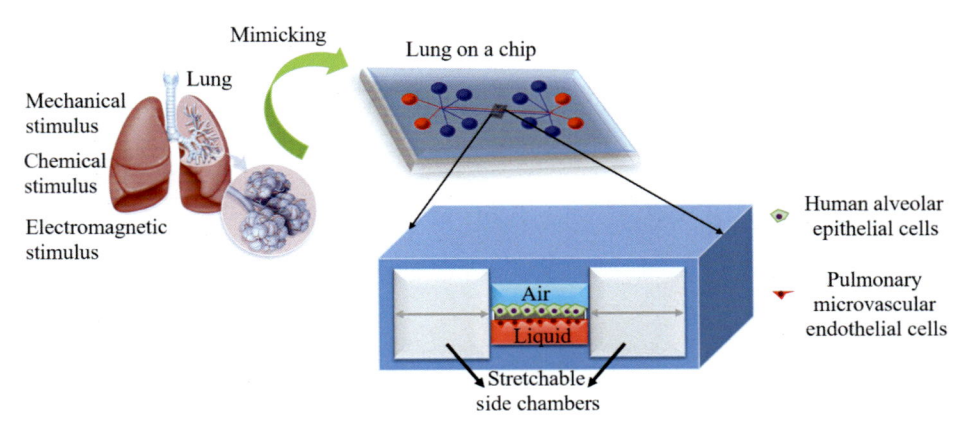

FIGURE 3.3   Lung on a chip. Fabricated microdevice with physiological breathing movement [157].

system of the living human lung. The *Escherichia coli (E. coli)* bacterial and inflammatory on the upper chamber resulted to activate the endothelial cells and expressed with adhesion molecules. Additionally, human neutrophils were dispersed in the lower section which emigrated and engulfed the *E. coli*. Further comparative study with living mouse model with ventilation-perfusion system presented the significance between fabricated breathing lung on a chip. In conclusion, pulmonary edema design was also micro engineered, in specific similar to vascular leakage syndrome were culture media was leaked to alveolar chamber.

3.8.2 Heart on a chip

Being a vital organ in human system tremendous efforts for heart regeneration and to screen cardio therapeutic drugs via biomimetic approaches are recently progressing [158]. Annabi et al., developed a method to coat the PDMS based microfluidic channel with hydrogel layer to enhance the cell adhesion and to mimic biological response similar to relevant tissue [159]. They have observed and demonstrated the effect of biocompatible hydrogels in responds to cardio myocytes cells and effect of suitable elasticity using soft tropoelastin scaffold to support elastic tissue. Inside the closed microfluidic channels with 50 μm width were coated by continuous flow of hydrogels made of photoresist bovine gelatin and human tropoelastin as shown in Fig. 3.4.

Coating of hydrogel with different thickness was succeeded using ultraviolet (UV) light with perfusion system by varying the flow rate the devices are exposed for 3 min at 14.6 mW/cm^2 and 6.90 mW/cm^2 for gelatin and tropoelastin, respectively. Further, before cell seeding each device were washed by Dulbecco's Phosphate Buffer Saline (DPBS) to remove the uncross linked polymer and photoinitator residue and stored at 4 °C. Cardio myocytes cells were suspended inside the coated microfluidic channel, different thickness in coating presented tunable substrate stiffness and thickness. In conclusion, tropoelastin coating presented improved cardio myocyte cell adhesion, alignment and proliferation

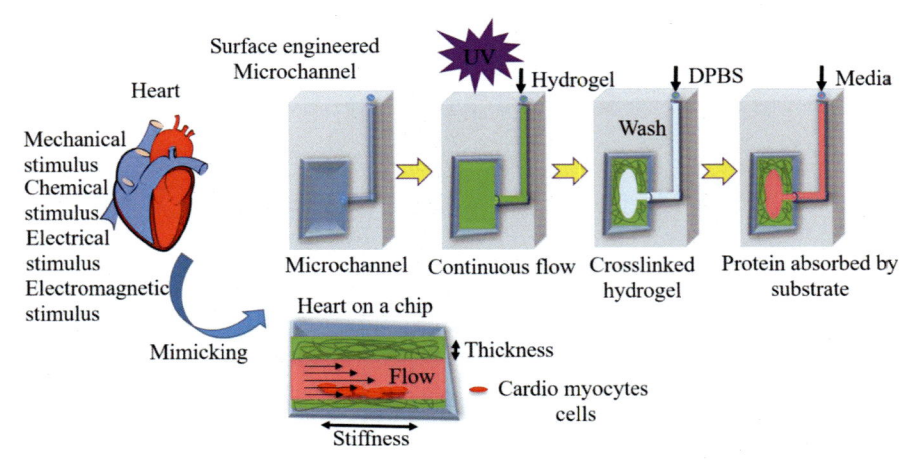

FIGURE 3.4   Heart on a chip. Coating procedure of the closed microfluidic channel [159].

then the gelatin coated channels. Therefore, designed system could help for cardiac drug and therapeutic studies and using such simple approach coating the microchannel could enhance the biophysical property.

Later, the study conducted by incorporating graphene oxide (GO) nanoparticles with elastic tropoelastin substituted by methacryloyl. The hydrogel was photo cross-linked where the substrate presented elasticity, conducting property and biocompatibility. By seeding cardiomyocytes cells along with electrical stimulus on the electro-active hydrogel substrate inside electrically insulated PDMS mold presented enhanced cell growth and functions supporting regeneration [160]. Such electroactive coated microfluidic channel with electrical stimulus could potentially screen and present micro device with advanced function. Mccain et al., replicated the diseased heart model with similar mechanical and structural microenvironment in failing myocardium which resulted with depicting the function as well as genetically. Such, diseased model can be designed for future drug screening for particular or specific diseases through mechanotransduction [161].

3.8.3 Liver on a chip

Being one of the fast growing organ, liver on a chip could detect toxicity and drug screening. Lee et al., designed an artificial liver sinusoidal using microfluidic endothelial like barrier along with huge transport channel that carried nutrient and drugs as shown in Fig. 3.5A [162]. Primary hepatocyte cells were seeded in the PDMS based microfluidic channel along with perfusion system without any surface modification and result showed sustained cell proliferation and viability. The result with mass transport of nutrient was proved whereas, favorable *in vitro* growth of primary cells was achieved with the designed model. The result showed no hepatotoxicity with the metabolic mediated with diclofenac.

Recently, Jang et al., designed similar microfluidic channel but with ECM coated channels for liver toxicity analysis and prediction (Fig. 3.5B) for phenotypes such as

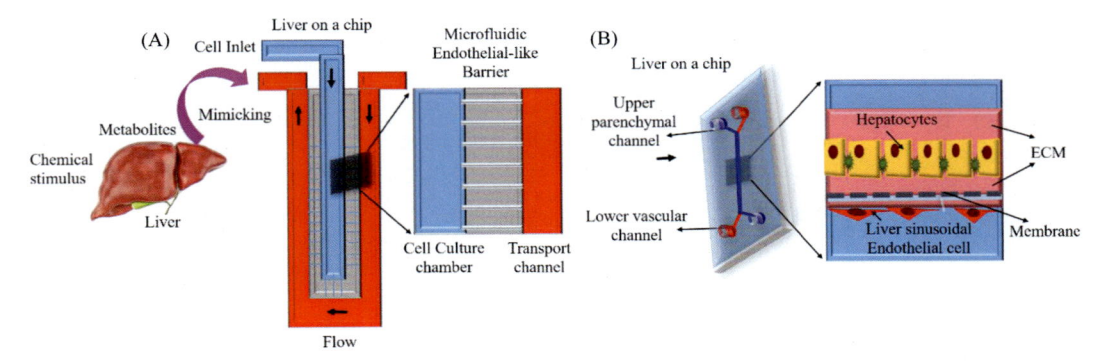

FIGURE 3.5   liver on a chip (A) micro device to mimic liver sinusoid [162]; (B) 3D view of liver on a chip and cross section of its complex liver cytoarchitecture [163].

hepatocellular injury, cholestasis, steatosis and fibrosis along with species specific toxicities [163]. The functionalized surface was premeditated by Emulate's proprietary protocols and reagents; whereas mixture of fibronectin with collagen type 1 and/or collagen type IV was used for human hepatocytes, rat liver chip and dog cells for species-specific toxic prediction. The upper section was seeded with species specific primary hepatocyte cells with or without liver sinusoidal endothelial cells. For multi-species liver on a chip provided toxicity and safety of its uses in relevance to animal studies which provided reduced human risk. Finally, the designed liver on a chip with functionalized surface proved its potential by presenting a disease modeling, biomarker identification, cytokine release and metabolic action determination to support drug screening for human hepatotoxicity's along with idiosyncratic responses. Therefore, such drug induced liver injury could be predicted without animal trial in an effective and precise way.

Ma et al., designed a multiplayer PDMS based reversibly assembled liver on chip for spheroids cell culture with 3D perfusion system, offering desirable native microenvironment [164]. The designed perfusion system showed its potential with remarkable metabolic activity, convenient, robust and exhibited long term conservation for drug related toxicity screening and to design diseased liver system. With hepatic spheroids the platform demonstrated and effective system with liver specific functions, hepatic polarity and metabolic activity where the designed concave micro well chip presented reproducible and rapid analysis for effective drug toxicity screening.

3.8.4 Kidney on a chip

One of the other major drug testing is nephrotoxicity, therefore mimicking the kidney function on a chip could support effective drug testing and significant improvement in curing diseases [165]. Designing kidney on a chip could be crucial but by mimicking the blood tissue barrier function, transport, physical, absorptive and physical property with metabolic activity would give a high throughput in drug screening industry [166]. Jang et al., designed a microfluidic channel coated with porous extracellular matrix and cultured with human kidney

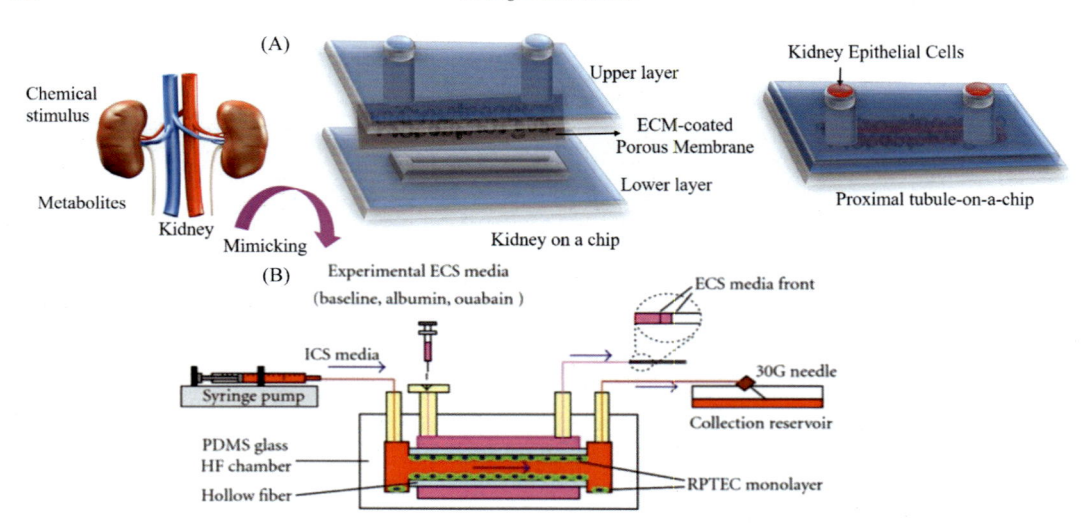

FIGURE 3.6 Kidney on a chip (A) design of a proximal tubule on a chip with ECM porous coated substrate [167]; (B) bio-artificial kidney device therapy [168] (Https://creativecommons.org/licenses/by/4.0/).

epithelial cells to mimic the living kidney proximal tubule as shown in Fig. 3.6A [167]. In comparison to traditional well plate the microfluidic channel resulted with enhanced epithelial proliferation and cilia formation along with high albumin conveyed, glucose reabsorption and alkaline phosphate activity. Further, comparison with *in vivo* model designed microdevice presented its potential by closely mimicking the function and response of cisplatin toxicity and P-glycoprotein (Pgp) efflux transporter activity.

Ng et al. engineered the tunable hollow fiber membrane with immunoprotected skin layer for cells and to closely mimic bio artificial kidney with enhanced renal substitution therapy [168]. The inner channel was coated with fibrin hydrogel to improve the cell adhesion and renal proximal tubule epithelial cells (RPTEC) were seeded. As shown in Fig. 3.6B the microfluidic channel with flow inlet and outlet was designed for continuous perfusion system. The immunofluorescence staining and other proximal tubule marker proved the successful growth of monolayer inside the channel. Further to understand the ability of monolayer with respect to exhibit a transport capability, inulin study was done with creatinine and glucose which proved the competency of designed microchip.

3.8.5 Brain on a chip

Being one of the most important and performs complex function in the human system, much research and advance technologies are required to understand the function and diseases of the brain as well for effective drug screening analysis [169]. Wevers et al., developed a microdevice which mimic human blood brain barrier for accessing delivery of large molecules into brain for therapeutics and antibodies drug [150]. The perfusion based organo plate with 96 or 40 tissue culture chip was designed by two and three-lane were, 384-well microliter on top with gel channel and microchannel on the bottom for media channel (Fig. 3.7A). The ECM gel are secured in the well with phase guide whereas, the

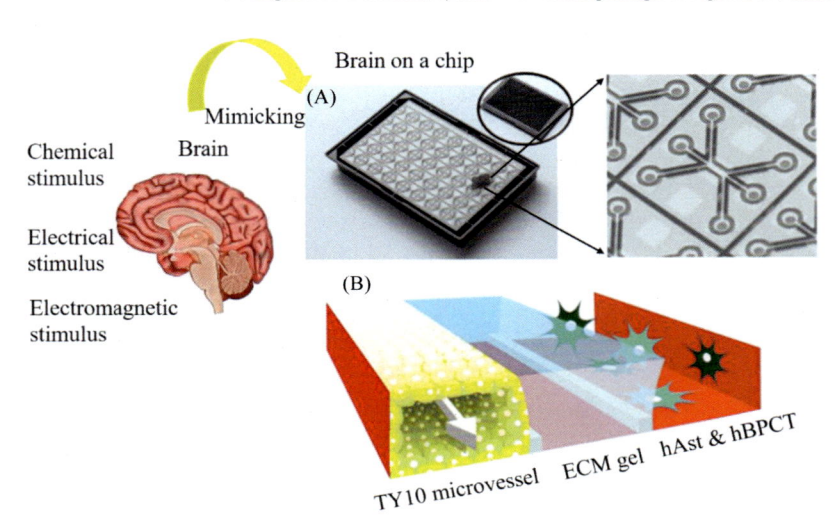

FIGURE 3.7 brain-on-a-chip (A) two-lane micro device with 384 well on upper chamber and 96 culture chip in the bottom section; (B) 3D pictoration of the microdevice with coculture to mimic the blood brain barrier (Http://creativecommons.org/licenses/by/4.0/).

cells are seeded after the gelation. Human Endothelial cells (TY10) are cultured alone for brain vasculature and also cocultured with human astrocyte cell line (hAst) and human brain pericyte cell line (hBPCT) are to mimic the blood brain basrrier function (Fig. 3.7B). With two different antibodies the transcytosis analysis showed perfusion system with lumen of endothelial layer penetrated to the basal chamber and quantified using the mesoscale discovery assay. The plate based microchip demonstrated its potential to mimic the blood brain barrier and antibody effects with different antibody and its penetration which could effectively support the drug screening.

Organ on a chip is a promising technology for effective discovery in drug screening for précised medicine and fast-track prospect in tissue engineering. The healthy and diseased models are developed for the precise and rapid analysis. Surface modification enhanced the tissue functionality in micro device which further presented the normal human like model. The surface property of the microchannel and well plays a vital role whereas, surface treatment, modified surface (e.g., micro pillars) or functionalized surface with electrical and electromagnetic properties could precisely support and enhance the analysis as well as tissue-tissue interaction in micro device. However, the choice of surface modification depends on the optimal for drug on specific organ model to enhance the cell adhesion and proliferation supporting tissue and drug effect analysis in the developed micro device.

The efficacy of an engineered matrix or micro device is not only determined from its cytocompatibility or bioactivity but also from its potential to inhibit or protect the matrix or growing tissue from microbial invasion.

3.9 Engineered bactericidal systems — a modern paradigm to regenerative micro devices

The field of engineered surfaces for medical devices has been an astounding success globally but the stumbling block from bench side to bedside transition still remains, thanks to the omnipresent microbes that lead to the infection of medical implants. The challenge becomes all the more daunting with the advent of widespread drug-resistant and nosocomial pathogens. According to the World Health Orgnization (WHO) joint news release (April 29, 2019), at least 700,000 people die each year due to drug-resistant diseases and if we are unable to contain the current trend, the statistics will explode to a catastrophe of 10 million deaths each year by 2050. The crisis is deepening with therapeutic void arising from decline in the number of new antibiotic drug approvals. WHO statistics also reveal that out of every 100 hospitalized patients at any given time, 7 in developed and 10 in developing countries will acquire at least one health care-associated infection. CDCs annual National and State Healthcare-Associated Infections Progress Report (HAI Progress Report) states that the potential danger lies in bloodstream infections, catheter-associated urinary tract infections, select surgical site infections, hospital-onset *Clostridium difficile* infections, and hospital-onset methicillin-resistant and vancomycin resistant *Staphylococcus aureus* (MRSA, VRSA) causing bacteremia. The causative agents of nosocomial as well as medical device related infections are *Acinetobacter*, *Burkholderia cepacia*, *Candida auris*, *Clostridiodes difficile*, *Clostridium sordellii*, carbapenem resistant *Enterobacteriaceae*, Human Immunodeficiency Virus, *Klebsiella* species, nontuberculous mycobacteria, *Mycobacterium tuberculosis*, *Pseudomonas aeruginosa*, vancomycin resistant *Enterococci*. The infection develops into a biofilm in three stages: adhesion, maturation, and detachment. Adhesion stage is the most important chance to prevent the infection since the susceptibility of the planktonic bacteria towards immune and antibiotic attack weans off with progression into the stationary biofilm stage [170].

In order to address the same, various strategies were adopted engineering the microdevices in a way so as to inhibit microbial adhesion, growth and biofilm formation on medical device surfaces. Few effective strategies for doing the same has been listed below:

3.9.1 Polymers

The first strategy to prevent microbial growth on the medical devices is to design the device out of the material that reduces adherence and further growth of the microbes. The second strategy is to kill the adhering microbes. Accordingly, the polymers are categorized as passive and active polymers [171] (Fig. 3.8).

3.9.1.1 Passive antimicrobial polymers (microbial resistive polymers)

Most microbes have a hydrophobic surface with net negative surface charge which arises due to the presence of phosphodiester bonds between teichoic acid monomers in the cell wall of Gram positive bacteria and lipopolysaccharides in the Gram negative bacteria. It has been found that quorum sensing causes an increase in the negative charge on cell surface of *E. coli*, which may facilitate interaction of bacteria with surfaces during the initial stages of biofilm formation [172]. Hence the 'passive' approach or the 'bacteria-resistive' approach is to design a hydrophilic or negatively-charged polymer with low surface free energy that will repel the microbes by preventing adhesion on the surface [171,173]. These include polymers used for antiadhesive coatings, for example Poly(ethylene glycol) (PEG), polypoly(*n*-vinyl-pyrrolidone), polypeptoid, heparin and others. Among these, the highly hydrated polymer chains of PEG exhibit remarkable antifouling properties due to a large exclusion volume and steric

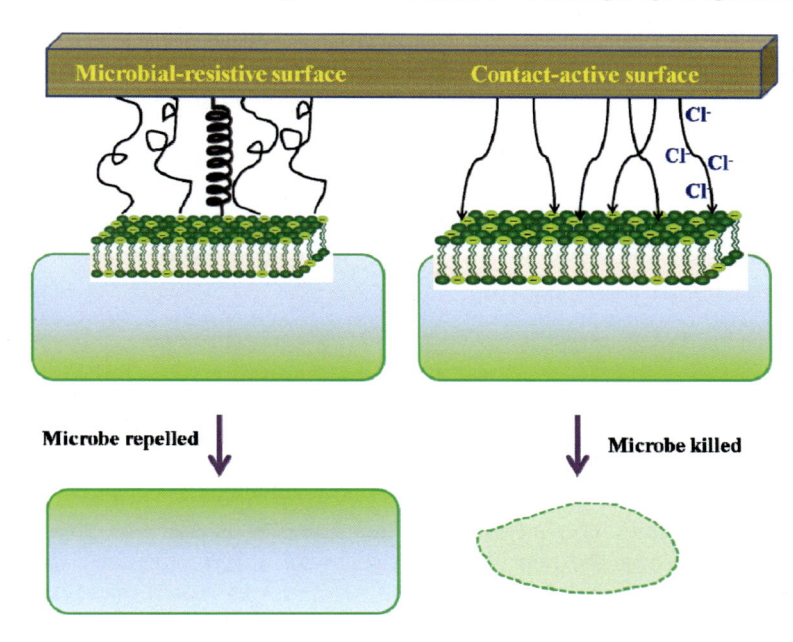

FIGURE 3.8 Microbial resistive surfaces are designed with polymers that reduce adherence and further growth of the microbes while contact-active surfaces have polymers (cationic or halogen-containing) that kill the adhering microbes.

hindrance effects [174] of highly hydrated layer which inhibits protein adsorption and bacterial adhesion [175]. Polyampholytes, zwitterionic polymers like phosphobetaine, sulfobetaine, poly(sulfobetaine methacrylate) (pSBMA) [176] and phospholipid polymers such as phosphorylcholine (PC)-based polymers, 2-methacryloyloxyethyl phosphorylcholine (MPC), positively charged ones like peptide-functionalized poly(L-lysine)-grafted poly(ethylene glycol) (PLL-g-PEG/PEG-RGD) are the charged polymers.

Hydrophobic polymers or slippery liquids like poly(dimethyl siloxane) (PDMS) and superhydrophobic molecule fluorosiloxane are the prominent members of this category [177]. Antifouling and antimicrobial mechanism of tethered quaternary ammonium salts in cross-linked PDMS matrices or coatings appears to be related to the length of the alkyl chain spanning between the nitrogen atom and silicon atom of the polymer and also to the length of the alkyl chain attached to the quaternary ammonium silane (QAS) nitrogen [178]. Heterogeneous surface morphologies arising out of self-assembly of moieties at the coating/air interface have been observed to provide highest antimicrobial activity particularly towards *E. coli*, *S. aureus* and *Candida albicans* [179].

3.9.1.2 Active antimicrobial polymers (contact-active biocides)

The strategies for designing vast expanse of antimicrobial polymers to kill the adhering microbes involve using intrinsically antibacterial bulk materials (Fig. 3.8) *i.e.*, biocidal polymers, functionalizing the polymers with cationic biocides, antimicrobial peptides, or antibiotics [171], or by incorporating these with bioactive materials that interfere with biofilm formation, and more recently, by fabricating micro- or nano-scale topographical patterns on the polymer surfaces. Chitosan being the perfect example of cationic polymer. It disrupts lipid—protein membrane assembly of the microbe gets disarrayed leading to the leakage of the intracellular components ultimately leading to microbial cell death [180]. Cationic polymers have an advantage as biocidal polymers for designing medical devices, since upon encounter with the cationic polymer, the. The antimicrobial activity of these polymers is embodied by the entire macromolecule rather than a specific site [171]. Thin films and coatings of chitosan serve as wound dressings and scaffolds for tissue and bone engineering [181]. 100% nonwoven, chitosan microfiber wound dressing 'Maxiocel' is highly absorbent as well as antimicrobial which is used for moderate to heavily exudating wounds of varied depths, shapes and sizes.

3.9.2 Metals

Metals can be bonded to at least one carbon atom of a polymer or an organic molecule. The modes of action of metals within the microbial cell which have been discussed below have been summarized in Fig. 3.9. Amidst all metals like silver, gold, copper, titanium etc., silver 'shines' through due to its most prominent antimicrobial property. Thus many surgical tools, catheters, wound dressings have silver based technologies as Ag^+ ion. Silver has multimodal activities like perforating the peptidoglycan cell wall, interfering with the negatively charged moieties on the cell membrane and microbial DNA to halt replication, forming insoluble compounds with phosphoryl and thiol groups of proteins and inhibiting respiratory enzymes [182]. Silver-based wound dressings are available under brand names with different compositions, such as Mepilex® Ag (silver foam), Acticoat™ (nanocrystalline silver on polyethylene mesh) and others. Silver can be applied as coatings, hydrogels, nanofibers, and nanoparticles (NPS) however, silver NPs have proven to be more efficacious owing to its ability to release silver ions, generate reactive oxygen species (ROS) and

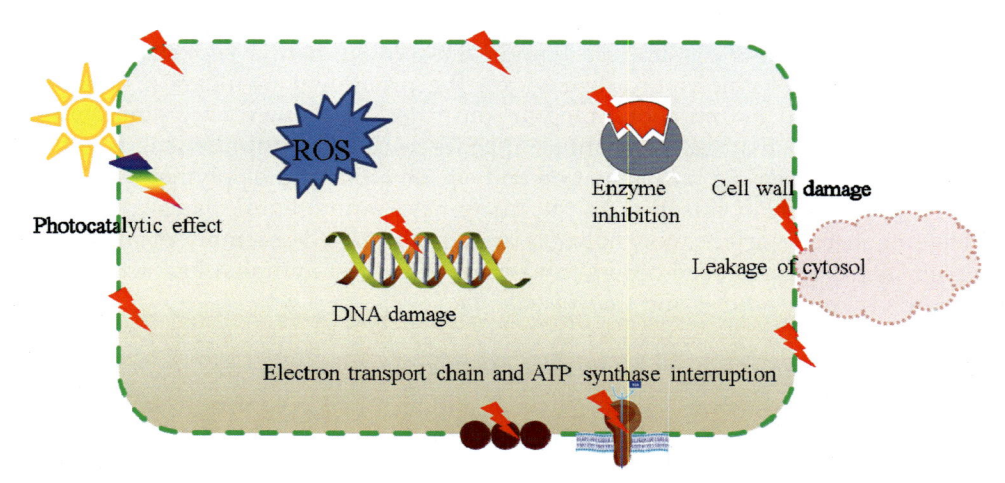

FIGURE 3.9 The metal ions damage the microbial cell wall, generate reactive oxygen species to destroy the cell membrane leading to the leakage of the cytosol, interrupt the electron transport chain and ATP synthase, inhibit glycolytic and other enzymes and bring about damage to DNA and its replication.

directly damage the cell membrane at tenfold lower concentrations than Ag$^+$ ions alone. Adherence and survival of Gram-positive methicillin-resistant *S. aureus* (MRSA) and the Gram-negative opportunistic pathogen *P. aeruginosa* (PAO-1) populations were significantly reduced on nanosilver/poly (DL-lactic-*co*-glycolic acid)-coated titanium with simultaneous enhancement of osteoinductivity [183].

Other metals like gold and copper also demonstrate antimicrobial activity though less pronounced than silver. Ultrasmall gold NPs (<2 nm) show high antibacterial activity against both Gram-positive (*S. aureus*) and Gram-negative bacteria (*E. coli*) and fungus (*C. albicans*) [184]. The gold NPs have the capability to dysregulate the tRNA function and modify the membrane potential and inhibit the Adenosine Triphosphate (ATP) synthase so as to decrease the ATP level leading to the bacterial cell collapse [185]. Further, Copper nanoparticles (CuNPs) also have the antimicrobial potential through oxidative stress and disturbance of the membrane integrity of the microorganisms thereby leading to the loss of vital nutrients and eventually hastening death [186]. CuNPs incorporated into medical grade polymers - polyurethane and silicone show potent antibacterial activity against MRSA and *E. coli* within 6 h [187]. Addition of CuNPs to the mixture of anionic carboxymethyl chitosan (CMC) and alginate (Alg) polymers eradicates clinical bacterial infection and also promotes bone formation. Furthermore, *in vivo* studies demonstrated that CMC/Alg/Cu scaffolds could induce the formation of vascularized new bone tissue in 4 weeks while avoiding clinical bacterial infection even when the implantation sites were challenged with clinically relevant *S. aureus* bacteria [188]. Another well explored metal is Zinc (Zn). Application of Zn nanoparticles at appropriate concentration, shape and size not only bearded a wide spectrum of bactericidal effects including gram negative bacteria like *E. coli*, *K. pneumoniae*, or *P. aeruginosa* and Gram-positive bacteria like *S. aureus* or *Bacillus subtilis*. The promising effects of titanium and its alloys as an antimicrobial material arises due to photocatalytic effect resulting in the generation of superoxide and hydroxyl radicals [189].

3.9.3 Antibiotics

Antibiotics are the most significant weapons in the antiinfective biomaterial armory. Antibiotic elution systems like films, coatings, incorporated biomaterials like hydrogels, surfaces, microspheres, nanoparticles, nanofibers etc. are the most common strategies used, many a times in conjunction with the above-mentioned strategies too. Some of the most common antibiotics used in this regard are beta-lactams, vancomycin (glycopeptide), erythromycin (macrolide), tetracycline, oxazolidinones, gentamycin (aminoglycoside), amoxicillin, ciprofloxacin, amikacin. An important study revealed that the most common Gram-negative MDR and extensively drug resistant (XDR) pathogens encountered in prosthetic joint infection are *E. coli* (33.6%), *P. aeruginosa* (25.2%), *K. pneumoniae* (21.4%) and *Enterobacter cloacae* (17.6%). *P. aeruginosa* predominated in XDR cases. Isolates were carbapenem-resistant, fluoroquinolone-resistant and extended beta lactamase producers [190]. Local delivery of amikacin and vancomycin from chitosan sponges prevents orthopedic hardware-associated polymicrobial (*S. aureus* and *E. coli*) biofilm [191]. The angiogenic and tissue repair properties of natural rubber latex has been augmented by incorporation of moxifloxacin, a broad spectrum orally administered antibiotic, to be exploited as treatment of infected wounds [192]. Further, Implant-associated osteomyelitis is a chronic infection that complicates orthopedic surgeries. Once infected, 50% of the patients suffer from treatment failure. As compared to the regular intramedullary nail, gentamicin-impregnated intramedullary interlocking nail prevents infection in Gustilo type I and II open tibia fractures [193]. Expert Tibia Nail (ETN) PROtect™ coated with a biodegradable gentamicin-laden polymer used against implant related infections for the prophylaxis of osteomyelitis demonstrated no deep infections after the placement of the gentamicin-coated nail and no side effects were linked to the implant coating [194]. Being uncharged molecules, quinolones can diffuse easily through the biofilm matrix. Hence due to the superior antibiofilm as well as tissue penetration effects of fluoroquinolones, they are frequently used in combination therapy for the treatment of bone and joint implant-associated biofilm infections and chronic lung infections [195]. Nano spray dried antibacterial coatings of PLGA and norfloxacin on titanium disc dental implant resulted in 99.83% reduction in the number of viable bacterial colonies [196].

3.9.4 Antiseptics

Antiseptics are very commonly used in hospital and other medical settings and also on skin during surgeries and other medical procedures to arrest microbial growth leading to infection and sepsis. Triclosan, chlorhexidine, benzalkonium chloride, povidine—iodine, octenidine are most frequently used antiseptics. Polyethylene terephthalate textile covered by layer-by-layer coating of genipin cross linked chitosan and loaded with chlorhexidine serves as a wound dressing with antibacterial effect against *S. aureus*. The number of layers applied to the textile and the thermal post treatment was able to control the release of the chlorhexidine while having antibacterial activity until 45 days [197]. A central venous catheter coated with benzalkonium chloride was able to prevent catheter-related microbial colonization [198]. Povidone-iodine-functionalized fluorinated copolymer, poly (hexafluorobutyl methacrylate-*co*-N-vinyl-2-pyrrolidone) has dual-functional rapid antibacterial activities against Gram-negative *E. coli* and Gram-positive *S. aureus*. It is thus suitable for coating porous fabrics and metal surfaces like gauzes and surgical instruments [199]. Sodium alginate/povidone iodine film-treated animals show a significantly higher wound closure compared to untreated animals considering the bactericidal and fungicidal properties of povidone iodine, providing a controlled antiseptic release [200]. Povidone—iodine swabstick, saturated with a 10% povidone—iodine, is used as a broad-range, first aid antimicrobial that helps prevent infection in minor cuts, scrapes and burns.

3.9.5 Antimicrobial peptides

Antimicrobial peptides (AMPs) are short amphiphilic sequences of 5—50 amino acids with a net positive charge. The sequence and propensity of cationic (arginine and lysine) and hydrophobic residues (valine, leucine, isoleucine, alanine, methionine, phenylalanine, tyrosine and tryptophan) govern the antimicrobial property. The positive charge elicits electrostatic interaction with the anionic lipoteichoic acids and/or lipopolysaccharide component of the cell wall to depolarize and destabilize it while the membrane permeabilization is brought about by the partitioning of the AMPs within the membrane lipid layer [201] (Fig. 3.10). Fabrication of KR-12 peptide-containing hyaluronic acid immobilized fibrous

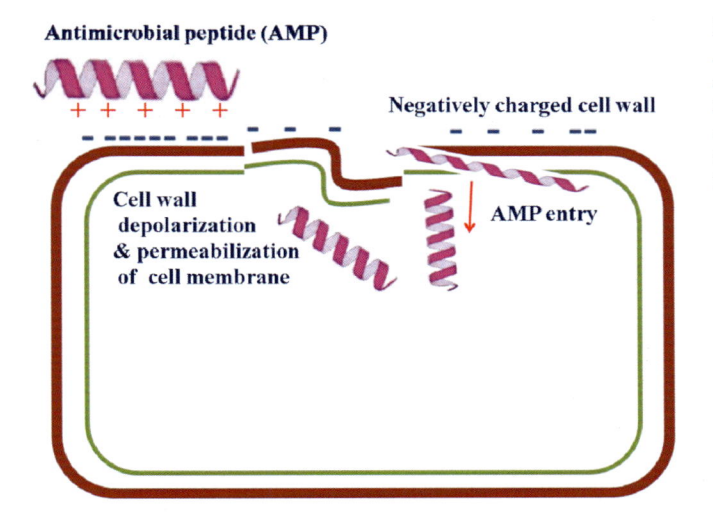

Antimicrobial peptide (AMP)

Negatively charged cell wall

Cell wall depolarization & permeabilization of cell membrane

AMP entry

FIGURE 3.10 The positively charged AMP depolarizes the cell wall by electrostatically interacting with the anionic lipoteichoic acids and/or lipopolysaccharide component and partitions into the membrane lipid layer to permeabilize the membrane.

eggshell membrane effectively kills MDR bacteria, promotes angiogenesis and accelerates re-epithelialization. KR-12 (KRIVQRIKDFLR-NH$_2$), a truncated derivative of LL-37 peptide, displays a selective toxic effect on bacteria but not on human cells [202]. Biocompatible antimicrobial electrospun nanofibers functionalized with antimicrobial polypeptide ε-poly-L-lysine impairs bacterial colonization of *S. epidermidis*, *S. aureus* and *E. coli* by damaging cell membranes and the formation of intracellular ROS [203]. RADA16-AMP (RARADADA)$_2$, a self-assembling peptide has twin effect - excellent bone formation and inhibition of *S. aureus* proliferation [204]. However, AMPs suffer issues of *in vivo* stability, salt sensitivity, and high toxicity towards mammalian cells which can be resolved by conjugation with functional polymers, e.g., PEGylation, which shields the positive charge of AMPs to escape immune system attack and prolong the circulation time in blood [205]. Lim et al. (2017), tethered an AMP, CWR11 (CWFWKWWRRRRR-NH$_2$) onto polydopamine functionalized polydimethylsiloxane (PDMS) catheter which imparted antimicrobial and antibiofilm properties against catheter associated urinary tract infections [206]. Surface tethering of GZ3.27 peptide (containing an added N-terminal cysteine), on titanium, glass, silicon surface demonstrated surface bactericidal activity against *P. aeruginosa* and *E. coli in vitro* [207].

Li et al. evaluated the feasibility of immobilizing two engineered arginine/lysine/tryptophan-rich AMPs, RK1 (RWKRWWRRKK) and RK2 (RKKRWWRRKK) with broad antimicrobial spectra and salt-tolerant properties on silicone surfaces to tackle catheter associated urinary tract infections. The peptides were successfully immobilized on PDMS and urinary catheter surfaces via an allyl glycidyl ether (AGE) polymer brush interlayer. Excellent antimicrobial and antibiofilm activities against *E. coli*, *S. aureus* and *C. albicans were observed without any* toxicity towards smooth muscle cells [208]. Contact-killing surfaces have been designed by creating chimeric peptides containing both a titanium binding domain and an antimicrobial motif [209]. The titanium-binding domain (RPRENRGRERG), binds the peptide on the implant surface which is separated from the freely exposed leucine-lysine rich antimicrobial domain (LKLLKKLLKLLKKL) that attacks the invading bacteria. Compared to bare titanium, this surface reduces adhesion of *S. mutans*, *S. epidermidis* and *E. coli* that affect oral and orthopedic implant related surgeries. Multifunctional coating developed by combining the cell adhesion sequence RGD with lactoferrin derived peptide enhanced *in vitro* cell integration along with inhibition of colonization by *S. aureus* and *S. sanguine* [210]. In conjuction to its membrane-lysis mechanism, magainin, a cationic amphipathic peptide, also displays antifungal action mode, by interfering with the DNA integrity of fungal cells [211].

A few AMPs and AMP derivates are at the preclinical stage and in clinical trials as reviewed nicely by Naafs [212] *viz.*, PL-5 (against skin infections), POL 7080 (for exacerbations of noncystic fibrosis bronchiectasis), DPK 060 (for atopic dermatitis) [213] and LL-37 (for chronic leg ulcers), SGX 942, Brilacidin and CTIX 1278 active against broad spectrum microbes. Among these POL 7080 and CTIX 1278 possess outstanding *in vivo* efficacy due to their potential against MDR superbugs like *P. aeruginosa* and *K. pneumonia*.

3.9.6 Antimicrobial surfaces

Antimicrobial surfaces with surface topographies impact bacterial attachment and hence biofilm formation. Nature-inspired surface topographies like nanopillars mimicking cicada and dragon-fly wing surfaces are bactericidal for *P. aeruginosa* [214]. Bacteria 'lying on the bed of thorns' get deformed [215] and ripped by the shear forces caused by the movement of the cells affixed on the nanopillars [216] (Fig. 3.11). A dragon-fly wing inspired synthetic, mechano-responsive, biomimetic antimicrobial surface analog termed as 'black silicon' has been designed with silicon substrates with sharper and more discreetly arranged nanoprotrusions. The diameters of the high-aspect-ratio nano-protrusions of the nanopillars ranged between ~20and 80 nm with a height of ~500 nm, which was approximately twice that of the dragonfly wing nano-pillars. Interestingly, the surface was found to be highly bactericidal against both Gram negative, Gram positive bacteria and endospores, exhibiting killing rates of up to ~450,000 cells/min/cm^2 [217].

In spite of potential strides in the field of medical technology with regards to the development and engineering of modern antibacterial systems to impart bactericidal efficacy to regenerative scaffolds, the field is still spotted with the nightmare of invading pathogenic microorganisms on the surgical equipment, medical implants, hospital and furniture surfaces. Hence, a vast array of strategies ranging from antimicrobial polymers, antibiotics, antiseptics, antimicrobial peptides to antimicrobial surfaces has been devised to contain the microbial growth.

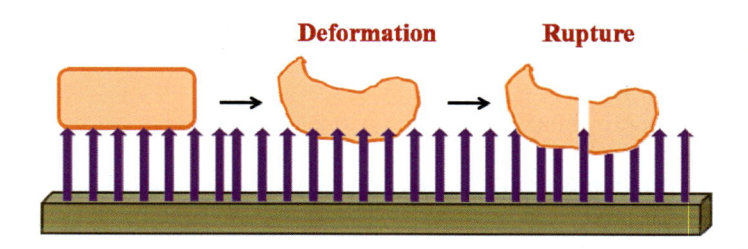

Deformation   **Rupture**

FIGURE 3.11 Surfaces with nano-indentations cause deformation and rupture of the microbial cell.

3.10 Future perspectives and challenges

The present study highlights the importance of surface tailoring in controlling bio-affinitive reactions. The growing demand in the field of tissue engineering opens up a huge scope for scientists to develop better devices with excellent reproducibility for acquiring better results. Surface functionalization not only helps in tuning the surface properties but also allows grafting biological molecules, chemical moieties or nanoparticles on the surface which may in turn effect any bio-affinitive reaction.

The grafting of physico-chemical, topographical or biological cues onto fabricated matrices have helped create biomimetic scaffolds with excellent cell—cell interaction or cell—tissue interaction. However, developing tissue constructs or organs possessing exact architectural or dimensional tribology is still a challenge. Development of hybrid scaffolds constituting of nano/micro/macro architecture with components of decellularized matrices and functional cues could further facilitate the engineers to design futuristic regenerative devices for next generation bio-medical application. Moreover, bioengineering has created the conducive environment for culturing of organoids for controlled and reproducible organoid development, but still there are other outstanding factors besides growth matrix that play an important role in cell differentiation process. Hence, there is need to have a control over all aspects of organoid formation. There are many opportunities for bioengineers to contribute in this field, and require collaboration between engineers, biologists, and users such as pharmaceutical companies. On the other hand, a vast array of strategies ranging from antimicrobial polymers, antibiotics, antiseptics, antimicrobial peptides to antimicrobial surfaces has been devised to contain the microbial growth. Thus, biomedical devices resistant to microbial adhesion and proliferation to some extent, are in use. However, developments are still required to contain the potential threat of upcoming infections, that are slowly but steadily becoming MDR or XDR varieties. Hence, biocidal material with excellent contact time, potent activity against adhering microbes with minimal toxicity towards host cells could be designed.

References

[1] Shi J, Votruba AR, Farokhzad OC, Langer R. Nanotechnology in drug delivery and tissue engineering: from discovery to applications. Nano Lett 2010;10:3223—30.

[2] Chen G, Roy I, Yang C, Prasad PN. Nanochemistry and nanomedicine for nanoparticle-based diagnostics and therapy. Chem Rev 2016;116:2826—85.

[3] Jain RK, Stylianopoulos T. Delivering nanomedicine to solid tumors. Nat Rev Clin Oncol 2010;7:653.

[4] Abudula T, Mohammed H, Joshi Navare K, Colombani T, Bencherif S, Memic A. Latest progress in electrospun nanofibers for wound healing applications. ACS Appl Bio Mater 2019.

[5] Zhang S, Geryak R, Geldmeier J, Kim S, Tsukruk VV. Synthesis, assembly, and applications of hybrid nanostructures for biosensing. Chem Rev 2017;117:12942—3038.

[6] Venditto VJ, Szoka Jr FC. Cancer nanomedicines: so many papers and so few drugs!. Adv Drug Delivery Rev 2013;65:80—8.

[7] Pillai G. Nanomedicines for cancer therapy: an update of FDA approved and those under various stages of development. SOJ Pharm Pharm Sci 2014;1:13.

[8] Wilhelm S, Tavares AJ, Dai Q, Ohta S, Audet J, Dvorak HF, et al. Analysis of nanoparticle delivery to tumours. Nat Rev Mater 2016;1:16014.

[9] Nichols JW, Bae YH. EPR: evidence and fallacy. J Controlled Rel 2014;190:451—64.

[10] Nel AE, Mädler L, Velegol D, Xia T, Hoek EM, Somasundaran P, et al. Understanding biophysicochemical interactions at the nano-bio interface. Nat Mater 2009;8:543.

[11] Cedervall T, Lynch I, Lindman S, Berggård T, Thulin E, Nilsson H, et al. Understanding the nanoparticle–protein corona using methods to quantify exchange rates and affinities of proteins for nanoparticles. Proc Natl Acad Sci 2007;104:2050–5.

[12] Lysaght MJ, Nguy NA, Sullivan K. An economic survey of the emerging tissue engineering industry. Tissue Eng 1998;4:231–8.

[13] Croisier F, Jérôme C. Chitosan-based biomaterials for tissue engineering. Eur Polym J 2013;49:780–92.

[14] Ricard-Blum S. The collagen family. Cold Spring Harb Perspect Biol 2011;3:a004978.

[15] Kasoju N, Bora U. Silk fibroin in tissue engineering. Adv Healthc Mater 2012;1:393–412.

[16] Cheng Z, Teoh S-H. Surface modification of ultra thin poly (ε-caprolactone) films using acrylic acid and collagen. Biomaterials 2004;25:1991–2001.

[17] Wang K. The use of titanium for medical applications in the USA. Mater Sci Eng: A 1996;213:134–7.

[18] Zhang LC, Liu Y, Li S, Hao Y. Additive manufacturing of titanium alloys by electron beam melting: a review. Adv Eng Mater 2018;20:1700842.

[19] Gepreel MA-H, Niinomi M. Biocompatibility of Ti-alloys for long-term implantation. J Mech Behav Biomed Mater 2013;20:407–15.

[20] Lancaster MA, Knoblich JA. Generation of cerebral organoids from human pluripotent stem cells. Nat Protoc 2014;9:2329.

[21] Lancaster MA, Knoblich JA. Organogenesis in a dish: modeling development and disease using organoid technologies. Science 2014;345 1247125.

[22] Nakano T, Ando S, Takata N, Kawada M, Muguruma K, Sekiguchi K, et al. Self-formation of optic cups and storable stratified neural retina from human ESCs. Cell Stem Cell 2012;10:771–85.

[23] Eiraku M, Sasai Y. Mouse embryonic stem cell culture for generation of three-dimensional retinal and cortical tissues. Nat Protoc 2012;7:69.

[24] Spence JR, Mayhew CN, Rankin SA, Kuhar MF, Vallance JE, Tolle K, et al. Directed differentiation of human pluripotent stem cells into intestinal tissue in vitro. Nature 2011;470:105.

[25] Takebe T, Sekine K, Suzuki Y, Enomura M, Tanaka S, Ueno Y, et al. Self-organization of human hepatic organoid by recapitulating organogenesis in vitro. Transplantation proceedings. Elsevier; 2012. p. 1018–20.

[26] Morizane R, Lam AQ, Freedman BS, Kishi S, Valerius MT, Bonventre JV. Nephron organoids derived from human pluripotent stem cells model kidney development and injury. Nat Biotechnol 2015;33:1193.

[27] Takasato M, Er P, Becroft M, Vanslambrouck JM, Stanley E, Elefanty AG, et al. Directing human embryonic stem cell differentiation towards a renal lineage generates a self-organizing kidney. Nat Cell Biol 2014;16:118.

[28] Ulman A. An introduction to ultrathin organic films: from Langmuir–Blodgett to self-assembly. Academic Press; 2013.

[29] Tsukruk VV. Molecular lubricants and glues for micro-and nanodevices. Adv Mater 2001;13:95–108.

[30] Lim DF, Fan J, Peng L, Leong K, Tan CS. Cu–Cu hermetic seal enhancement using self-assembled monolayer passivation. J Electron Mater 2013;42:502–6.

[31] Song M, Kang J-W, Kim D-H, Kwon J-D, Park S-G, Nam S, et al. Self-assembled monolayer as an interfacial modification material for highly efficient and air-stable inverted organic solar cells. Appl Phys Lett 2013;102:59.

[32] Luzinov I, Minko S, Tsukruk VV. Responsive brush layers: from tailored gradients to reversibly assembled nanoparticles. Soft Matter 2008;4:714–25.

[33] Minko S. Responsive polymer brushes. J Macromol Sci Part C: Polym Rev 2006;46:397–420.

[34] Senaratne W, Andruzzi L, Ober CK. Self-assembled monolayers and polymer brushes in biotechnology: current applications and future perspectives. Biomacromolecules 2005;6:2427–48.

[35] Krishnan S, Weinman CJ, Ober CK. Advances in polymers for anti-biofouling surfaces. J Mater Chem 2008;18:3405–13.

[36] Uhlmann P, Ionov L, Houbenov N, Nitschke M, Grundke K, Motornov M, et al. Surface functionalization by smart coatings: stimuli-responsive binary polymer brushes. Prog Org Coat 2006;55:168–74.

[37] Johnston AP, Cortez C, Angelatos AS, Caruso F. Layer-by-layer engineered capsules and their applications. Curr Opcolloid Interface Sci 2006;11:203–9.

[38] Shchepelina O, Kozlovskaya V, Singamaneni S, Kharlampieva E, Tsukruk VV. Replication of anisotropic dispersed particulates and complex continuous templates. J Mater Chem 2010;20:6587–603.

[39] Cui J, van Koeverden MP, Müllner M, Kempe K, Caruso F. Emerging methods for the fabrication of polymer capsules. Adv Colloid Interface Sci 2014;207:14–31.

[40] Richardson JJ, Teng D, Björnmalm M, Gunawan ST, Guo J, Cui J, et al. Fluidized bed layer-by-layer microcapsule formation. Langmuir 2014;30:10028–34.

[41] Yeung T, Georges PC, Flanagan LA, Marg B, Ortiz M, Funaki M, et al. Effects of substrate stiffness on cell morphology, cytoskeletal structure, and adhesion. Cell Motil Cytoskeleton 2005;60:24–34.

[42] Balgude A, Yu X, Szymanski A, Bellamkonda R. Agarose gel stiffness determines rate of DRG neurite extension in 3D cultures. Biomaterials 2001;22:1077–84.

[43] Engler AJ, Griffin MA, Sen S, Bönnemann CG, Sweeney HL, Discher DE. Myotubes differentiate optimally on substrates with tissue-like stiffness: pathological implications for soft or stiff microenvironments. J Cell Biol 2004;166:877–87.

[44] Wong JY, Velasco A, Rajagopalan P, Pham Q. Directed movement of vascular smooth muscle cells on gradient-compliant hydrogels. Langmuir 2003;19:1908–13.

[45] Tzvetkova-Chevolleau T, Stéphanou A, Fuard D, Ohayon J, Schiavone P, Tracqui P. The motility of normal and cancer cells in response to the combined influence of the substrate rigidity and anisotropic microstructure. Biomaterials 2008;29:1541–51.

[46] Sundaram HS, Han X, Nowinski AK, Brault ND, Li Y, Ella-Menye JR, et al. Achieving one-step surface coating of highly hydrophilic poly (carboxybetaine methacrylate) polymers on hydrophobic and hydrophilic surfaces. Adv Mater Interfaces 2014;1:1400071.

[47] Croll TI, O'Connor AJ, Stevens GW, Cooper-White JJ. Controllable surface modification of poly (lactic-co-glycolic acid)(PLGA) by hydrolysis or aminolysis I: physical, chemical, and theoretical aspects. Biomacromolecules 2004;5:463–73.

[48] Barragán F, Guardián R, Menchaca C, Rosales I, Uruchurtu J. Electrochemical corrosion of hot pressing titanium coated steels for biomaterial applications. Int J Electrochem Sci 2010;5:1799–809.

[49] van Kooten TG, Spijker HT, Busscher HJ. Plasma-treated polystyrene surfaces: model surfaces for studying cell–biomaterial interactions. Biomaterials 2004;25:1735–47.

[50] Yang J, Khan M, Zhang L, Ren X, Guo J, Feng Y, et al. Antimicrobial surfaces grafted random copolymers with REDV peptide beneficial for endothelialization. J Mater Chem B 2015;3:7682–97.

[51] Khan M, Yang J, Shi C, Lv J, Feng Y, Zhang W. Surface tailoring for selective endothelialization and platelet inhibition via a combination of SI-ATRP and click chemistry using Cys–Ala–Gly-peptide. Acta Biomater 2015;20:69–81.

[52] Xiong J-P, Stehle T, Zhang R, Joachimiak A, Frech M, Goodman SL, et al. Crystal structure of the extracellular segment of integrin $\alpha V\beta 3$ in complex with an Arg-Gly-Asp ligand. Science 2002;296:151–5.

[53] Humphries JD, Askari JA, Zhang X-P, Takada Y, Humphries MJ, Mould AP. Molecular basis of ligand recognition by integrin $\alpha 5\beta 1$ II. Specificity of Arg-Gly-Asp binding is determined by Trp157 of the α subunit. J Biol Chem 2000;275:20337–45.

[54] Kanie K, Narita Y, Zhao Y, Kuwabara F, Satake M, Honda S, et al. Collagen type IV-specific tripeptides for selective adhesion of endothelial and smooth muscle cells. Biotechnol Bioeng 2012;109:1808–16.

[55] Hubbell JA, Massia SP, Desai NP, Drumheller PD. Endothelial cell-selective materials for tissue engineering in the vascular graft via a new receptor. Biotechnology 1991;9:568.

[56] Lin Q-K, Hou Y, Ren K-F, Ji J. Selective endothelial cells adhesion to Arg-Glu-Asp-Val peptide functionalized polysaccharide multilayer. Thin Solid Films 2012;520:4971–8.

[57] Kim Y-Y, Li H, Song YS, Jeong H-S, Yun H-Y, Baek KJ, et al. Laminin peptide YIGSR enhances epidermal development of skin equivalents. J Tissue Viability 2018;27:117–21.

[58] dos Santos BP, Garbay B, Pasqua M, Chevron E, Chinoy ZS, Cullin C, et al. Production, purification and characterization of an elastin-like polypeptide containing the Ile-Lys-Val-Ala-Val (IKVAV) peptide for tissue engineering applications. J Biotechnol 2019;298:35–44.

[59] Hristov M, Weber C. Endothelial progenitor cells: characterization, pathophysiology, and possible clinical relevance. J Cell Mol Med 2004;8:498–508.

[60] Melchiorri A, Hibino N, Yi T, Lee Y, Sugiura T, Tara S, et al. Contrasting biofunctionalization strategies for the enhanced endothelialization of biodegradable vascular grafts. Biomacromolecules 2015;16:437–46.

[61] Zhang M, Wang K, Wang Z, Xing B, Zhao Q, Kong D. Small-diameter tissue engineered vascular graft made of electrospun PCL/lecithin blend. J Mater Sci: Mater Med 2012;23:2639–48.

[62] Lu S, Zhang P, Sun X, Gong F, Yang S, Shen L, et al. Synthetic ePTFE grafts coated with an anti-CD133 antibody-functionalized heparin/collagen multilayer with rapid in vivo endothelialization properties. ACS Appl Mater Interfaces 2013;5:7360−9.

[63] Yu Q, Zhang Y, Wang H, Brash J, Chen H. Anti-fouling bioactive surfaces. Acta Biomater 2011;7:1550−7.

[64] Wei Y, Ji Y, Xiao L, Lin Q, Ji J. Different complex surfaces of polyethyleneglycol (PEG) and REDV ligand to enhance the endothelial cells selectivity over smooth muscle cells. Colloids Surf B: Biointerfaces 2011;84:369−78.

[65] Liu P, Chen Q, Li L, Lin S, Shen J. Anti-biofouling ability and cytocompatibility of the zwitterionic brushes-modified cellulose membrane. J Mater Chem B 2014;2:7222−31.

[66] Andorko JI, Jewell CM. Designing biomaterials with immunomodulatory properties for tissue engineering and regenerative medicine. Bioeng Transl Med 2017;2:139−55.

[67] Vishwakarma A, Bhise NS, Evangelista MB, Rouwkema J, Dokmeci MR, Ghaemmaghami AM, et al. Engineering immunomodulatory biomaterials to tune the inflammatory response. Trends Biotechnol 2016;34:470−82.

[68] Zhang K, Zheng J, Bian G, Liu L, Xue Q, Liu F, et al. Polarized macrophages have distinct roles in the differentiation and migration of embryonic spinal-cord-derived neural stem cells after grafting to injured sites of spinal cord. Mol Ther 2015;23:1077−91.

[69] Zhang Q, Hwang JW, Oh J-H, Park CH, Chung SH, Lee Y-S, et al. Effects of the fibrous topography-mediated macrophage phenotype transition on the recruitment of mesenchymal stem cells: an in vivo study. Biomaterials 2017;149:77−87.

[70] Wilkinson C, Riehle M, Wood M, Gallagher J, Curtis A. The use of materials patterned on a nano-and micrometric scale in cellular engineering. Mater Sci Eng: C 2002;19:263−9.

[71] Ber S, Köse GT, Hasırcı V. Bone tissue engineering on patterned collagen films: an in vitro study. Biomaterials 2005;26:1977−86.

[72] Loesberg W, Te Riet J, Van Delft F, Schön P, Figdor C, Speller S, et al. The threshold at which substrate nanogroove dimensions may influence fibroblast alignment and adhesion. Biomaterials 2007;28:3944−51.

[73] Berry CC, Campbell G, Spadiccino A, Robertson M, Curtis AS. The influence of microscale topography on fibroblast attachment and motility. Biomaterials 2004;25:5781−8.

[74] Turner A, Dowell N, Turner S, Kam L, Isaacson M, Turner J, et al. Attachment of astroglial cells to microfabricated pillar arrays of different geometries. J Biomed Mater Res: Off J Soc Biomater 2000;51:430−41. The Japanese Society for Biomaterials, and The Australian Society for Biomaterials and the Korean Society for Biomaterials.

[75] Ding Y, Yang Z, Bi CW, Yang M, Xu SL, Lu X, et al. Directing vascular cell selectivity and hemocompatibility on patterned platforms featuring variable topographic geometry and size. ACS Appl Mater Interfaces 2014;6:12062−70.

[76] Guo Y, Caslaru R. Fabrication and characterization of micro dent arrays produced by laser shock peening on titanium Ti−6Al−4V surfaces. J Mater Process Technol 2011;211:729−36.

[77] Lim JY, Donahue HJ. Cell sensing and response to micro-and nanostructured surfaces produced by chemical and topographic patterning. Tissue Eng 2007;13:1879−91.

[78] Srivas PK, Kapat K, Das B, Pal P, Ray PG, Dhara S. Hierarchical surface morphology on Ti6Al4V via patterning and hydrothermal treatment towards improving cellular response. Appl Surf Sci 2019;478:806−17.

[79] Sasikumar Y, Indira K, Rajendran N. Surface modification methods for titanium and its alloys and their corrosion behavior in biological environment: a review. J Bio- And Tribo-Corrosion 2019;5:36.

[80] Müeller WD, Gross U, Fritz T, Voigt C, Fischer P, Berger G, et al. Evaluation of the interface between bone and titanium surfaces being blasted by aluminium oxide or bioceramic particles. Clin Oral Implant Res 2003;14:349−56.

[81] Jemat A, Ghazali MJ, Razali M, Otsuka Y. Surface modifications and their effects on titanium dental implants. Biomed Res Int 2015;2015.

[82] Li H, Wang Y, Zheng Y, Lin J. Osteoblast response on Ti-and Zr-based bulk metallic glass surfaces after sand blasting modification. J Biomed Mater Res Part B: Appl Biomater 2012;100:1721−8.

[83] Bahraminasab M, Sahari B, Edwards K, Farahmand F, Arumugam M. Aseptic loosening of femoral components − materials engineering and design considerations. Mater Des 2013;44:155−63.

[84] De Nardo L, Raffaini G, Ebramzadeh E, Ganazzoli F. Titanium oxide modeling and design for innovative biomedical surfaces: a concise review. Int J Artif Organs 2012;35:629−41.

[85] Bjursten LM, Rasmusson L, Oh S, Smith GC, Brammer KS, Jin S. Titanium dioxide nanotubes enhance bone bonding in vivo. J Biomed Mater Res Part A: Off J Soc Biomater 2010;92:1218−24. The Japanese Society for Biomaterials, and The Australian Society for Biomaterials and the Korean Society for Biomaterials.

[86] Sobieszczyk S. Surface modifications of Ti and its alloys. Adv Mater Sci 2010;10:29−42.

[87] Ban S, Iwaya Y, Kono H, Sato H. Surface modification of titanium by etching in concentrated sulfuric acid. Dental Mater 2006;22:1115−20.

[88] Tavares MG, Tambasco de Oliveira P, Nanci A, Hawthorne AC, Rosa AL, Xavier SP. Treatment of a commercial, machined surface titanium implant with H2SO4/H2O2 enhances contact osteogenesis. Clin Oral Implant Res 2007;18:452−8.

[89] Bearinger JP, Orme CA, Gilbert JL. Effect of hydrogen peroxide on titanium surfaces: in situ imaging and step-polarization impedance spectroscopy of commercially pure titanium and titanium, 6-aluminum, 4-vanadium. J Biomed Mater Res Part A: Off J Soc Biomater 2003;67:702−12. The Japanese Society for Biomaterials, and The Australian Society for Biomaterials and the Korean Society for Biomaterials.

[90] Macak JM, Tsuchiya H, Taveira L, Ghicov A, Schmuki P. Self-organized nanotubular oxide layers on Ti-6Al-7Nb and Ti-6Al-4V formed by anodization in NH4F solutions. J Biomed Mater Res Part A: Off J Soc Biomater 2005;75:928−33. The Japanese Society for Biomaterials, and The Australian Society for Biomaterials and the Korean Society for Biomaterials.

[91] Burns K, Yao C, Webster TJ. Increased chondrocyte adhesion on nanotubular anodized titanium. J Biomed Mater Res Part A: Off J Soc Biomater 2009;88:561−8. The Japanese Society for Biomaterials, and The Australian Society for Biomaterials and the Korean Society for Biomaterials.

[92] Kulkarni M, Mazare A, Schmuki P, Iglič A. Biomaterial surface modification of titanium and titanium alloys for medical applications. Nanomedicine 2014;111:111.

[93] Thull R, Grant D. Physical and chemical vapor deposition and plasma-assisted techniques for coating titanium. Titanium in medicine. Springer; 2001. p. 283−341.

[94] Zhao X, Liu X, Ding C. Acid-induced bioactive titania surface. J Biomed Mater Res Part A: Off J Soc Biomater 2005;75:888−94. The Japanese Society for Biomaterials, and The Australian Society for Biomaterials and the Korean Society for Biomaterials.

[95] Yavari SA, van der Stok J, Chai YC, Wauthle R, Birgani ZT, Habibovic P, et al. Bone regeneration performance of surface-treated porous titanium. Biomaterials 2014;35:6172−81.

[96] Mendonça G, Mendonça DB, Aragao FJ, Cooper LF. The combination of micron and nanotopography by H2SO4/H2O2 treatment and its effects on osteoblast-specific gene expression of hMSCs. J Biomed Mater Res Part A: Off J Soc Biomater 2010;94:169−79. The Japanese Society for Biomaterials, and The Australian Society for Biomaterials and the Korean Society for Biomaterials.

[97] Diniz M, Soares G, Coelho M, Fernandes M. Surface topography modulates the osteogenesis in human bone marrow cell cultures grown on titanium samples prepared by a combination of mechanical and acid treatments. J Mater Sci: Mater Med 2002;13:421−32.

[98] Takeuchi M, Abe Y, Yoshida Y, Nakayama Y, Okazaki M, Akagawa Y. Acid pretreatment of titanium implants. Biomaterials 2003;24:1821−7.

[99] Nanci A, Wuest J, Peru L, Brunet P, Sharma V, Zalzal S, et al. Chemical modification of titanium surfaces for covalent attachment of biological molecules. J Biomed Mater Res: Off J Soc Biomater 1998;40:324−35. The Japanese Society for Biomaterials, and the Australian Society for Biomaterials.

[100] Wang XX, Hayakawa S, Tsuru K, Osaka A. A comparative study of in vitro apatite deposition on heat-, H2O2-, and NaOH-treated titanium surfaces. J Biomed Mater Res: Off J Soc Biomater Japanese Soc Biomater 2001;54:172−8.

[101] Kokubo T, Miyaji F, Kim HM, Nakamura T. Spontaneous formation of bonelike apatite layer on chemically treated titanium metals. J Am Ceram Soc 1996;79:1127−9.

[102] Sollazzo V, Pezzetti F, Scarano A, Piattelli A, Massari L, Brunelli G, et al. Anatase coating improves implant osseointegration in vivo. J Craniofacial Surg 2007;18:806−10.

[103] Delplancke J-L, Winand R. Galvanostatic anodization of titanium—I. Structures and compositions of the anodic films. Electrochim Acta 1988;33:1539−49.

[104] Jonášová L, Müller FA, Helebrant A, Strnad J, Greil P. Biomimetic apatite formation on chemically treated titanium. Biomaterials 2004;25:1187−94.

[105] Popat KC, Leoni L, Grimes CA, Desai TA. Influence of engineered titania nanotubular surfaces on bone cells. Biomaterials 2007;28:3188–97.

[106] Yao C, Webster TJ. Anodization: a promising nano-modification technique of titanium implants for orthopedic applications. J Nanosci Nanotechnol 2006;6:2682–92.

[107] Kapat K, Maity PP, Rameshbabu AP, Srivas PK, Majumdar P, Dhara S. Simultaneous hydrothermal bioactivation with nano-topographic modulation of porous titanium alloys towards enhanced osteogenic and antimicrobial responses. J Mater Chem B 2018;6:2877–93.

[108] Dominik M, Leśniewski A, Janczuk M, Niedziółka-Jönsson J, Hołdyński M, Godlewski M, et al. Titanium oxide thin films obtained with physical and chemical vapour deposition methods for optical biosensing purposes. Biosens Bioelectron 2017;93:102–9.

[109] de Jonge LT, Leeuwenburgh SC, Wolke JG, Jansen JA. Organic–inorganic surface modifications for titanium implant surfaces. Pharm Res 2008;25:2357–69.

[110] Mansour A, Mezour MA, Badran Z, Tamimi F. Extracellular matrices for bone regeneration: a literature review. Tissue Eng Part A 2017;23:1436–51.

[111] De Lange G, Donath K. Interface between bone tissue and implants of solid hydroxyapatite or hydroxyapatite-coated titanium implants. Biomaterials 1989;10:121–5.

[112] Jansen J, Van De Waerden J, Wolke J, Groot K De. Histologic evaluation of the osseous adaptation to titanium and hydroxyapatite-coated titanium implants. J Biomed Mater Res 1991;25:973–89.

[113] Sigrist H, Gao H, Wegmüller B. Light–dependent, covalent immobilization of biomolecules on 'inert'surfaces. Biotechnology 1992;10:1026.

[114] Hanawa T. A comprehensive review of techniques for biofunctionalization of titanium. J Periodontal Implant Sci 2011;41:263–72.

[115] Dettin M, Bagno A, Gambaretto R, Iucci G, Conconi MT, Tuccitto N, et al. Covalent surface modification of titanium oxide with different adhesive peptides: surface characterization and osteoblast-like cell adhesion. J Biomed Mater Res Part A: Off J Soc Biomater 2009;90:35–45. The Japanese Society for Biomaterials, and The Australian Society for Biomaterials and the Korean Society for Biomaterials.

[116] Morra M, Cassinelli C, Cascardo G, Cahalan P, Cahalan L, Fini M, et al. Surface engineering of titanium by collagen immobilization. Surface characterization and in vitro and in vivo studies. Biomaterials 2003;24:4639–54.

[117] Morra M. Biochemical modification of titanium surfaces: peptides and ECM proteins. Eur Cell Mater 2006;12:15.

[118] Michael J, Schönzart L, Israel I, Beutner R, Scharnweber D, Worch H, et al. Oligonucleotide – RGD peptide conjugates for surface modification of titanium implants and improvement of osteoblast adhesion. Bioconjugate Chem 2009;20:710-–718.

[119] Xiao S-J, Textor M, Spencer ND, Sigrist H. Covalent attachment of cell-adhesive, (Arg-Gly-Asp)-containing peptides to titanium surfaces. Langmuir 1998;14:5507–16.

[120] Yang W, Guo D, Harris MA, Cui Y, Gluhak-Heinrich J, Wu J, et al. Bmp2 in osteoblasts of periosteum and trabecular bone links bone formation to vascularization and mesenchymal stem cells. J Cell Sci 2013;126:4085–98.

[121] Gao W, Feng B, Lu X, Wang J, Qu S, Weng J. Characterization and cell behavior of titanium surfaces with PLL/DNA modification via a layer-by-layer technique. J Biomed Mater Res Part A 2012;100:2176–85.

[122] Cai K, Hu Y, Jandt KD, Wang Y. Surface modification of titanium thin film with chitosan via electrostatic self-assembly technique and its influence on osteoblast growth behavior. J Mater Sci: Mater Med 2008;19:499.

[123] Edmondson R, Broglie JJ, Adcock AF, Yang L. Three-dimensional cell culture systems and their applications in drug discovery and cell-based biosensors. Assay Drug Dev Technol 2014;12:207–18.

[124] Cukierman E, Pankov R, Yamada KM. Cell interactions with three-dimensional matrices. Curr Opcell Biol 2002;14:633–40.

[125] Mahoney ZX, Stappenbeck TS, Miner JH. Laminin $\alpha 5$ influences the architecture of the mouse small intestine mucosa. J Cell Sci 2008;121:2493–502.

[126] Shyer AE, Tallinen T, Nerurkar NL, Wei Z, Gil ES, Kaplan DL, et al. Villification: how the gut gets its villi. Science 2013;342:212–18.

[127] Finkbeiner SR, Hill DR, Altheim CH, Dedhia PH, Taylor MJ, Tsai Y-H, et al. Transcriptome-wide analysis reveals hallmarks of human intestine development and maturation in vitro and in vivo. Stem Cell Rep 2015;4:1140–55.

[128] Huch M, Knoblich JA, Lutolf MP, Martinez-Arias A. The hope and the hype of organoid research. Development 2017;144:938–41.

[129] Gjorevski N, Sachs N, Manfrin A, Giger S, Bragina ME, Ordóñez-Morán P, et al. Designer matrices for intestinal stem cell and organoid culture. Nature 2016;539:560.

[130] Cruz-Acuña R, Quirós M, Farkas AE, Dedhia PH, Huang S, Siuda D, et al. Synthetic hydrogels for human intestinal organoid generation and colonic wound repair. Nat Cell Biol 2017;19:1326.

[131] Broguiere N, Isenmann L, Hirt C, Ringel T, Placzek S, Cavalli E, et al. Growth of epithelial organoids in a defined hydrogel. Adv Mater 2018;30 1801621.

[132] Yui S, Azzolin L, Maimets M, Pedersen MT, Fordham RP, Hansen SL, et al. YAP/TAZ-dependent reprogramming of colonic epithelium links ECM remodeling to tissue regeneration. Cell Stem Cell 2018;22:35–49 e37.

[133] Capeling MM, Czerwinski M, Huang S, Tsai Y-H, Wu A, Nagy MS, et al. Nonadhesive alginate hydrogels support growth of pluripotent stem cell-derived intestinal organoids. Stem Cell Rep 2019;12:381–94.

[134] Cruz-Acuña R, Quirós M, Huang S, Siuda D, Spence JR, Nusrat A, et al. PEG-4MAL hydrogels for human organoid generation, culture, and in vivo delivery. Nat Protoc 2018;13:2102.

[135] Poling HM, Wu D, Brown N, Baker M, Hausfeld TA, Huynh N, et al. Mechanically induced development and maturation of human intestinal organoids in vivo. Nat Biomed Eng 2018;2:429.

[136] Eyckmans J, Boudou T, Yu X, Chen CS. A hitchhiker's guide to mechanobiology. Dev Cell 2011;21:35–47.

[137] Lee DA, Knight MM, Campbell JJ, Bader DL. Stem cell mechanobiology. J Cell Biochem 2011;112:1–9.

[138] Pin C, Parker A, Gunning AP, Ohta Y, Johnson IT, Carding SR, et al. An individual based computational model of intestinal crypt fission and its application to predicting unrestrictive growth of the intestinal epithelium. Integr Biol 2014;7:213–28.

[139] Drost J, Van Jaarsveld RH, Ponsioen B, Zimberlin C, Van Boxtel R, Buijs A, et al. Sequential cancer mutations in cultured human intestinal stem cells. Nature 2015;521:43.

[140] Woo D-H, Chen Q, Yang T-LB, Glineburg MR, Hoge C, Leu NA, et al. Enhancing a Wnt-telomere feedback loop restores intestinal stem cell function in a human organotypic model of dyskeratosis congenita. Cell Stem Cell 2016;19:397–405.

[141] Schwank G, Koo B-K, Sasselli V, Dekkers JF, Heo I, Demircan T, et al. Functional repair of CFTR by CRISPR/Cas9 in intestinal stem cell organoids of cystic fibrosis patients. Cell Stem Cell 2013;13:653–8.

[142] Bigorgne AE, Farin HF, Lemoine R, Mahlaoui N, Lambert N, Gil M, et al. TTC7A mutations disrupt intestinal epithelial apicobasal polarity. J Clin Invest 2014;124:328–37.

[143] Wiegerinck CL, Janecke AR, Schneeberger K, Vogel GF, van Haaften–Visser DY, Escher JC, et al. Loss of syntaxin 3 causes variant microvillus inclusion disease. Gastroenterology 2014;147:65–8 e10.

[144] Bhise NS, Ribas J, Manoharan V, Zhang YS, Polini A, Massa S, et al. Organ-on-a-chip platforms for studying drug delivery systems. J Controlled Rel 2014;190:82–93.

[145] Sosa-Hernández JE, Villalba-Rodríguez AM, Romero-Castillo KD, Aguilar-Aguila-Isaías MA, García-Reyes IE, Hernández-Antonio A, et al. Organs-on-a-chip module: a review from the development and applications perspective. Micromachines 2018;9:536.

[146] Younan X, Whitesides GM. Soft lithography. Annu Rev Mater Sci 1998;28:153–84.

[147] Gale BK, Jafek AR, Lambert CJ, Goenner BL, Moghimifam H, Nze UC, et al. A review of current methods in microfluidic device fabrication and future commercialization prospects. Inventions 2018;3:60.

[148] Duffy DC, McDonald JC, Schueller OJ, Whitesides GM. Rapid prototyping of microfluidic systems in poly (dimethylsiloxane). Anal Chem 1998;70:4974–84.

[149] Sin A, Chin KC, Jamil MF, Kostov Y, Rao G, Shuler ML. The design and fabrication of three-chamber microscale cell culture analog devices with integrated dissolved oxygen sensors. Biotechnol Prog 2004;20:338–45.

[150] Wevers NR, Kasi DG, Gray T, Wilschut KJ, Smith B, van Vught R, et al. A perfused human blood–brain barrier on-a-chip for high-throughput assessment of barrier function and antibody transport. Fluids Barriers CNS 2018;15:23.

[151] Chan CY, Huang P-H, Guo F, Ding X, Kapur V, Mai JD, et al. Accelerating drug discovery via organs-on-chips. Lab A Chip 2013;13:4697–710.

[152] Zhang B, Korolj A, Lai BFL, Radisic M. Advances in organ-on-a-chip engineering. Nat Rev Mater 2018;3:257.

[153] Wong I, Ho C-M. Surface molecular property modifications for poly (dimethylsiloxane)(PDMS) based microfluidic devices. Microfluidics Nanofluidics 2009;7:291.

[154] Gökaltun A, Kang YBA, Yarmush ML, Usta OB, Asatekin A. Simple surface modification of poly (dimethylsiloxane) via surface segregating smart polymers for biomicrofluidics. Sci Rep 2019;9:7377.

[155] Doryab A, Amoabediny G, Salehi-Najafabadi A. Advances in pulmonary therapy and drug development: lung tissue engineering to lung-on-a-chip. Biotechnol Adv 2016;34:588—96.

[156] Kuddannaya S, Chuah YJ, Lee MHA, Menon NV, Kang Y, Zhang Y. Surface chemical modification of poly (dimethylsiloxane) for the enhanced adhesion and proliferation of mesenchymal stem cells. ACS Appl Mater Interfaces 2013;5:9777—84.

[157] Huh D. A human breathing lung-on-a-chip. Ann Am Thorac Soc 2015;12:S42—4.

[158] Zhang YS, Aleman J, Arneri A, Bersini S, Piraino F, Shin SR, et al. From cardiac tissue engineering to heart-on-a-chip: beating challenges. Biomed Mater 2015;10:034006.

[159] Annabi N, Selimović Š, Cox JPA, Ribas J, Bakooshli MA, Heintze D, et al. Hydrogel-coated microfluidic channels for cardiomyocyte culture. Lab A Chip 2013;13:3569—77.

[160] Annabi N, Shin SR, Tamayol A, Miscuglio M, Bakooshli MA, Assmann A, et al. Highly elastic and conductive human-based protein hybrid hydrogels. Adv Mater 2016;28:40—9.

[161] McCain ML, Sheehy SP, Grosberg A, Goss JA, Parker KK. Recapitulating maladaptive, multiscale remodeling of failing myocardium on a chip. Proc Natl Acad Sci 2013;110:9770—5.

[162] Lee PJ, Hung PJ, Lee LP. An artificial liver sinusoid with a microfluidic endothelial-like barrier for primary hepatocyte culture. Biotechnol Bioeng 2007;97:1340—6.

[163] Jang K-J, Otieno MA, Ronxhi J, Lim H-K, EwartL, Kodella K, et al. Liver-chip: reproducing human and cross-species toxicities. bioRxiv 2019: 631002.

[164] Ma L-D, Wang Y-T, Wang J-R, Wu J-L, Meng X-S, Hu P, et al. Design and fabrication of a liver-on-a-chip platform for convenient, highly efficient, and safe in situ perfusion culture of 3D hepatic spheroids. Lab A Chip 2018;18:2547—62.

[165] Paoli R, Samitier J. Mimicking the kidney: a key role in organ-on-chip development. Micromachines 2016;7:126.

[166] Wilmer MJ, Ng CP, Lanz HL, Vulto P, Suter-Dick L, Masereeuw R. Kidney-on-a-chip technology for drug-induced nephrotoxicity screening. Trends Biotechnol 2016;34:156—70.

[167] Jang K-J, Mehr AP, Hamilton GA, McPartlin LA, Chung S, Suh K-Y, et al. Human kidney proximal tubule-on-a-chip for drug transport and nephrotoxicity assessment. Integr Biol 2013;5:1119—29.

[168] Ng CP, Zhuang Y, Lin AWH, Teo JCM. A fibrin-based tissue-engineered renal proximal tubule for bioartificial kidney devices: development, characterization and in vitro transport study. Intj Tissue Eng 2013;2013.

[169] Li X, Ming G-l. Using chips to simulate the brain as a tool to investigate brain development. Expert Rev Neurotherapeutics 2008;8:1001—4.

[170] Flemming H-C, Wingender J, Szewzyk U, Steinberg P, Rice SA, Kjelleberg S. Biofilms: an emergent form of bacterial life. Nat Rev Microbiol 2016;14:563.

[171] Huang K-S, Yang C-H, Huang S-L, Chen C-Y, Lu Y-Y, Lin Y-S. Recent advances in antimicrobial polymers: a mini-review. Int J Mol Sci 2016;17:1578.

[172] Eboigbodin K, Newton J, Routh A, Biggs C. Bacterial quorum sensing and cell surface electrokinetic properties. Appl Microbiol Biotechnol 2006;73:669—75.

[173] Francolini I, Donelli G, Crisante F, Taresco V, Piozzi A. Antimicrobial polymers for anti-biofilm medical devices: state-of-art and perspectives. Biofilm-based healthcare-associated infections. Springer; 2015. p. 93—117.

[174] Jeon S, Lee J, Andrade J, De Gennes P. Protein—surface interactions in the presence of polyethylene oxide: I. Simplified theory. J Colloid Interface Sci 1991;142:149—58.

[175] Ostuni E, Chapman RG, Holmlin RE, Takayama S, Whitesides GM. A survey of structure—property relationships of surfaces that resist the adsorption of protein. Langmuir 2001;17:5605—20.

[176] Chang Y, Liao S-C, Higuchi A, Ruaan R-C, Chu C-W, Chen W-Y. A highly stable nonbiofouling surface with well-packed grafted zwitterionic polysulfobetaine for plasma protein repulsion. Langmuir 2008;24:5453—8.

[177] Liang J, Barnes K, Akdag A, Worley SD, Lee J, Broughton RM, et al. Improved antimicrobial siloxane. Ind Eng Chem Res 2007;46:1861—6.

[178] Ye S, Majumdar P, Chisholm B, Stafslien S, Chen Z. Antifouling and antimicrobial mechanism of tethered quaternary ammonium salts in a cross-linked poly (dimethylsiloxane) matrix studied using sum frequency generation vibrational spectroscopy. Langmuir 2010;26:16455—62.

[179] Majumdar P, Lee E, Gubbins N, Christianson DA, Stafslien SJ, Daniels J, et al. Combinatorial materials research applied to the development of new surface coatings XIII: an investigation of polysiloxane antimicrobial coatings containing tethered quaternary ammonium salt groups. J Combin Chem 2009;11:1115—27.

[180] Rekha Deka S, Kumar Sharma A, Kumar P. Cationic polymers and their self-assembly for antibacterial applications. Curr Top Med Chem 2015;15:1179—95.

[181] Fei Liu X, Lin Guan Y, Zhi Yang D, Li Z, De Yao K. Antibacterial action of chitosan and carboxymethylated chitosan. J Appl Polym Sci 2001;79:1324—35.

[182] Dakal TC, Kumar A, Majumdar RS, Yadav V. Mechanistic basis of antimicrobial actions of silver nanoparticles. Front Microbiol 2016;7:1831.

[183] Zeng X, Xiong S, Zhuo S, Liu C, Miao J, Liu D, et al. Nanosilver/poly (DL-lactic-co-glycolic acid) on titanium implant surfaces for the enhancement of antibacterial properties and osteoinductivity. Int J Nanomed 2019;14:1849.

[184] Zheng K, Setyawati MI, Leong DT, Xie J. Antimicrobial gold nanoclusters. ACS Nano 2017;11:6904—10.

[185] Shamaila S, Zafar N, Riaz S, Sharif R, Nazir J, Naseem S. Gold nanoparticles: an efficient antimicrobial agent against enteric bacterial human pathogen. Nanomaterials 2016;6:71.

[186] Kamaruzzaman NF, Tan LP, Hamdan RH, Choong SS, Wong WK, Gibson AJ, et al. Antimicrobial polymers: the potential replacement of existing antibiotics? Int J Mol Sci 2019;20:2747.

[187] Sehmi SK, Noimark S, Weiner J, Allan E, macrobert AJ, Parkin IP. Potent antibacterial activity of copper embedded into silicone and polyurethane. ACS Appl Mater Interfaces 2015;7:22807—13.

[188] Lu Y, Li L, Zhu Y, Wang X, Li M, Lin Z, et al. Multifunctional copper-containing carboxymethyl chitosan/alginate scaffolds for eradicating clinical bacterial infection and promoting bone formation. ACS Appl Mater Interfaces 2017;10:127—38.

[189] Liou J-W, Chang H-H. Bactericidal effects and mechanisms of visible light-responsive titanium dioxide photocatalysts on pathogenic bacteria. Archivum Immunol Therapiae Exp 2012;60:267—75.

[190] Papadopoulos A, Ribera A, Mavrogenis AF, Rodriguez-Pardo D, Bonnet E, Salles MJ, et al. Multidrug-resistant and extensively drug-resistant Gram-negative prosthetic joint infections: role of surgery and impact of colistin administration. Int J Antimicrob Agents 2019;53:294—301.

[191] Boles LR, Awais R, Beenken KE, Smeltzer MS, Haggard WO, Jessica AJ. Local delivery of amikacin and vancomycin from chitosan sponges prevent polymicrobial implant-associated biofilm. Military Med 2018;183:459—65.

[192] Garms BC, Borges FA, de Barros NR, Marcelino MY, Leite MN, Del Arco MC, et al. Novel polymeric dressing to the treatment of infected chronic wound. Appl Microbiol Biotechnol 2019;103:4767—78.

[193] Pinto D, Manjunatha K, Savur AD, Ahmed NR, Mallya S, Ramya V. Comparative study of the efficacy of gentamicin-coated intramedullary interlocking nail versus regular intramedullary interlocking nail in Gustilo type I and II open tibia fractures. Chin J Traumatol 2019;.

[194] Moghaddam A, Graeser V, Westhauser F, Dapunt U, Kamradt T, Woerner SM, et al. Patients' safety: is there a systemic release of gentamicin by gentamicin-coated tibia nails in clinical use? Therapeutics Clin Risk Manag 2016;12:1387.

[195] Høiby N, Bjarnsholt T, Moser C, Bassi G, Coenye T, Donelli G, et al. ESCMID guideline for the diagnosis and treatment of biofilm infections 2014. Clin Microbiol Infect 2015;21:S1—25.

[196] Baghdan E, Raschpichler M, Lutfi W, Pinnapireddy SR, Pourasghar M, Schäfer J, et al. Nano spray dried antibacterial coatings for dental implants. Eur J Pharmaceutics Biopharmaceutics 2019;139:59—67.

[197] Aubert-Viard F, Mogrovejo-Valdivia A, Tabary N, Maton M, Chai F, Neut C, et al. Evaluation of antibacterial textile covered by layer-by-layer coating and loaded with chlorhexidine for wound dressing application. Mater Sci Eng: C 2019;100:554—63.

[198] Moss H, Tebbs S, Faroqui M, Herbst T, Isaac J, Brown J, et al. A central venous catheter coated with benzalkonium chloride for the prevention of catheter-related microbial colonization. Eur J Anaesthesiol 2000;17:680—7.

[199] Borjihan Q, Yang J, Song Q, Gao L, Xu M, Gao T, et al. Povidone-iodine-functionalized fluorinated copolymers with dual-functional antibacterial and antifouling activities. Biomater Sci 2019.

[200] Summa M, Russo D, Penna I, Margaroli N, Bayer IS, Bandiera T, et al. A biocompatible sodium alginate/povidone iodine film enhances wound healing. Eur J Pharmaceutics Biopharmaceutics 2018;122:17—24.

[201] Pasupuleti M, Schmidtchen A, Malmsten M. Antimicrobial peptides: key components of the innate immune system. Crit Rev Biotechnol 2012;32:143—71.

[202] Liu M, Liu T, Zhang X, Jian Z, Xia H, Yang J, et al. Fabrication of KR-12 peptide-containing hyaluronic acid immobilized fibrous eggshell membrane effectively kills multi-drug-resistant bacteria, promotes angiogenesis and accelerates re-epithelialization. Int J Nanomed 2019;14:3345.

[203] Amariei G, Kokol V, Vivod V, Boltes K, Letón P, Rosal R. Biocompatible antimicrobial electrospun nanofibers functionalized with ε-poly-l-lysine. Int J Pharmaceutics 2018;553:141—8.

[204] Yang G, Huang T, Wang Y, Wang H, Li Y, Yu K, et al. Sustained release of antimicrobial peptide from self-assembling hydrogel enhanced osteogenesis. J Biomater Sci Polym Ed 2018;29:1812–24.

[205] Sun H, Hong Y, Xi Y, Zou Y, Gao J, Du J. Synthesis, self-assembly, and biomedical applications of antimicrobial peptide—polymer conjugates. Biomacromolecules 2018;19:1701–20.

[206] Lim K, Chua RRY, Ho B, Tambyah PA, Hadinoto K, Leong SSJ. Development of a catheter functionalized by a polydopamine peptide coating with antimicrobial and antibiofilm properties. Acta Biomater 2015;15:127–38.

[207] De Zoysa GH, Sarojini V. Feasibility study exploring the potential of novel battacin lipopeptides as antimicrobial coatings. ACS Appl Mater Interfaces 2017;9:1373–83.

[208] Li X, Li P, Saravanan R, Basu A, Mishra B, Lim SH, et al. Antimicrobial functionalization of silicone surfaces with engineered short peptides having broad spectrum antimicrobial and salt-resistant properties. Acta Biomater 2014;10:258–66.

[209] Yazici H, O'Neill MB, Kacar T, Wilson BR, Oren EE, Sarikaya M, et al. Engineered chimeric peptides as antimicrobial surface coating agents toward infection-free implants. ACS Appl Mater Interfaces 2016;8:5070–81.

[210] Hoyos-Nogués M, Velasco F, Ginebra M-P, Manero JMA, Gil FJ, Mas-Moruno C. Regenerating bone via multifunctional coatings: the blending of cell integration and bacterial inhibition properties on the surface of biomaterials. ACS Appl Mater Interfaces 2017;9:21618–30.

[211] Matejuk A, Leng Q, Begum M, Woodle M, Scaria P, Chou S, et al. Peptide-based antifungal therapies against emerging infections. Drugs Future 2010;35:197.

[212] Naafs M. The antimicrobial peptides: ready for clinical trials. Biomed J Sci Tech Res 2018;7:6038–42.

[213] Mahlapuu M, Håkansson J, Ringstad L, Umerska A, Andersson T, Boge L, et al. Characterization of the in vitro, ex vivo, and in vivo efficacy of the antimicrobial peptide DPK-060 used for topical treatment. Front Cell Infect Microbiol 2019;9:174.

[214] Ivanova EP, Hasan J, Webb HK, Truong VK, Watson GS, Watson JA, et al. Natural bactericidal surfaces: mechanical rupture of Pseudomonas aeruginosa cells by cicada wings. Small 2012;8:2489–94.

[215] Cheng Y, Feng G, Moraru CI. Micro-and nanotopography sensitive bacterial attachment mechanisms: a review. Front Microbiol 2019;10.

[216] Bandara CD, Singh S, Afara IO, Wolff A, Tesfamichael T, Ostrikov K, et al. Bactericidal effects of natural nanotopography of dragonfly wing on Escherichia coli. ACS Appl Mater Interfaces 2017;9:6746–60.

[217] Ivanova EP, Hasan J, Webb HK, Gervinskas G, Juodkazis S, Truong VK, et al. Bactericidal activity of black silicon. Nat Commun 2013;4:2838.

4

Strategies of 3D bioprinting and parameters that determine cell interaction with the scaffold - A review

Greeshma Ratheesh[1], Cedryck Vaquette[2], Prashant Sonar[3] and Yin Xiao[1,4]

[1]Institute of Health and Biomedical Innovation, Queensland University of Technology, Brisbane, QLD, Australia [2]School of Dentistry, The University of Queensland, Brisbane, QLD, Australia [3]School of Chemistry and Physics, Queensland University of Technology, Brisbane, QLD, Australia [4]Australia-China Centre for Tissue Engineering and Regenerative Medicine, Centre for Biomedical Technologies, Queensland University of Technology, Brisbane, QLD, Australia

4.1 Introduction

According to a classical Greek mythology, a Titan called Prometheus stole the fire from Zeus and gave it to humans. As a punishment for this crime, he was chained to rock and a great eagle was sent to feed on the Titan's liver. The shocking incident was the ability of Prometheus's liver to regenerate every day. Biomedical researches nowadays are investigating to achieve the same potential as Promethean to regenerate organs. The Promethean ability of regenerating the tissue by the use of human cells which restore, maintain or repair human tissue is called tissue engineering.

Tissue engineering by definition is the application of scientific postulates to the design, construction, modification and growth of living tissues using a suitable biomaterials, cells and growth factors, alone or in combination [1−3]. The process helps to eliminate problems such as donor site scarcity, immune rejection and pathogen transfer. Rapid prototyping is one of the method in the advance scaffold fabrication process. The use of additive

81

manufacturing is advancing due to its rapid fabrication based on a computer assisted/aided design. Charles W. Hull in 1986 first patented the 3D printing process, which is the rapid fabrication process of physical, three-dimensional structure [4]. Bioprinting a type of additive manufacturing technique is a promising approach in the biofabrication process. Unlike other methods the material used in bioprinting is living cells and biomaterial which is commonly referred as bioink. The process of bioprinting helps to bypass the cell seeding step.

Based on tissue engineering and regenerative medicine, bioprinting process emerged as a tool that enables to create 3D functional tissues with tailored biological and mechanical properties. The term bioprinting is referred as the use of computer aided technology to pattern and assemble living cells and biomaterial in two dimensional or three dimensional orientation, so as to produce a bioengineered structure that could serve in tissue engineering and regenerative medicine, pharmacokinetics and cell biology studies. Bioprinting process has a wide range of application such as regeneration of tissues like bone [5], cartilage [6], liver [7], and skin [8]. The process helps in protein coating [9], cell deposition for drug delivery [10,11], and also to study tumor growth [12], vascular network [13,14] and stem cell differentiation [15,16].

In this review, we focus on the different strategies of bioprinting process and the commonly used material for bioprinting. We also specify on the properties of an ideal bioink, the cell response to the printed scaffold, the current challenges and the future perspective of 3D organ printing.

4.2 Types of bioprinting

Biomimicry, autonomous self-assembly and mini-tissue building blocks are the main approaches in 3D bioprinting process. There are a number of bioprinting process that are commonly employed [17,18] (Fig. 4.1, Table 4.1).

4.2.1 Extrusion-based bioprinting

Extrusion based printing, often referred as direct-write printing is one of the cheapest and most widely used method of non-biological printing [34]. Moreover, the method

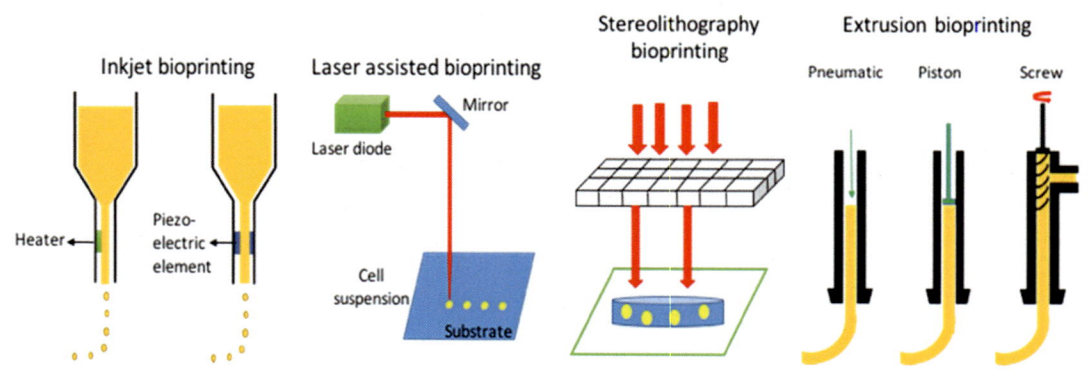

FIGURE 4.1   Schematic illustration of the different types of bioprinting process.

TABLE 4.1 Comparison between the different bioprinting processes.

	Laser assisted bioprinting	Inkjet bioprinting	Extrusion bioprinting	Reference
Viscosity of bioink	1−300 mPa	<10 mPa	$30-6 \times 10^{-7}$ mPa	[19−22]
Method of gelation	Chemical, photo-crosslinking	Chemical, photo-crosslinking	Chemical, photo-crosslinking, temperature	[23−25]
Cell density	Medium (10^8 cells/mL)	Low <10 cells/mL	High	[26−28]
Resolution	10−100 μm	10−50 μm	200−1000 μm	[15,26,29]
Single cell control	Medium	Low	Medium	[15,29,30]
Speed	Medium (200−1600 mm/s)	Fast (100,000 droplet/s)	Slow (700 mm/s−10 μm/s)	[26,30,31]
Cell viability	>95%	>85%	80−90%	[27,32,33]

has a better control over the resulting scaffold. The main components of the printer is the xyz axis which is controlled using a CNC software and the dispense head [17]. The material can be extruded either mechanically or pneumatically. High viscous material can be printed using a pneumatic system, wherein high pressure air pushes the material through the nozzle and the extruded material is deposited on the plotting table based on the designed pattern in a layer-by-layer fashion till the 3D geometry is achieved. Whereas motor-based method is complex and difficult to print material with high viscosity [4].

The printing resolution of extrusion based method is in the range of 10−200 μm based on the bioinks used [35]. High cell density can be printed by this method since air pressure is used to extrude the material; and cell viability post-printing is in the range of 45%−98% [18,36]. The cell viability is significantly low as 45% or below due to the use of high pressure to push the material through a small nozzle, this can be overcome by the use of high gauge size nozzle; however the use of high gauge size nozzle will tend to affect the resolution and speed of printing [32]. Khalil et al. demonstrated the use of multiple extrusion head, which facilitates the printing of multiple material and multi-cell printing [37]. Similarly, Xu et al. employed the use of dual ejector to print two different cell type, OVCAR-5 cancer cells and MRC-5 fibroblasts cells to print a 3D coculture model [38]. Duan et al. 2013 fabricated aortic valve by extrusion method using alginate/gelatin hydrogel by direct incorporation of aortic root sinus smooth muscle cells (SMC) and aortic valve leaflet interstitial cells (VIC). The cell viability of the cells was estimated to be 81.4 ± 3.4% in SMC and 83.2 ± 4.0% in VIC [39].

4.2.2 Inkjet bioprinting

Yet another commonly used technique for 3D bioprinting is the inkjet printing process.

The process is a digital printing technology that was developed for graphics and text printing method. The presence and absence of each drop of ink determines the image construct on the substrate. The printing is carried out by two processes; firstly, droplet formation on a specific location of the substrate and secondly, interaction of the droplet and the target substrate. The inkjet printing can be classified into two methods: (1) continuous inkjet printing (CIJ) and (2) drop-on-demand printing (DOD). In CIJ electrically conductive ink discharge charged droplet which is deposited on the desired location on the target substrate subjected to electric or magnetic field; whereas in DOD drop generator positions just above the desired location and the droplet is dispensed [40]. The cell viability for inkjet printer is in the range of 70–90% with a printing resolution of 0.5–50 μm. Some of the greatest demerits of this approach is the low cell density (<1 million cells/mL) and restricted range of material viscosity to prevent nozzle clogging or cell damage due to the force exerted [18,41].

4.2.3 Laser-assisted bioprinting (LAB)

Laser-assisted bioprinting technique has great advantages over the other bioprinting techniques in the absence of nozzle clogging, speed, high resolution and the process enables a broad spectrum of rheological properties of the material. However, LAB is not used commonly pertaining to the complex set-up, minimum material usage, and the difficulty to change the set-up parameters [42]. The bioprinter is composed of three main component: (1) a laser source, (2) a ribbon which is composed of the biological material and (3) a target substrate where the printed material is deposited. The method uses laser pulse which is focussed on the ribbon made up of absorbing metal like gold or titanium onto which biological material is layered. A high pressure bubble is created by the laser on the ribbon which forces the biological material to be deposited on the target substrate [43,44]. The printing resolution is dependent on certain parameters like the bioink thickness on the ribbon, viscosity and surface tension of the bioink, surface wettability, hypnotic power of the laser, gap between the ribbon and the collector [26,43]. LAB is reported to possess a high viability of 95% with a high seeding density of 10^8 cells/mL [18].

4.2.4 Stereolithography and projection pattern bioprinting

Traditionally stereolithography technique used laser and a *xyz* control to fabricate structures using a photocurable polymer or resin. Recently, the technique has been modified by the incorporation of micromirror array which helps to fine tune the intensity of light for each pixel on the printing area that polymerizes photopolymerisable material and thereby facilitate the fabrication of cell free [45,46] and cell- laden scaffold [47,48]. One of the greatest advantages of the stereolithography technique is the ability to print light-sensitive hydrogel. Moreover, the technique is said to reduce the printing time as it depends on the thickness of the structure [47]. However, the development of complicated system with the ability to polymerize the entire 3D structure tend to influence on the printing speed [49]. Since the process is nozzle free printing, there is high cell viability of 90% with a resolution of 200 μm [16,50].

4.3 Hydrogels for 3D bioprinting

Hydrogels are an ideal biomaterial for tissue engineering that can be of natural origin, artificially synthesized, or a combination of other materials. Hydrogel are built by a network of polymers or peptide chains. The ability of hydrogel to hold high water content that allow cells to migration, nutrient/oxygen transfer, and waste diffusion, makes it an important candidate in the 3D bioprinting process. Some of the commonly used hydrogel for 3D bioprinting are as follows.

4.3.1 Alginate

Alginate is a commonly used naturally occurring hydrogel. They are made up of alternating units of β-D-mannuronate (M) and α-L-guluronate (G) units. They tend to undergo ionic crosslinking in which negatively charged carboxylate group interacts with positively charged ions (Ca^{2+}) forming a hydrogel network [51]. In the past alginate was used for encapsulating cells, for instance O'shea G.M and Sun A.M., used sodium alginate to microencapsulate Langerhan cell which was transplanted to a diabetic rat [52]. Xu et al., studied on the use of alginate to fabricate heterogeneous tissue constructs using cells such as human amniotic fluid-derived stem cells (hAFSCs), canine smooth muscle cells (dSMCs), and bovine aortic endothelial cells (bECs) [53]. Alginate in combination with other biomaterial, serves as an excellent tool for the fabrication of scaffold. For example, a blend of low melting agarose and low viscosity alginate was used to bioprint bifurcated vascular tissue using liquid fluorocarbon by Blaeser's group [54]. Sithole et al., used alginate as bio-ink to fabricate scaffold for bone tissue engineering application, in which the alginate interacted with the poly(ethyleneimine) that forms a polyelectrolyte complex through ionic bond. An inorganic support to the complex was provided by the addition of silica gel which also enhanced the osteoinduction [55].

4.3.2 Collagen type I

Collagen is a type of fibrous protein that is found in multicellular animals, which is found to be secreted into the extracellular matrix. They are the main components of bone and skin, and are known to contribute about 25% of the total protein. Collagen is a triple helical polypeptide chain [56]; this thermosensitive hydrogel has been widely used in the field of tissue engineering because of their biocompatibility and cell adhesion property [57]. In a recent study conducted by Yang et al., collagen type I or agarose was mixed with sodium alginate which was used as a bioink for cartilage tissue engineering [58]. Roth et al. investigated on the cell adhesion and proliferation using a collagen- coated cell repellent substrate [57]. In another study conducted by Skardal et al., used fibrin-collagen bioink carrying different cell type to print on the wound site in order to test the surgical skin wound [59].

4.3.3 Methacrylated gelatin

Methacrylated gelatin is nothing but the denatured collagen protein in which the methacrylate group is conjugated with the amine side chain. They have gained a lot of interest in the field of tissue engineering in the recent years pertaining to its tuneable mechanical properties and biocompatibility [60]. Methacrylated gelatin in the presence of a photoinitiator can be photocrosslinked with UV light thereby making it biomimetic, biodegradable and mechanically strong. For example, Stratesteffen et al. used a tailored hydrogel blend of methacrylated gelatin and collagen type I to investigate on the rheological properties using 3D drop-on-demand printing process [61]. In another study conducted by Duchi et al. wherein they employed core/shell printing using GelMA/HAMa printed using a handheld printer called 'Biopen'. The group claims to have achieved a stiffness of 200KPa at 10 seconds crosslinking with $700 \, mW/cm^2$ of 365 nm UV-A; and with a cell viability of 90% [62].

4.3.4 Fibrin

Fibrin are hydrogel that are produced by reaction of thrombin and fibrinogen. Fibrin is known for its ability to support cell growth and proliferation. Their application in wound healing and skin graft fabrication are quite significant [59,63]. Cui et al., employed TIJ bioprinting process, in which HMVECs (human dermal microvascular endothelial cells) was printed on crosslinked fibrin. Crosslinking was achieved by the use of thrombin and Ca^{2+} on fibrinogen substrate. They results suggest that after 21 days of culture HMVECs formed an extensive capillary network by aligning on the hydrogel [64]. The same group in another study fabricated a viable neural construct by bioprinting fibrin and neural cells alternatively [33]. Hakam et al., fabricated a biopaper made from hybrid combination of fibrin-gelatin for its application in skin bioprinting. In another study conducted by lee et al., constructed an artificial neural tissue built with murine neural stem cells (C17.2), collagen loaded with vascular endothelial growth factor (VEGF) releasing fibrin gel. The group observed a sustained release of growth factor from the construct [65]. Xu et al., fabricated cartilage tissue by employing a hybrid construction method using electrospinning and micro-valve bioprinting in which fibrin was used that enhanced the mechanical and functional properties of the tissue [66].

4.3.5 Polyethylene glycol

Yet another commonly used material for bioprinting is the polyethylene glycol. It is a polyether-based material that when functionalized by the addition of diacrylate (DA) or methacrylate (MA) yields photo-labile polymers. PEG is extensively used in droplet-based bioprinting pertaining to its high mechanical strength when compared to other naturally derived polymers [67]. The two important properties of PEG is the mechanical stability and solubility in water. It is said by altering the composition and crosslinking with photoinitiator helps to tune the structural, functional and mechanical properties of the fabricated tissue construct. Cui et al., functionalized PEG with methacryloyl chloride for cartilage tissue engineering application using TIJ bioprinting by photocrosslinking using I2959 photoinitiator in which human articular chondrocytes where cultured; and suggests that the hydrogel not only induced production of ECM but also enhance growth and proliferation of chondrocytes [68]. Gao et al., bioprinted acrylated-PEG with acrylated peptide made up of Arg-Gly-Asp (RGD) sequence that are essential for cell adhesion, this was photopolymerization at each layer. This construct carried hMSCs which differentiated to osteogenic and chondrogenic lineages [69]. In a recent study conducted by zheng et al., fabricated self-standing bioink using silk fibroin protein and polyethylene (PEG). The mixing of silk fibroin and PEG creates a physical crosslinking due to the β sheet structure formation inducted by the silk. The group claim that the cell (human bone marrow mesenchymal cells) loaded construct could retain the shape for over 12 weeks [70]. Bone marrow-derived hMSCs in PEGDMA (polyethylene glycol dimethacrylate) combined with bioactive glass (BG) and nano- particles of hydroxyapatite (HA) was bioprinted to control spatial delivery of hMSCs and bioactive ceramic materials for the fabrication bone tissue [71].

4.4 Properties of a bioink

Bioink is defined as the materials that has the ability to mimic the extracellular matrix environment in-order to enhance the proliferation and differentiation of cells. Development of an optimal biomaterial that possess a suitable microenvironment for cell response is the main challenge in regenerative engineering. The two most important properties that an ideal bioink has to possess are mechanical requirement and the cell survival within the construct [72,73]. This was addressed by the use of hydrogel with a tuneable physicochemical properties and the ability to mimic body's ECM. Hydrogel are composed of chains of polymers and peptides which has the ability to absorb high content of water, nutrient and oxygen and thereby enabling cell migration. Synthetic and naturally derived hydrogel are been used for the bioprinting process. However, the use of synthetic material is more advantages when compared to the natural hydrogel with the ability to modify the chemical structure [17,74].

The hydrogel can be crosslinked by either physical method or chemical method. Physical crosslinking of the hydrogel is achieved by the non-covalent interaction like van der Waals forces of interaction, hydrogen bonding, hydrophobic attraction, electrostatic forces, and hence such kind of interaction procedures a hydrogel with low mechanical properties that has the ability to disintegrate faster due to the non-covalent linking [75,76]. On the other hand chemical crosslinking produces hydrogel with high mechanical properties due to the strong covalent bonding [77]. However, such hydrogel are non-biocompatibility due to the organic solvents and the polymerization initiators used in the process. In some instance dual-stage crosslinking approach is used such as a low-viscous alginate and gelatin methacryloyl (GelMA) that improve the printing and enhance cell viability [78].

4.5 Parameters that determines the cell responses on tissue engineered scaffold

The very crucial challenge in the field of 3D bioprinting is the cell viability and the ability to form blood vessels within the construct. Viability of the cell is dependent on various aspects such as (1) preprocessing, the cell viability before printing; (2) processing, cell viability during the printing; and (3) postprocessing, cell viability after printing. There is a misunderstanding that high initial seeding density will help in rapid tissue formation. However, researches have proved the presence of cell has an influence on the printability. Cell response to the surface and architecture of engineered scaffold is dependent on various factors such as chemistry, topographical, hydrophobicity, protein adsorption, surface charge, surface roughness, porosity, and other surface characteristics of the material [79]. These characteristics play a vital role in cell proliferation, differentiation and initial cell adheresion and spreading. The cell behavior at the initial stage of culture determines its growth and spreading at the later stage. Cells communicate among themselves and with the surface of ECM through various means. They possess specific interaction between the ligand and the adhesion molecule. The various adhesion molecules include selectins, immunoglobulin superfamily, cadherins and integrin; among which integrins guide the adhesion between the cells and the substrate and cadherins guides cell to cell adhesion.

Surface hydrophobicity of a material is a very critical parameter that determines the initial cell adhesion. The surface hydrophobicity of a material can be measured by the water contact angle of the material. Higher the contact angle, higher is the hydrophobicity of the material (Fig. 4.2). Various studies have reported that the cell adhesion on the surface of the material is dependent on its hydrophilicity; that is if the material is more hydrophilic, there is an increase in cell attachment [80,81]. For instance, it is reported in few studies that there is a decrease in osteoblast adhesion if the contact angle is from 0° to 106°. Similarly fibroblast showed increased cell attachment when the contact angle was in the range on 60°−80° [82]. In a recent study conducted by Moroi et al. the impact of UV treatment on the hydrophobicity of unsintered hydroxyapatite/ poly-L-lactic acid revealed that the contact angle of the UV-treated samples have reduced when compared to that of the untreated sample, which significantly enhanced the cell adhesion [83]. The adhesion of the cell to the surface of a material is dependent on various aspect. It requires a series of cytoplasmic, transmembranal and extracellular protein that assemble on to the stable contact sites. Proteins like immunoglobulins, vitronectin, fibrinogen and fibonectin tend to adsorb on the surface of the material and thereby trigger the inflammatory response. This lead to the activation of neutrophils, macrophages and other inflammatory cells. Studies have revealed that the hydrophobicity of the material has an influence on the protein adsorption. In other words the hydrophilic surface tend to resist protein adsorption, whereas hydrophobic material attract more protein in its surface [81]. For instance Tamada et al. studied that the use of bovine serum albumin, bovine γ globulin and plasma Fn to investigate on the protein absorption on polymer surface. The surface which possess a water contact angle in the range of 60°−80° was reported with maximum protein absorption.

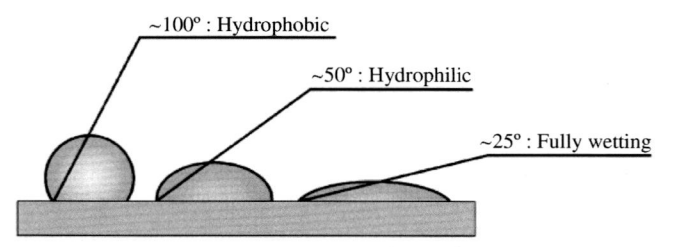

~100° : Hydrophobic

~50° : Hydrophilic

~25° : Fully wetting

FIGURE 4.2   Schematic of contact angle [80].

This suggest that the surface wettability is directly related to the protein adsorption on the material [84].

The other important parameter which determines the cell response to the construct is the surface charge. In a study conducted by Schneider et al. HEMA (2-hydroxyethyl methacrylate) hydrogel incorporated with positive charge was reported to possess increased cell attachment and cell spreading of osteoblast and fibroblast [85]. On the other hand in one of the study conducted by Dadsetan et al. negatively charged hydrogel (oligo poly ethylene glycol fumarate) was reported to possess high chondrocyte differentiation and expression of glycosaminoglycan when compared with that of neutrally or positively charged scaffold [86]. Gopikrishna J. Pillai et al. suggested that the surface charge of the polymeric nanoparticle plays a very critical role in the interplay between the plasma protein adsorption and cellular interaction. The study suggest that positively charged particles showed an increase in size variation and maximum surface protein adsorption [87]. Schmidt et al. reported that the surface charge of the polymer material can be tuned by the use of various functional groups that in turn modify the cell behavior (Table 4.2) [88].

Yet another important parameter that regulates the cell response on the surface and architecture of scaffold is the surface roughness and porosity. The biological response of the tissue and implant is regulated by the surface roughness of the material [79]. Reports suggest that cell tends to show high proliferation and differentiation on microrough surface when compared to smooth surface. For instance, in a study conducted by Hatano et al. the alkaline phosphatase (ALP) activity and the expression of osteocalcin was observed to be increased by rat primary osteoblast cells on a rough surface with porosity of 0.81 μm when compared to that of smooth surface [89]. In a recent study conducted by Wang et al. cold atmospheric plasma (CAP) was used to modify 3D printed PLA scaffold so as to improve the surface roughness and the chemical composition of the scaffold. The results suggest that the surface modification with CAP on PLA scaffold increased the hydrophilicity and nanoscale surface roughness which in turn enhanced the bone cells and mesenchymal stem cell attachment [90]. Lee et al. studied on the behavior of MG63 osteoblast like cells that was cultured on polycarbonate (PC) membrane with different pore size (0.2–0.8 μm diameter). The study suggest that increase in the micropore size would show an increase in ALP activity. In a nut shell the study suggest that the increase in pore size have a negative impact on the cell number and cell differentiation whereas there is an increase in matrix production (Fig. 4.3).

TABLE 4.2 The effect of function group on protein and cell attachment [88].

Functional groups	Properties	Effect on cells
-CH$_3$	Neutral, hydrophobic	Promotes increased leukocyte adhersion and phagocyte migration
-OH	Neutral, hydrophilic	Increases osteoblast differentiation
-COOH	Negative, hydrophilic	Increase osteoblast attachment
-NH$_2$	Positive, hydrophilic	Promotes myoblast and endothelial proliferation and osteoblast differentiation
-CH$_2$NH$_2$	Neutral, hydrophilic	Enhance CHO attachment of Chinese hamster ovary cells

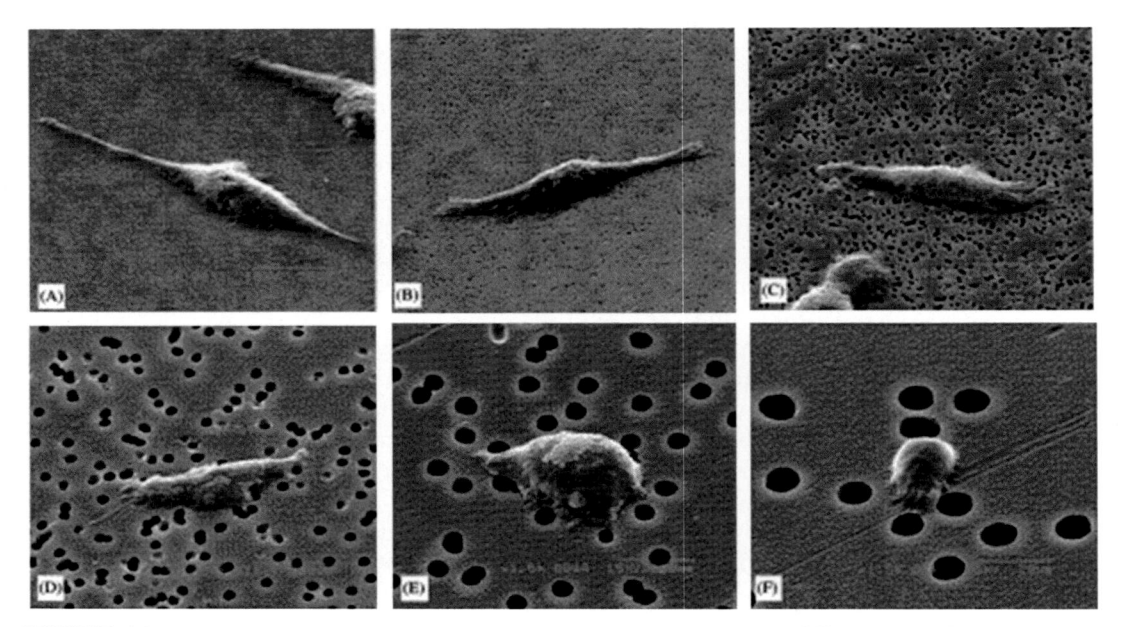

FIGURE 4.3   SEM image of MG63 cells attached on the PC membrane surfaces with different micropore sizes: (A) 0.2, (B) 0.4, (C) 1.0, (D) 3.0, (E) 5.0, and (F) 8.0 μm in diameter [91].

Similar to porosity and surface roughness; inter-connectivity and pore tortuosity (ratio of actual path length through connected pores to the shortest linear distance) play an important role in cell attachment and differentiation [79]. Silva et al. suggested that cell penetration and infiltration to the interior of the scaffold is enhanced by the aligned channels in hydroxyapatite (HA) and poly (D,L-lactic acid) scaffold [92]. Ark et al. fabricated a soft tissue scaffold with Rose Bengal/Chitosan solution impregination. The group proves that the scaffold possess high porosity (86.46 ± 2.95%) and high connectivity that tends to enhance rapid fluid movement, which in-turn shows high cell survivability [93]. Rose et al. claims that the channel diameter has a critical role on cell coverage. These is an increase in cell coverage with increase in channel diameter to 38% in 420 μm channel when compared to 170 μm diameter channel which was 22% [94]. Apart from all this parameter there are certain critical challenges that are to be considered in the field of tissue/organ printing. Table 4.3 gives a details of some of the important challenges that has to be address in the future tissue engineering studies.

4.6 Conclusion

3D bioprinting is an emerging field in the tissue and organ regeneration. Unlike other printing process, bioprinting is a system that can simultaneously construct a scaffold and deposit cells, growth factors and other biological cue in a controlled fashion. The process

TABLE 4.3 Challenges in 3D bioprinting.

Challenges	Summary	Reference
• Type and properties of material: • Degradation pattern	Degradation is determined by hydophilicity, degree of crystallinity, catalysts, porosity, surface area	[95,96]
• Degradation products	Possibility of acid based by-product formation dependent on the degradation rate	[95–97]
• Mechanical strength	Depends on the type of material used (polymer, hydrogel, metals)	[95,98]
• Organ blue print	Post-processing fusion, retraction, remodeling and compaction of the soft tissue construct.	[99–101]
• Scaffold architecture • Pore size	cell growth, attachment, vascularization	[95,102]
• scaffold morphology	Cell adhesion and migration	[95,103]
• surface topography	Cell matrix interaction	[95,96]
Bioink	Biocompatibility, biodegradability, stimuli sensitive, fast solidification, cost	[95,101]
Cell viability and vascularization	• Cell seeding, cell survival during loading, processing and post processing • Vascularization in thick tissue construct	[95,101,104]

has promising use for the accuracy in the placement of cells for tissue engineering and promote regenerative medicine. The popularity in the printing technology had led to the development of cheaper systems and therefore increase the accessibility. However, one of the gap that has to be filled for the successful translation of printed organ or tissue is the speed at which they are printed. The bioprinting process should not only aim in maintaining high cell viability, but also to scale up and fabricate enough scaffold to meet the clinical demand. Although challenges remain in this field, there is a lot of room for development in the area of stem cell research, the development of bioprintable material, new printing strategies and engineering designs.

References

[1] Langer R, Vacanti J. Tissue engineering. Science 1993;260:920–6 *TISSUE ENGINEERING: THE UNION OF BIOLOGY AND ENGINEERING* 98.
[2] Sittinger M, et al. Tissue engineering and autologous transplant formation: practical approaches with resorbable biomaterials and new cell culture techniques. Biomaterials 1996;17:237–42.
[3] Rose FR, Oreffo RO. Bone tissue engineering: hope vs hype. Biochem Biophys Res Commun 2002;292:1–7.
[4] DeSimone E, Schacht K, Jungst T, Groll J, Scheibel T. Biofabrication of 3D constructs: fabrication technologies and spider silk proteins as bioinks. Pure Appl Chem 2015;87:737–49.
[5] Loozen LD, Wegman F, Öner FC, Dhert WJA, Alblas J. Porous bioprinted constructs in BMP-2 non-viral gene therapy for bone tissue engineering. J Mater Chem B 2013;1:6619. Available from: https://doi.org/10.1039/c3tb21093f.
[6] Cui X, Breitenkamp K, Finn MG, Lotz M, D'Lima DD. Direct human cartilage repair using three-dimensional bioprinting technology. Tissue Eng Part A 2012;18:1304–12. Available from: https://doi.org/10.1089/ten. TEA.2011.0543.

[7] Chang R, Emami K, Wu H, Sun W. Biofabrication of a three-dimensional liver micro-organ as an in vitro drug metabolism model. Biofabrication 2010;2:045004. Available from: https://doi.org/10.1088/1758-5082/2/4/045004.

[8] Michael S, et al. Tissue engineered skin substitutes created by laser-assisted bioprinting form skin-like structures in the dorsal skin fold chamber in mice. PLoS One 2013;8:e57741. Available from: https://doi.org/10.1371/journal.pone.0057741.

[9] Derby B. Bioprinting: inkjet printing proteins and hybrid cell-containing materials and structures. J Mater Chem 2008;18:5717. Available from: https://doi.org/10.1039/b807560c.

[10] Snyder JE, et al. Bioprinting cell-laden matrigel for radioprotection study of liver by pro-drug conversion in a dual-tissue microfluidic chip. Biofabrication 2011;3:034112. Available from: https://doi.org/10.1088/1758-5082/3/3/034112.

[11] Rodriguez-Devora JI, Zhang B, Reyna D, Shi ZD, Xu T. High throughput miniature drug-screening platform using bioprinting technology. Biofabrication 2012;4:035001. Available from: https://doi.org/10.1088/1758-5082/4/3/035001.

[12] Zhao Y, et al. Three-dimensional printing of Hela cells for cervical tumor model in vitro. Biofabrication 2014;6:035001.

[13] Kolesky DB, et al. 3D bioprinting of vascularized, heterogeneous cell-laden tissue constructs. Adv Mater 2014;26:3124—30.

[14] Bertassoni LE, et al. Hydrogel bioprinted microchannel networks for vascularization of tissue engineering constructs. Lab A Chip 2014;14:2202—11.

[15] Phillippi JA, et al. Microenvironments engineered by inkjet bioprinting spatially direct adult stem cells toward muscle-and bone-like subpopulations. Stem Cell 2008;26:127—34.

[16] Wang Z, et al. A simple and high-resolution stereolithography-based 3D bioprinting system using visible light crosslinkable bioinks. Biofabrication 2015;7:045009.

[17] Mehrban N, Teoh GZ, Birchall MA. 3D bioprinting for tissue engineering: stem cells in hydrogels. Int J Bioprinting 2016;2.

[18] Murphy SV, Atala A. 3D bioprinting of tissues and organs. Nat Biotechnol 2014;32:773—85.

[19] Guillemot F, et al. High-throughput laser printing of cells and biomaterials for tissue engineering. Acta Biomater 2010;6:2494—500.

[20] Kim JD, Choi JS, Kim BS, Choi YC, Cho YW. Piezoelectric inkjet printing of polymers: stem cell patterning on polymer substrates. Polymer 2010;51:2147—54.

[21] Chang CC, Boland ED, Williams SK, Hoying JB. Direct-write bioprinting three-dimensional biohybrid systems for future regenerative therapies. J Biomed Mater Res Part B: Appl Biomater 2011;98:160—70.

[22] Guillotin B, Guillemot F. Cell patterning technologies for organotypic tissue fabrication. Trends Biotechnol 2011;29:183—90.

[23] Murphy SV, Skardal A, Atala A. Evaluation of hydrogels for bio-printing applications. J Biomed Mater Res Part A 2013;101:272—84.

[24] Smith CM, Christian JJ, Warren WL, Williams SK. Characterizing environmental factors that impact the viability of tissue-engineered constructs fabricated by a direct-write bioassembly tool. Tissue Eng 2007;13:373—83.

[25] Koch L, et al. Laser printing of skin cells and human stem cells. Tissue Eng Part C: Methods 2009;16:847—54.

[26] Guillotin B, et al. Laser assisted bioprinting of engineered tissue with high cell density and microscale organization. Biomaterials 2010;31:7250—6.

[27] Xu T, Jin J, Gregory C, Hickman JJ, Boland T. Inkjet printing of viable mammalian cells. Biomaterials 2005;26:93—9.

[28] Mironov V, Kasyanov V, Markwald RR. Organ printing: from bioprinter to organ biofabrication line. Curr Opbiotechnol 2011;22:667—73.

[29] Campbell PG, Miller ED, Fisher GW, Walker LM, Weiss LE. Engineered spatial patterns of FGF-2 immobilized on fibrin direct cell organization. Biomaterials 2005;26:6762—70.

[30] Smith CM, et al. Three-dimensional bioassembly tool for generating viable tissue-engineered constructs. Tissue Eng 2004;10:1566—76.

[31] Demirci U, Montesano G. Single cell epitaxy by acoustic picolitre droplets. Lab A Chip 2007;7:1139—45.

[32] Chang R, Nam J, Sun W. Effects of dispensing pressure and nozzle diameter on cell survival from solid freeform fabrication—based direct cell writing. Tissue Eng Part A 2008;14:41—8.

[33] Xu T, et al. Viability and electrophysiology of neural cell structures generated by the inkjet printing method. Biomaterials 2006;27:3580—8.

[34] Hockaday L, et al. Rapid 3D printing of anatomically accurate and mechanically heterogeneous aortic valve hydrogel scaffolds. Biofabrication 2012;4:035005.

[35] Hanson Shepherd JN, et al. 3D microperiodic hydrogel scaffolds for robust neuronal cultures. Adv Funct Mater 2011;21:47—54.

[36] Schacht K, et al. Biofabrication of cell-loaded 3D spider silk constructs. Angew Chem Int Ed 2015;54:2816—20.

[37] Khalil S, Nam J, Sun W. Multi-nozzle deposition for construction of 3D biopolymer tissue scaffolds. Rapid Prototyp J 2005;11:9—17.

[38] Xu F, et al. A three-dimensional in vitro ovarian cancer coculture model using a high-throughput cell patterning platform. Biotechnol J 2011;6:204—12.

[39] Duan B, Hockaday LA, Kang KH, Butcher JT. 3D bioprinting of heterogeneous aortic valve conduits with alginate/gelatin hydrogels. J Biomed Mater Res Part A 2013;101:1255—64.

[40] Saunders RE, Derby B. Inkjet printing biomaterials for tissue engineering: bioprinting. Int Mater Rev 2014;59:430—48.

[41] Moon S, et al. Layer by layer three-dimensional tissue epitaxy by cell-laden hydrogel droplets. Tissue Eng Part C: Methods 2009;16:157—66.

[42] Guillotin B, et al. Rapid prototyping of complex tissues with laser assisted bioprinting (LAB). Woodhead Publ Ser Biomater 2014;70:156—75.

[43] Guillemot F, Souquet A, Catros S, Guillotin B. Laser-assisted cell printing: principle, physical parameters versus cell fate and perspectives in tissue engineering. Nanomedicine 2010;5:507—15.

[44] Serra P, Duocastella M, Fernández-Pradas J, Morenza J. Liquids microprinting through laser-induced forward transfer. Appl Surf Sci 2009;255:5342—5.

[45] Seck TM, Melchels FP, Feijen J, Grijpma DW. Designed biodegradable hydrogel structures prepared by stereolithography using poly (ethylene glycol)/poly (D, L-lactide)-based resins. J Controlled Release 2010; 148:34—41.

[46] Ronca A, Ambrosio L, Grijpma D. Preparation of designed poly (D, L-lactide)/nanosized hydroxyapatite composite structures by stereolithography. Acta Biomater 2013;9:5989—96.

[47] Grogan SP, et al. Digital micromirror device projection printing system for meniscus tissue engineering. Acta Biomater 2013;9:7218—26.

[48] Gauvin R, et al. Microfabrication of complex porous tissue engineering scaffolds using 3D projection stereolithography. Biomaterials 2012;33:3824—34.

[49] Gou M, et al. Bio-inspired detoxification using 3D-printed hydrogel nanocomposites. Nat Commun 2014;5.

[50] Lin H, et al. Application of visible light-based projection stereolithography for live cell-scaffold fabrication with designed architecture. Biomaterials 2013;34:331—9.

[51] Blandino A, Macias M, Cantero D. Formation of calcium alginate gel capsules: influence of sodium alginate and $CaCl_2$ concentration on gelation kinetics. J Biosci Bioeng 1999;88:686—9.

[52] O'Shea GM, Sun A. Encapsulation of rat islets of Langerhans prolongs xenograft survival in diabetic mice. Diabetes 1986;35:943—6.

[53] Xu T, et al. Complex heterogeneous tissue constructs containing multiple cell types prepared by inkjet printing technology. Biomaterials 2013;34:130—9.

[54] Blaeser A, et al. Biofabrication under fluorocarbon: a novel freeform fabrication technique to generate high aspect ratio tissue-engineered constructs. Bioresearch Open Access 2013;2:374—84.

[55] Sithole MN, et al. A 3D bioprinted in situ conjugated-co-fabricated scaffold for potential bone tissue engineering applications. J Biomed Mater Res Part A 2018;106:1311—21.

[56] Park JY, et al. A comparative study on collagen type I and hyaluronic acid dependent cell behavior for osteochondral tissue bioprinting. Biofabrication 2014;6:035004.

[57] Roth E, et al. Inkjet printing for high-throughput cell patterning. Biomaterials 2004;25:3707—15.

[58] Yang X, et al. Collagen-alginate as bioink for three-dimensional (3D) cell printing based cartilage tissue engineering. Mater Sci Eng: C 2018;83:195—201.

[59] Skardal A, et al. Bioprinted amniotic fluid-derived stem cells accelerate healing of large skin wounds. Stem Cell Transl Med 2012;1:792—802.

[60] Nichol JW, et al. Cell-laden microengineered gelatin methacrylate hydrogels. Biomaterials 2010;31:5536—44.

[61] Stratesteffen H, et al. GelMA-collagen blends enable drop-on-demand 3D printablility and promote angiogenesis. Biofabrication 2017;9:045002.

[62] Duchi S, et al. Handheld co-axial bioprinting: application to in situ surgical cartilage repair. Sci Rep 2017;7:5837.

[63] Yanez M, et al. In vivo assessment of printed microvasculature in a bilayer skin graft to treat full-thickness wounds. Tissue Eng Part A 2014;21:224—33.

[64] Cui X, Boland T. Human microvasculature fabrication using thermal inkjet printing technology. Biomaterials 2009;30:6221—7.

[65] Lee Y-B, et al. Bio-printing of collagen and VEGF-releasing fibrin gel scaffolds for neural stem cell culture. Exp Neurol 2010;223:645—52.

[66] Xu T, et al. Hybrid printing of mechanically and biologically improved constructs for cartilage tissue engineering applications. Biofabrication 2012;5:015001.

[67] Zalipsky S, Harris JM. ACS Publications; 1997.

[68] Cui X, Breitenkamp K, Lotz M, D'Lima D. Synergistic action of fibroblast growth factor-2 and transforming growth factor-beta1 enhances bioprinted human neocartilage formation. Biotechnol Bioeng 2012;109:2357—68.

[69] Gao G, Yonezawa T, Hubbell K, Dai G, Cui X. Inkjet-bioprinted acrylated peptides and PEG hydrogel with human mesenchymal stem cells promote robust bone and cartilage formation with minimal printhead clogging. Biotechnol J 2015;10:1568—77.

[70] Zheng Z, et al. 3D bioprinting of self-standing silk-based bioink. Adv Healthc Mater 2018;7:e1701026.

[71] Gao G, et al. Bioactive nanoparticles stimulate bone tissue formation in bioprinted three-dimensional scaffold and human mesenchymal stem cells. Biotechnol J 2014;9:1304—11.

[72] Levato R, et al. Biofabrication of tissue constructs by 3D bioprinting of cell-laden microcarriers. Biofabrication 2014;6:035020.

[73] Skardal A, et al. A hydrogel bioink toolkit for mimicking native tissue biochemical and mechanical properties in bioprinted tissue constructs. Acta Biomater 2015;25:24—34.

[74] Tan H, Li H, Rubin JP, Marra KG. Controlled gelation and degradation rates of injectable hyaluronic acid-based hydrogels through a double crosslinking strategy. J Tissue Eng Regener Med 2011;5:790—7.

[75] Hennink W, Van Nostrum CF. Novel crosslinking methods to design hydrogels. Adv Drug Deliv Rev 2012;64:223—36.

[76] Bae KH, Wang L-S, Kurisawa M. Injectable biodegradable hydrogels: progress and challenges. J Mater Chem B 2013;1:5371—88.

[77] Wang H, Heilshorn SC. Adaptable hydrogel networks with reversible linkages for tissue engineering. Adv Mater 2015;27:3717—36.

[78] Colosi C, et al. Microfluidic bioprinting of heterogeneous 3D tissue constructs using low-viscosity bioink. Adv Mater 2016;28:677—84.

[79] Chang H-I, Wang Y. Regenerative medicine and tissue engineering-cells and biomaterials. InTech; 2011.

[80] Goddard JM, Hotchkiss J. Polymer surface modification for the attachment of bioactive compounds. Prog Polym Sci 2007;32:698—725.

[81] Xu L-C, Siedlecki CA. Effects of surface wettability and contact time on protein adhesion to biomaterial surfaces. Biomaterials 2007;28:3273—83.

[82] Wei J, et al. Influence of surface wettability on competitive protein adsorption and initial attachment of osteoblasts. Biomed Mater 2009;4:045002.

[83] Moroi A, et al. Effect on surface character and mechanical property of unsintered hydroxyapatite/poly-l-lactic acid (uHA/PLLA) material by UV treatment. J Biomed Mater Res - Part B Appl Biomater 2018;106:191—200. Available from: https://doi.org/10.1002/jbm.b.33833.

[84] Tamada Y, Ikada Y. Effect of preadsorbed proteins on cell adhesion to polymer surfaces. J Colloid Interface Sci 1993;155:334—9.

[85] Schneider GB, et al. The effect of hydrogel charge density on cell attachment. Biomaterials 2004;25:3023—8.

[86] Dadsetan M, et al. The effects of fixed electrical charge on chondrocyte behavior. Acta Biomater 2011;7:2080—90. Available from: https://doi.org/10.1016/j.actbio.2011.01.012.

[87] Pillai GJ, Greeshma MM, Menon D. Impact of poly(lactic-co-glycolic acid) nanoparticle surface charge on protein, cellular and haematological interactions. Colloids Surf B Biointerfaces 2015;136:1058—65. Available from: https://doi.org/10.1016/j.colsurfb.2015.10.047.

[88] Schmidt DR, Waldeck H, Kao WJ. Biological interactions on materials surfaces. Springer; 2009. p. 1–18.

[89] Hatano K, et al. Effect of surface roughness on proliferation and alkaline phosphatase expression of rat calvarial cells cultured on polystyrene. Bone 1999;25:439–45.

[90] Wang M, et al. Cold atmospheric plasma (CAP) surface nanomodified 3D printed polylactic acid (PLA) scaffolds for bone regeneration. Acta Biomater 2016;46:256–65. Available from: https://doi.org/10.1016/j.actbio.2016.09.030.

[91] Lee SJ, et al. Response of MG63 osteoblast-like cells onto polycarbonate membrane surfaces with different micropore sizes. Biomaterials 2004;25:4699–707.

[92] Silva M, et al. The effect of anisotropic architecture on cell and tissue infiltration into tissue engineering scaffolds. Biomaterials 2006;27:5909–17.

[93] Ark M, et al. Characterisation of a novel light activated adhesive scaffold: potential for device attachment. J Mech Behav Biomed Mater 2016;62:433–45. Available from: https://doi.org/10.1016/j.jmbbm.2016.05.029.

[94] Rose FR, et al. In vitro assessment of cell penetration into porous hydroxyapatite scaffolds with a central aligned channel. Biomaterials 2004;25:5507–14.

[95] Ratheesh G, et al. 3D fabrication of polymeric scaffolds for regenerative therapy. ACS Biomater Sci & Eng 2017;3:1175–94.

[96] Yeong W-Y, Chua C-K, Leong K-F, Chandrasekaran M. Rapid prototyping in tissue engineering: challenges and potential. Trends Biotechnol 2004;22:643–52.

[97] Sung H-J, Meredith C, Johnson C, Galis ZS. The effect of scaffold degradation rate on three-dimensional cell growth and angiogenesis. Biomaterials 2004;25:5735–42.

[98] Wang N, et al. Mechanical behavior in living cells consistent with the tensegrity model. Proc Natl Acad Sci 2001;98:7765–70.

[99] Jakab K, Neagu A, Mironov V, Markwald RR, Forgacs G. Engineering biological structures of prescribed shape using self-assembling multicellular systems. Proc Natl Acad Sci USA 2004;101:2864–9.

[100] Napolitano AP, Chai P, Dean DM, Morgan JR. Dynamics of the self-assembly of complex cellular aggregates on micromolded nonadhesive hydrogels. Tissue Eng 2007;13:2087–94.

[101] Mironov V, Kasyanov V, Drake C, Markwald RR. Organ printing: promises and challenges. 2008.

[102] Ranucci CS, Kumar A, Batra SP, Moghe PV. Control of hepatocyte function on collagen foams: sizing matrix pores toward selective induction of 2-D and 3-D cellular morphogenesis. Biomaterials 2000;21:783–93.

[103] Yin L, Bien H, Entcheva E. Scaffold topography alters intracellular calcium dynamics in cultured cardiomyocyte networks. Am J Physiol-Heart Circulatory Physiol 2004;287:H1276–85.

[104] Saunders RE, Gough JE, Derby B. Delivery of human fibroblast cells by piezoelectric drop-on-demand inkjet printing. Biomaterials 2008;29:193–203.

5

Multipotent nature of dental pulp stem cells for the regeneration of varied tissues — A personalized medicine approach

V.P. Sivadas[1], D.P. Rahul[2] and Prabha D. Nair[1]

[1]Division of Tissue Engineering and Regeneration Technologies, Biomedical Technology Wing, Sree Chitra Tirunal Institute for Medical Sciences and Technology (SCTIMST), Poojapura, Thiruvananthapuram, India [2]Department of Orthodontics and Dentofacial Orthopedics, School of Dentistry, Amrita Vishwa Vidyapeetham, Amrita Institute of Medical Sciences, Kochi, India

5.1 Introduction

Dental Pulp Stems cells (DPSCs) are ectomesenchymal derivatives of cephalic neural crest cells. The neural crest cells are structures formed at the fusion of neural folds during embryonic development. The neural crest cells predominantly produce neural tissue along with few other ectomesenchymal derivatives [1]. Postnatal human dental pulp stem cells (DPSC) were first identified by Gronthos et al. [2], and reported for proliferative and osteogenic properties. Gronthos et al. [2] isolated and characterized DPSCs from impacted third molar tooth. Later, Miura et al. [3] isolated dental pulp stem cells from human exfoliated deciduous teeth and described the higher yield of stems capable of multilineage differentiation. DPSCs could be considered as a preosteogenic mesenchymal stem cell (MSC). Since then studies were extensively conducted for characterization of DPSCs as MSCs [4,5].

The International Society for Cellular Therapy has set minimum criteria for any cell to be described as a MSC. The minimum criteria for recognition of Multipotent MSCs under standard culture conditions include the potential to adhere to plastic in *in vitro* culture and further, MSCs must express the surface markers CD105, CD73 and CD90, and should lack the expression of CD45, CD34, CD14 or CD11b, CD79α or CD19 and HLA-DR. Third condition for a cell to be regarded as MSC is its differentiation potential to osteoblasts, adipocytes and chondroblasts *in vitro* [6]. DPSCs that satisfy these criteria, can be regarded as multi-potent MSCs. Recent research has even shown a wider spectrum of differentiation of DPSCs into osteoblasts, odontontoblasts, adipocytes, chondrocytes, endothelial cells, muscle tissue, islet cells, and neural cells [7,8]. DPSCs do express surface antigens like STRO-1, CD13, CD24, CD29, CD44, CD73, CD90, CD105, CD106, CD146 and the stem cell markers like Oct4, Nanog and β2 integrin. Further, DPSCs are negative for CD14, CD34, CD45 and HLA-DR. Notably, they are exclusively negative for CD45 and CD34 that suggest these cells are not of hematopoietic origin [2,9]. Detailed information regarding immunophenotypic characteristics and localization of dental MSCs is given by Bakopoulou and About [10] and beyond the scope of this chapter.

5.2 Importance of DPSCs in personalized regenerative medicine

Regenerative medicine has the potential to heal or replace tissues and organs damaged by age, disease, or trauma, as well as to normalize congenital defects. Regenerative medicine substitutes for or regenerates damaged human cells, tissues and/or organs in order to restore their normal functioning [11]. Tissue engineering is an integral part of modern regenerative medicine. Tissue engineering involves the application of adult and/or stem cells, usage of cellular regeneration enhancing scaffolds and microenvironments, and important bioactive molecules and growth factors [12,13]. The success of tissue engineering and cellular regeneration is dependent on the biocompatibility of the scaffolds/molecules used, management of immune rejection and chronic inflammation and control of bacterial infections [13,14]. Recently, Dental Stem Cells (DSCs) are gaining more attention as a stem cell source in regenerative medicine due to its higher clonality, proliferation potential and the capacity to retain stemness even after long-term cryopreservation [15]. Several studies have provided evidence that human dental pulp contains precursor cells, named dental pulp stem cells (hDPSC). These cells have self-renewal potential and multilineage differentiation capacity. As these cell cells can be easily isolated, cultured and cryopreserved, they form an attractive stem cell source for futuristic tissue engineering purposes [16].

Dental Stem Cells (DSCs) are mesenchymal cell populations that exhibit self-renewal capacity and multidifferentiation potential [17,18]. As mentioned earlier, Dental Pulp Stem Cells (DPSCs) are the first identified and characterized DSCs [2]. Currently, there are five main types of DSCs [19,20]. They are: stem cells from exfoliated deciduous teeth (SHED) [3], periodontal ligament stem cells (PDLSCs) [21], and dental follicle precursor cells (DFPCs) [22], stem cells from apical papilla (SCAP) [23]. All these stem cells except SHED are capable of forming permanent teeth [19]. Since these cells are easily accessible, and they prevail throughout the lifetime of human beings, they are widely studied in regenerative medicine as a source of autologous stem cells. These cells find applications in regenerative therapies including oro-facial, neurologic, ocular, cardiovascular, diabetic, renal, muscular dystrophy and autoimmune conditions [19,20]. In this chapter, we aim to highlight the recent developments and findings in the field of DPSC mediated regenerative medicine. Indeed, DPSCs can be used for clinical applications in a wide array of diseases. But, only the most relevant findings with regards to regenerative medicine associated with DPSCs is discussed in the current chapter.

5.3 Usefulness of DPSCs in osteogenic regeneration therapy

A tooth for bone is a concept, which has not been exploited to its full potential. Perhaps, the current trend in dental pulp stem cell research is more oriented towards their tooth regeneration potential [4]. Some of the initial studies have also concluded that the osteogenic differentiation of DPSCs is formation of dentin [24]. However, dentin is acellular and differs from bone by absence of osteocyte cells. Moreover, further development in this area provided evidence for the ability of DPSCs to produce lamellar bone containing osteocytes [25]. At present, bone regeneration and dental regeneration potential of DPSCs are widely studied areas, because DPSCs show inherent potential for dental and bone regeneration related applications. Apart from being defined as multi-potent MSCs, the potential of DPSCs for osteogenic differentiation have further been confirmed through many osteogenic gene expression studies. Gene expression analysis of osteogenic-induced DPSCs exhibit expression of typical osteoblast marker genes like ALP, Osteocalcin, osteopontin, RUNX2 and collagen type I. In addition, micro array analysis experiments have shown DPSCs to express osteoblast phenotype factors like IGFBP-5, JunB and NURR1 [5].

A recent study reported how the seeding density of DPSCs during initial expansion can affect the property and differentiation potential of DPSCs [26]. Two seeding densities were evaluated: Sparse seeding (sDPSCs; 5×10^3 cells/cm^2) or dense seeding (dDPSCs; 1×10^5 cells/cm^2) conditions for initial expansion of 4 days were compared for their properties and differentiation potential. Notably, the population of CD73 + and CD105 + cells were much less in dDPSCs group as compared to sDPSCs group. Though both groups retained multi-differentiation potential, the dDPSCs showed higher number of mineralized nodule formation. Further, when dDPSCs were implanted into mouse bone cavities, more mineralized tissue was formed when compared to sDPSCs and control. This study indicates that dense seeding conditions can influence the properties of DPSCs promoting osteogenic-lineage commitment.

In order to get a clear understanding on the *in vivo* functions and therapeutic potential of human DPSCs, one must understand the importance of selective cell markers in the progenitor cells. Important information regarding the surface markers in human DPSCs for bone tissue engineering was provided by Yasui et al. [27]. As per their findings, the side-populations of DPSCs with higher osteogenesis potential are LNGFR(Low +)THY-1(High +) cells. These cells have much higher clonogenic capacity, do express the well-known mesenchymal stem cell markers and are able to heal critical-size calvarial defects in mouse model to a better extent than other DPSCs *in vivo*. These findings put forward that LNGFR(Low +)THY-1 (High +) DPSC side population cells can be considered as an excellent cell source for bone regenerative therapies. Another study used a sub-population of human dental pulp mesenchymal cells (DPMSCs) called the dental pulp pluripotent stem cells (DPPSCs). These cells can undergo differentiation into mesodermal, ectodermal and endodermal cell lineages [28]. DPPSCs when cultured in 3D Cell Carrier glass scaffolds could form bone-like tissue and showed trabecular host bone structure, with high interconnectivity. Further, DPPSCs after osteogenic differentiation showed higher levels of expression of bone markers, calcium deposition and ALP activity as compared to DPMSCs. Thus DPPSCs could act as a potential cell source for the use in bone tissue engineering applications.

Another important area of concern is the generation of suitable scaffolds and osteoinductive molecules in DPSC mediated bone regeneration. Osteogenic differentiation of DPSCs was studied on a variety of scaffolds and DPSCs were found to be compatible with an array of scaffolds including bioglass (BG), commercially available Titanium implants (like Ti6A14V and biomimetic BAS™) and synthetic bone grafts such as Bonelike® [29–31]. Notably, DPSCs showed decreased osteogenic activity when compared with BMSCs in Hydroxy apatite (HA)—poly caprolactone (PCL) composites (HA—PCL) scaffolds [32]. Osteogenic activity of DPSC in HA/collagen/PLCL scaffolds were enhanced in comparison to PLCL alone scaffolds indicating a role of HA/collagen. Osteogenic activity of DPSC have also been detected on sponge like bioactive glass — porcine gelatin composite scaffolds which were confirmed by alkaline phosphatase (ALP) staining and immunohistochemical studies [33]. Silk fibroin scaffold constructs made with DPSC and human amniotic fluid derived stem cell were compared for bone regeneration to repair critical size calvarial defects in immuno compromised rats [34]. Both constructs showed healing of defect without infection or other complications. Human amniotic fluid derived stem cells constructs however, showed better bone formation than DPSC constructs. However when collagen scaffold constructs made with DPSC and human amniotic fluid derived stem cell were compared for bone regeneration in the repair of critical size calvarial defects in immuno-compromised rats, both constructs showed healing of bone defects [35]. Vascularization was also observed in both constructs indicating angiogenic potential of both DPSC and amniotic fluid derived stem cells. Akkouch et al. [36] studied the proliferation and osteogenic differentiation of DPSC in three different scaffolds viz. collagen, bioglass—collagen and bioglass—collagen—PCL. Higher proliferation and osteogenic differentiation of DPSCs on bioglass—collagen and bioglass—collagen—PCL than collagen alone were observed indicating an osteoinductive effect of the bioglass. Further, DPSCs were shown to be viable and proliferating on β-tricalciumphosphate(TCP)-/poly(L-lactic/caprolactone)(PLL/PC) (TCP/PLL/PCL) composite scaffolds. Osteogenic activity on these scaffolds was confirmed by ALP staining and mRNA expression of ALP and Osteocalcin (OC) [37].

Annibali et al. [38] evaluated the usefulness of DPSCs seeded along with Granular Deproteinized Bovine Bone (GDPB) or Beta-Tricalcium Phosphate (ß-TCP) in healing a athymic T-cell deficient nude rats of calvarial "critical size" defect. They created two bilateral critical-size circular defects (5 mm x 1 mm size) on the parietal bone. One cranial defect for each rat was filled with the scaffold alone and the second defect was filled with the scaffold seeded with DPSCs. Notably, GDPB group showed higher percentage of lamellar bone as compared to GDPB/DPSC, ß-TCP and ß-TCP/DPSC. However, the addition of DPSCs to scaffolds could significantly increase the woven bone formation at the defect size.

Notably, some of the studies used DPSCs for bone regeneration without using any scaffolds. Tatsuhiro et al. [39] developed a scaffold-free tissue construct based on human DPSCs for the regeneration of tissue defects. As an initial step, they created basal sheets of human DPSCs by 4-week culture. These basal sheets were then subjected to 1-week 3D culture, with or without osteogenic induction. This method generated spherical shaped scaffold-free constructs of human DPSCs with a calcified matrix that are absent in the control. The expression levels of bone-related genes were found to be significantly upregulated in these constructs as compared to the control. These results suggest that scaffold-free constructs based on human DPSC could be useful for bone regeneration.

Similarly, the helioxanthin derivative 4-(4-methoxyphenyl)pyrido[40,30:4,5]thieno[2,3-b] pyridine-2-carboxamide (TH), can be used as an inducer of osteogenic differentiation of mesenchymal stem cells and preosteoblastic cells. Fujii et al. [40] assessed the effect of TH on the osteogenic differentiation of human DPSCs, and the bone formation ability of TH-induced DPSCs *in vivo*. DPSCs treated with TH were transplanted into mouse calvarial defects using cell-sheet technology. Notably, TH induced the osteogenic differentiation of DPSCs more efficiently than bone morphogenetic protein-2 treated group and TH untreated control group. Successful bone regeneration was observed *in vivo*, using DPSC sheets with TH treatment. Therefore, TH-induced DPSCs cell sheet transplantation is a useful scaffold-free method for bone regeneration *in vivo*.

Though many studies demonstrated the superior nature of DPSCs for usage in bone regeneration, the study by Jin et al. [41] yielded contradictory results. This study compared the bone regeneration capacity of DPSCs and adipose derived mesenchymal stem cells (ADSCs) both *in vitro* and *in vivo*. DPSCs showed superior colony-forming capacities, better migration and higher expression levels of proangiogenic genes. However, ADSCs exhibited greater osteogenic differentiation potential, strong expression of bone-related marker genes, and better mineral deposition *in vitro*. Furthermore, they implanted ADSCs and DPSCs into a mandibular defect of a rat to assess the bone regeneration potential of these cells *in vivo*. Interestingly, ADSCs promoted faster bone regeneration *in vivo*, as compared DPSCs. These results suggested that ADSCs may be more useful than DPSCs for bone regeneration, at least for the repair of mandibular defects. The recent developments in this field suggests that DPSCs could be useful in providing personalized therapy for persons suffering from bone defects, including cranio-facial deformity.

5.4 DPSCs for the regeneration of neuronal tissues and central nervous system

Human adult dental pulp stem cells (DPSCs) reside within the perivascular niche of dental pulp and are thought to originate from migrating cranial neural crest (CNC) cells [42]. Various groups have reported on the neurogenic differentiation potential of DPSCs. Since availability of live neurons is the major limiting factor in studying neurogenetic syndromes, the neurogenic differentiation potential of DPSCs makes them a valuable resource for translational neuronal research. Studies have shown that neurons from DPSCs exhibit similar morphological, gene expression patterns and electrophysiological properties of native neurons [43—49]. Though there are several studies on the neurogenic potential of DPSCs, only relevant studies with regards to translational research are discussed in this section.

In one of such initial studies on the neurogenic differentiation potential of DPSCs, Arthur et al. [47] showed DPSCs can differentiate into functionally active neurons, under appropriate cellular cues. Human DPSCs were able to differentiate to neuron-like cell that expressed neuronal-specific markers both *in vitro* and *in vivo*. Human DPSCs expressed neuronal markers and attained a neuronal morphology after xenotransplantation into the mesencephalon of day-2 chicken embryo, whereas human foreskin fibroblasts failed to do so.

The helix-loop-helix transcription factor Olig2 is an essential differentiation inducer of the oligodendrogenic pathway. In a notable study, Askari et al. [50] differentiated human DPSCs in to oligodendrocyte progenitor cell (OPC) lineage by transfecting with the human Olig2 gene. The OPCs on transplantation were able to actively remyelinate and repair the local sciatic nerve damage induced by lysolecithin in a mouse model of for demyelination. Therefore, differentiation of DPSCs into oligodendrocytes using Olig2 and transplanting them on to the affected site can be regarded as a valuable strategy for the treatment of different types demyelination diseases [50]. However, this study used ectopic expression of Olig2 gene, raising ethical and safety concerns and thus making this strategy less preferred. In a comparatively recent development, El Ayachi et al. [51] developed a method for generating transgene-free-induced pluripotent stem cells (TF-iPSCs) from dental stem cells (DSCs). They generated functional neurons *in vitro*, from TF-iPSCs derived from both DPSCs and stem cells of apical papilla (SCAP) through two methods-embryoid body-mediated and direct induction. Both differentiation methods produced neuron-like cells *in vitro*, with characteristic sodium and potassium current and action potential (or spontaneous excitatory postsynaptic potential). Therefore, TF-DPSC iPSCs could be regarded as an alternative cell source for neural regeneration.

Spinal cord injuries (SCI) have complicated pathophysiology and often lead to life-long functional deficits due limited axonal regeneration. In a commendable study, Sakai et al. [52] transplanted human DPSCs into a completely transected spinal cord of adult rat to assess the neuronal regeneration potential of human DPSCs. After DPSC transplantation, the rats showed marked improvement in locomotor function when compared to controls. This study reported three major neuroregenerative activities of human DPSCs. First, human DPSCs can inhibit the SCI-induced apoptosis of neurons and supporting cells, to preserve neuronal filaments and myelin sheaths. Secondly, DPSCs promoted fast regeneration of transected axons by inhibiting axon growth inhibitors, via paracrine mechanisms. Third, they compensated for the loss of neurons cells by differentiating into mature oligodendrocytes. This study points towards the possibility of developing DPSC transplantation as an effective regimen for treating SCIs. However, detailed studies are required before reaching a conclusion.

In another CNS system based study, Kiraly et al. [53] transplanted predifferentiated DPSCs into the cerebrospinal fluid of injured newborn rats' brains. The engrafted DPSCs differentiated in to neuron-like cells in rat brain. Further, they expressed neuron-specific markers and voltage dependent Na + and K + channels. Results of this study suggested that DPSCs can act as a potential cell source for treating brain injuries, since they can compensate for neuronal loss by differentiating in to neuronal and glial cell lineages *in vivo*. Similarly, in a co-culture system comprising DPSCs and adult mouse hippocampal slices on matrigel *in vitro*, DPSCs differentiated in to and promoted the growth of neuronal cells in both the CA1 zone and the edges of the hippocampal slices. Interestingly, DPSCs expressed and secreted brain-derived neurotrophic factor (BDNF) in co-culture system. Therefore, DPSCs are able to stimulate neuroregeneration in hippocampal region of adult mouse, through neurotrophic support *in vitro* [54]. From these studies, it is clear that DPSCs can be used for neuronal tissue regeneration, Further studies in this field could prove beneficial for providing personalized stem cell therapy for patients suffering from neurodegenerative disorders.

5.5 Applicability of DPSCs as a stem cell source for the regeneration of myocardial and vascular tissues

Angiogenesis is regarded as a key process in the regeneration and restoration of different tissues through tissue engineering. The induction of de novo blood vessels seems to be vital to the implant site, as there is a high chance for necrosis of the implanted tissue/cells due to insufficient oxygen supply and nutrient transport. Though some initial studies have shown some promising results on the angiogenic abilities of the DPSCs, there is a lack of knowledge on the mechanisms of DPSC-mediated angiogenesis and conditions required for DPSC-mediated angiogenesis. In one of the early studies, Bronckaers et al. [16] evaluated the angiogenic profile of human DPSC using an antibody array. The array identified various pro-and anti-angiogenic factors like vascular endothelial growth factor (VEGF), plasminogen activator inhibitor-1 (PAI-1), monocyte chemotactic protein-1 (MCP-1) and endostatin. Further, this study indicated that human DPSCs significantly induce HMEC-1 migration *in vitro* through modulating Akt and ERK pathways. The chemotactic action of hDPSCs is also dependent on its VEGF secreting properties. Notably, hDPSC did not influence the proliferation of human microvascular endothelial cells (HMEC-1), indicating the vasculature inducing abilities of hDPSCs is mainly based on its ability to induce HMEC-1 migration. This study also confirmed the ability of hDPSCs to induce angiogenesis through chicken chorioallantoic membrane (CAM) assay. Thus hDPSCs is of great importance not only for tissue engineering, but also for the treatment of chronic wounds and myocardial infarction.

In a comparatively recent study, Martínez-Sarrà et al. [55] has shown that human DPSCs could be differentiated to various lineages including endothelial, smooth and skeletal muscle cells. Most importantly, they assessed the therapeutic potential of hDPSCs using a wound healing mouse model and in two genetic mouse models of muscular dystrophy (Scid/mdx and Sgcb-null Rag2-null γc-null). DPSC transplantation in mice resulted in complete re-epithelialization of wounds as compared to only 40% of re-epithelialization in the PBS-treated mice. Moreover, DPSC improved collagen deposition in healing wounds. Another important finding was that engraftment of the DPSCs in the skeletal muscle (tibialis anterior) of both dystrophic murine models, resulted in integration in muscular fibers and vessels. Further, reduced fibrosis and infiltration of proangiogenic CD206 + macrophages was observed on the DPSC implant site, indicating its' angiogenesis enhancing potential. Since DPPSC shows muscular integration potential and re-vascularization potential in muscular dystrophy models, they could be used for slowing down the dystrophic muscle degeneration.

Endothelial progenitor cells (EPCs) are known to stimulate vasculogenesis and are good candidates for the treatment of ischemic diseases. However, the availability of autologous EPCs is limited and invasive biopsy is needed for collecting EPCs. As an alternative source to EPCs, Iohara et al. [56] isolated a highly vasculogenic side population cells from porcine DPSCs, based on CD31 and CD146 expression status. They showed that the CD31 (−)/CD146(−)/CD34(+)/VEGFR2(+)/Flk1(+) side population of DPSCs, are similar to EPCs. These cells can be distinguished from the hematopoietic cell lineages by the absence of CD11b, CD14, and CD45 expression. When this side population of DPSCs were transplanted on to the models of mouse hind limb ischemia, they showed successful tissue integration and high density of capillary formation at the ischemic site. Notably, the DPSCs were found in close proximity of the newly formed vasculature, indicating their role in promoting angiogenesis. Further, these DPSCs were found to express proangiogenic factors, like VEGF-A, G-CSF, GM-CSF, and MMP3. Hence, this side population of DPSCs can be regarded as a stem cell source for stimulating angiogenesis/vasculogenesis during tissue regeneration.

In a comparative study, Ishizaka et al. [57] studied regenerative potential of CD31(−) side population (SP) cells isolated from dental pulp, bone marrow and adipose tissue of porcine origin. Notably, the dental pulp CD31(−) side population cells showed highest expression levels of angiogenic/neurotrophic factors. Transplantation of dental pulp CD31(−) SP cells in a mouse hindlimb ischemia model generated higher levels of capillary bed formation as compared to bone marrow and adipose CD31(−) SP cells transplantation. Further, dental pulp CD31(−) SP cells showed better recovery of the motor function recovery and infarct size reduction. Last but not least, the pulp CD31(−) SP cells were able to induced angiogenesis, and neurogenesis in ectopic transplantation models at much higher levels, as compared to bone marrow and adipose CD31(−) SP cells. Thus, the CD31(−) SP cells derived from dental pulp can be considered as a superior stem cell source for angiogenic and neurogenic regeneration compared to CD31(−) SP cells from bone marrow and adipose tissue.

In a commendable study, Gandia et al. [58] investigated therapeutic potential of human DPSCs for the repair of myocardial infarction (MI). They used green fluorescent protein (GFP) tagged human DPSCs for this study. The GFP-DPSCs were injected intra-myocardially to nude rats, in which myocardial infarction was induced by coronary artery ligation. Four weeks post-injection of GFP-DPSCs, the animals showed improved cardiac function, anterior wall thickening, left ventricular fractional area change, and reduction in infarct size compared to control animals. However, histologic evaluation showed that GFP-DPSCs have not differentiated to form endothelial cells or cardiac muscle cells. This could be due to the fact that the cells used in this study are of exogenous origin, making the integration and differentiation difficult to happen. However, there was an increased angiogenesis in GFP-DPSC treated animals as compared to control animals. One reason for this could be that hDPSCs are known to secrete several proangiogenic factors. Thus, human DPSCs could be a promising cell population for cell based cardiac repair.

As described earlier, transplantation of DPSCs is a well-demonstrated technique to promote the regeneration of tissues and organs. Zhang et al. [59] explored the therapeutic propensity of DPSCs transplantation in rat models of acute radiation-induced esophageal injury. After transplantation on to the injured esophagus, the PKH26-labeled DPSCs showed co-localized with esophageal stem cells. Further, the esophageal stem cell markers like PCNA, CK14, CD71, and integrin α6 showed significantly upregulation after DPSC transplantation. It is suggested that the transplanted DPSCs trans-differentiated in to esophageal stem cells *in vivo*. Further, the injured esophagus recovered fast, inflammation subsided, and the walls of DPSC transplanted esophagus showed a greater thickness. This study demonstrated how beneficial DPSCs are, for regeneration and repair of the damaged esophageal tissue.

5.6 Dental pulp stem cells as mediators of optic system regeneration

Though the studies on the utility of DPSCs for the regeneration of optic system are not adequate, the preliminary results are quite promising and support further studies. Mead et al. [60] studied the significance of intravitreal transplants of DPSCs, as compared to that of bone marrow-derived mesenchymal stem cells (BMSCs). ELISA based assessment of conditioned medium showed that rat DPSCs do secrete significantly higher levels of nerve growth factor (NGF), brain-derived neurotrophic factor (BDNF), and neurotrophin-3 (NT-3) in to the culture medium as compared to rat BMMSCs. In co-culturing systems, DPSCs promoted better survival of βIII-tubulin(+) retinal cells *in vitro* as compared to BMMSCs. Further, the intravitreal transplants of DPSCs in rat optic nerve injury models promoted significantly higher levels of retinal ganglion cells (RGCs) survival and axon regeneration after optic nerve injury. This action was found to be mediated through neurotrophin, since the effects were abolished after TrK receptor blockade. An *in vitro* study conducted by Gomes et al. [61] explored the utility of tissue-engineered cell sheet composed of human undifferentiated immature dental pulp stem cells (hIDPSC) for ocular surface reconstruction. They used a rabbit model of total limbal stem cell deficiency (LSCD) for this purpose. Transplantation of the tissue-engineered hIDPSC sheet onto the corneal bed LSCD rabbit resulted in improved corneal transparency as compared to control corneas, which developed total conjunctivalization and opacification. The rabbits which received hIDPSC implantation showed clearer corneas with less neovascularization and with uniform corneal epithelium. Thus, transplantation of tissue-engineered hIDPSC sheet could be useful for the regeneration of damaged corneal epithelium. Further exploration in this regard is necessary. Similarly, Syed-Picard et al. [62] reported the applicability of DPSCs in corneal stromal blindness. After *in vitro* differentiation into keratinocytes of central cornea, DPSCs expressed characteristic markers of keratocytes like keratocan, and keratansulfate proteoglycans. In addition, DPSC-derived keratinocytes cultured on aligned nanofiber substrates were able to mimic a corneal stromal-like structure. Further, after injection into mouse corneal stroma, human DPCs were able to produce corneal stromal extracellular matrix. Notably, expression of human Type I collagen and Keratocan neither affected corneal transparency in mice nor did it induce immunological rejection. Recently, Kushnerev et al. [63] delivered DPSCs onto debrided human cornea using contact lenses. This study provided evidence that DPSCs can migrate onto the cornea, and establish a barrier to prevent the conjunctivalization of cornea. Thus, DPSCs could be useful for regeneration of corneal epithelium and stromal blindness. Further, these findings demonstrate the great potential of DPSCs in personalized stem cell based therapies for human optical system degeneration.

5.7 Regenerative therapeutic potential of DPSCs in diabetes

Diabetes mellitus (DM) comprises a group of endocrine disorders described by erroneous carbohydrate metabolism, which causes hyperglycemia and associated complications. Primarily, diabetes can be induced by the autoimmune destruction of insulin-producing pancreatic β cells (type 1 diabetes) or by resistance to insulin (type 2 diabetes), which was characterized by initial overexpression of insulin, followed by insulin deficiency [64]. Autologous islet cells transplantation is one of the highly preferred strategies for treating type 1 diabetes, since it can ward off lifelong insulin replacement and also can avoid immune rejection to a significant level [65].

Recently, several studies have shown that DPSCs have the potential to differentiate into pancreatic islet like clusters (ICCs). Govindasamy et al. [66] explored the efficiency of DPSCs to differentiate in to the pancreatic islet-like cell aggregates (ICAs). The DPSC derived ICAs were positive for C-peptide, Pdx-1, Pax4, Pax6, Ngn3, and Isl-1. After 10 days of differentiation, ICAs exhibited *in vitro* functionality and released insulin and C-peptide in a glucose-dependent manner. These results demonstrated that DPSCs could be differentiated into pancreatic cell lineage, making it a useful candidate for autologous stem cell therapy in diabetes. Currently, there are well established protocols for the differentiation of DPSCs in to ICAs. Recently, Yagi Mendoza et al. [67] demonstrated a 3-step 3D culture system, by which human CD117 + DPSCs are magnetically selected and successfully differentiated in to ICAs. Their method derived ICAs with enhanced insulin and C-peptide secretion and expression of pancreatic markers. Further, they showed glucose-dependent secretion of insulin and upregulation of important pancreatic endocrine markers, transcriptional factors, and the PI3K/AKT and WNT pathways. This 3-step 3D system is very useful for the differentiation of DPSC, for usage in stem cell based regenerative medicine in diabetes.

In one of our studies, we have demonstrated that islet-like cell clusters (ICCs) can be generated from human dental pulp stem cells from permanent teeth (DPSCs) and exfoliated deciduous teeth (SHED) [7]. Since the ICCs were of human origin, we packed ICCs in immuno-isolation macro-capsules during the transplantation of ICCs into streptozotocin (STZ)-induced diabetic mice to avoid immune rejection. These biocompatible macro-capsules are selectively permeable to glucose and insulin, but are impermeable to immunoglobulins [68]. Though no significant difference was observed with regards to the size and morphology of ICCs produced by DPSCs and SHED, the number of ICCs produced was higher in SHED group. Our results demonstrate that ICCs derived from human DPSCs and SHED in immuno-isolation macro-capsules can reverse STZ-induced diabetes in mice, without the need for immunosuppression. Thus, SHED and DPSCs can be considered as an autologous source of human tissue for stem cell therapy in type 1 diabetes.

In a notable study, Guimarães et al. [69] has demonstrated that the endovenous injection of mouse dental pulp stem cell (mDPSC) in to a streptozotocin (STZ)-induced type 1 diabetes C57BL/6 mice could contribute to pancreatic β-cell renewal. They reported an increase in pancreatic islets and insulin as well as normalized urea and proteinuria levels, 30 days after mDPSC transplantation in diabetic mice. Moreover, diabetic mice after mDPSC transplantation exhibited improved renal function and nociceptive thresholds similar to that of non-diabetic control mice. Therefore, endovenous administration of autologous DPSCs may be useful with regards to pancreatic β-cell renewal, improved renal function, and long-lasting anti-nociceptive effect in diabetic mice. This study should be explored further in larger animal models to confirm the efficacy of DPSCs in controlling diabetes complications and diabetic neuropathic pain.

5.8 DPSCs as a therapeutic tool for the regeneration of cartilage and tendon

Another important area of regenerative medicine in which DPSCs are useful is the tissue engineering of cartilage and tendon. DPSCs have the ability to differentiate into multiple cell lineages including chondrocytes and tenocytes. Chen et al. [70] demonstrated that DPSCs can be differentiated in to tenocyte-like cells, using mechanical loading conditions. Under mechanical loading, DPSCs differentiated in to tendon-like cells and showed enhanced expression of tendon-related markers such as scleraxis, tenascin-C, tenomodulin, Type I and Type VI collagens, and eye absent homolog 2. This study applied static mechanical stimulation on to

Long polyglycolic acid (PGA) fiber scaffolds seeded with DPSCs, using a custom-fabricated spring, made of stainless steel frame under static tension. Furthermore, mature tendon-like tissue with significant extra cellular matrix was formed after transplantation of these DPSC seeded-PGA constructs to a mouse model. These results suggest that DPSCs can be regarded as a useful source of stem cells for tissue engineering of tendon-like tissue.

High incidence of articular cartilage defects resulting from age-related degeneration or trauma injuries is a major problem worldwide. Studies has provided evidence that DPSCs has rapid *ex vivo* expansion capacity and chondrogenic differentiation potential, which makes them an encouraging stem cell type for therapeutic application in cartilage tissue engineering. A recent study assessed the effect of a hypoxia mimicking agent, cobalt chloride (CoCl2), on chondrogenic differentiation of human DPSCs using pellet culture system [71]. Notably, CoCl2 treatment resulted in increased pellet size, structural integrity and matrix deposition and organizations mimicking native cartilage. Further, enhanced glycosaminoglycans (GAGs) content and type II collagen II expression was also observed. Most importantly, CoCl2 prevented hypertrophy, as verified from decreased collagen X expression in CoCl2 treated pellets. This study suggests that, usage of CoCl2 may be a useful for obtaining hyaline cartilage, while using DPSCs as a cell source for articular cartilage tissue engineering.

Dai et al. [72] reported that the co-culture system comprising Dental pulp stem cells (DPSCs) and Costal chondrocytes (CCs) could prevent the hypertrophic development of DPSCs that has undergone chondrogenic differentiation. They used human CCs and human DPSCs co-culture system articular cartilage. Notably, CCs served a chondro-inductive niche in which DPSCs underwent chondrogenic differentiation to produce hyaline cartilage. It is well known that CCs alone could not prevent the hypertrophy and mineralization. Furthermore, they usage of FGF9 in this study may also have served as an inhibitor of mineralization by binding to FGFR3 and also by the phosphorylation of ERK1/2 in DPSCs. However, the co-culture system comprising CCs and DPSCs in medium containing FGF9 can be useful in preventing the ossification in tissue engineered cartilage.

A study by Mata et al. [73], showed the superior nature of human dental pulp stem cells (hDPSCs) as compared to rabbit chondrocytes for the regeneration of rabbit cartilage defects. The authors used 3% alginate hydrogels for the implantation of both hDPSCs and rabbit chondrocytes in rabbit models for cartilage damage. Higher degree of cartilage regeneration and healing was observed in animals implanted with hDPSCs as compared to rabbit chondrocyte implanted animals. It is not clear that how a cell type of xenogenic origin could provide better results without immunosuppression, than cells from the same species. It was previously reported that mesenchymal stem cells can modulate the host immune system and alleviate inflammatory responses as well [74]. This could be one reason why hDPSCs could outperform the allotransplantation using rabbit chondrocytes in this study.

A study conducted by Rizk and Rabie [75] showed that human DPSCs are useful for creating larger three-dimensional cartilage-like constructs. They induced hDPSCs to differentiate in to chondrogenic lineage using recombinant transforming growth factor β3 (TGFβ3). Notably, recombinant adenoviral vector system encoding human TGFβ3 was used for used for transducing DPSCs. These transduced cells were seeded on to a poly-L-lactic acid/polyethylene glycol (PLLA/PEG) electrospunfiber scaffold and then implanted in nude mice. This system could continuously supply TGFβ3 up to 48 days in nude mice and resulted in the differentiation of hDPSCs and generation of cartilage-like matrix.

Though this system could successfully differentiate hDPSCs toward chondrogenic lineage, the usage of adenoviral based transducing system makes it less favorable for future treatment applications of cartilage defects.

In addition to hyaline cartilage regeneration studies, DPSCs were also used for the regeneration of degenerated vertebral disks. A recent study by Yao and Flynn [76] assessed the migration potential of chondrogenic cells derived from DPSCs in type 1 and type 2 collagen gels. Since the normal nucleus pulposus (NP) cells has similar phenotype to chondrocytes, they presumed that a combination of DPSC derived chondrocytes and collagen gels could heal disk damage. The motility of transplanted chondrogenic cells is crucial because the cells should migrate away from the hydrogels and disperse throughout the NP tissue after implantation to heal it. Notably, the migration of DPSC-derived chondrocytes was slightly higher in type I collagen hydrogels than in type II collagen hydrogels. However, crosslinking of type I collagen with poly(ethylene glycol) ether tetrasuccinimidylglutarate (4S-StarPEG) significantly reduced the cell migration. This study provided evidence that application of DPSC-derived chondrogenic cells could migrate through native collagen gels and thus could be beneficial for regeneration of damaged vertebral disks.

In brief, the future of DPSC-based personalized regenerative therapy for several disease states could be regarded as promising. A brief outline of the important studies which evaluated the multilineage differentiation potential and regenerative medicine propensity of DPSCs is given in Table 5.1.

5.9 Future perspectives and conclusions

Stem cell therapy is a promising strategy for regeneration of damaged organs, tissues or functions through the transplantation of stem cells. Stem cells are known to supplement the organ function through differentiation into specific cell lineages and by inducing regeneration through paracrine mechanisms. Mesenchymal stem cells are the most widely implicated stem cell type in tissue engineering and regenerative medicine. Over the last decade, dental stem cells have evolved as a favorite source of cells in tissue engineering and regenerative medicine due to high clonality, ease of obtaining and their multipotent nature. Further, they are special as compared to other mesenchymal cells, because they show much higher potential to differentiate to neuronal, vascular and myocardial cells [15,19,20]. Though research in this area has made considerable advances over the last decade, many key areas are still needs to be addressed, which can be done only through clinical research.

DPSCs are a heterogeneous population of mesenchymal stem cells, comprising side-populations of stem cells which express different types of surface antigens in addition to the common mesenchymal markers [10,17]. Some studies have used these side populations of DPSCs to get much better regeneration of bone and vascular tissue, as compared to the unsorted DPSCs [27,56]. Results of these studies suggest that the sub-populations in the DPSCs have different potentials of differentiation towards a specific adult cell lineage. Thus, detailed combinations of surface antigens of the DPSC sub-populations are required, to define the sub-populations of DPSCs that has the highest potential to differentiate into a specific cell type. Such specific sub-populations, when used in stem cell therapy could

TABLE 5.1 A brief outline of recent research studies evaluating the usage of DPSCs for regeneration of various types of tissues.

Sl. no.	Tissue/cell type generated	Method used	Conditions/molecules/ animal model	Outcome	Reference
1.	Tendon	Polyglycolic acid (PGA) fiber scaffolds	Static mechanical stimulation, mice model	Tendon-like tissue was formed *in vivo*, with extra cellular matrix showing enhanced expression of tendon markers such as scleraxis, tenascin-C, tenomodulin, etc.	[70]
2.	Pancreatic Islet-like cells (ICCs)	3-step *in vitro* differentiation protocol	Glucagon-like peptide (GLP)-1	Pancreatic islet-like cell aggregates (ICAs) were obtained, which were positive for pancreatic markers like C-peptide, Pdx-1, Pax4, Pax6, Ngn3, and Isl-1	[66]
3.	Pancreatic Islet-like cells (ICCs)	3D culture system using magnetically separated CD117 + DPSCs	Glucagon-like peptide (GLP)-1	ICAs with glucose-dependent secretion of insulin, upregulation of important pancreatic endocrine markers, transcriptional factors, and PI3K/AKT and WNT pathways	[67]
4.	Pancreatic Islet-like cells (ICCs)	ICCs encapsulated in immuno-isolation macro-capsules	Streptozotocin (STZ)-induced diabetic mice	Reversal of STZ-induced diabetes in mice	[7]
5.	Pancreatic Islet-like cells (ICCs)	Endovenous injection of mouse dental pulp stem cell (mDPSC)	Streptozotocin (STZ)-induced type 1 diabetes C57BL/6 mice	Pancreatic β-cell renewal *in vivo*, improved renal function, and long-lasting anti-nociceptive effect in diabetic mice	[69]
6.	Cartilage-like tissue construct	Adeno-associated viral vector encoding human TGFβ3 and poly-L-lactic acid/polyethylene glycol (PLLA/PEG) electrospunfiber scaffolds	Nude mice and pellet culture	hDPSCs were successfully differentiated in nude mice and generation of cartilage-like matrix	[75]
7.	Cartilage damage repair *in vivo*	3% alginate hydrogels	Rabbit model of cartilage damage	Higher degree of cartilage regeneration was observed in animals implanted with hDPSCs as compared to rabbit chondrocyte implanted animals	[73]

(Continued)

TABLE 5.1   (Continued)

Sl. no.	Tissue/cell type generated	Method used	Conditions/molecules/ animal model	Outcome	Reference
8.	Cartilage	Co-culture system of DPSCs and Costal chondrocytes	FGF9	Co-culture system comprising DPSCs and Costal chondrocytes (CCs) prevented the hypertrophic development of DPSCs that has undergone chondrogenic differentiation	[72]
9.	Cartilage	Pellet culture system	CoCl2	CoCl2 prevented hypertrophy in DPSC-derived chondrogenic cells, as verified from decreased collagen X expression	[71]
10.	Nucleus Pulposus tissue	Type I and Type II collagen gels	Poly(ethylene glycol) ether tetrasuccinimidylglutarate (4S-StarPEG)	DPSC-derived chondrogenic cells are able to migrate through native collagen gels and could be beneficial for regeneration of damaged vertebral disks	[76]
11.	Vascular	Antibody array	chicken chorioallantoic membrane (CAM) assay	confirmed the ability of hDPSCs to induce angiogenesis through chemotaxis of human microvascular endothelial cells (HMEC-1)	[16]
12.	Endothelial, smooth muscle and skeletal muscle	Direct engraftment of DPSCs *in vivo*	Wound healing mouse model and genetic mouse models of muscular dystrophy (Scid/mdx and Sgcb-null Rag2-null γc-null).	Transplantation of DPSC in wound model mice resulted in complete re-epithelialization of wounds. Engraftment of DPSCs in the skeletal muscle (tibialis anterior) of dystrophic murine models reduced fibrosis and infiltration of proangiogenic CD206 + macrophages	[55]

(*Continued*)

TABLE 5.1 (Continued)

Sl. no.	Tissue/cell type generated	Method used	Conditions/molecules/ animal model	Outcome	Reference
13.	Endothelial progenitor-like cells	Direct engraftment of CD31(−)/CD146 (−)/CD34(+)/VEGFR2 (+)/Flk1(+) side population of DPSCs *in vivo*	Hind limb ischemia mice model	Side population of DPSCs showed successful tissue integration and high density of capillary formation at the ischemic site	[56]
14.	Vasculature/ capillary bed formation	CD31(−) side population (SP) of DPSCs	mouse hind limb ischemia model	CD31(−) side population (SP) of DPSCs showed better recovery of the motor function recovery, infarct size reduction and capillary formation *in vivo*	[57]
15.	Vasculature/ capillary bed formation	GFP-DPSCs were injected intra-myocardially to MI model nude rats	MI model nude rats	intra-myocardial injection of DPSCs improved cardiac function, left ventricular fractional area change, and reduction in infarct size. Further, there was increased vasculature	[58]
16.	Esophageal muscle	Direct transplantation of DPSCs on to the esophageal injury site	rat models of acute radiation-induced esophageal injury	This study suggested the trans-differentiation of transplanted DPSCs into esophageal stem cells *in vivo*, resulting in the fast recovery of the injured esophagus	[59]
17.	Optic system/ retinal ganglion cells	intravitreal transplantation of DPSCs	rat optic nerve injury models	DPSCs promoted Neurotrophin-mediated retinal ganglion cells (RGCs) survival and axon regeneration after optic nerve injury	[60]
18.	Optic system/ corneal epithelium	Tissue-engineered cell sheet with DPSCs	Rabbit model of total limbal stem cell deficiency (LSCD)	Tissue-engineered cell sheet composed of human undifferentiated immature dental pulp stem cells (hIDPSC) promoted repair of damaged corneal epithelium	[61]

(Continued)

TABLE 5.1   (Continued)

Sl. no.	Tissue/cell type generated	Method used	Conditions/molecules/ animal model	Outcome	Reference
19.	Optical system/ corneal keratinocytes	Direct injection of human DPSCs into mouse cornea.	Mouse model of Corneal stromal blindness	Newly formed corneal stromal extracellular matrix by human DPSCs in mouse cornea, without causing opacity or immune rejection	[62]
20.	Neuronal cells	Xenotransplantation into the mesencephalon of chicken embryo	Embryonic day-2 chicken embryo	Human DPSCs expressed neuronal markers and showed neuronal morphology after xenotransplanatation	[47]
21.	Neuronal (ligodendrocyte progenitor cell (OPC))	Transfection of human DPSCs with Olig2 gene	Mouse model of for demyelination / Olig2 gene	The OPCs on transplantation were able to remyelinate and repair the local sciatic nerve damage	[50]
22.	Neuronal cells	embryoid body-mediated and direct induction		Generated functional neurons *in vitro*, from iPSCs derived from both DPSCs and stem cells of apical papilla (SCAP)	[51]
23.	Neuronal cells/ spinal cord	Direct transplantation of human DPSCs into a completely transected spinal cord of adult rat	SCI model rat with completely transected spinal cord	After DPSC transplantation, the rats showed marked improvement in locomotor function and regeneration of spinal cord	[52]
24.	Neuronal cells of brain	Transplantation of predifferentiated DPSCs into the cerebrospinal fluid of injured newborn rat's brains	Rat brain injury model	DPSCs differentiated into neuron-like cells in rat brain showing neuron-specific marker expression and voltage dependent Na + and K + channels	[53]
25.	Bone	Transplantation of LNGFR(Low +)THY-1 (High +) DPSC side population *in vivo*.	mouse model with critical-size calvarial defect.	The LNGFR(Low +) THY-1(High +) side population of DPSCs successfully healed the critical size calvarial defects in mice model	[27]

(*Continued*)

TABLE 5.1   (Continued)

Sl. no.	Tissue/cell type generated	Method used	Conditions/molecules/ animal model	Outcome	Reference
26.	Bone	3D Cell Carrier glass scaffolds	Dental pulp pluripotent stem cells (DPPSCs)	Bone-like tissue showing trabecular host bone structure, with high interconnectivity was formed by DPPSCs *in vitro*	[28]
27.	Bone	Silk fibroin scaffold constructs made with DPSC and human amniotic fluid derived stem cells (AFSCs)	Critical size calvarial defects in immuno compromised rat model	Though both AFSCs and DPSCs formed bone formation, AFSC showed superior bone formation than DPSC constructs	[34]
28.	Bone	Scaffold-free tissue engineered construct derived from human DPSCs based basal sheets		Scaffold-free constructs of human DPSCs showed a calcified matrix and increased expression levels of bone-related genes	[39]

provide the best results in personalized regeneration therapy. Further, Karamzadeh et al. [77] reported heterogeneous sub-populations among hDPSCs and hDFSCs with regards to stemness, differentiation fate, and cell cycle phases. Usage of such heterogeneous population of stem cells may result in clinical treatment variations. Therefore, specific studies for getting better understanding on the phenotypic and genotypic heterogeneity among different DPSCs subpopulations could be beneficial for obtaining optimal results in personalized regeneration therapy.

Recent studies have shown the involvement of microRNAs in modulating and defining the differentiation potential of dental stem cells towards different adult cell lineages [78—87]. Therefore, detailed information on the different microRNAs involved in the DPSC differentiation pathways could be beneficial for deciding optimized personalized therapy in future. DPSCs can be cryopreserved for comparatively longer times without losing their stemness [88,89]. However, maintaining the stemness and differentiation potential of DPSCs in *in vitro* culture is a major concern. Usage of small molecules like Pluripotin (SC1) and 6-bromoindirubin-3-oxime has shown to preserve the stemness of DPSCs up to 70 days in culture [90]. However, the safety of these small molecules for human use should be clearly established before using in conjunction with stem cell therapy regimen in clinical scenarios.

Another issue is that DPSCs or exfoliated deciduous tooth stem cells may not available throughout a patient's lifetime. Though DSC banking can offer a potential solution by cryopreserving them for future use, the cost of long-term cryopreservation could limit their application in regenerative medicine [88]. In addition, Chen et al. [91] reported different properties and age related bias in differentiation potential of DPSCs. These issues are to be addressed in future to optimize the results in DPSC-based regeneration therapy. To conclude, the future of DPSC-based personalized regenerative medicine is bright provided that the aforementioned concerns are meticulously addressed.

References

[1] Baroffio A, Dupin E, Le Douarin NM. Common precursors for neural and mesectodermal derivatives in the cephalic neural crest. Development 1991;112(1):301−5.

[2] Gronthos S, Mankani M, Brahim J, Robey PG, Shi S. Postnatal human dental pulp stem cells (DPSCs) in vitro and in vivo. Proc Natl Acad Sci USA 2000;97(25):13625−30.

[3] Miura M, Gronthos S, Zhao M, Lu B, Fisher LW, Robey PG, et al. SHED: stem cells from human exfoliated deciduous teeth. Proc Natl Acad Sci USA 2003;100(10):5807−12.

[4] Gronthos S, Brahim J, Li W, Fisher LW, Cherman N, Boyde A, et al. Stem cell properties of human dental pulp stem cells. J Dent Res 2002;81(8):531−5.

[5] Mori G, Brunetti G, Oranger A, Carbone C, Ballini A, Lo Muzio L, et al. Dental pulp stem cells: osteogenic differentiation and gene expression. Ann NY Acad Sci 2011;1237:47−52. Available from: https://doi.org/10.1111/j.1749-6632.2011.06234.x Nov.

[6] Dominici M, Le Blanc K, Mueller I, Slaper-Cortenbach I, Marini F, Krause D, et al. Minimal criteria for defining multipotent mesenchymal stromal cells. The International Society for Cellular Therapy position statement. Cytotherapy 2006;8(4):315−17.

[7] Kanafi MM, Rajeshwari YB, Gupta S, Dadheech N, Nair PD, Gupta PK, et al. Transplantation of islet-like cell clusters derived from human dental pulp stem cells restores normoglycemia in diabetic mice. Cytotherapy 2013;15(10):1228−36. Available from: https://doi.org/10.1016/j.jcyt.2013.05.008.

[8] Sedgley CM, Botero TM. Dental stem cells and their sources. Dent Clin North Am 2012;56(3):549−61. Available from: https://doi.org/10.1016/j.cden.2012.05.004.

[9] D'Aquino R, De Rosa A, Laino G, Caruso F, Guida L, Rullo R, et al. Human dental pulp stem cells: from biology to clinical applications. J Exp Zool Part B: Mol Dev Evol 2009;312(5):408−15 (July 2009), ISSN 1552-5007.

[10] Bakopoulou A, About I. Stem cells of dental origin: current research trends and key milestones towards clinical application Stem Cell Int 2016;2016Article ID 4209891. Available from: http://dx.doi.org/10.1155/2016/4209891.

[11] Mason C, Dunnill P. A brief definition of regenerative medicine. Regen Med 2008;3(1):1−5.

[12] Langer R, Vacanti JP. Tissue engineering. Science 1993;260(5110):920−6.

[13] Lanza R, Langer R, Vacanti JP. Principles of tissue engineering. Fourth Sci Direct 2013;1476.

[14] Heil M, Ziegelhoeffer T, Mees B, Schaper W. A different outlook on the role of bone marrow stem cells in vascular growth: bone marrow delivers software not hardware. Circ Res 2004;94(5):573−4.

[15] Botelho J, Cavacas MA, Machado V, Mendes JJ. Dental stem cells: recent progresses in tissue engineering and regenerative medicine. Ann Med 2017;49(8):644−51. Available from: https://doi.org/10.1080/07853890.2017.1347705.

[16] Bronckaers A, Hilkens P, Fanton Y, Struys T, Gervois P, Politis C, et al. Angiogenic properties of human dental pulp stem cells. PLoS One 2013;8(8):e71104. Available from: https://doi.org/10.1371/journal.pone.0071104 eCollection 2013.

[17] Huang GT, Gronthos S, Shi S. Mesenchymal stem cells derived from dental tissues vs. those from other sources: their biology and role in regenerative medicine. J Dent Res 2009;88(9):792−806. Available from: https://doi.org/10.1177/0022034509340867.

[18] Rodríguez-Lozano FJ, Bueno C, Insausti CL, Meseguer L, Ramírez MC, Blanquer M, et al. Mesenchymal stem cells derived from dental tissues. Int Endod J 2011;44(9):800−6. Available from: https://doi.org/10.1111/j.1365-2591.2011.01877.x.

[19] Barbara Z, Eriberto B, Stefano S, Giulia B, Chiara G, Ferrarese N, et al. Dental pulp stem cells and tissue engineering strategies for clinical application on odontoiatric field In: Pignatello R, editor. Biomaterials science and engineering. IntechOpen; 2011. Available from: http://dx.doi.org/10.5772/24871Available from. Available from: https://www.intechopen.com/books/biomaterials-science-and-engineering/dental-pulp-stem-cells-and-tissue-engineering-strategies-for-clinical-application-on-odontoiatric-fi.

[20] Al-Habib M, George T-J, Huang. Dental mesenchymal stem cells: dental pulp stem cells, periodontal ligament stem cells, apical papilla stem cells, and primary teeth stem cells—isolation, characterization, and expansion for tissue engineering. In: Papagerakis P, editor. Odontogenesis: methods and protocols, methods in molecular biology, 1922. © Springer Science + Business Media; 2019. LLC, part of Springer Nature.

[21] Seo BM, Miura M, Gronthos S, Bartold PM, Batouli S, Brahim J, et al. Investigation of multipotent postnatal stem cells from human periodontal ligament. Lancet 2004;364(9429):149–55.

[22] Morsczeck C, Götz W, Schierholz J, Zeilhofer F, Kühn U, Möhl C, et al. Isolation of precursor cells (PCs) from human dental follicle of wisdom teeth. Matrix Biol 2005;24(2):155–65.

[23] Sonoyama W, Liu Y, Fang D, Yamaza T, Seo BM, Zhang C, et al. Mesenchymal stem cell-mediated functional tooth regeneration in swine. PLoS One 2006;1:e79 Dec 20.

[24] Batouli S, Miura M, Brahim J, Tsutsui TW, Fisher LW, Gronthos S, et al. Comparison of stem-cell-mediated osteogenesis and dentinogenesis. J Dent Res 2003;82(12):976–81.

[25] Laino G, d'Aquino R, Graziano A, Lanza V, Carinci F, Naro F, et al. A new population of human adult dental pulp stem cells: a useful source of living autologous fibrous bone tissue (LAB). J Bone Min Res 2005;20(8):1394–402.

[26] Noda S, Kawashima N, Yamamoto M, Hashimoto K, Nara K, Sekiya I, et al. Effect of cell culture density on dental pulp-derived mesenchymal stem cells with reference to osteogenic differentiation. Sci Rep 2019;9(1):5430. Available from: https://doi.org/10.1038/s41598-019-41741-w.

[27] Yasui T, Mabuchi Y, Toriumi H, Ebine T, Niibe K, Houlihan DD, et al. Purified human dental pulp stem cells promote osteogenic regeneration. J Dent Res 2016;95(2):206–14. Available from: https://doi.org/10.1177/0022034515610748.

[28] Atari M, Caballé-Serrano J, Gil-Recio C, Giner-Delgado C, Martínez-Sarrà E, García-Fernández DA, et al. The enhancement of osteogenesis through the use of dental pulp pluripotent stem cells in 3D. Bone 2012;50(4):930–41. Available from: https://doi.org/10.1016/j.bone.2012.01.005.

[29] Gholami, Labbaf S, Houreh AB, Ting H, Jones JR, Esfahani MN. Long term effects of bioactive glass particulates on dental pulp stem cells in vitro. Biomed Glasses 2017;3:96–103. Available from: https://doi.org/10.1515/bglass-2017-0009.

[30] Irastorza I, Luzuriaga J, Martinez-Conde R, Ibarretxe G, Unda F. Adhesion, integration and osteogenesis of human dental pulp stem cells on biomimetic implant surfaces combined with plasma derived products. Eur Cell Mater 2019;38:201–14. Available from: https://doi.org/10.22203/eCM.v038a14 Nov 4.

[31] Campos JM, Sousa AC, Caseiro AR, Pedrosa SS, Pinto PO, Branquinho MV, et al. Dental pulp stem cells and Bonelike® for bone regeneration in ovine model. Regen Biomater 2019;6(1):49–59. Available from: https://doi.org/10.1093/rb/rby025.

[32] D'Antò V, Raucci MG, Guarino V, Martina S, Valletta R, Ambrosio L. Behaviour of human mesenchymal stem cells on chemically synthesized HA-PCL scaffolds for hard tissue regeneration. J Tissue Eng Regen Med 2016;10(2):E147–54. Available from: https://doi.org/10.1002/term.1768.

[33] Nadeem D, Kiamehr M, Yang X, Su B. Fabrication and in vitro evaluation of a sponge-like bioactive-glass/gelatin composite scaffold for bone tissue engineering. Mater Sci Eng C Mater Biol Appl 2013;33(5):2669–78. Available from: https://doi.org/10.1016/j.msec.2013.02.021.

[34] Riccio M, Maraldi T, Pisciotta A, La Sala GB, Ferrari A, Bruzzesi G, et al. Fibroin scaffold repairs critical-size bone defects in vivo supported by human amniotic fluid and dental pulp stem cells. Tissue Eng Part A 2012;18(9-10):1006–13. Available from: https://doi.org/10.1089/ten.TEA.2011.0542 Epub 2012 Jan 26.

[35] Maraldi T, Riccio M, Pisciotta A, Zavatti M, Carnevale G, Beretti F, et al. Human amniotic fluid-derived and dental pulp-derived stem cells seeded into collagen scaffold repair critical-size bone defects promoting vascularization. Stem Cell Res Ther 2013;4(3):53. Available from: https://doi.org/10.1186/scrt203.

[36] Akkouch A, Zhang Z, Rouabhia M. Engineering bone tissue using human dental pulp stem cells and an osteogenic collagen-hydroxyapatite-poly (L-lactide-co-ε-caprolactone) scaffold. J Biomater Appl 2014;28(6):922–36. Available from: https://doi.org/10.1177/0885328213486705.

[37] Khanna-Jain R, Mannerström B, Vuorinen A, Sándor GK, Suuronen R, Miettinen S. Osteogenic differentiation of human dental pulp stem cells on β-tricalcium phosphate/poly (l-lactic acid/caprolactone) three-dimensional scaffolds. J Tissue Eng 2012;3(1). Available from: https://doi.org/10.1177/2041731412467998 2041731412467998.

[38] Annibali S, Quaranta R, Scarano A, Pilloni A, Cicconetti A, Cristalli MP, et al. Histomorphometric evaluation of bone regeneration induced by biodegradable scaffolds as carriers for dental pulp stem cells in a rat model of calvarial "critical size" defect. Stem Cell Res Ther 2016;6:1. Available from: http://dx.doi.org/10.4172/2157-7633.1000322.

[39] Tatsuhiro F, Seiko T, Yusuke T, Reiko TT, Kazuhito S. Dental pulp stem cell-derived, scaffold-free constructs for bone regeneration. Int J Mol Sci 2018;19(7):E1846. Available from: https://doi.org/10.3390/ijms19071846 pii:.

[40] Fujii Y, Kawase-Koga Y, Hojo H, Yano F, Sato M, Chung UI, et al. Bone regeneration by human dental pulp stem cells using a helioxanthin derivative and cell-sheet technology. Stem Cell Res Ther 2018;9(1):24. Available from: https://doi.org/10.1186/s13287-018-0783-7.

[41] Jin Q, Yuan K, Lin W, Niu C, Ma R, Huang Z. Comparative characterization of mesenchymal stem cells from human dental pulp and adipose tissue for bone regeneration potential. Artif Cell Nanomed Biotechnol 2019;47(1):1577—84. Available from: https://doi.org/10.1080/21691401.2019.1594861.

[42] Thesleff I, Aberg T. Molecular regulation of tooth development. Bone 1999;25(1):123—5.

[43] Bonnamain V, Thinard R, Sergent-Tanguy S, Huet P, Bienvenu G, Naveilhan P, et al. Human dental pulp stem cells cultured in serum-free supplemented medium. Front Physiol 2013;11(4):357. Available from: https://doi.org/10.3389/fphys.2013.00357 eCollection 2013.

[44] Ebrahimi B, Yaghoobi MM, Kamal-abadi AM, Raoof M. Human dental pulp stem cells express many pluripotency regulators and differentiate into neuronal cells. Neural Regen Res 2011;6(34):2666—72.

[45] Isobe Y, Koyama N, Nakao K, Osawa K, Ikeno M, Yamanaka S, et al. Comparison of human mesenchymal stem cells derived from bone marrow, synovial fluid, adult dental pulp, and exfoliated deciduous tooth pulp. Int J Oral Maxillofac Surg 2016;45(1):124—31. Available from: https://doi.org/10.1016/j.ijom.2015.06.022.

[46] Nosrat IV, Smith CA, Mullally P, Olson L, Nosrat CA. Dental pulp cells provide neurotrophic support for dopaminergic neurons and differentiate into neurons in vitro; implications for tissue engineering and repair in the nervous system. Eur J Neurosci 2004;19(9):2388—98.

[47] Arthur A, Rychkov G, Shi S, Koblar SA, Gronthos S. Adult human dental pulp stem cells differentiate toward functionally active neurons under appropriate environmental cues. Stem Cell 2008;26(7):1787—95. Available from: https://doi.org/10.1634/stemcells.2007-0979.

[48] Urraca N, Memon R, El-Iyachi I, Goorha S, Valdez C, Tran QT, et al. Characterization of neurons from immortalized dental pulp stem cells for the study of neurogenetic disorders. Stem Cell Res 2015;15 (3):722—30. Available from: https://doi.org/10.1016/j.scr.2015.11.004.

[49] Victor AK, Reiter LT. Dental pulp stem cells for the study of neurogenetic disorders. Hum Mol Genet 2017;26(R2):R166—71. Available from: https://doi.org/10.1093/hmg/ddx208.

[50] Askari N, Yaghoobi MM, Shamsara M, Esmaeili-Mahani S. Human dental pulp stem cells differentiate into oligodendrocyte progenitors using the expression of Olig2 transcription factor. Cell Tissues Organs 2014;200 (2):93—103. Available from: https://doi.org/10.1159/000381668.

[51] El Ayachi I, Zhang J, Zou XY, Li D, Yu Z, Wei W, et al. Human dental stem cell derived transgene-free iPSCs generate functional neurons via embryoid body-mediated and direct induction methods. J Tissue Eng Regen Med 2018;12(4):e1836—51. Available from: https://doi.org/10.1002/term.2615.

[52] Sakai K, Yamamoto A, Matsubara K, Nakamura S, Naruse M, Yamagata M, et al. Human dental pulp-derived stem cells promote locomotor recovery after complete transection of the rat spinal cord by multiple neuro-regenerative mechanisms. J Clin Invest 2012;122(1):80—90. Available from: https://doi.org/10.1172/JCI59251.

[53] Kiraly M, Kádár K, Horváthy DB, Nardai P, Rácz GZ, Lacza Z, et al. Integration of neuronally predifferentiated human dental pulp stem cells into rat brain in vivo. Neurochem Int 2011;59(3):371—81. Available from: https://doi.org/10.1016/j.neuint.2011.01.006.

[54] Xiao L, Ide R, Saiki C, Kumazawa Y, Okamura H. Human dental pulp cells differentiate toward neuronal cells and promote neuroregeneration in adult organotypic hippocampal slices in vitro. Int J Mol Sci 2017;18 (8):E1745. Available from: https://doi.org/10.3390/ijms18081745 pii:.

[55] Martínez-Sarrà E, Montori S, Gil-Recio C, Núñez-Toldrà R, Costamagna D, Rotini A, et al. Human dental pulp pluripotent-like stem cells promote wound healing and muscle regeneration. Stem Cell Res Ther 2017;8 (1):175. Available from: https://doi.org/10.1186/s13287-017-0621-3.

[56] Iohara K, Zheng L, Wake H, Ito M, Nabekura J, Wakita H, et al. A novel stem cell source for vasculogenesis in ischemia: subfraction of side population cells from dental pulp. Stem Cell 2008;26(9):2408—18. Available from: https://doi.org/10.1634/stemcells.2008-0393.

[57] Ishizaka R, Hayashi Y, Iohara K, Sugiyama M, Murakami M, Yamamoto T, et al. Stimulation of angiogenesis, neurogenesis and regeneration by side population cells from dental pulp. Biomaterials 2013;34(8):1888—97. Available from: https://doi.org/10.1016/j.biomaterials.2012.10.045.

[58] Gandia C, Armiñan A, García-Verdugo JM, Lledó E, Ruiz A, Miñana MD, et al. Human dental pulp stem cells improve left ventricular function, induce angiogenesis, and reduce infarct size in rats with acute myocardial infarction. Stem Cell 2008;26(3):638−45.

[59] Zhang C, Zhang Y, Feng Z, Zhang F, Liu Z, Sun X, et al. Therapeutic effect of dental pulp stem cell transplantation on a rat model of radioactivity-induced esophageal injury. Cell Death Dis 2018;9(7):738. Available from: https://doi.org/10.1038/s41419-018-0753-0.

[60] Mead B, Logan A, Berry M, Leadbeater W, Scheven BA. Intravitreally transplanted dental pulp stem cells promote neuroprotection and axon regeneration of retinal ganglion cells after optic nerve injury. Invest Ophthalmol Vis Sci 2013;54(12):7544−56. Available from: https://doi.org/10.1167/iovs.13-13045.

[61] Gomes JA, Geraldes Monteiro B, Melo GB, Smith RL, Cavenaghi Pereira da Silva M, Lizier NF, et al. Corneal reconstruction with tissue-engineered cell sheets composed of human immature dental pulp stem cells. Invest Ophthalmol Vis Sci 2010;51(3):1408−14. Available from: https://doi.org/10.1167/iovs.09-4029.

[62] Syed-Picard FN, Du Y, Lathrop KL, Mann MM, Funderburgh ML, Funderburgh JL. Dental pulp stem cells: a new cellular resource for corneal stromal regeneration. Stem Cell Transl Med 2015;4(3):276−85. Available from: https://doi.org/10.5966/sctm.2014-0115.

[63] Kushnerev E, Shawcross SG, Sothirachagan S, Carley F, Brahma A, Yates JM, et al. Regeneration of corneal epithelium with dental pulp stem cells using a contact lens delivery system. Invest Ophthalmol Vis Sci 2016;57(13):5192−9. Available from: https://doi.org/10.1167/iovs.15-17953.

[64] American Diabetes Association. Diagnosis and classification of diabetes mellitus. Diabetes Care 2010;33 (Suppl 1):S62−9. Available from: https://doi.org/10.2337/dc10-S062.

[65] Lysy PA, Weir GC, Bonner-Weir S. Concise review: pancreas regeneration: recent advances and perspectives. Stem Cell Transl Med 2012;1(2):150−9. Available from: https://doi.org/10.5966/sctm.2011-0025.

[66] Govindasamy V, Ronald VS, Abdullah AN, Nathan KR, Ab Aziz ZA, Abdullah M, et al. Differentiation of dental pulp stem cells into islet-like aggregates. J Dent Res 2011;90(5):646−52. Available from: https://doi.org/10.1177/0022034510396879.

[67] Yagi Mendoza H, Yokoyama T, Tanaka T, Ii H, Yaegaki K. Regeneration of insulin-producing islets from dental pulp stem cells using a 3D culture system. Regen Med 2018;13(6):673−87. Available from: https://doi.org/10.2217/rme-2018-0074.

[68] Kadam SS, Bhonde RR. Islet neogenesis from the constitutively nestin expressing human umbilical cord matrix derived mesenchymal stem cells. Islets 2010;2(2):112−20. Available from: https://doi.org/10.4161/isl.2.2.11280.

[69] Guimarães ET, Cruz Gda S, Almeida TF, Souza BS, Kaneto CM, Vasconcelos JF, et al. Transplantation of stem cells obtained from murine dental pulp improves pancreatic damage, renal function, and painful diabetic neuropathy in diabetic type 1 mouse model. Cell Transpl 2013;22(12):2345−54. Available from: https://doi.org/10.3727/096368912X657972.

[70] Chen YY, He ST, Yan FH, Zhou PF, Luo K, Zhang YD, et al. Dental pulp stem cells express tendon markers under mechanical loading and are a potential cell source for tissue engineering of tendon-like tissue. Int J Oral Sci 2016;8(4):213−22. Available from: https://doi.org/10.1038/ijos.2016.33.

[71] Khajeh S, Razban V, Talaei-Khozani T, Soleimani M, Asadi-Golshan R, Dehghani F, et al. Enhanced chondrogenic differentiation of dental pulp-derived mesenchymal stem cells in 3D pellet culture system: effect of mimicking hypoxia Biologia 2018;73:715−26(2018). Available from: https://doi.org/10.2478/s11756-018-0080-z.

[72] Dai J, Wang J, Lu J, Zou D, Sun H, Dong Y, et al. The effect of co-culturing costal chondrocytes and dental pulp stem cells combined with exogenous FGF9 protein on chondrogenesis and ossification in engineered cartilage. Biomaterials 2012;33(31):7699−711. Available from: https://doi.org/10.1016/j.biomaterials.2012.07.020.

[73] Mata M, Milian L, Oliver M, Zurriaga J, Sancho-Tello M, de Llano JJM, et al. In vivo articular cartilage regeneration using human dental pulp stem cells cultured in an alginate scaffold: a preliminary study. Stem Cell Int 2017;2017:8309256. Available from: https://doi.org/10.1155/2017/8309256.

[74] Li H, Shen S, Fu H, Wang Z, Li X, Sui X, et al. Immunomodulatory functions of mesenchymal stem cells in tissue engineering. Stem Cell Int 2019;2019:9671206. Available from: https://doi.org/10.1155/2019/9671206 Jan 13, eCollection 2019.

[75] Rizk A, Rabie AB. Human dental pulp stem cells expressing transforming growth factor β3 transgene for cartilage-like tissue engineering. Cytotherapy 2013;15(6):712−25. Available from: https://doi.org/10.1016/j.jcyt.2013.01.012.

[76] Yao L, Flynn N. Dental pulp stem cell-derived chondrogenic cells demonstrate differential cell motility in type I and type II collagen hydrogels. Spine J 2018;18(6):1070—80. Available from: https://doi.org/10.1016/j.spinee.2018.02.007.

[77] Karamzadeh R, Baghaban Eslaminejad M, Sharifi-Zarchi A. Comparative in vitro evaluation of human dental pulp and follicle stem cell commitment. Cell J 2017;18(4):609—18 Winter.

[78] Zhang P, Yang W, Wang G, Li Y. miR-143 suppresses the osteogenic differentiation of dental pulp stem cells by inactivation of NF-κB signaling pathway via targeting TNF-α. Arch Oral Biol. 2018;87:172—9. Available from: https://doi:10.1016/j.archoralbio.2017.12.031.

[79] Sun DG, Xin BC, Wu D, Zhou L, Wu HB, Gong W, et al. miR-140-5p-mediated regulation of the proliferation and differentiation of human dental pulp stem cells occurs through the lipopolysaccharide/toll-like receptor 4 signaling pathway. Eur J Oral Sci 2017;125(6):419—25. Available from: https://doi.org/10.1111/eos.12384.

[80] Vasanthan P, Govindasamy V, Gnanasegaran N, Kunasekaran W, Musa S, Abu Kasim NH. Differential expression of basal microRNAs' patterns in human dental pulp stem cells. J Cell Mol Med 2015;19(3):566—80. Available from: https://doi.org/10.1111/jcmm.12381.

[81] Huang X, Liu F, Hou J, Chen K. Inflammation-induced overexpression of microRNA-223-3p regulates odontoblastic differentiation of human dental pulp stem cells by targeting SMAD3. Int Endod J 2019;52(4):491—503. Available from: https://doi.org/10.1111/iej.13032.

[82] Ke Z, Qiu Z, Xiao T, Zeng J, Zou L, Lin X, et al. Downregulation of miR-224-5p promotes migration and proliferation in human dental pulp stem cells. Biomed Res Int 2019;2019:4759060. Available from: https://doi.org/10.1155/2019/4759060 Jul 18, eCollection 2019.

[83] Lu X, Chen X, Xing J, Lian M, Huang D, Lu Y, et al. miR-140-5p regulates the odontoblastic differentiation of dental pulp stem cells via the Wnt1/β-catenin signaling pathway. Stem Cell Res Ther 2019;10(1):226. Available from: https://doi.org/10.1186/s13287-019-1344-4.

[84] Qiu Z, Lin S, Hu X, Zeng J, Xiao T, Ke Z, et al. Involvement of miR-146a-5p/neurogenic locus notch homolog protein 1 in the proliferation and differentiation of STRO-1 + human dental pulp stem cells. Eur J Oral Sci 2019;127(4):294—303. Available from: https://doi.org/10.1111/eos.12624.

[85] Wang BL, Wang Z, Nan X, Zhang QC, Liu W. Downregulation of microRNA-143-5p is required for the promotion of odontoblasts differentiation of human dental pulp stem cells through the activation of the mitogen-activated protein kinases 14-dependent p38 mitogen-activated protein kinases signaling pathway. J Cell Physiol 2019;234(4):4840—50. Available from: https://doi.org/10.1002/jcp.27282.

[86] Yao S, Li C, Budenski AM, Li P, Ramos A, Guo S. Expression of microRNAs targeting heat shock protein B8 during in vitro expansion of dental pulp stem cells in regulating osteogenic differentiation. Arch Oral Biol 2019;107:104485. Available from: https://doi.org/10.1016/j.archoralbio.2019.104485 Nov.

[87] Qiao W, Li D, Shi Q, Wang H, Wang H, Guo J. miR-224-5p protects dental pulp stem cells from apoptosis by targeting Rac1. Exp Ther Med 2020;19(1):9—18. Available from: https://doi.org/10.3892/etm.2019.8213.

[88] Chalisserry EP, Nam SY, Park SH, Anil S. Therapeutic potential of dental stem cells. J Tissue Eng 2017;8. Available from: https://doi.org/10.1177/2041731417702531 Jan-Dec, 2041731417702531.

[89] Alsulaimani RS, Ajlan SA, Aldahmash AM, Alnabaheen MS, Ashri NY. Isolation of dental pulp stem cells from a single donor and characterization of their ability to differentiate after 2 years of cryopreservation. Saudi Med J 2016;37(5):551—60. Available from: https://doi.org/10.15537/smj.2016.5.13615.

[90] Al-Habib M, Yu Z, Huang GT. Small molecules affect human dental pulp stem cell properties via multiple signaling pathways. Stem Cell Dev 2013;22(17):2402—13. Available from: https://doi.org/10.1089/scd.2012.0426.

[91] Chen L, Jiang Y, Du Z. Molecular differences between mature and immature dental pulp cells: bioinformatics and preliminary results. Exp Ther Med 2018;15(4):3362—8. Available from: https://doi.org/10.3892/etm.2018.5847.

Cardiovascular System

Regenerating the heart: The past, present, & future

Aditya Sengupta and Raghav A. Murthy

Department of Cardiovascular Surgery, Icahn School of Medicine at Mount Sinai, New York, NY, United States

6.1 Introduction

Cardiovascular disease, specifically ischemic heart disease (IHD), is the leading cause of global mortality. Patients with IHD, including those who suffer an acute myocardial infarction, often go on to develop devastating sequelae such as ventricular dysfunction and ischemic cardiomyopathy. This is a consequence not only of fibrous scar tissue replacing damaged myocardial cells, but also of the limited regenerative potential of adult cardiomyocytes [1]. A remodeling process then ensues that results in further fibrosis, loss of viable cardiomyocytes, ventricular dilatation, and heart failure [2]. Current therapeutic approaches to IHD aim to prevent this progression. For instance, β-blockers and angiotensin-converting enzyme (ACE) inhibitors have been shown to at least slow remodeling following an ischemic insult, with resultant reductions in mortality and hospitalizations from heart failure [3,4]. Similarly, revascularization by percutaneous coronary intervention (PCI) or coronary artery bypass surgery (CABG), when indicated, serves to improve blood supply and salvage injured myocardium [2]. For patients already in irreversible or end-stage heart failure, options such as left ventricular assist device (LVAD) implantation and cardiac transplantation exist [5]. However, the latter cannot be accepted as standard therapy due to the lack of donors worldwide [6].

Given the limited number of therapeutic options for patients with advanced heart failure, researchers have long focused their efforts at promoting cardiac regeneration [7]. Over the past three decades, various strategies have been attempted, including first-generation cell-based therapies (transplantation of non-cardiac cells such as skeletal myoblasts and bone-marrow derived cells) [8], use of resident cardiac cells with stem cell-like properties [9], transplantation of functional cardiomyocytes that were previously generated in vitro [10], and cell-free-based approaches [11]. However, preclinical outcomes of the aforementioned regenerative strategies have yet to be translated effectively to clinical trials, with unique advantages and shortcomings associated with each approach [1]. In this chapter, we provide a thorough review of the various therapeutic approaches for cardiac regeneration that have been attempted in the past, as well as an assessment of current and future strategies.

6.2 Cardiomyocyte regenerative potential

Although cardiac muscle cells are considered to be terminally differentiated, the proliferative and regenerative capacity of such cells vary by species and life stages. For instance, cardiomyocytes from newts and zebrafish are able to proliferate for the entirety of the animal's life span, whereas adult mammalian cardiomyocytes are permanently quiescent [12,13]. Interestingly, studies have demonstrated that the heart of a mouse at one day of age retains its regenerative capacity following an ischemic injury, but that at seven days of age or greater does not [14]. This regenerative time window also correlates with the postnatal period when human cardiac muscle cells permanently exit the cell cycle [15]. Although increasing evidence shows that cardiomyocyte turnover does in fact occurs in the adult human heart, the rate of turnover is unfortunately far too insufficient to restore contractility and replace the vast numbers of cardiac muscle cells lost following an acute myocardial infarction [1,7,16]. This further validates the need for novel regenerative therapies for patients with IHD.

6.3 Cell-based strategies

6.3.1 Non-cardiac cells

Cells other than cardiomyocytes have been the principal source for cell-based strategies in ischemic cardiomyopathy. Skeletal myoblasts were the first such cells to be used. Transplantation of such cells in animal models led to improvements in ejection fraction, and preliminary clinical trials demonstrated that the transplanted myoblasts resulted in improved cardiac function in patients with heart failure [8,17,18]. However, such beneficial effects were not observed at long-term follow-up. Furthermore, failure of the skeletal myoblasts to integrate with the host's cardiomyocytes led to electromechanical discordance and deleterious effects such as arrhythmogenesis, thus precluding their use in future studies [19].

Researchers then turned their attention to bone marrow-derived mononuclear cells (BMCs). Studies such as the BOOST and REPAIR-AMI trials showed improvements in ejection in patients with acute myocardial infarction following transplantation of BMCs [20]. However, numerous follow-up trials with larger patient populations, including the TIME and FOCUS-CCTRN trials, have been unable to replicate these early results [21,22].

Finally, regeneration strategies involving mesenchymal stem cells (MSCs) have been attempted given the ability of MSCs to self-renew and differentiate into a variety of cell types, including cardiomyocytes [23]. Once again, despite early promising results, studies such as the POSEIDON and MSC-HF clinical trials have failed to demonstrate a sustained benefit from transplantation of MSCs in patients with ischemic heart disease [24,25].

6.4 Cardiac stem cells

Given the lack of consistent results seen with non-cardiac cells, investigators turned their attention to cardiac stem cells (CSCs), or resident cardiac cells with self-renewal properties and the ability to differentiate into cardiac muscle, smooth muscle, and endothelial cells. Preclinical studies involving the autologous transplantation of cardiosphere-derived cells showed promise with improvements in ejection fraction at eight weeks [26,27]. Since then, two clinical trials involving CSCs have been conducted, namely the SCIPIO and CADUCEUS trials [28,29]. Even though the former showed slight improvements in ejection fraction and infarct size, questions remained about the cardiomyogenic potential and engraftment rates of CSCs. Furthermore, concern over the integrity of data used in this trial have led to the retraction of its results [30]. The CADUCEUS trial showed that CSCs transplanted into patients following a myocardial infarction led to an increase in viable tissue at six months and one year. However, there was no evident improvement in ventricular function, casting a shadow over CSCs as viable therapeutic options [29].

6.5 Pluripotent stem cells

Given the limited capacity of transplanted stem cells to differentiate into cardiac muscle cells, scientists turned their attention to generating de novo cardiomyocytes in vitro. Two such cell populations have been extensively studied with regards to cardiac regeneration: embryonic stem cells (ESCs) and induced pluripotent stem cells (iPSCs). ESCs can easily be made to differentiate into cardiomyocytes in vitro, and transplantation of such cells into ischemic hearts in animal models have been shown to improve ventricular function and integrate electromechanically with resident cardiac cells [31]. Even though complications such as arrhythmias, associated risks such as tumorigenesis and immune rejection, and ethical considerations have limited their therapeutic use [32], early results from the ESCORT trial (designed to assess the feasibility of ESCs as cardiac regenerators in humans) are encouraging [1,33].

In 2006, Yamanaka and colleagues introduced the world to iPSCs, or fibroblasts reprogrammed to a pluripotent state via the forced expression of four key genes [34], thus offering a novel approach to cardiac regeneration while avoiding the ethical dilemmas associated with the use of ESCs. A recent preclinical study involving the intramyocardial transplantation of allogenic iPSC-cardiomyocytes into macaque models two weeks following a myocardial infarction has shown improvements in ventricular function at twelve weeks, albeit a noticeable incidence of ventricular tachycardia [35]. Even though there are risks of tumorigenesis, arrhythmias, and low engraftment rates with iPSC-derived cardiomyocytes, considerable progress has been made towards integrating such cells into the host upon transplantation. For instance, mature cardiomyocytes can now be generated from iPSCs with enhanced purity and large numbers, thus improving cell retention rates [36]. Furthermore, scaffolds such as hydrogels and cell sheets are paving the way for more complete biointegration [37]. However, much research is needed before this technology can be applied to humans.

6.6 Cell-free strategies

In contrast to cell-based approaches, cell-free strategies involve the manipulation of key secretory factors. Examples of such modalities include growth factors, microRNAs (miRNAs), and exosomes, with resultant downstream paracrine effects such as angiogenesis and cardiomyocyte proliferation with concomitant inhibition of fibrosis, apoptosis, and oxidative stress [1]. These cardioprotective factors can thus be harnessed for cardiac repair and regeneration.

6.7 Growth factors

Signaling molecules such as neuregulin 1 (NRG1) and its receptors (receptor tyrosine-protein kinase ERBB2 and ERBB4) are crucial in the early stages of embryonic cardiac development. These growth factors have also been shown to induce cardiac muscle proliferation and improve ventricular function. For instance, systemic delivery of human recombinant NRG1 led to decreases in end-systolic and end-diastolic volumes in patients with heart failure [38]. These results, however, stand in stark contrast to findings seen in mouse models where NRG1 treatment failed to enhance cardiomyocyte proliferation [39]. The converse has also been seen with certain other growth factors: results from animal studies have not been translated to success in clinical trials. For instance, vascular endothelial growth factor A (VEGFA), a potent pro-angiogenic molecule, has been shown to improve coronary blood flow and restore cardiomyocyte function in animal models of chronic myocardial ischemia [40]. However, as the EUROINJECT-ONE trial demonstrated, these findings could not be replicated in human studies [41]. Similar outcomes have been reported for fibroblast growth factor 2 (FGF2) [42]. The lack of sufficient quantities of therapeutic factors at the target sites, either due to ineffective delivery or to a short half-life of the factors themselves, has often been identified as the culprit for the aforementioned inconsistent results. However, the recent use of scaffolding materials and improved gene-based therapeutics have led to enhanced delivery and sustained local release of therapeutic growth factors [43].

6.8 Exosomes & microRNA technology

Exosomes refer to extracellular vesicles that are less than 100 nm in diameter. They generally have specific surface markers that define their function, are secreted by a variety of cells, and may contain various products [1]. Research has shown that exosomes may play a pivotal role in cellular communication by transporting cell-specific mRNAs and miRNAs. This has been exploited in the laboratory for cardiac repair. For example, injection of murine cardiosphere-derived exosomes carrying miR-451 has been shown to inhibit cardiac muscle apoptosis [44]. Human CSC- and MSC-derived exosomes have also demonstrated beneficial functions such as attenuation of adverse cardiac remodeling and the damaging effects of ischemia—reperfusion injury [45]. However, much research is needed before exosomes can be utilized for cardiac regeneration.

miRNAs have also been investigated as a therapeutic modality. These are small, single-strand, non-coding RNAs that are involved in post-transcriptional gene expression and paracrine signaling. miRNAs have been implicated in cardiac development and disease progression given their ability to interfere with numerous target RNAs [46]. For instance, inhibition of miR-15 molecules has been shown to induce cardiac muscle cell proliferation and improve cardiac function in ischemic mouse models [47]. Similarly, cardiac-specific overexpression of miRNAs such as miR-199a and miR-590 induces in vivo proliferation of postnatal and adult cardiomyocytes along with attenuation of post-infarction fibrosis [48]. As with exosomes, although these preclinical studies are encouraging, much investigation is needed before miRNAs can be used for sustained cardiomyocyte proliferation in humans.

6.9 Direct reprogramming

Direct reprogramming is a novel strategy for cardiac repair and regeneration whereby resident fibroblasts are directly converted into de novo cardiomyocytes. This is a promising option given that ischemic injury is often followed by a fibrotic response, and that fibroblasts make up a sizeable proportion of all cardiac cells. Furthermore, this technology could theoretically allow one to bypass the issues of tumorigenesis and low engraftment rates seen with iPSCs [1]. The three main strategies employed are (1) in vitro direct reprogramming of mouse fibroblasts into cardiomyocytes, (2) in vitro direct reprogramming of human cells, and (3) in vivo direct reprogramming.

Forced expression of multiple transcription factors related to heart development (specifically *Gata4*, *Mef2c*, *Tbx5*, and *Hand2*, or GHMT) can successfully transform murine fibroblasts into cardiomyocyte-like cells with sarcomeres, spontaneous intracellular calcium oscillations, and beating contractions [49]. However, certain issues currently preclude the use of this technology in clinical trials. For one, reprogramming efficiency is a major hurdle. Modification of the reprogramming strategy with miRNAs and growth factors has been investigated as a potential solution, but with varied results [50]. Investigators have also attempted to reprogram fibroblasts into cardiac progenitor cells via the forced expression of five key genes encoding early developmental factors (*Mesp1*, *Gata4*, *Tbx5*, *Nkx2−5*, and *Baf60c*, or *Smarcd3*). Murine fibroblasts can now be transformed into multipotent cardiac progenitor cells. Following transplantation into post-ischemic murine hearts, these progenitor cells differentiated into cardiac muscle and endothelial cells with some improvement in survival. However, as always, the disadvantages of transplanting cells remain with this strategy [1,51].

Scientists have also attempted to directly reprogram human cells in vitro. Studies have shown that human dermal fibroblasts can be transformed into cardiac progenitor cells via the expression of the transcription factors c-ETS2 (ETS2) and mesoderm posterior protein 1 (MESP1) [52].

Cardiomyocytes thus created display similar gene expression profiles to native cardiac muscle cells, and also exhibit appropriate sarcomere structure [53]. Transplantation of human reprogrammed cardiomyocyte-like cells has shown viability in mouse models, but numerous challenges remain in applying this technology to humans [54]. Perhaps most importantly, human reprogramming is less efficient when compared to murine models, and human cells generally take a much longer time to exhibit the properties of cardiac muscle cells. This has been attributed to the longer developmental time observed in humans, as well as to the varying epigenetic landscapes between the mouse and human fibroblasts [1,55].

Direct reprogramming of fibroblasts into cardiomyocytes in vivo has also been demonstrated. GMT- and GHMT-based reprogramming strategies and retroviral delivery techniques have been successfully used in mice to generate cardiomyocyte-like cells. As with the previous two strategies, these newly generated cells display similar gene expression profiles, sarcomere structure, contractility patterns, and calcium transients to endogenous cardiac muscle cells [49,56]. Although this technology has been shown to improve ventricular function in post-ischemic murine models, a few challenges remain before eventual clinical translation can be advocated [57]. For one, induction of reprogramming is currently limited by the small therapeutic time window after coronary ligation. Heterogeneity of the newly generated cardiomyocytes and low reprogramming efficiency are additional issues. Finally, there is a dearth of feasible delivery methods, although work with a Sendai virus vector shows promise [58].

6.10 Endogenous repair & regeneration

In contrast to cell-based and cell-free strategies, endogenous cardiac repair and regeneration can be evoked by modulating cardiomyocyte proliferation and targeting the cardiac fibrotic response. Numerous strategies have been employed to achieve the former, including modulation of cell cycle regulators such as cyclins and cyclin-dependent kinases (CDKs) [59]. More recently, genetic ablation of *Meis1*, a transcription factor homeobox protein that regulates cardiomyocyte cell cycle arrest, has been shown to prolong postnatal cardiac muscle cell proliferation. These induced cardiomyocytes are also able to re-enter the cell cycle in adult mice [1,60]. Scientists have also targeted the Hippo signaling pathway for stimulation of endogenous cardiac repair [61]. This pathway plays a pivotal role in controlling cell proliferation and organ size, and is activated via the phosphorylation of a number of factors (mammalian STE20-like protein kinase 1 or MST1, MST2, protein salvador homologue 1 or SAV1, large tumor suppressor homologue 1 or LATS1, and LATS2) [1,62]. Genetic ablation of *Mst1*, *Mst2*, *Lats2*, or *Sav1* has been shown to promote cardiomyocyte proliferation and improve cardiac function in murine models [63,64]. However, Hippo signaling inhibition also leads to off-target cellular proliferation, and further research is needed before this endogenous repair technique can be used clinically.

The fibrotic process can also be targeted for cardiac repair. Following an ischemic insult, myofibroblasts and inflammatory cells are activated that produce a secretome conducive to the formation of a fibrotic scar. This reparative fibrosis continues until pathological cardiac remodeling results [65]. Numerous signaling pathways have been implicated in this process, including the renin—angiotensin—aldosterone system (RAAS) and the TGFβ and WNT pathways, thus providing targets for inhibiting fibrotic remodeling [66]. For instance, inhibition of chymase, a protease that activates RAAS and TGFβ, leads to improved cardiac function and attenuated fibrosis in post-infarction rats [67]. In addition, clinical trials that demonstrated the potency of inhibiting angiotensin II in counteracting the fibrotic response [68].

Pharmacologic inhibition of the remodeling process, however, is complicated by the fact that the initial fibrotic response is necessary for proper repair and regeneration. In murine models, inhibition of TGFβ in the first day following a myocardial infarction leads to increased mortality, while TGFβ inhibition at a later time leads to reversal of pathologic remodeling [69]. Additionally, in the post-infarction heart, cells other than fibroblasts also play an active role in the fibrotic remodeling process. These include endothelial cells, epicardial cells, and perivascular cells. Holistic inhibition of fibrosis, therefore, also has to take these other non-fibroblast cells into consideration. Studies have shown that pigment epithelium-derived factor slows cardiac remodeling by halting endothelial-to-mesenchymal conversion [1,70]. A more thorough understanding of the post-infarction fibrotic response is required before these strategies are implemented in the clinical setting.

6.11 The future

Therapeutic strategies for cardiac repair and regeneration will undoubtedly continue to expound upon the aforementioned technologies. Cell-based and cell-free strategies, along with techniques that enhance and modulate cardiac repair, will evolve with an eventual transition to clinical studies as the mechanisms underlying cardiac fibrotic remodeling and regeneration are further elucidated.

There have been recent developments in basic science that show promise for future therapies for cardiac repair. For instance, in postnatal murine cardiac development, reactive oxygen species (ROS) have been identified as a potential inducer of cell cycle arrest via the DNA damage response (DDR) pathway. Pharmacologic inhibition of the DDR pathway leads to prolongation of this postnatal proliferative window [71]. More interestingly, in mice, gradual exposure to systemic hypoxia leads to a reduction in ROS synthesis and oxidative DNA damage, which then reactivates the mitotic process and leads to repair following an ischemic insult [72]. Thus, there may be a role for modulation of environmental signals that scavenge ROS or modify the DDR pathway as a therapeutic approach for cardiac regeneration.

Genome editing also has potential and is actively being investigated as a therapeutic modality for improving cardiac function. For instance, targeting the dystrophin gene with CRISPR-Cas9 technology leads to improved cardiomyocyte function in murine models of Duchenne muscular dystrophy (DMD) [73]. Furthermore, genome editing of iPSC-derived cardiomyocytes from patients with DMD results in restoration of dystrophic expression with enhanced ventricular performance in vitro [74,75]. Investigators have also exploited this technology for xenotransplantation. Genome-edited xenografts theoretically carry a smaller risk of immune-mediated rejection and post-transplantation infection, thus opening the door for primed stem cells that do no elicit undesired host responses [76]. Finally, CRISPR may additionally be used to modulate the endogenous repair mechanisms of the heart [77].

A key challenge moving forward will be to reconcile the encouraging outcomes seen in preclinical studies with the suboptimal results of recent clinical trials. Numerous explanations have been put forth, but the most plausible mechanism seems to be that of inefficient delivery of therapeutic factors to the desired target [1]. To tackle this issue, researchers have turned to biointegration, or the process of coalescing the mechanical and functional properties of a given biomaterial, such as an implant, with living tissue. This integration, or biologization, provides a scaffold upon which native tissue is allowed to regenerate, thus essentially converting biomaterials into a dynamic component of living beings that react and adapt to the environment. Here, decellularized matrices have the capacity to promote endothelial cell migration and proliferation have come to the forefront. However, much work needs to be done before this technology can be effectively applied to human cardiac regeneration [78].

Recent evidence also suggests that a paradigm shift may be imminent in the field of cardiovascular repair and regeneration. Rather than focusing on a single therapeutic modality such as iPSC technology or genome editing, investigators have turned their attention to a more holistic approach that combines the strengths of each individual component. For instance, Bertero and Murry have proposed a paradigm for understanding cardiac repair based on the classic report by Hanahan and Weinberg in 2000 [79]. Here, the authors identify five key "hallmarks" of the regenerating heart that have to be simultaneously addressed by any lasting therapeutic modality: angiogenesis and arteriogenesis, remuscularization, electromechanical stability, resolution of fibrosis, and immunological balance [80]. With the evolution of stem cell technology and genomics, along with the popularization of biointegration, perhaps this future is not too distant.

6.12 Conclusion

Much progress has been made since the early decades of regenerative medicine. Both cell-based and cell-free strategies for cardiac repair have evolved over the years, with some success seen in clinical trials. However, issues such as insufficient delivery of therapeutic factors and tumorigenesis still plague outcomes. Newer approaches such as modulation of endogenous cardiac repair and genome editing show promise, but it remains to be seen if such technologies can provide durable reparative potential in large-scale human studies. Moving forward, regenerating the heart will likely require a holistic approach that combines the various aforementioned therapeutic modalities.

References

[1] Hashimoto H, Olson EN, Bassel-Duby R. Therapeutic approaches for cardiac regeneration and repair. Nat Rev Cardiol 2018;15:585—600.
[2] Aimo A, Gaggin HK, Barison A, Emdin M, Januzzi Jr. JL. Imaging, biomarker, and clinical predictors of cardiac remodeling in heart failure with reduced ejection fraction. JACC Heart Fail 2019;7:782—94.
[3] Investigators S, Yusuf S, Pitt B, Davis CE, Hood WB, Cohn JN. Effect of enalapril on survival in patients with reduced left ventricular ejection fractions and congestive heart failure. N Engl J Med 1991;325:293—302.
[4] Packer M, Coats AJ, Fowler MB, et al. Effect of carvedilol on survival in severe chronic heart failure. N Engl J Med 2001;344:1651—8.

[5] Mehra MR, Uriel N, Naka Y, et al. A fully magnetically levitated left ventricular assist device — final report. N Engl J Med 2019;380:1618—27.

[6] Yacoub M. Cardiac donation after circulatory death: a time to reflect. Lancet 2015;385:2554—6.

[7] Laflamme MA, Murry CE. Heart regeneration. Nature 2011;473:326—35.

[8] Menasche P, Alfieri O, Janssens S, et al. The Myoblast Autologous Grafting in Ischemic Cardiomyopathy (MAGIC) trial: first randomized placebo-controlled study of myoblast transplantation. Circulation 2008;117:1189—200.

[9] Beltrami AP, Barlucchi L, Torella D, et al. Adult cardiac stem cells are multipotent and support myocardial regeneration. Cell 2003;114:763—76.

[10] Mummery CL, Zhang J, Ng ES, Elliott DA, Elefanty AG, Kamp TJ. Differentiation of human embryonic stem cells and induced pluripotent stem cells to cardiomyocytes: a methods overview. Circ Res 2012;111:344—58.

[11] Xin M, Kim Y, Sutherland LB, et al. Hippo pathway effector Yap promotes cardiac regeneration. Proc Natl Acad Sci USA 2013;110:13839—44.

[12] Porrello ER, Olson EN. A neonatal blueprint for cardiac regeneration. Stem Cell Res 2014;13:556—70.

[13] Poss KD, Wilson LG, Keating MT. Heart regeneration in zebrafish. Science 2002;298:2188—90.

[14] Porrello ER, Mahmoud AI, Simpson E, et al. Transient regenerative potential of the neonatal mouse heart. Science 2011;331:1078—80.

[15] Soonpaa MH, Kim KK, Pajak L, Franklin M, Field LJ. Cardiomyocyte DNA synthesis and binucleation during murine development. Am J Physiol 1996;271:H2183—9.

[16] Bergmann O, Bhardwaj RD, Bernard S, et al. Evidence for cardiomyocyte renewal in humans. Science 2009;324:98—102.

[17] Durrani S, Konoplyannikov M, Ashraf M, Haider KH. Skeletal myoblasts for cardiac repair. Regen Med 2010;5:919—32.

[18] Povsic TJ, O'Connor CM, Henry T, et al. A double-blind, randomized, controlled, multicenter study to assess the safety and cardiovascular effects of skeletal myoblast implantation by catheter delivery in patients with chronic heart failure after myocardial infarction. Am Heart J 2011;162:654—62 e1.

[19] Fouts K, Fernandes B, Mal N, Liu J, Laurita KR. Electrophysiological consequence of skeletal myoblast transplantation in normal and infarcted canine myocardium. Heart Rhythm 2006;3:452—61.

[20] Schachinger V, Erbs S, Elsasser A, et al. Intracoronary bone marrow-derived progenitor cells in acute myocardial infarction. N Engl J Med 2006;355:1210—21.

[21] Perin EC, Willerson JT, Pepine CJ, et al. Effect of transendocardial delivery of autologous bone marrow mononuclear cells on functional capacity, left ventricular function, and perfusion in chronic heart failure: the FOCUS-CCTRN trial. JAMA 2012;307:1717—26.

[22] Traverse JH, Henry TD, Pepine CJ, et al. Effect of the use and timing of bone marrow mononuclear cell delivery on left ventricular function after acute myocardial infarction: the TIME randomized trial. JAMA 2012;308:2380—9.

[23] Planat-Benard V, Menard C, Andre M, et al. Spontaneous cardiomyocyte differentiation from adipose tissue stroma cells. Circ Res 2004;94:223—9.

[24] Hare JM, Fishman JE, Gerstenblith G, et al. Comparison of allogeneic vs autologous bone marrow-derived mesenchymal stem cells delivered by transendocardial injection in patients with ischemic cardiomyopathy: the POSEIDON randomized trial. JAMA 2012;308:2369—79.

[25] Mathiasen AB, Qayyum AA, Jorgensen E, et al. Bone marrow-derived mesenchymal stromal cell treatment in patients with severe ischaemic heart failure: a randomized placebo-controlled trial (MSC-HF trial). Eur Heart J 2015;36:1744—53.

[26] Bolli R, Tang XL, Sanganalmath SK, et al. Intracoronary delivery of autologous cardiac stem cells improves cardiac function in a porcine model of chronic ischemic cardiomyopathy. Circulation 2013;128:122—31.

[27] Johnston PV, Sasano T, Mills K, et al. Engraftment, differentiation, and functional benefits of autologous cardiosphere-derived cells in porcine ischemic cardiomyopathy. Circulation 2009;120:1075—83, 7 p following 1083.

[28] Bolli R, Chugh AR, D'Amario D, et al. Cardiac stem cells in patients with ischaemic cardiomyopathy (SCIPIO): initial results of a randomised phase 1 trial. Lancet 2011;378:1847—57.

[29] Makkar RR, Smith RR, Cheng K, et al. Intracoronary cardiosphere-derived cells for heart regeneration after myocardial infarction (CADUCEUS): a prospective, randomised phase 1 trial. Lancet 2012;379:895—904.

[30] The Lancet E. Retraction-Cardiac stem cells in patients with ischaemic cardiomyopathy (SCIPIO): initial results of a randomised phase 1 trial. Lancet 2019;393:1084.

[31] Mummery C, Ward-van Oostwaard D, Doevendans P, et al. Differentiation of human embryonic stem cells to cardiomyocytes: role of coculture with visceral endoderm-like cells. Circulation 2003;107:2733—40.

[32] Chong JJ, Yang X, Don CW, et al. Human embryonic-stem-cell-derived cardiomyocytes regenerate non-human primate hearts. Nature 2014;510:273—7.

[33] Menasche P, Vanneaux V, Hagege A, et al. Transplantation of human embryonic stem cell-derived cardiovascular progenitors for severe ischemic left ventricular dysfunction. J Am Coll Cardiol 2018;71:429—38.

[34] Takahashi K, Yamanaka S. Induction of pluripotent stem cells from mouse embryonic and adult fibroblast cultures by defined factors. Cell 2006;126:663—76.

[35] Shiba Y, Gomibuchi T, Seto T, et al. Allogeneic transplantation of iPS cell-derived cardiomyocytes regenerates primate hearts. Nature 2016;538:388—91.

[36] Tohyama S, Hattori F, Sano M, et al. Distinct metabolic flow enables large-scale purification of mouse and human pluripotent stem cell-derived cardiomyocytes. Cell Stem Cell 2013;12:127—37.

[37] Chow A, Stuckey DJ, Kidher E, et al. Human induced pluripotent stem cell-derived cardiomyocyte encapsulating bioactive hydrogels improve rat heart function post myocardial infarction. Stem Cell Rep 2017;9:1415—22.

[38] Gao R, Zhang J, Cheng L, et al. A Phase II, randomized, double-blind, multicenter, based on standard therapy, placebo-controlled study of the efficacy and safety of recombinant human neuregulin-1 in patients with chronic heart failure. J Am Coll Cardiol 2010;55:1907—14.

[39] Reuter S, Soonpaa MH, Firulli AB, Chang AN, Field LJ. Recombinant neuregulin 1 does not activate cardiomyocyte DNA synthesis in normal or infarcted adult mice. PLoS One 2014;9:e115871.

[40] Harada K, Friedman M, Lopez JJ, et al. Vascular endothelial growth factor administration in chronic myocardial ischemia. Am J Physiol 1996;270:H1791—802.

[41] Gyongyosi M, Khorsand A, Zamini S, et al. NOGA-guided analysis of regional myocardial perfusion abnormalities treated with intramyocardial injections of plasmid encoding vascular endothelial growth factor A-165 in patients with chronic myocardial ischemia: subanalysis of the EUROINJECT-ONE multicenter double-blind randomized study. Circulation 2005;112:I157—65.

[42] Simons M, Annex BH, Laham RJ, et al. Pharmacological treatment of coronary artery disease with recombinant fibroblast growth factor-2: double-blind, randomized, controlled clinical trial. Circulation 2002;105:788—93.

[43] Zangi L, Lui KO, von Gise A, et al. Modified mRNA directs the fate of heart progenitor cells and induces vascular regeneration after myocardial infarction. Nat Biotechnol 2013;31:898—907.

[44] Chen L, Wang Y, Pan Y, et al. Cardiac progenitor-derived exosomes protect ischemic myocardium from acute ischemia/reperfusion injury. Biochem Biophys Res Commun 2013;431:566—71.

[45] Arslan F, Lai RC, Smeets MB, et al. Mesenchymal stem cell-derived exosomes increase ATP levels, decrease oxidative stress and activate PI3K/Akt pathway to enhance myocardial viability and prevent adverse remodeling after myocardial ischemia/reperfusion injury. Stem Cell Res 2013;10:301—12.

[46] Liu N, Olson EN. MicroRNA regulatory networks in cardiovascular development. Dev Cell 2010;18:510—25.

[47] Porrello ER, Mahmoud AI, Simpson E, et al. Regulation of neonatal and adult mammalian heart regeneration by the miR-15 family. Proc Natl Acad Sci USA 2013;110:187—92.

[48] Chen J, Huang ZP, Seok HY, et al. mir-17-92 cluster is required for and sufficient to induce cardiomyocyte proliferation in postnatal and adult hearts. Circ Res 2013;112:1557—66.

[49] Song K, Nam YJ, Luo X, et al. Heart repair by reprogramming non-myocytes with cardiac transcription factors. Nature 2012;485:599—604.

[50] Muraoka N, Yamakawa H, Miyamoto K, et al. MiR-133 promotes cardiac reprogramming by directly repressing Snail and silencing fibroblast signatures. EMBO J 2014;33:1565—81.

[51] Lalit PA, Salick MR, Nelson DO, et al. Lineage reprogramming of fibroblasts into proliferative induced cardiac progenitor cells by defined factors. Cell Stem Cell 2016;18:354—67.

[52] Islas JF, Liu Y, Weng KC, et al. Transcription factors ETS2 and MESP1 transdifferentiate human dermal fibroblasts into cardiac progenitors. Proc Natl Acad Sci USA 2012;109:13016—21.

[53] Fu JD, Stone NR, Liu L, et al. Direct reprogramming of human fibroblasts toward a cardiomyocyte-like state. Stem Cell Rep 2013;1:235—47.

[54] Cao N, Huang Y, Zheng J, et al. Conversion of human fibroblasts into functional cardiomyocytes by small molecules. Science 2016;352:1216–20.

[55] Zhou JX, Huang S. Understanding gene circuits at cell-fate branch points for rational cell reprogramming. Trends Genet 2011;27:55–62.

[56] Qian L, Huang Y, Spencer CI, et al. In vivo reprogramming of murine cardiac fibroblasts into induced cardiomyocytes. Nature 2012;485:593–8.

[57] Nam YJ, Song K, Olson EN. Heart repair by cardiac reprogramming. Nat Med 2013;19:413–15.

[58] Miyamoto K, Akiyama M, Tamura F, et al. Direct in vivo reprogramming with sendai virus vectors improves cardiac function after myocardial infarction. Cell Stem Cell 2018;22:91–103 e5.

[59] Pasumarthi KB, Nakajima H, Nakajima HO, Soonpaa MH, Field LJ. Targeted expression of cyclin D2 results in cardiomyocyte DNA synthesis and infarct regression in transgenic mice. Circ Res 2005;96:110–18.

[60] Mahmoud AI, Kocabas F, Muralidhar SA, et al. Meis1 regulates postnatal cardiomyocyte cell cycle arrest. Nature 2013;497:249–53.

[61] Flinn MA, Link BA, O'Meara CC. Upstream regulation of the Hippo-Yap pathway in cardiomyocyte regeneration. Semin Cell Dev Biol 2019.

[62] Wang J, Liu S, Heallen T, Martin JF. The Hippo pathway in the heart: pivotal roles in development, disease, and regeneration. Nat Rev Cardiol 2018;15:672–84.

[63] Heallen T, Morikawa Y, Leach J, et al. Hippo signaling impedes adult heart regeneration. Development 2013;140:4683–90.

[64] Heallen T, Zhang M, Wang J, et al. Hippo pathway inhibits Wnt signaling to restrain cardiomyocyte proliferation and heart size. Science 2011;332:458–61.

[65] Weber KT, Sun Y, Bhattacharya SK, Ahokas RA, Gerling IC. Myofibroblast-mediated mechanisms of pathological remodelling of the heart. Nat Rev Cardiol 2013;10:15–26.

[66] Cittadini A, Monti MG, Isgaard J, et al. Aldosterone receptor blockade improves left ventricular remodeling and increases ventricular fibrillation threshold in experimental heart failure. Cardiovasc Res 2003;58:555–64.

[67] Kanemitsu H, Takai S, Tsuneyoshi H, et al. Chymase inhibition prevents cardiac fibrosis and dysfunction after myocardial infarction in rats. Hypertens Res 2006;29:57–64.

[68] Pitt B, Poole-Wilson PA, Segal R, et al. Effect of losartan compared with captopril on mortality in patients with symptomatic heart failure: randomised trial – the Losartan Heart Failure Survival Study ELITE II. Lancet 2000;355:1582–7.

[69] Ikeuchi M, Tsutsui H, Shiomi T, et al. Inhibition of TGF-beta signaling exacerbates early cardiac dysfunction but prevents late remodeling after infarction. Cardiovasc Res 2004;64:526–35.

[70] Zeisberg EM, Tarnavski O, Zeisberg M, et al. Endothelial-to-mesenchymal transition contributes to cardiac fibrosis. Nat Med 2007;13:952–61.

[71] Tao G, Kahr PC, Morikawa Y, et al. Pitx2 promotes heart repair by activating the antioxidant response after cardiac injury. Nature 2016;534:119–23.

[72] Nakada Y, Canseco DC, Thet S, et al. Hypoxia induces heart regeneration in adult mice. Nature 2017;541:222–7.

[73] Long C, McAnally JR, Shelton JM, Mireault AA, Bassel-Duby R, Olson EN. Prevention of muscular dystrophy in mice by CRISPR/Cas9-mediated editing of germline DNA. Science 2014;345:1184–8.

[74] Long C, Li H, Tiburcy M, et al. Correction of diverse muscular dystrophy mutations in human engineered heart muscle by single-site genome editing. Sci Adv 2018;4:eaap9004.

[75] Strong A, Musunuru K. Genome editing in cardiovascular diseases. Nat Rev Cardiol 2017;14:11–20.

[76] Yang L, Guell M, Niu D, et al. Genome-wide inactivation of porcine endogenous retroviruses (PERVs). Science 2015;350:1101–4.

[77] Chakraborty S, Ji H, Kabadi AM, Gersbach CA, Christoforou N, Leong KW. A CRISPR/Cas9-based system for reprogramming cell lineage specification. Stem Cell Rep 2014;3:940–7.

[78] Moroni F, Mirabella T. Decellularized matrices for cardiovascular tissue engineering. Am J Stem Cell 2014;3:1–20.

[79] Hanahan D, Weinberg RA. The hallmarks of cancer. Cell 2000;100:57–70.

[80] Bertero A, Murry CE. Hallmarks of cardiac regeneration. Nat Rev Cardiol 2018;15:579–80.

Engineered cardiac tissue: Concepts and future

Soumya K. Chandrasekhar[1,2], *Finosh G. Thankam*[3], *Joshi C. Ouseph*[2] *and Devendra K. Agrawal*[3]

[1]Department of Zoology, KKTM Government College, Pullut, Calicut University, Kerala, India
[2]Department of Zoology, Christ College, Irinjalakuda, Calicut University, Kerala, India
[3]Department of Translational Research, Western University of Health Sciences, Pomona, CA, United States

7.1 Introduction

Significant increase of 14.5% in the global mortality due to CVDs especially myocardial infarction (MI) occurred during 2006—16. Heart transplantation has been considered to be the lifesaving management strategy for end stage cardiac failure; however, the lack of sufficient donors offers a major constrain. In USA, heart failure is tormenting 6 million people and 10% of them are in the late stages requiring heart transplantation. Also, this figure continues to grow if global scenario is taken into consideration. In addition, the difficulties in the instantaneous procurement of healthy donor hearts for transplantation is a major challenge. Millions of patients are in the waiting list and can be saved if transplantation is feasible. There is an increase of 51% in the number of adult candidates waiting for a transplant, globally since 2004. However, the relatively static nature of donor pool opposed by the increased surge in the number of sufferers requiring transplants has faded the hope of survival of many. Sadly, many of the MI patients die on the waiting list aspiring a chance for survival [1].

The increased chances of immunological rejection of the transplanted heart, principally with the discontinuation of immuno-suppressants, is another serious threat for the success of heart transplantation [2]. In addition to the chronic immune suppression, the post transplantation complications/comorbidities such as diabetes mellitus, kidney disorders, and hypertension affects the performance of the transplanted heart. Allograft vasculopathy and malignancy following the transplantation, age and ethnic factors, infections, trauma, and cost effectiveness also offer hurdles [3]. Even though there is an increase of

20% patients in the waiting list, the relatively less number of transplants being performed in the last decade has been accounted by these obstacles [4].

The multitude of such barriers for heart transplantation, whilst the increasing demand has led the biomedical researchers over the globe to seek alternative strategies. Regenerative therapy/Tissue engineering (TE) has been emerged as a relatively novel approach wherein functional cells, supporting scaffolds or physiological signals are administered simultaneously in a controlled manner to the site of injury to accelerate repair and regeneration [5]. The TE strategies aim to overcome the challenges of limited organ donors and serious complications of lifelong immune suppression and the graft tissue rejection. Cardiac tissue engineering (CTE) specifically aims in the in vitro manipulation of high fidelity tissue mimics containing cardiac cells which are either injected or implanted into to infarcted heart to improve the repair to regenerate a functional myocardium [6]. The major focus of this chapter is to throw light to the recent advancements in the CTE and to reveal the scope and concepts of next-generation CTE.

7.2 Cardiac ECM

Understanding regarding the various physiological and anatomical aspects of cardiac tissue is essential for the development of tissue engineered cardiac constructs. In this discussion, the cardiac tissue has been demonstrated as two parts cardiac ECM and cell phenotypes giving importance to their role in normal vs infarcted heart. The cardiac ECM (cECM) is a complex meshwork of fibers comprising diverse matrix proteins in which cardiac myocytes, fibroblasts, leukocytes, and cardiac vascular cells are homed. Apart from its structural role in cardiac tissue, cECM exhibits nonstructural properties by entangling versatile proteins including growth factors, signaling mediators and several binding (cell—cell and cell—cECM) cues. In general, the cECM composed of structural proteins consisting of fibrillar molecules (e.g. collagen types I and III), basement membrane (pericellular) proteins that interface between the interstitial tissue and cardiomyocytes, and nonstructural (interstitial) proteins such as proteoglycans, glycosaminoglycans, glycoproteins and nonproteoglycans [7].

The beating function of cardiac tissue defines the critical role of cECM in maintaining the myocardial homeostasis which creates the balance in dynamic remodeling of the cardiac niche. Additionally, the cECM facilitates the interaction, behavior and response of the cardiac cells in the dynamic micro-niche. Consequently, the sustained cell—cECM interactions reflect the biophysical and biochemical associations driving the cell fate and performance which in turn determine the myocardial function [7]. Interestingly, the cECM and cardiac cells are arranged in a heterogeneous fashion throughout the heart. Around 70% of myocardial tissue is constituted by cardiac fibroblasts; however, approximately 70% of myocardial tissue volume is occupied by cardiomyocytes (CM), the beating elements of the heart. The endothelial cells decorate the remaining part of cardiac tissue. The biomechanical and mechano-transduction signaling are responsible for the elasticity and mechanical strength favouring the beating function of the heart. To cite, the collagen—integrin—cytoskeleton—myofibril alliance between the cECM and CM assist theproper transmission of mechanosensory cues which modulate the gene expression and signaling for maintaining the cECM homeostasis [8]. Growth factor components of cECM especially the vascular endothelial growth factor (VEGF), fibroblast growth factor (FGF), and transforming growth factor-β (TGF-β) are heterogeneously dispensed throughout cECM and are sequestered from fibrillar ECM components to act as a repository for to trigger cECM remodeling/repair following an injury. FGF and VEGF are masked by binding to the cECM proteoglycan heparan sulfate [7]. Also, the binding of VEGF with the fibronectin type III domains of tenascin-C results in the sequestration [9]. The degradation of cECM due to the upregulation of various MMPs following the injury results in the release of these growth factors which is pivotal for tissue repair/remodeling. However, the hyperactivity of the MMPs results in excessive ECM degradation and hurdles the tissue repair. Hence, the MMP expression is tightly regulated and the MMP activities are controlled by tissue inhibitors of metalloproteinases (TIMPs) [7,10].

7.3 Post-MI scarring

Myocardial infarction (MI) is the annihilation of cardiac myocytes due to prolonged hypoxic episodes following the obstruction/s in coronary artery. The wound healing phase at the infarct zone following MI is progressively leads to the fibrous tissue or collagenous scar. Also, the scarring zone is functionally inert and fails to perform the beating function effectively as a healthy heart. Moreover, the size and composition of the scar are crucial in deciding the fate of patients subsisting after the first infarction. Scar formation is one of the highly coordinated phases of wound healing process initiated in a post infarct heart and the other phases being inflammation, resolution of inflammation, fibroblast proliferation and neovascularization. In general, the inflammatory events mark the onset of reparative responses in a post infarct heart which is followed by the reparative signaling. Any imbalances between these phases result in the expansion of infarct zone and sustain the tissue damage, ultimately leading to contractile dysfunction. The process of inflammation begins with the infiltration of neutrophils to the injury site triggered by the activation of damage associated molecular patterns (DAMPs) released from the injured cells.

The primary structural element of cECM is constituted by fibrillar collagen which is organized into distinct networks including the endomysium (the fibers surrounding individual CM), the perimysium (constitutes the matrix for major bundles) and the epimysium (covers the entire cardiac muscle) [11]. In mammalian system, 85–90% of the cECM is constituted by type I collagen located mainly at epimysium and perimysium and 5–11% by type III collagen which is concentrated on the endomysium [12,13]. Even though, the collagenous cECM is involved in signal transduction, the molecular interactions between the matrix components and cellular elements has not been established. The early inflammatory phase demonstrates the increased level of bioactive cECM degradation fragments called matrikines which acts as DAMPs for immune system activation. In addition, a provisional matrix is orchestrated by the deposition of extravasated plasma proteins that facilitates the infiltration of inflammatory cells [14]. The phagocytic clearance of tissue debris and the upregulation of antiinflammatory mediators marks the transition of inflammatory phase to the proliferative phase of cardiac repair. The enrichment of cECM by the deposition of matricellular proteins (ECM proteins without a primary structural role) facilitates the phenotype switch of cells, activates the MMPs and growth factors, and transduces the signaling pathways [14,15]. The activation of myofibroblasts marks the crucial event of proliferative phase of cardiac repair characterizing the increased deposition of structural components of cECM. The crosslinking of cECM components, maturation of myoblasts and the vascular maturation results in collagenous scar tissue; which represent the maturation phase [14].

The ischemic insult associated with MI triggers the rapid activation of MMPs, both collagenases (MMP1, and gelatinases (MMP2 and MMP9)), and the subsequent generation of matrikines/matrix fragments [16,17]. It has been reported that, the MMP activation peaks within ten minutes of the coronary event; possibly due to the ischemia-driven ROS generation and oxidative stress [18,19]. The pool of MMPs at the infarct zone are being contributed from CM, fibroblasts, endothelial cells and various immune cells and the necrotic CM accentuates the degradation of cECM [20]. Apart from the matrix degradation functions, MMPs are involved in the proteolytic processing of cytokines and growth factors and possess intra-CM targets including myosin, α-actinin, and titin [21−24]. Although the MMP activation has been associated with MI, the bioactive proinflammatory matrikines has not been well-established. However, the fragments of type I and IV collagens and the fragments of noncollagenous have been detected in the infarct zone following the MI [25,26]. Also, the low-molecular weight hyaluronan fragments, endostatin (fragment of collagen XVIII), and tumstatin (MMP9-mediated cleavage fragment of collagen IV) were reported to exert potent proinflammatory effects at the infarct zone [27−29].

MMP mediated degradation of cECM triggers the formation of a provisional matrix driven by the plasma components, especially by establishing the interactions between the fibrinogen and fibronectin accompanied by the surge of VEGF to increase the vascular permeability at the infarct zone [30,31]. The provisional matrix acts as a template for the recruitment and assembly of repair machinery and is replaced by fibrotic scar tissue, triggers the trans-differentiation of fibroblasts to myofibroblasts, and functions as a reservoir for the growth factors and cytokines as the heparin-binding domain of fibrin exhibit higher affinity towards a diverse family of growth factors including FGF, VEGF, TGF and PDGF [32,33]. Fibrinolysis of the provisional matrix and the phagocytic clearance of the fragments by $CCR2^+$ macrophages precede the deposition of cECM components such as collagen, hyaluronan and versican [14,34]. Activation of the TGF-β1 signaling in the provisional matrix results in the recruitment and activation of resident fibroblasts and subsequent trans-differentiation to myofibroblast phenotype characterized with the expression of α-smooth muscle actin (α-SMA) and increased collagen secretion [35]. Eventually, the myoblasts undergo apoptosis and the deposited collagen at the infarct zone undergo extensive cross linking leading to the maturation of fibrotic scar tissue [14]. The scar tissue affects the contractile function of the heart leading to complications.

7.4 Cardiac tissue engineering

Tissue engineering and regenerative medicine emerged as an alternative approach to restore the myocardial function following MI. TE is multidisciplinary by combining the elements of engineering, biology, medicine, biochemistry, and pharmacology aiming to create suitable tissue substitutes. The concept of engineered cardiac patch is to mimic the native cECM, provide mechanical support and ensure the delivery of cells and mediator to the infarct zone. The major focus of CTE is to limit LV remodeling, prevent dilatation and thinning of the infarct zone, enhance mechanical properties of ventricle, and prevent the apoptosis in CM. In addition, it aids to retain viable transplanted stem cells, which stimulate the formation of vasculature, trans-differentiation of myofibroblasts, and differentiation of CM [36]. The biomaterial scaffolds are the inevitable for CTE and the ideal characteristics of the scaffolds applicable for CTE has been well-defined [37]. The general principle of CTE is displayed in Fig. 7.1. This section throws light to the some of the pivotal features need to be considered for the better performance CTE of scaffolds.

Mechanical properties of the CTE scaffolds are to be addressed carefully while designing any cardiac tissue engineering scaffold. Mechano-compatibility of the scaffolds drive for effective cardiac tissue regeneration. And, the biomaterial scaffolds need to be designed with appropriate strength to support attachment, migration and beating of CM. Moreover, optimum stiffness of the scaffolds facilitates the proper alignment of the seeded cells to form the cardiac tissue architecture and to generate and conduct electrical forces. Precisely, a biomaterial scaffold should match cardiac ECM in its molecular composition and mechanical properties [38,39]. Moreover, the mechano-compatibility of cardiac scaffolds are crucial as the beating function of the surviving heart should not interfere the mechanical integrity of the biomaterial. Human myocardium exhibits the stiffness of

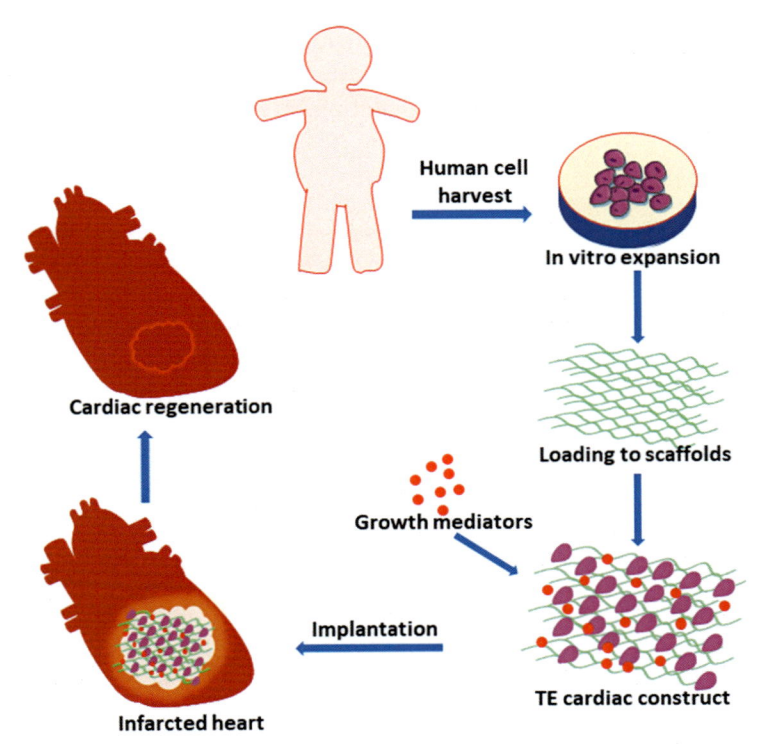

FIGURE 7.1 The general principle of CTE.

20–500 kPa (end of diastole to systole). Therefore, the CTE scaffolds require the stiffness greater than the native myocardium [40].

Another basic attribute of any CTE scaffold is the properly oriented interconnected network of pores. The size, shape, interconnectivity and density are the different variables of pores that needs to be considered while designing the CTE scaffolds. These factors drive the transport of nutrients and oxygen into the interstices of the scaffold, traffic the removal of metabolic exhausts, facilitate neovascularization, and promote cell migration, attachment and orientation. A pore size threshold is required for effective the better performance of the scaffolds. The smaller porosity hurdles the cell migration and attachment whereas larger pore size triggers the rapid hydrolytic degradation of the scaffolds as the biomaterial absorb more water. Generally, the CTE scaffolds with desired pore architecture can be created with the selection of with appropriate fabrication techniques such as solvent casting, particle leaching, freeze drying, and gas foaming [41,42].

Biocompatibility is the most desired character of any biomaterial intended for TE applications. Biocompatibility is defined as the ability of a biomaterial to interact and perform without evoking any adverse host responses. The scaffold as such or the degradation products should be beneficial or neutral in the host system. Biocompatibility encompasses various aspects including hemo-compatibility, cyto-compatibility, and immuno-compatibility. In addition, the biomaterial is expected to function in the host body until the processes of repair and regeneration is completed [43]. The evolving concept of immunocompatibility is based on biomaterials-induced sterile inflammation (BISI) triggered by DAMP release at the interface between the host tissue and biomaterial surface, subsequently leading to the activation of inflammasomes [44]. Hence, the ideal CTE scaffold elicits no considerable DAMP release and inflammasome activation at the interface while retaining the capability of cell survival and function. The scaffold should resist the blood clotting, prevents infections and supports M2 macrophages over M1 phenotypes [40].

Biodegradation refers to the breakdown of scaffolds due to the cleavage of specific bonds there by releasing unique fragments. Biodegradation occurs via various mechanisms including bioerosion (degradation by hydrolytic mechanisms and results in both surface and bulk erosion), bioresorption (degradation by cellular activity) and through enzymatic activity. Importantly, the degradation products should be nontoxic and nonimmunogenic. An ideal scaffold material for CTE should be fully degraded within 6–8 weeks as the pathological remodeling following MI requires approximately 6 weeks to complete. The general principle is that the degradation of cardiac scaffolds should be in accordance with the extent of myocardial regeneration which needs to be immediately replaced by functional tissues [40]. The major advantages of biodegradable cardiac scaffolds are (1) they do not have to be removed after use by secondary surgery because degradation products formed can be excreted from the body via general circulation and (2) progressive loss of degradable implant material accelerate the regeneration of heart tissue [37].

Clinically relevant thickness for the tissue engineered cardiac construct has been reported to be up to 10 mm for full thickness cardiac graft; however, the cardiac patch implants can be thinner [45]. In general, the oxygen diffusion in a metabolically active tissue (10^8 cells/cm^3 cell density) has been found to be difficult beyond 200 μm warranting the requirement of channeling system while considering thicker scaffolds [6]. In addition, the time consideration for the implantation raises concerns as the rapid cell death and infarct zone expansion aggravates the pathology immediately after the coronary occlusion. Hence, the CTE based strategies are recommended shortly after the cardiac events to prevent the cardiac remodeling and subsequent scar formation. In addition, the quality and performance of the biomaterials employed for the design of CTE scaffolds determine the success of cardiac regeneration. A bird's eye view of various CTE-biomaterials has been discussed in the following section.

7.5 Biomaterials for CTE

Since the delivery of cells and mediators to the infarct zone form the major approach for CTE, the biomaterials for cardiac regeneration have been featured into in vitro preengineered cardiac constructs and injectable biomaterials. The performance of cardiac constructs depends on the stability and response of the biomaterial scaffolds employed which depends on the chemistry and the availability of free functional groups to promote the tissue-biomaterial interaction. The field of CTE indepthly owe to versatile biomaterial types derived diverse sources. This section overviews various classes of biomaterials employed for CTE.

Natural biomaterials including proteins, and polysaccharides derived either directly or indirectly from plants or animals and possess a multitude of desirable features ideal for CTE applications. Collagen, chitosan, hyaluronic acid, alginate, matrigel, gelatin, and fibrin are the common natural polymers used in the fabrication of CTE scaffolds. Inspired from cECM, the natural polymeric biomaterials especially collagen subtypes, fibrin and gelatin are benefitted by their native cellular interactions [46]. Biocompatibility, biodegradability, intrinsic cellular interactions, controlled degradability, low toxicity by-products and non-immunogenicity are the beneficial features which escalates the appeal for natural polymers as scaffold materials. The excellent water absorption and swelling property inherent with natural polymers help in nutrient and waste trafficking across the scaffold material which further improves the adhesion, cell survival, migration and differentiation. However, rapid degradation, long gelation times, poor mechanical properties, insufficient electrical conductivity are undesirable traits of natural biomaterials.

Synthetic biomaterials are superior to the natural polymers owing to the mechanical characters, minimal batch variations and low risk of infections. They are tunable beginning from the synthesis stage with respect to their shape, architecture and chemistry to generate templates that matches specific requirements. In addition, the degradation profile, porosity and mechanical properties can easily be tailored for specific TE applications. Poly(ethyleneglycol) (PEG), polyvinyl alcohol (PVA), poly(caprolactone) (PCL), polypropylene, polyester, and poly(N-isopropylacrylamide) (PNIPAM) have been extensively utilized for CTE [47]. Contrastingly, the low biodegradability and the toxicity/immunogenity of degradation products offers challenge for synthetic polymers. The increased resistance of synthetic polymers to degradation owe to the strength of their strong carbon bonding [48]. And, the factors governing their decomposition include hydrophobicity/hydrophilicity, molecular mass, crystallinity and the physical conditions of the environment such as pH, enzyme concentration, water content and ionic strength [49]. The toxic products released into the body from the degrading biomaterial may induce secondary complications and triggers immune responses [50]. In addition, the molecular cues essential for attachment and biological functions of the cell generally lack in synthetic polymers [51].

The hybrid biomaterials address the demerits associated with the individual applications of natural and synthetic biomaterials and improve mechanical properties, biocompatibility and regenerative function. The hybrid materials are easy to fabricate and process and exhibits minimal batch-to-batch variation. However, the selection of fabrication techniques is crucial as the chemical treatments may alter the native conformation of the natural biomaterials [52]. In other words, the biological performance of natural biomaterials can be improved by reinforcement with biocompatible synthetic polymers. Hybrid biomaterials have been hailed to possess exceptional biocompatibility and appreciable mechanical strength for cardiac applications. Hydrophilicity, cell attachment, and biodegradability properties of the hybrid polymers are superior when compared to the individual natural and synthetic polymeric components used for their fabrication (D.P. [53]).

7.6 Decellularized ECM

Decellularization refers to the removal of cell component of the tissue by solely retaining ECM and the decellularized ECM functions as scaffold for CTE [54]. Allogenic or xenogenic heart are decellularized using physical, chemical or enzymatic treatments which are later seeded with cardiomyocytes or stem cells to fabricate regeneration patch [55,56]. The decellularized scaffold is beneficial as it retains the three-dimensional structure of the parent tissue and provides a precise microenvironment required for the seeded cells to thrive, differentiate and function [32]. Since the composition of the decellularized scaffold simulates the native tissue, the regenerative cues are inherently available in the scaffold which considerably minimize the adverse immune responses in the host body following implantation. However, there exist certain challenges for the effective translation of this strategy [57]. A major hurdle is the partial or complete loss of composition, structure and integrity of ECM during the process of decellularization which may result in aberrant immunological responses in the host tissue [58]. The method for decellularization is determined by several factors such as cellularity of the tissue, thickness, and density and generally rely on the lysis of the cells in the tissue leaving behind while retaining the ECM [55]. However, the cell disruption techniques may alter the chemical and microarchitecture of the ECM. Thus, the techniques need to be refined in order to maximize cell removal with minimum damage to the physico-chemical parameters of the scaffold [59]. Also, the lack of effective technique to ensure the uniform reseeding of cells on the scaffold is challenging as the uneven distribution of cells in the scaffold severely disrupt the force generation and electrical synchronization within the surviving tissue causing graft failure [60]. Standard techniques for decellularization and adequate cell distribution are warranting for the fabrication of scaffolds with desirable attributes and successful translation.

7.7 Tissue—biomaterial interaction

Blood is the first tissue to interact with the implant which happens simultaneously with implant placement which is characterized by the spontaneous interaction with the water and the adhesion of plasma components including ions, minerals, proteins sugars and lipids towards the surface of the biomaterial [61]. Generally, the protein components are the major macromolecule adhered to the biomaterials surface determining the fate of the implant. Among the plasma proteins, fibrinogen and albumin are the major players determining the biocompatibility of the implant [62]. However, the process of adsorption strictly depends on the surface characteristics of the implant such as roughness, topography, surface functional groups and hydrophilicity. In fact, surface hydrophobicity or hydrophilicity, which is a function of functional groups, is the primary determinant of albumin or fibrinogen binding [63]. Although albumin binding favors both hydrophilic and hydrophobic surfaces, studies have shown that albumin exhibits adsorption strength and favorable molecular conformations with hydrophilic surface than a hydrophobic exterior. Fibrinogen, on the other hand, prefers a hydrophobic surface whereas repelled by hydrophilic surface [64].

In general, the protein molecules in a solution tend to assemble spontaneously on biomaterial interfaces depending on their molecular weight and abundance. Smaller molecular weight proteins having appreciable abundance tend to adsorb faster compared to larger molecules [65]. Yet, these initially adsorbed proteins are replaced by other higher affinity proteins which are recruited to the interface owing to their lower mobility. This process of protein competition and displacement is referred as Vroman effect [66]. Albumin, owing to its smaller molecular weight and abundance, tend to be easily adsorbed to any surface of contact than any other plasma proteins. However, at hydrophilic interface the albumin exhibits stronger interactions and the lower mobility fibrinogen often fails to displace albumin. Moreover, the fibrinogen is repelled from the surface as they prefer hydrophobic interactions. Hence, the albumin layer formed on the biomaterial interface prevents the immunological responses against biomaterial by inhibiting the platelet adhesion and subsequent clotting cascade and following inflammatory events [62].

Contrastingly, the biomaterial surfaces rich in hydrophobic moieties display weaker interactions with albumin. Hence, the fibrinogen easily desorbs albumin and displace by Vroman effect. The biochemical events associated with Vroman effect is depicted in Fig. 7.2. The adhered fibrinogen layer subsequently facilitates other adhesion proteins such as fibronectin, vitronectin, complement proteins, coagulation proteins, and platelets [67]. This consortium of proteins and platelets forms a provisional matrix around the implant material. Once provisional matrix is established, fibrinogen and complement proteins of the provisional matrix trigger the activation and subsequent aggregation of platelets [68]. Parallel to this, fibrinogen polymerizes to from fibrin under the influence of thrombin, leading to initial clot formation and the subsequent onset of acute and chronic phases of inflammatory responses [69]. This is marked by the recruitment of immunological cells mainly neutrophils and macrophages to the implant under the influence of provisional matrix proteins and other local proinflammatory cytokines, chemokines, damage associated molecular pattern (DAMPs) molecules, and growth factors released at the implantation site [50]. Integrin signals the attachment of macrophages and neutrophils on to the surface of the biomaterial and facilitates the biodegradation mediated through the activation of phagocytosis, hydrolysis, ROS, and degradative enzymes released from their cytoplasmic granules. In addition, the neutrophil extracellular traps (NETs) which is a sticky meshwork of proteins released from neutrophils interferes with the normal healing response and forms an encapsulation around the biomaterial preventing its integration with normal tissue.

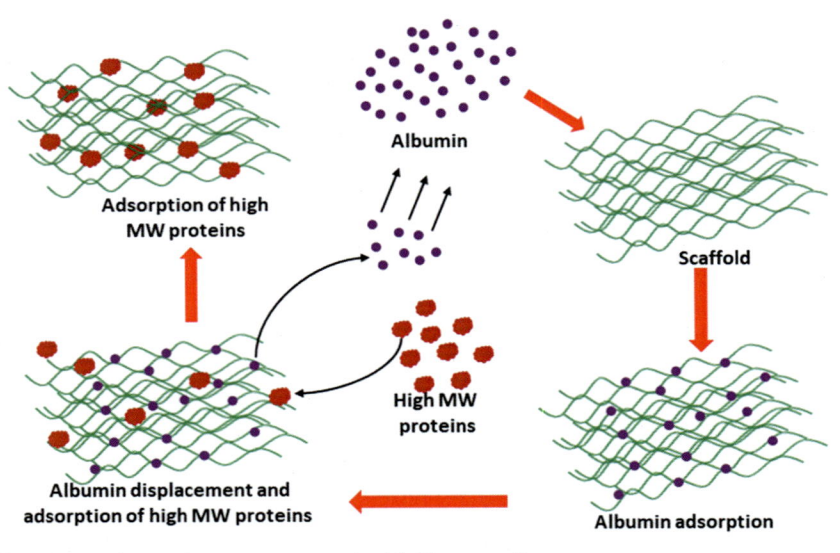

FIGURE 7.2   The biochemical events associated with Vroman effect.

The cells seeded on the biomaterial network facilitates the scaffolds restructuring through multiple biochemical mechanisms which is vital for the maintenance of healthy tissue and the repair of the damage. The cell surface adhesion molecules especially integrins find binding ligands in the biomaterial surface which marks the initial event of cell–biomaterial interaction [70]. However, the availability and accessibility of the cell binding cues largely depend on the chemistry of the biomaterials employed. For example, the natural biomaterials including collagen, fibrin and hyaluronic acid display ample distribution of inherent binding sites for cell adhesion. On the other hand, the synthetic polymers and nonprotein biomaterials lack the binding cues which improves the cell adhesion by incorporating the oligopeptides especially arginine–glycine–aspartic acid (RGD; [71]). The spacing and the density of the binding sites also influences the cytoskeletal architecture and cell morphology [72]. Interestingly, the cells seeded on the biomaterials devoid of binding cues also facilitate cell adhesion via the secretion of ECM to form a pericellular matrix for survival and function until the formation of a stable ECM [73].

The cell mobility through a TE scaffold relies largely on contractile/adhesive forces, availability of binding cues, lysis/controlled degradation of matrix component and biocompatibility of the biomaterial [74,75]. The cellular protrusions facilitating the movement are governed by the polymerization of the actin cytoskeleton as the cell attach to the ligands. The myosin II is responsible for the generation of the forces required for the cellular translocation through the scaffolds and the movement progresses by the disassembly of adhesions at trailing end [70]. Hence, the incorporation of chemotactic gradient across the scaffolds promotes cell migration in a specified direction [76]. Alternatively, durotaxis mechanism operates based on the rigidity of the scaffold surface to regulate the direction of cell migration. For example, the adherent cells prefer rigid surface than softer ones and the manipulation of scaffold rigidity facilitates the migration of cells to and from the hydrogel to/from the host tissue [77].

The cell contractions tend to remodel the scaffold surface influencing the cell migration and interactions. For example, the collagen-based hydrogels facilitate the migration of fibroblasts via the coordinated binding and release of α2β1 integrin with the motifs present in the collagen fibers [78]. In addition, the rate of scaffold degradation is influenced by the mechanical force exerted by the cells to the scaffolds as well as to the adjacent cells [7,79]. Contrastingly, the ECM formation and/mineralization increase the stiffness and affect the bulk properties of the scaffolds which offers beneficial as well as detrimental effects depending on the type of tissue [80]. In addition, the changes in the scaffold environment elicit cellular responses including phenotype switch, cytoskeletal reorganization, proliferation, migration and function [70]. The hydrophilic biomaterials retain bulk of water and improves the viscoelasticity which in turn influences the interactions and behavior of the seeded cells [70].

The formation of fibrous encapsulations hurdles the cell—biomaterial interactions and often creates a fluid filled space at the interface which affects the cellular performance [81]. However, the biomaterials with increased wettability facilitate the early stage cellular adhesions and possess greater biocompatibility to prevent the activation of macrophages and subsequent giant cell formation [82]. In general, the cell—material interaction is a complex process that requires the integration of many biochemical and mechanical signals to establish a proper communication between the cells and the scaffolds. The present knowledge regarding the cell—biomaterial interactions are largely based on the adhesion proteins and mechano-transduction [83]. Hence, a better understanding of cell behavior in native cardiac tissue is inevitable for developing scaffolds for CTE applications. Moreover, the strategies highlighting to improve the cell—biomaterial interactions are essential to be considered for the biofabrication of CTE scaffolds; however, warrants further research.

7.8 Functional modifications for CTE scaffolds

Biomaterials for CTE are tuneable and favors physical, chemical, mechanical and biological modifications. While physical modifications alter the topography of biomaterials chemical modifications changes the chemistry of the material by altering the surface functional groups or by introducing new functional groups. The goal of modifying the surface of a biomaterial is to create a specific chemical and physical environment that offers a favorable cellular response cardiac tissue regeneration [51]. In cases where tissue integration is desired the physical environment includes macro, micro, and/or nano scale features facilitating the cells to adhere, proliferate, and migrate [53]. However, the textured surfaces can be detrimental to the function of the device such as articulating surfaces or cardiovascular devices [84]. Diverse agents including growth factors, bioactive compounds, components of ECM and synthetic chemicals have been employed for decorating scaffold surface to manipulate the surface chemistry and to create a cardiac tissue simulated surface [51].

The bioactive molecules, such as chemicals in the form of ionic dissolution products, drugs, peptide sequences and/or growth factors (GFs) impregnated onto the CTE scaffolds were reported to improve the biological performance [85]. Bioactive molecules programmed to be discharged from the scaffold by controlled release, diffusion or network breakdown, can interact with cells and signal to activate cardiac regeneration. In addition, the cells produce additional GFs which in turn stimulates multiple generations of growing/proliferating cells to align/orient properly for better integration with host tissue [86]. The published literature demonstrates several molecular cues tailored onto versatile biomaterial surface which improved the cell adhesion, migration, proliferation and differentiation for CTE applications [87,37]. This section overviews the widely employed extra cellular cues to decorate the surface of cardiac constructs and their biological responses.

Among the various candidates, peptides and protein molecules isolated from ECM are more preferred for mosaicking surfaces of cardiac constructs as they act as physiochemical signals regulating intracellular pathways that direct cell differentiation [88]. Various studies have been undertaken with collagen coated biomaterial scaffolds for cardiac regeneration and displayed encouraging results. Two subtypes of collagen "(Collage I and Collagen III)" are essential for sustaining the integrity of cardiac ECM. Nearly 80% of collagen in an intact heart is collagen I; however, the proportion of collagen III increases drastically following an injury [89]. As these collagen subtypes play significant role in maintaining structural integrity of cardiac myocytes, cECM, wall stiffness and regulating force transmission, the CTE scaffolds impregnated with type I and III collagens improve the performance of the construct [90]. In addition, the excellent biocompatibility, biodegradability and weak antigenicity makes collagen a preferred candidate for CTE applications. Also, the biological functions of collagen in tissue repair by promoting angiogenesis, reducing fibrosis and attracting native cells are beneficial for successful CTE strategies [41].

Laminins are present in basal membrane and communicates with the cells through integrin signals and are composed of three polypeptide subunits namely α, β, and γ. In the embryonic phase, laminin is critical for proper development of heart. It has been established that laminin deficient embryos compromised the pericardial cell integrity and heart formation was severely impaired as the myotubes fail to expand properly to reach their sites of attachments [91]. Changes in laminin subtypes are observed in post infarct heart impairing multiple aspects of cardiac function and organization [92]. Laminins are highly cell-type specific similar to collagen, promote adhesion migration and differentiation of cells revealing the potential application in CTE. A recent study reported that cardiac cell specific markers were upregulated in human embryonic stem cells when grown on a culture coated with laminin [93]. The study suggests the importance of laminin in maintaining phenotype stability of cells. In addition, it has been demonstrated that laminin coated scaffolds have the potential to direct the proper alignment of cardiomyocytes in the CTE patch and contract as a single unit [94].

Fibronectins have also been investigated as bioactive leads for coating the biomaterial scaffolds for the development of cardiac patches. Being an ECM protein, the fibronectin has profound influence on physiological and morphological aspects of cell function. Fibronectin expression remains elevated during the early developmental phases and in a post infarct heart whereas in the normal heart the expression level remains low. The presence of fibronectin in cardiac stem cell niches and the fact that they are being secreted by stem cells suggesting their role in stem cell differentiation towards cardiac lineage [95]. A seminal study demonstrates that the human embryonic stem cells cultured in a matrix containing fibronectin and laminin in the ratio 70:30 promoted the differentiation of the stem cell to cardiomyocytes through integrin-mediated MEK/ERK signaling [96]. Another study revealed that the cardiac progenitor cells (CPC) cultured on matrix containing fibronectin had optimal growth factor secretion, proliferation, Cnx43 expression, and CPC alignment required for cardiac regeneration [97].

Nephronectin is an RGD bearing cECM protein expressed by CM and secreted mainly to cardiac jelly. The scaffolds incorporated with nephronectin displayed increased cell—cell interaction, cell-scaffold interactions, sarcomere maturation and cECM alignment and higher beating frequency. Similarly, the hydrogels decorated with gelatin, the denaturation product of collagen, facilitated to maintain the characteristic features of CM [98]. However, the potential of nephronectin in CTE is superior to fibronectin and gelatin [51]. The RGD immobilized scaffolds exhibited superior viability, differentiation and contractile function of CM suggesting its potential application in CTE [99]. In addition, the incorporation of growth factors including VEGF, G-CSF, IGF, EGF, and others in the scaffolds revealed encouraging outcomes for CTE [51]. However, the lower half-lives of the growth factors and cytokines demands continuous/sustained bioavailability for better performance.

7.9 Bottlenecks and future

The advancements in the knowledge regarding the underlying molecular mechanisms of CVDs especially the MI pathology drives the focus of medical research for the development of cardiac regenerative therapies and the concept of bioartificial heart aiming to alleviate the shortage of organ donors. Despite the developments achieved in regenerative cardiology and CTE, the attempts for the complete reversal of end-stage cardiac failure have not been successful so far and heart transplantation remains the ideal treatment for end stage heart failure. In addition, the present-day clinical medicine succeeded in prolonging the lives of MI patients; however, striving to cure the failing heart. The concept of bioartificial heart would be the target for next generation cardiac therapies which solely depends on the principles of CTE. The exploitation of live cells as raw material and creation of scaffolds competent for endogenous cardiac repair has considered to be the unique presentation of CTE; however, appropriate sizing, electrical integration, interdependency between the left and right sides of the organ, antithrombogenicity and durability offers challenge. The use of diverse stem cell population for the in vitro populating cardiac cells for TE and autologous CTE are promising; however, the need of billions of beating CM for replenishing the infarct zone is challenging. The goal of next generation CTE would be focusing to address such hurdles.

The simultaneous formation and integration of cECM with the host myocardium to form a functional unit warrants further attention regarding CTE. The retention of cardiac complexity under macro, micro and nano scale and the creation of vascular channeling system in the cardiac construct is inevitable for full thickness myocardial regeneration. More importantly, the entanglement of bioactive molecules to facilitate cell migration, proliferation, and differentiation and their controlled/sustained release on demand remain as unattained goal in CTE. In addition, the incorporation of multiple mediators such as growth factors and cytokines and their specific release pattern from a cell seeded CTE construct have not been attempted. Also, smart/intelligent scaffolds regulating the simultaneous release kinetics of diverse mediators depending on the specific need of loaded cells and host tissue is wanting in the field of CTE. The future of CTE research relies on addressing such potential research challenges.

The strategies to improve the perfusion of nutrients and metabolite trafficking in CTE have been emerged many of which have significantly increased the survival and performance of the seeded cells for cardiac regeneration. The upgradation and automation of bioreactor systems to ensure active perfusion in combination with the strategies to accelerate angiogenesis offer promising future for successful CTE. The growth, survival and maintenance of ideal proportion/density of cocultured cells in the engineered cardiac constructs are ideal to promote cardiac regeneration; however, are not achieved yet. The advent of 3D bioprinting technologies has raised the hope of CTE; but is currently hurdled with unsuitable bioinks, lack of multimaterial scaffolding, insufficient tissue-mimetics, improper cell density and impaired cECM formation. None-the-less, the integration of conventional scaffold fabrication techniques and advanced strategies such as micropatterning, and microfluidics along with 3D bioprinting would be ideal to CTE to ensure cardiac muscle maturity and global contractile function.

The greater heterogeneity in the cellular organization and orientation of CTE constructs often result in the increased chance of reentry arrhythmia which impairs the transmission of electrical waves from the native myocardium through the construct. This results in the blockage of electrical conduction of the heart and significantly affect the electrophysiological micro niche. The epicardial patches were evolved as a potent alternative to overcome the complications associated with the electrical integrity of the infarcted myocardium. However, the complete regeneration of myocardium remains unresolved. In addition, the advent of CRISPR/Cas9 technology has raised the scope of regenerative cardiology and the looking forward for the integration with CTE strategies.

A long journey ahead is waiting for the successful transition of CTE from conceptual to experimental evidence and then to therapeutic arena. Thousands of researchers across the globe invest their time, efforts and knowledge for upgrading CTE to a reality. The future of CTE relies on the development of in vitro beating CTE construct decorated with the promising features of native cardiac tissue which supports electrical integration upon implantation to infarct zone. The advancements in the understanding regarding the cardiac tissue organization and signaling at the molecular and atomic levels inspires the ex vivo/ in vitro simulation of native cardiac features. It is logical to believe that the next generation cardiology will be depending on CTE which gives pleasant hope for the millions of MI patients throughout the globe.

Thanks to Cardiac Tissue Engineering

References

[1] Yancy Clyde W, et al. 2017 ACC/AHA/HFSA focused update of the 2013 ACCF/AHA guideline for the management of heart failure: a report of the American College of Cardiology/American Heart Association Task Force on Clinical Practice Guidelines and the Heart Failure Society of America. Circulation 2017;136:e137−61. Available from: https://doi.org/10.1161/CIR.0000000000000509.

[2] Choudhury M. Post-cardiac transplant recipient: implications for anaesthesia. Indian J Anaesth 2017;61:768. Available from: https://doi.org/10.4103/ija.IJA_390_17.

[3] Szyguła-Jurkiewicz B, Szczurek W, Gąsior M, Zembala M. Risk factors of cardiac allograft vasculopathy. Kardiochirurgia Torakochirurgia Pol. Pol. J Cardio-Thorac Surg 2015;12:328—33. Available from: https://doi.org/10.5114/kitp.2015.56783.

[4] Tonsho M, Michel S, Ahmed Z, Alessandrini A, Madsen JC. Heart transplantation: challenges facing the field. Cold Spring Harb Perspect Med. 2014;4. Available from: https://doi.org/10.1101/cshperspect.a015636.

[5] De Witte T-M, Fratila-Apachitei LE, Zadpoor AA, Peppas NA. Bone tissue engineering via growth factor delivery: from scaffolds to complex matrices. Regen Biomater. 2018;5:197—211. Available from: https://doi.org/10.1093/rb/rby013.

[6] Radisic M. Biomaterials for cardiac tissue engineering. Biomed Mater 2015;10:030301. Available from: https://doi.org/10.1088/1748-6041/10/3/030301.

[7] Chang CW, Dalgliesh AJ, López JE, Griffiths LG. Cardiac extracellular matrix proteomics: challenges, techniques, and clinical implications. Proteom Clin Appl 2016;10:39—50. Available from: https://doi.org/10.1002/prca.201500030.

[8] Takawale A, Sakamuri SSVP, Kassiri Z. Extracellular matrix communication and turnover in cardiac physiology and pathology. Compr Physiol 2015;5:687—719. Available from: https://doi.org/10.1002/cphy.c140045.

[9] Barallobre-Barreiro J, Didangelos A, Yin X, Doménech N, Mayr M. A sequential extraction methodology for cardiac extracellular matrix prior to proteomics analysis. Methods Mol Biol Clifton NJ 2013;1005:215—23. Available from: https://doi.org/10.1007/978-1-62703-386-2_17.

[10] DeCoux A, Lindsey ML, Villarreal F, Garcia RA, Schulz R. Myocardial matrix metalloproteinase-2: inside out and upside down. J Mol Cell Cardiol 2014;77:64—72. Available from: https://doi.org/10.1016/j.yjmcc.2014.09.016.

[11] Leonard BL, Smaill BH, LeGrice IJ. Structural remodeling and mechanical function in heart failure. Microsc. Microanal. J Microsc Soc Am Microbeam Anal Soc Microsc Soc Can. 2012;18:50—67. Available from: https://doi.org/10.1017/S1431927611012438.

[12] Bashey RI, Martinez-Hernandez A, Jimenez SA. Isolation, characterization, and localization of cardiac collagen type VI. Associations with other extracellular matrix components. Circ Res 1992;70:1006—17. Available from: https://doi.org/10.1161/01.res.70.5.1006.

[13] Weber KT. Cardiac interstitium in health and disease: the fibrillar collagen network. J Am Coll Cardiol 1989;13:1637—52. Available from: https://doi.org/10.1016/0735-1097(89)90360-4.

[14] Frangogiannis NG. The extracellular matrix in myocardial injury, repair, and remodeling. J Clin Invest 2017;127:1600—12. Available from: https://doi.org/10.1172/JCI87491.

[15] Murphy-Ullrich JE, Sage EH. Revisiting the matricellular concept. Matrix Biol J Int Soc Matrix Biol 2014;37:1—14. Available from: https://doi.org/10.1016/j.matbio.2014.07.005.

[16] Danielsen CC, Wiggers H, Andersen HR. Increased amounts of collagenase and gelatinase in porcine myocardium following ischemia and reperfusion. J Mol Cell Cardiol 1998;30:1431—42. Available from: https://doi.org/10.1006/jmcc.1998.0711.

[17] Whittaker P, Boughner DR, Kloner RA. Role of collagen in acute myocardial infarct expansion. Circulation 1991;84:2123—34. Available from: https://doi.org/10.1161/01.cir.84.5.2123.

[18] Etoh T, Joffs C, Deschamps AM, Davis J, Dowdy K, Hendrick J, et al. Myocardial and interstitial matrix metalloproteinase activity after acute myocardial infarction in pigs. Am J Physiol Heart Circ Physiol 2001;281:H987—94. Available from: https://doi.org/10.1152/ajpheart.2001.281.3.H987.

[19] Wang Y, Yin P, Bian G-L, Huang H-Y, Shen H, Yang J-J, et al. The combination of stem cells and tissue engineering: an advanced strategy for blood vessels regeneration and vascular disease treatment. Stem Cell Res Ther. 2017;8. Available from: https://doi.org/10.1186/s13287-017-0642-y.

[20] Alfonso-Jaume MA, Bergman MR, Mahimkar R, Cheng S, Jin ZQ, Karliner JS, et al. Cardiac ischemia-reperfusion injury induces matrix metalloproteinase-2 expression through the AP-1 components FosB and JunB. Am J Physiol Heart Circ Physiol 2006;291:H1838—1846. Available from: https://doi.org/10.1152/ajpheart.00026.2006.

[21] Ali MAM, Cho WJ, Hudson B, Kassiri Z, Granzier H, Schulz R. Titin is a target of matrix metalloproteinase-2: implications in myocardial ischemia/reperfusion injury. Circulation 2010;122:2039—47. Available from: https://doi.org/10.1161/CIRCULATIONAHA.109.930222.

[22] Cauwe B, Opdenakker G. Intracellular substrate cleavage: a novel dimension in the biochemistry, biology and pathology of matrix metalloproteinases. Crit Rev Biochem Mol Biol 2010;45:351−423. Available from: https://doi.org/10.3109/10409238.2010.501783.

[23] Cox JH, Dean RA, Roberts CR, Overall CM. Matrix metalloproteinase processing of CXCL11/I-TAC results in loss of chemoattractant activity and altered glycosaminoglycan binding. J Biol Chem 2008;283:19389−99. Available from: https://doi.org/10.1074/jbc.M800266200.

[24] Denney H, Clench MR, Woodroofe MN. Cleavage of chemokines CCL2 and CXCL10 by matrix metalloproteinases-2 and -9: implications for chemotaxis. Biochem Biophys Res Commun 2009;382:341−7. Available from: https://doi.org/10.1016/j.bbrc.2009.02.164.

[25] Trial J, Baughn RE, Wygant JN, McIntyre BW, Birdsall HH, Youker KA, et al. Fibronectin fragments modulate monocyte VLA-5 expression and monocyte migration. J Clin Invest 1999;104:419−30. Available from: https://doi.org/10.1172/JCI4824.

[26] Villarreal F, Omens J, Dillmann W, Risteli J, Nguyen J, Covell J. Early degradation and serum appearance of type I collagen fragments after myocardial infarction. J Mol Cell Cardiol 2004;36:597−601. Available from: https://doi.org/10.1016/j.yjmcc.2004.01.004.

[27] Hamano Y, Zeisberg M, Sugimoto H, Lively JC, Maeshima Y, Yang C, et al. Physiological levels of tumstatin, a fragment of collagen IV alpha3 chain, are generated by MMP-9 proteolysis and suppress angiogenesis via alphaV beta3 integrin. Cancer Cell 2003;3:589−601. Available from: https://doi.org/10.1016/s1535-6108(03)00133-8.

[28] Okada M, Oba Y, Yamawaki H. Endostatin stimulates proliferation and migration of adult rat cardiac fibroblasts through PI3K/Akt pathway. Eur J Pharmacol 2015;750:20−6. Available from: https://doi.org/10.1016/j.ejphar.2015.01.019.

[29] Taylor KR, Trowbridge JM, Rudisill JA, Termeer CC, Simon JC, Gallo RL. Hyaluronan fragments stimulate endothelial recognition of injury through TLR4. J Biol Chem 2004;279:17079−84. Available from: https://doi.org/10.1074/jbc.M310859200.

[30] Andersson L, Scharin Täng M, Lundqvist A, Lindbom M, Mardani I, Fogelstrand P, et al. Rip2 modifies VEGF-induced signalling and vascular permeability in myocardial ischaemia. Cardiovasc Res 2015;107:478−86. Available from: https://doi.org/10.1093/cvr/cvv186.

[31] Dobaczewski M, Bujak M, Zymek P, Ren G, Entman ML, Frangogiannis NG. Extracellular matrix remodeling in canine and mouse myocardial infarcts. Cell Tissue Res 2006;324:475−88. Available from: https://doi.org/10.1007/s00441-005-0144-6.

[32] Chan BP, Leong KW. Scaffolding in tissue engineering: general approaches and tissue-specific considerations. Eur Spine J 2008;17:467−79. Available from: https://doi.org/10.1007/s00586-008-0745-3.

[33] Martino MM, Briquez PS, Ranga A, Lutolf MP, Hubbell JA. Heparin-binding domain of fibrin(ogen) binds growth factors and promotes tissue repair when incorporated within a synthetic matrix. Proc Natl Acad Sci USA 2013;110:4563−8. Available from: https://doi.org/10.1073/pnas.1221602110.

[34] Motley MP, Madsen DH, Jürgensen HJ, Spencer DE, Szabo R, Holmbeck K, et al. A CCR2 macrophage endocytic pathway mediates extravascular fibrin clearance in vivo. Blood 2016;127:1085−96. Available from: https://doi.org/10.1182/blood-2015-05-644260.

[35] Serini G, Bochaton-Piallat ML, Ropraz P, Geinoz A, Borsi L, Zardi L, et al. The fibronectin domain ED-A is crucial for myofibroblastic phenotype induction by transforming growth factor-beta1. J Cell Biol 1998;142:873−81. Available from: https://doi.org/10.1083/jcb.142.3.873.

[36] Domenech M, Polo-Corrales L, Ramirez-Vick JE, Freytes DO. Tissue engineering strategies for myocardial regeneration: acellular versus cellular scaffolds? Tissue Eng Part B Rev 2016;22:438−58. Available from: https://doi.org/10.1089/ten.teb.2015.0523.

[37] Finosh GT, Jayabalan M. Regenerative therapy and tissue engineering for the treatment of end-stage cardiac failure: new developments and challenges. Biomatter 2012;2:1−14. Available from: https://doi.org/10.4161/biom.19429.

[38] Thankam FG, Muthu J. Influence of physical and mechanical properties of amphiphilic biosynthetic hydrogels on long-term cell viability. J Mech Behav Biomed Mater 2014;35:111−22. Available from: https://doi.org/10.1016/j.jmbbm.2014.03.010.

[39] Vunjak Novakovic G, Eschenhagen T, Mummery C. Myocardial tissue engineering: in vitro models. Cold Spring Harb Perspect Med 2014;4:a014076. Available from: https://doi.org/10.1101/cshperspect.a014076.

[40] Reis LA, Chiu LLY, Feric N, Fu L, Radisic M. Biomaterials in myocardial tissue engineering. J Tissue Eng Regen Med 2016;10:11—28. Available from: https://doi.org/10.1002/term.1944.

[41] Geng X, Liu B, Liu J, Liu D, Lu Y, Sun X, et al. Interfacial tissue engineering of heart regenerative medicine based on soft cell-porous scaffolds. J Thorac Dis 2018;10:S2333—45. Available from: https://doi.org/10.21037/jtd.2018.01.117.

[42] Gnanaprakasam Thankam F, Muthu J. Alginate based hybrid copolymer hydrogels--influence of pore morphology on cell-material interaction. Carbohydr Polym 2014;112:235—44. Available from: https://doi.org/10.1016/j.carbpol.2014.05.083.

[43] Kaiser NJ, Coulombe KLK. Physiologically inspired cardiac scaffolds for tailored *in vivo* function and heart regeneration. Biomed Mater 2015;10:034003. Available from: https://doi.org/10.1088/1748-6041/10/3/034003.

[44] Williams DF. Biocompatibility pathways in tissue-engineering templates. Engineering 2018;4:286—90. Available from: https://doi.org/10.1016/j.eng.2018.03.007.

[45] Chiu LLY, Radisic M. Scaffolds with covalently immobilized VEGF and Angiopoietin-1 for vascularization of engineered tissues. Biomaterials 2010;31:226—41. Available from: https://doi.org/10.1016/j.biomaterials.2009.09.039.

[46] Rodrigues ICP, Kaasi A, Maciel Filho R, Jardini AL, Gabriel LP. Cardiac tissue engineering: current state-of-the-art materials, cells and tissue formation. Einstein Sao Paulo Braz 2018;16:eRB4538. Available from: https://doi.org/10.1590/S1679-45082018RB4538.

[47] Khan F, Tanaka M. Designing smart biomaterials for tissue engineering. Int J Mol Sci. 2017;19:17. Available from: https://doi.org/10.3390/ijms19010017.

[48] Diez-Pascual AM. Tissue engineering bionanocomposites based on poly(propylene fumarate). Polymers 2017;9:260. Available from: https://doi.org/10.3390/polym9070260.

[49] Kamaly N, Yameen B, Wu J, Farokhzad OC. Degradable controlled-release polymers and polymeric nanoparticles: mechanisms of controlling drug release. Chem Rev 2016;116:2602—63. Available from: https://doi.org/10.1021/acs.chemrev.5b00346.

[50] Mariani E, Lisignoli G, Borzì RM, Pulsatelli L. Biomaterials: foreign bodies or tuners for the immune response? Int J Mol Sci 2019;20. Available from: https://doi.org/10.3390/ijms20030636.

[51] Tallawi M, Rosellini E, Barbani N, Cascone MG, Rai R, Saint-Pierre G, et al. Strategies for the chemical and biological functionalization of scaffolds for cardiac tissue engineering: a review. J R Soc Interface 2015;12:20150254. Available from: https://doi.org/10.1098/rsif.2015.0254.

[52] Coenen AMJ, Bernaerts KV, Harings JAW, Jockenhoevel S, Ghazanfari S. Elastic materials for tissue engineering applications: natural, synthetic, and hybrid polymers. Acta Biomater 2018;79:60—82. Available from: https://doi.org/10.1016/j.actbio.2018.08.027.

[53] Bhattarai DP, Aguilar LE, Park CH, Kim CS. A review on properties of natural and synthetic based electrospun fibrous materials for bone tissue engineering. Membranes 2018;8:62. Available from: https://doi.org/10.3390/membranes8030062.

[54] Li Y, Zhang D. Artificial cardiac muscle with or without the use of scaffolds. Biomed Res Int. 2017;2017:1—15. Available from: https://doi.org/10.1155/2017/8473465.

[55] Crapo PM, Gilbert TW, Badylak SF. An overview of tissue and whole organ decellularization processes. Biomaterials 2011;32:3233—43. Available from: https://doi.org/10.1016/j.biomaterials.2011.01.057.

[56] Gilpin A, Yang Y. Decellularization strategies for regenerative medicine: from processing techniques to applications. Biomed Res Int 2017;2017:1—13. Available from: https://doi.org/10.1155/2017/9831534.

[57] Bejleri D, Davis ME. Decellularized extracellular matrix materials for cardiac repair and regeneration. Adv Healthc Mater 2019;81801217. Available from: https://doi.org/10.1002/adhm.201801217.

[58] Aamodt JM, Grainger DW. Extracellular matrix-based biomaterial scaffolds and the host response. Biomaterials 2016;86:68—82. Available from: https://doi.org/10.1016/j.biomaterials.2016.02.003.

[59] Huang S, Yang Y, Yang Q, Zhao Q, Ye X. Engineered circulatory scaffolds for building cardiac tissue. J Thorac Dis 2018;10:S2312—28. Available from: https://doi.org/10.21037/jtd.2017.12.92.

[60] Zheng MH, Chen J, Kirilak Y, Willers C, Xu J, Wood D. Porcine small intestine submucosa (SIS) is not an acellular collagenous matrix and contains porcine DNA: possible implications in human implantation. J Biomed Mater Res B Appl Biomater 2005;73:61—7. Available from: https://doi.org/10.1002/jbm.b.30170.

[61] Xu L-C, Bauer JW, Siedlecki CA. Proteins, platelets, and blood coagulation at biomaterial interfaces. Colloids Surf B Biointerfaces 2014;124:49—68. Available from: https://doi.org/10.1016/j.colsurfb.2014.09.040.

[62] Gnanaprakasam Thankam F, Muthu J. Influence of plasma protein—hydrogel interaction moderated by absorption of water on long-term cell viability in amphiphilic biosynthetic hydrogels. RSC Adv 2013;3:24509. Available from: https://doi.org/10.1039/c3ra43710h.

[63] Roach P, Farrar D, Perry CC. Interpretation of protein adsorption: surface-induced conformational changes. J Am Chem Soc 2005;127:8168—73. Available from: https://doi.org/10.1021/ja042898o.

[64] Recek N. Biocompatibility of plasma-treated polymeric implants. Materials 2019;12:240. Available from: https://doi.org/10.3390/ma12020240.

[65] Dee KC, Puleo DA, Bizios R. An introduction to tissue—biomaterial interactions. Hoboken, NJ: Wiley-Liss; 2002.

[66] Jung S-Y, Lim S-M, Albertorio F, Kim G, Gurau MC, Yang RD, et al. The Vroman effect: a molecular level description of fibrinogen displacement. J Am Chem Soc 2003;125:12782—6. Available from: https://doi.org/10.1021/ja037263o.

[67] de Mel A, Cousins BG, Seifalian AM. Surface modification of biomaterials: a quest for blood compatibility. Int J Biomater 2012;2012:1—8. Available from: https://doi.org/10.1155/2012/707863.

[68] Parker TJ, Broadbent JA, McGovern JA, Broszczak DA, Parker CN, Upton Z. Provisional matrix deposition in hemostasis and venous insufficiency: tissue preconditioning for nonhealing venous ulcers. Adv Wound Care 2015;4:174—91. Available from: https://doi.org/10.1089/wound.2013.0462.

[69] Shiu HT, Goss B, Lutton C, Crawford R, Xiao Y. Formation of blood clot on biomaterial implants influences bone healing. Tissue Eng Part B Rev 2014;20:697—712. Available from: https://doi.org/10.1089/ten.teb.2013.0709.

[70] Ahearne M. Introduction to cell-hydrogel mechanosensing. Interface Focus 2014;4:20130038. Available from: https://doi.org/10.1098/rsfs.2013.0038.

[71] Yang F, Williams CG, Wang D-A, Lee H, Manson PN, Elisseeff J. The effect of incorporating RGD adhesive peptide in polyethylene glycol diacrylate hydrogel on osteogenesis of bone marrow stromal cells. Biomaterials 2005;26:5991—8. Available from: https://doi.org/10.1016/j.biomaterials.2005.03.018.

[72] Cavalcanti-Adam EA, Volberg T, Micoulet A, Kessler H, Geiger B, Spatz JP. Cell spreading and focal adhesion dynamics are regulated by spacing of integrin ligands. Biophys J 2007;92:2964—74. Available from: https://doi.org/10.1529/biophysj.106.089730.

[73] Steward AJ, Wagner DR, Kelly DJ. The pericellular environment regulates cytoskeletal development and the differentiation of mesenchymal stem cells and determines their response to hydrostatic pressure. Eur Cell Mater. 2013;25:167—78. Available from: https://doi.org/10.22203/ecm.v025a12.

[74] Gobin AS, West JL. Cell migration through defined, synthetic ECM analogs. FASEB J Publ Fed Am Soc Exp Biol. 2002;16:751—3. Available from: https://doi.org/10.1096/fj.01-0759fje.

[75] Zaman MH, Trapani LM, Sieminski AL, Siemeski A, Mackellar D, Gong H, et al. Migration of tumor cells in 3D matrices is governed by matrix stiffness along with cell-matrix adhesion and proteolysis. Proc Natl Acad Sci USA 2006;103:10889—94. Available from: https://doi.org/10.1073/pnas.0604460103.

[76] Cheng S-Y, Heilman S, Wasserman M, Archer S, Shuler ML, Wu M. A hydrogel-based microfluidic device for the studies of directed cell migration. Lab Chip 2007;7:763—9. Available from: https://doi.org/10.1039/b618463d.

[77] Vincent LG, Choi YS, Alonso-Latorre B, del Álamo JC, Engler AJ. Mesenchymal stem cell durotaxis depends on substrate stiffness gradient strength. Biotechnol J 2013;8:472—84. Available from: https://doi.org/10.1002/biot.201200205.

[78] Meshel AS, Wei Q, Adelstein RS, Sheetz MP. Basic mechanism of three-dimensional collagen fibre transport by fibroblasts. Nat Cell Biol 2005;7:157—64. Available from: https://doi.org/10.1038/ncb1216.

[79] Ellsmere JC, Khanna RA, Lee JM. Mechanical loading of bovine pericardium accelerates enzymatic degradation. Biomaterials 1999;20:1143—50. Available from: https://doi.org/10.1016/s0142-9612(99)00013-7.

[80] Hu JC, Athanasiou KA. Low-density cultures of bovine chondrocytes: effects of scaffold material and culture system. Biomaterials 2005;26:2001—12. Available from: https://doi.org/10.1016/j.biomaterials.2004.06.038.

[81] Biggs MJP, Richards RG, Dalby MJ. Nanotopographical modification: a regulator of cellular function through focal adhesions. Nanomed Nanotechnol Biol Med 2010;6:619—33. Available from: https://doi.org/10.1016/j.nano.2010.01.009.

[82] Cuvelier D, Théry M, Chu Y-S, Dufour S, Thiéry J-P, Bornens M, et al. The universal dynamics of cell spreading. Curr Biol CB 2007;17:694—9. Available from: https://doi.org/10.1016/j.cub.2007.02.058.

[83] Sanz-Herrera JA, Reina-Romo E. Cell-biomaterial mechanical interaction in the framework of tissue engineering: insights, computational modeling and perspectives. Int J Mol Sci. 2011;12:8217–44. Available from: https://doi.org/10.3390/ijms12118217.

[84] Parisi L, Toffoli A, Ghiacci G, Macaluso G. Tailoring the interface of biomaterials to design effective scaffolds. J Funct Biomater. 2018;9:50. Available from: https://doi.org/10.3390/jfb9030050.

[85] Kesireddy V, Kasper FK. Approaches for building bioactive elements into synthetic scaffolds for bone tissue engineering. J Mater Chem B Mater Biol Med. 2016;4:6773–86. Available from: https://doi.org/10.1039/C6TB00783J.

[86] Barthes J, Özçelik H, Hindié M, Ndreu-Halili A, Hasan A, Vrana NE. Cell microenvironment engineering and monitoring for tissue engineering and regenerative medicine: the recent advances. Biomed Res Int. 2014;2014. Available from: https://doi.org/10.1155/2014/921905.

[87] Ravichandran R, Venugopal JR, Sundarrajan S, Mukherjee S, Sridhar R, Ramakrishna S. Minimally invasive injectable short nanofibers of poly(glycerol sebacate) for cardiac tissue engineering. Nanotechnology 2012;23:385102. Available from: https://doi.org/10.1088/0957-4484/23/38/385102.

[88] Akhyari P, Kamiya H, Haverich A, Karck M, Lichtenberg A. Myocardial tissue engineering: the extracellular matrix☆. Eur J Cardiothorac Surg 2008;34:229–41. Available from: https://doi.org/10.1016/j.ejcts.2008.03.062.

[89] Pauschinger M, Doerner A, Remppis A, Tannhäuser R, Kühl U, Schultheiss H-P. Differential myocardial abundance of collagen type I and type III mRNA in dilated cardiomyopathy: effects of myocardial inflammation. Cardiovasc Res. 1998;37:123–9. Available from: https://doi.org/10.1016/S0008-6363(97)00217-4.

[90] Horn MA, Trafford AW. Aging and the cardiac collagen matrix: novel mediators of fibrotic remodelling. J Mol Cell Cardiol 2016;93:175–85. Available from: https://doi.org/10.1016/j.yjmcc.2015.11.005.

[91] Yarnitzky T, Volk T. Laminin is required for heart, somatic muscles, and gut development in the Drosophila embryo. Dev Biol 1995;169:609–18. Available from: https://doi.org/10.1006/dbio.1995.1173.

[92] Wagner JUG, Chavakis E, Rogg E-M, Muhly-Reinholz M, Glaser SF, Günther S, et al. Switch in laminin β2 to laminin β1 isoforms during aging controls endothelial cell functions—brief report. Arterioscler Thromb Vasc Biol 2018;38:1170–7. Available from: https://doi.org/10.1161/ATVBAHA.117.310685.

[93] Yap L, Wang J-W, Moreno-Moral A, Chong LY, Sun Y, Harmston N, et al. In vivo generation of post-infarct human cardiac muscle by laminin-promoted cardiovascular progenitors Cell Rep 2019;26:3231–45e9. Available from: https://doi.org/10.1016/j.celrep.2019.02.083.

[94] McDevitt TC, Woodhouse KA, Hauschka SD, Murry CE, Stayton PS. Spatially organized layers of cardiomyocytes on biodegradable polyurethane films for myocardial repair. J Biomed Mater Res 2003;66A:586–95. Available from: https://doi.org/10.1002/jbm.a.10504.

[95] Konstandin MH, et al. Fibronectin is essential for reparative cardiac progenitor cell response after myocardial infarction. Circ Res 2013;113:115–25. Available from: https://doi.org/10.1161/CIRCRESAHA.113.301152.

[96] Sa S, Wong L, McCloskey KE. Combinatorial fibronectin and laminin signaling promote highly efficient cardiac differentiation of human embryonic stem cells. Biores Open Access 2014;3:150–61. Available from: https://doi.org/10.1089/biores.2014.0018.

[97] French KM, Maxwell JT, Bhutani S, Ghosh-Choudhary S, Fierro MJ, Johnson TD, et al. Fibronectin and cyclic strain improve cardiac progenitor cell regenerative potential *in vitro*. Stem Cell Int. 2016;2016:1–11. Available from: https://doi.org/10.1155/2016/8364382.

[98] Choi S, Hong Y, Lee I, Huh D, Jeon T-J, Kim SM. Effects of various extracellular matrix proteins on the growth of HL-1 cardiomyocytes. Cell Tissues Organs 2013;198:349–56. Available from: https://doi.org/10.1159/000358755.

[99] Yu J, Du KT, Fang Q, Gu Y, Mihardja SS, Sievers RE, et al. The use of human mesenchymal stem cells encapsulated in RGD modified alginate microspheres in the repair of myocardial infarction in the rat. Biomaterials 2010;31:7012–20. Available from: https://doi.org/10.1016/j.biomaterials.2010.05.078.

Vascular regeneration and tissue engineering: Progress, clinical impact, and future challenges

Santanu Hati[1], Swati Agrawal[2] and Vikrant Rai[1]

[1]Department of Biomedical Science, Creighton University School of Medicine, Omaha, NE, United States [2]Department of Surgery, Creighton University School of Medicine, Omaha, NE, United States

8.1 Introduction

Cardiovascular disease, the principal cause of death in the western world, caused by occlusive arterial diseases, results in myocardial infarction, stroke, and lower limb ischemia. All these manifestations have a common denominator, atherosclerosis, a chronic inflammatory disease initiated by infiltration of low-density lipoprotein cholesterol (LDL) into the intimal layer of the artery. Lipid deposition in the arterial wall leads to the formation of fatty streak which with progression leads to the formation of atheroma, fibroatheroma, and plaque. The process of plaque formation is mediated by LDL deposition, endothelial dysfunction, recruitment of inflammatory cells (leukocytes), smooth muscle cell (SMC) proliferation, intimal inflammation, foam cell formation, apoptosis and necrosis, matrix synthesis, calcification, angiogenesis, and arterial remodeling. Inflammatory cells (macrophages, dendritic cells), proinflammatory cytokines (interleukin (IL)-6, IL-1β, tumor necrosis factor (TNF)-α), damage-associated molecular proteins (DAMPs) including high mobility group box protein (HMGB)-1, receptor for advanced glycation end products (RAGE), S100 proteins, cellular membrane receptors such as triggering receptor expressed on membrane (TREMs) and toll-like receptors (TLRs), vitamin D deficiency, and modulation of transcription factors play a role in the pathogenesis of atherosclerosis. Smoking, hypertension, hyperlipidemia, diabetes mellitus, male sex, and family history are the common risk factors of atherosclerosis [1−10].

Regenerated Organs
DOI: https://doi.org/10.1016/B978-0-12-821085-7.00008-7

Many therapeutic approaches to attenuate thrombus formation and atherosclerosis have been suggested in the literature, however, they remain elusive and no definitive treatment is known yet [1,3,4,9]. For instance, TREM-1 plays an important role in atherosclerosis and TREM-1 inhibitors might play a therapeutic role as evidenced by its use in animal models in other pathological conditions such as in alcoholic liver disease, the use of TREM-1 inhibitory peptide (GF9) was shown to suppress inflammation. GF9 also exhibits therapeutic effects in animal models of sepsis and rheumatoid arthritis. But the stability, half-life and the delivery of the peptide to the particular target was not suitable and further modification of GF9 peptide sequence resulting in the formation of 31 amino acid-long peptides GA31 and GE31 were shown to inhibit TREM-1 in vivo [11]. Similarly, the role of pravastatin, a derivative of compactin, was demonstrated in reducing atherosclerotic plaques and improving the plaque formation by reducing lipid deposits and alleviating plaque inflammatory responses by decreasing the level of TREM-1 [12]. Further, the probable use of nangibotide (LR12), a TREM-1 antagonist derived from the residues 94−105 of TREM-like transcript-1 and the use of TREM-1 sequence corresponding conserved domain (LQVTDSGLYRCVIYHPP) LP17 to suppress TREM-1 expression and inflammation has been discussed [13,14].

As discussed above, HMGB-1 also plays a role in the pathogenesis of atherosclerosis and it might be a therapeutic target. Glycyrrhizin or glycyrrhizinic acid, a constituent of Glycyrrhiza glabra root, acts as an HMGB-1 inhibitor and its inhibitory role on proliferation and migration of VSMCs and its role in decreasing oxidative stress and inflammation has been discussed [15]. Plant-derived small molecules such as tanshinone IIA, (−)-epigallocatechin-3-gallate, quercetin, lycopene and synthetic small molecules such as nafamostat, sivelestat, atorvastatin, simvastatin, gabexate-mesylate, and methotrexate in inhibiting HMGB1 pathological activity have been described in the literature [16]. The role of Ethyl pyruvate in mice for the reversal of the chronic unpredicted mild stress-induced atherosclerotic development and pro-inflammatory cytokines upregulation by HMGB1/TLR4 pathway has been documented [17]. Furthermore, TLRs play a crucial role in the development of atherosclerosis, targeting TLRs might be a promising approach. TLRs are components of the innate immune system and interaction of TLRs with their ligands can activate the downstream signaling pathways ultimately inducing an immune response by producing inflammatory cytokines and other inflammatory mediators leading to plaque formation and rupture. Thus, targeting TLRs might be considered. There are several TRL inhibitors known for cardiovascular diseases. Eritoran acts through TLR-4 inhibition, quinine drugs such as chloroquine, hydroxychloroquine, and quinacrine acts by masking the TLR binding epitope of nucleic acids, valsartan and candesartan act on TLR2 and TLR4, and small molecules such as fluvastatin, atorvastatin, ST2825, TAK-242 also inhibits TLR pathways [18]. These studies suggest that prevention of atherosclerotic plaque formation should be targeted to decrease the incidences, however, these researches need to be translated from lab to clinics. Thus, at present, replacement of the vessel is the only mode of treatment available for severely damaged vessels and in case of emergencies like rupture of the aortic aneurysm or aortic dissection which poses an emergency situation and need urgent surgical intervention to repair the vessel. In such cases, vessel transplantation and vascular regeneration are effective therapeutic options. We will expound various regenerative therapies and tissue engineering vascular grafts (TEVGs) formation using fabrication techniques with their merits and demerits.

8.2 Regenerative therapies

Vascular pathologies lead to clinical manifestations including ischemia, stroke, and gangrene (atherosclerosis and peripheral vascular and arterial diseases due to lumen obstruction), aortic dissection, and aortic aneurysm. Anti-atherosclerotic measures and allogeneic and autologous blood vessel transplantation are standard therapeutic options, however, are limited by medication-related side effects, autoimmune complications, and scarcity of donors and the patient's condition. Further, despite the development of endovascular therapy and bypass surgery, the prognosis of vascular diseases is very poor. Tissue engineering with regenerative medicine seems to be promising approach for vascular regeneration [19–24]. We will discuss gene therapy, stem cell therapy, and use of tissue engineering and regenerative medicine as an approach for vascular regeneration.

8.2.1 Gene therapy

Patients suffering from chronic critical limb ischemia (CLI), an end-stage manifestation of peripheral arterial disease (PVD), and coronary artery disease caused due to atherosclerosis, are at very high risk of amputation and cardiovascular complications. The manifestations and sequelae of PVD and limb ischemia result in severe morbidity and mortality. Revascularization involving angiogenesis, vasculogenesis, and arteriogenesis is the most commonly used treatment for CLI/PVD, however, many patients remain unsuitable for the surgical procedure. Hence, there is a need for an alternative treatment strategy to improve blood supply. Gene/protein therapy using proangiogenic growth factors/cytokines has been discussed, however, bone marrow (BM)-derived endothelial progenitor cells (EPCs) and mesenchymal stem/progenitor cells (MSCs) based therapy drastically developed the field of therapeutic angiogenesis and changed the treatment of ischemic limb. Although early phase clinical trials showed safety, feasibility, and potential effectiveness of stem cell therapy, only few late-phase clinical trials, as summarized by Fujita et al., have been conducted [21]. Vasculogenesis, a process of de novo formation of new vessels or capillaries, is initiated by the recruitment of endothelial progenitor cells (EPCs) which secrete various proangiogenic cytokines and growth factors including VEGF, HIF1-α, angiopoietin-1, SDF-1α, IGF-1, and endothelial nitric oxide synthase promoting proliferation and migration of endothelial cells and contribute to vascular regeneration. Angiogenic gene therapy using vascular endothelial growth factor (VEGF), hepatocyte growth factor (HGF), fibroblast growth factor (FGF), developmental endothelial locus-1 (Del-1), hypoxia-inducible factor 1-alpha (HIF-1α), stroma derived factor-1 alpha (SDF-1α), and recombinant Sendai virus vector with human FGF-2 gene (rSeV/dF-hFGF-2) resulted in neoangiogenesis characterized by capillary sprouting, endothelial cell migration, proliferation, and luminogenesis to generate new capillaries [19,21]. Angiogenic gene therapies for vascular regeneration have been confirmed to be safe and feasible in small-scale clinical study, however, the clinical use of these therapies in large-scale randomized comparative studies and various RCTs showed no clear benefit and warrants future studies with large clinical trials [19,21].

8.2.2 Stem cell therapy

Stem cell therapy consists of transplantation of stem or progenitor cells, capable of self-renewal and differentiation into an organ-specific cell type, at the site of injury to repair the damaged tissue. Stem cells have a paracrine effect through the release of pro-angiogenic growth factors or cytokines at the site of injury. Although stem cell therapy is limited by the poor survival and rapid removal of cells, it has advantage over gene therapy. The superiority of stem cell therapy over protein or gene therapy might be due to the direct vasculogenic properties of stem cells and the paracrine effects of these cells mediated by secretion of exosomes containing proteins, ribonucleic acids and microRNAs stimulating both receptor-mediated and genetic mechanisms needed for angiogenesis [22]. The ongoing clinical trial and pros and cons of the most commonly used cells including bone marrow-mononuclear cells (MNCs), G-CSF-mobilized peripheral blood MNCs, unmobilized peripheral blood-MNCs, adipose-derived stem cells, umbilical cord blood-derived stem cells, CD34 + Cells, and Endothelial Progenitor Cells (EPCs), and mesenchymal stem cells for stem cell therapy for vasculogenesis has been discussed [19,21]. Kang et al. [25] reported the superiority of adipose tissue-derived MSCs over adult stem cells for vascular regeneration in animal models and showed that synergistic use of three-dimensional stem cell clusters and angiopoietin-1 promotes vascular regeneration in ischemic region. A maltose-binding protein-linked basic fibroblast growth factor-immobilized polystyrene surface was used to culture the human adipose-derived stem cells and design the three-dimensional cell clusters having the capacity to differentiate into endothelial lineage cells and releasing various angiogenic factors such as FGF, IL-8 except angiopoietin-1. So, these clusters were combined with angiopoietin-1 and transplanted into the ischemic region. The combined therapy showed vessel regeneration with no presence of fibrotic collagen and exhibited limb salvage suggesting the increased efficiency of the combined therapy in blood vessel regeneration, tissue regeneration, and minimizing ischemic fibrosis and muscle degeneration [25]. The use of bone marrow MNCs, EPCs, and MSCs for vascularization has been confirmed in various animal and a few clinical studies, however, the efficacy and the effect of stem cell therapy is not fully sufficient and thus need further research or an alternative strategy for vascular regeneration [19].

8.2.3 Tissue engineering and regenerative medicine

Surgical intervention is needed to replace the damaged blood vessels in CAD and PVD and the options are to do endarterectomy procedures or use vascular patches. Limitations of stem cell therapy, gene therapy, autologous and allogenic endarterectomy with poor prognosis and morbidity make regenerative medicine and tissue engineering (replacing the damaged vessel with bio-fabricated constructs) with vascular graft by a tissue engineering-derived blood vessel a promising alternative [19]. The advancement and use of regenerative medicine and tissue engineering provide exciting new perspectives and offer an immediate cure and fast recovery with a normal functioning organ along with decreased medical and economic burden [24]. Vascular tissue engineering, which produces biomimetic constructs resembling normal tissues that replace the damaged tissues and restores the function by delivering living elements plays a crucial role in increasing life expectancy and preservation of extremities. However, the effectiveness of these techniques in clinics remains to be explicated [20,23].

The three basic components of tissue engineering consist of (a) cells (embryonic stem cells or adult stem cells derived from bone marrow, adipose tissue, hair follicle or induced pluripotent stem cells) expressing the appropriate genes to maintain appropriate phenotype and perform the specific function; (b) bioreactive agent or signal (growth factors/cytokines, adhesion factors, and bioreactors) inducing cells to perform their function; and (c) the biodegradable scaffold (synthetic, biological, or composite) housing the cells acting as a substitute for damaged vessel [26–29]. The stem cells in the biodegradable scaffold secrete extracellular matrix which pertains the mechanical stability and also secretes various growth factors necessary for vessel regeneration. Scaffold aids in holding the cells and the bioactive factors play a crucial role in angiogenesis [30]. In the following section, we will describe briefly about the stem cells, types of scaffolds, and microenvironment needed for neovascularization because the source of origin, potential use of stem cells in vasculogenesis, their characterization, angiogenic potential and differentiation towards VSMCs and endothelial cells, paracrine effect, signaling pathways and mechanisms involved in governing this differentiation, role of mechanical factors such as shear stress and stretch and multimodal mechanical stimulation and the ideal properties of the scaffold (mechanical and physiological properties, rigidity, patterning, biodegradability) needed for a successful angiogenesis or vasculogenesis, and various growth factors involved in three components of regeneration have been reviewed and discussed in detail in the existing literature [31–38].

8.2.3.1 Stem cells in tissue engineering

Downstream and upstream are the most promising tissue engineering approaches. The downstream approach involves implanting the pre-cultured cells and synthetic scaffold complexes into the defect area and involves techniques like cell aggregation, microfabrication, cell sheeting, cell stacking, and cell printing. While in upstream approach an engineered tissue created by combined culture of cells and biomaterial scaffolds is implanted or an acellular scaffold with desired bioactive molecules is delivered to the site of injury and these bioactive molecules recruits and promotes the proliferation and differentiation of the recruited progenitor cells to repair the injured tissue [39,40]. Among various issues related to use of scaffold, leakage of cells, and immune-mediated rejection, major obstacle in using the tissue-engineered graft is vascularization of the central/deeper part and it may undergo ischemia and necrosis due to poor blood supply and hypoxia. Autologous vein patches from saphenous or jugular veins are resistant to infections compared to synthetic analog, however, their use is associated with morbidity of the donor site [41]. Biological patches (bovine pericardium) might be an option due to its off-the-shelf availability, durability, and biocompatibility, but poses a high risk of infection or contamination with pathogens [42]. Various preclinical and clinical trials focusing on these issues and working on to develop technologies for integrating grafts with host vasculature and fabricated sophisticated grafts and tissue mimic, strategies to alter the microenvironment to enhance the intrinsic regenerative capacity of the host by immunomodulation have been discussed in the literature and these strategies have been proposed as a future direction to develop improved regenerative medicine therapies [43].

Stem cells, as discussed above, commonly used with the scaffold for regeneration prevent thrombosis through paracrine signaling to endogenous cells and helps to initiate scaffold-mediated remodeling and vascular regeneration. The current strategy of stem cells-scaffold based TEVGs (tissue-engineered vascular grafts) is based on using autologous stem cells, however, limited by the reduced regenerative capacity of autologous stem cells in certain groups of patients limiting the therapeutic potential of this approach [44]. This limitation of using stem cells prompts the need to investigate another strategy for vascular regeneration. Allogeneic stem cells or stem cell-secreted products such as extracellular vesicles (EVs — the cell-derived phospholipid membrane-based nanoparticles with functional surface/membrane proteins and protein and RNA species reflecting the parent cell and tissue) may provide the basis of TEVGs therapeutics. The use of EVs is exciting due to their potential use as a cell-free therapeutic base and other benefits including cell-specific targeting, reduced immunogenicity, stability, and cargo delivery to target cells under physiological conditions. Based on the density, size, precipitation, immunoaffinity, and microfluidic techniques EVs can be isolated from cell culture supernatant and are of three classes; (1) exosomes (30−200 nm) formed via invagination of the cell membrane, (2) micro-vesicles (200−1000 nm) released via direct outward budding of the cell membrane, and (3) apoptotic bodies (1000−5000 nm) released by cells via fragmentation of the plasma membrane. Exosomes and microvesicles used in TEVGs contain protein, mRNA, and miRNA reflective of the parent cellular membrane. EV isolation, standardization of stem cell culture conditions, and scaffold functionalization with EVs are the important points that must be taken into consideration and addressed before the use of EVs in TEVGs therapeutics [44].

Pericytes surrounding the smaller vessels and capillaries are another class of cells that have the potential for tissue repair, vascular homeostasis, and angiogenesis. These cells are in direct contact with endothelial cells (ECs) [45]. Pericytes have various markers such as CD44, CD90, CD105, CD34, CD146, vimentin, alpha-smooth muscle actin (α-SMA) and properties similar to SMCs, MSCs, and ECs and have been recently isolated and characterized for angiogenesis, but there is no common consensus till date [30,46]. Pericyte shares a common phenotype, however, their shape, size, distribution, attachment, and density are location dependent. Pericytes lack the expression of hematopoietic and endothelial markers (CD45 and CD31) and express neural/glial antigen2 (NG2) and platelet-derived growth factor receptor beta (PDGFRβ). A comparison of expression markers on SMCs, MSCs, ECs, and pericytes and expression markers of pericytes from different body parts (saphenous vein, cardiac, myocardial, skeletal muscle, brain, liver, and dental pulp) have been summarized in detail by Avolio et al. [30]. Pericytes or perivascular cells' capacity to stabilize blood vessels regulate angiogenesis and immunological response and their contribution to physiological and pathological repair processes make them an ideal candidate for tissue engineering [30]. To increase the efficiency, efficacy, and outcome of vascular regeneration, strategies such as retrograde coronary delivery, improved combinations such as combination therapy of stem cells and macrophages, preconditioning and pretreatment of stem cell, stem cell exosomes, use of mannitol, magnet, nanoparticles, and ultrasound-enhanced delivery, enhanced homing of stem cells with bioactive factors, drugs or other techniques, and stem cell modulation genetically, or with antibody, selectins, and peptides have been discussed widely in the literature and proposes an enhanced repair, remodeling, and angiogenesis [47].

8.2.3.2 Scaffolds in tissue engineering

The success and efficacy of tissue engineering and regeneration depend on the choice of stem cells, scaffold material and use of bioactive factors. The advantages and limitations of various natural and synthetic biomaterials to design and structure of scaffold in a vessel and other organ regeneration have been discussed elaborately in the literature [26,48]. Biocompatible and biodegradable tissue-engineered vascular patches or scaffolds derived from decellularized tissues (to remove all cellular and nuclear matter to minimize the adverse effects on the composition, biological activity, and mechanical integrity of the host ECM for the development of a new tissue) or made of natural (fibrin, elastin, collagen, silk fibroin, hyaluronan) or synthetic polymers (polyglycolic acid (PGA), polylactic acid (PLA), poly-caprolactone (PCL), and polyglycerolsebacate (PGS)) are component of the vessel regeneration in TE and regenerative medicine [48]. The uses of hybrid scaffolds made of natural and synthetic polymers, hydrogels, cardiogels, and matrigels have also been discussed in the literature [47,48]. The dense or porous scaffold made of chitosan complexed with alginate or pectin are being used as vascular graft [49] and they have their own merits and demerits. Chitosan-alginate scaffolds have higher culture medium uptake capacity while chitosan-pectin scaffold has longer-term stability in culture media and higher elastic modulus, longer-term stability upon degradation, enhanced mechanical properties, better blood-contact response with an ability to withstand larger deformations. Chitosan-pectin scaffolds have been shown to have better hemocompatibility, lower levels of platelet adhesion and activation, and better spread, adherence, proliferation, and survival of SMCs. These biocompatible chitosan-pectin based tridimensional structures have a high potential for the reconstruction and regeneration of vascular tissues [50]. Similarly, the use of tissue-engineered vascular graft (TEVGs) made of decellularized matrices and natural and/or biodegradable synthetic polymers and of polytetrafluoroethylene (ePTFE) prosthesis have been discussed. However, their use was limited due to neointimal hyperplasia, remarkable calcification, thrombosis, and foreign body reaction [48,51].

8.2.3.3 Three-dimensional (3D) printing and tissue engineering

Incorporation of a stable vascular network anastomosing with host vasculature to support the biological functions of stem cells incorporated in the scaffold is the basis of the success of TEVGs. Various strategies including the use of decellularized matrix, self-assembled structures, and cell sheets, laser, micro-molding, and electrospinning-based techniques facilitating the layer-by-layer fabrication of any geometry at the place of injury have significantly progressed TEVGs. Since perfusion of the vascular network is major problem, use of 3D fabrication to fabricate vascular networks with incorporated cells and bioactive factors including angiogenic factors, proteins, and/or peptides, cytokines, and growth factors might enhance the outcome [52].

Among various 3D fabrication approaches such as EB biofabrication, micro-pattern fabrication and assembly, laser-based fabrication, nanoscale fabrication, and natural matrix recellularization evolved to mimic the native vascular network, direct and indirect bioprinting approaches have proven promising for large 3D tissue constructs with an intricate vascular network. Direct bioprinting involves active bioprinting of hollow constructs with cell-loaded or cell-compatible bioinks whereas indirect bioprinting encompasses the bioprinting of a sacrificial biopolymers scaffold which is subsequently removed leaving behind the lumen. Indirect bioprinting is convenient in cases where scaffolding biopolymers have poor printability and manipulation complexity. Further, the use of coaxial needles in extrusion-based (EB) biofabrication has revolutionized the 3D biofabrication endowing the ability to print lumen-incorporated strands. Extrusion-, droplet-, and laser-based bioprinting techniques, the use of exogenous biomaterial-free cell aggregates such as tissue spheroids and cell pellets, and bioprinting of hydrogels have also been used in TE-derived vascular grafts [53]. Fabrication techniques such as layer-by-layer fabrication, laser degradation, bio-printing vascular networks, inkjet- and extrusion-based bioprinting, laser-based 3D printing, sources of biomaterials, cell seeding and grafting conditions, assembly of the scaffolds, and type of scaffolds to be used, the effect of prevascularization on the vascular network formed with their advantages and drawbacks with their clinical implications and outlook have been discussed and reviewed in the literature [52,54—56].

Scaffold-free tissue engineering approaches such as coculture system, sheet-based engineering, decellularization, direct cell injection, bioprinting, and biofabrication in a bioreactor system are additional techniques for vascular regeneration where the fabrication of the tissue construct is attached in the vital capability of the cells to prepare their own extracellular matrix along with the bioactive and growth factors/ bioreactors [57—59]. The use of rapid prototyping bioprinting method with bioactive factors/ bioreactors for scaffold-free small-diameter vascular reconstruction has been used for fully biological self-assembly approaches [57,60]. The success in developing vascular tissues in this approach is based on the development of biomimetic bioreactors with a combination of stress, strain, and perfusion stimulation designed on the basis of expansion and recoil properties of blood vessels [61]. A comparative discussion with merits and drawbacks of scaffold and scaffold-free approaches over each other by Demirbag et al. [28] deduce that these approaches should be considered complementary to each other because of the uniqueness of cells, tissues, nature of the vascular defects, and pathological condition of the patient. Further, to enhance the efficacy and efficiency of TEVGs, combination of various techniques to design better-engineered vessels and techniques used in other organ systems (High-Density Suspension System [62]) for regeneration might be a lead to expand stem cells in vascular regeneration.

Designing a completely scaffold-free technique will be beneficial because of the advantage of promoting a greater life expectancy of the individuals affected with vascular diseases and prevention of complications. Further, designing a completely biomimetic construct is a challenge because angiogenesis is a crucial step in vasculogenesis and monitoring the signaling pathways for vascularization is difficult. However, combining the bioactive and growth factors with automated cell tissue culture systems is a promising approach [54]. The common consensus is that the viability of large cell populations incorporated in TEVGs depends on the micro-scale capillary networks and for the success of TEVG the perfusion is a must.

This can be improved by various fabrication approaches such as plasma etching, laser ablation, soft lithography, and replica molding. Further, the formation of micro- and macro-blood vessels using nanofiber-based engineered constructs, use of decellularized matrix, 2D and 3D microfluidic systems have promoted the EC monolayer formation on the lumen of the vessel and a combination of inkjet and 3D bioplotter bio-printing system will be a promising approach [52].

8.2.3.4 Potential challenges and recent approaches for TEVGs

Increased cell senescence of VSMCs is a limiting factor in re-reendothelialization or TEGV and lifespan extension of the cells via telomerase expression in smooth muscle cells and endothelial cells in the elderly patients via ectopic expression of hTERT is a promising approach [63–65]. Grafts rejection, hypersensitivity, and immunological reaction are the major concern of TEVGs. With the development of TE, the focus is on to develop a biocompatible artificial vascular graft with less immunological reaction which also has the capability of allowing cellular adhesion and cellular proliferation for tissue regeneration. Recently, Chiu et al. [66] have discussed the use of digital light processing (DLP) 3D printing technology to manufacture near-optimal processing parameters to construct artificial vascular grafts with vascular characteristics closer to native vessels supporting cellular adhesion and proliferation. The study suggests that proper optimization of fabrication procedures and ratios of materials including amino resin (AR), 2-hydroxyethyl methacrylate (HEMA), dopamine, and curing durations enables successfully fabricated vascular grafts with good printing resolutions.

Improved and advanced approaches of TE and the biomaterial used for artificial vessels have enabled the improved larger diameter of artificial vessels with reduced adverse responses driving the associated complications, however, the problem still exist for small-diameter vessels and vascularization is a potential challenge in TEVGs. Recently, Wang et al. [37], using in vivo model, have reported the use of modified freeze-cast technique for lamellar nanotopography on the luminal surface to design the biomimetic design of an acellular small-diameter vascular graft having the features of inhibiting the adherence and activation of platelets, inducing oriented growth of ECs, promoting revascularization, and eventually remodeling a neovessel with long-term patency. Silk fibroin and gelatin were used to provide hemocompatibility and elasticity for integrating the graft with adjacent tissues and polycaprolactone (PCL) was used as a mechanical sheath to avoid graft rupture. It was also reported that compared with random topography, regularly lamellar nanopattern can manipulate blood flow to reduce the flow disturbances, the cause for rethrombosis. Similarly, the role of neuropeptide substance P (SP) in inducing endogenous tissue regeneration in cell-free grafts was demonstrated by Shafiq et al. [67]. The group showed the vascular regeneration potential of SP, which can accelerate tissue repair by endogenous cell mobilization and recruitment, with heparin co-tethered vascular grafts polycaprolactone (PCL), PCL/SP-conjugated poly(L-lactide-co-ε-caprolactone) (PLCL-SP) (SP), and PCL/PLCL-SP/heparin-conjugated PLCL (Hep/SP) vascular grafts. Improved hemocompatibility due to heparin and recruitment of mesenchymal stem cells by SP induced the endogenous tissue regeneration in cell-free grafts. Another potential challenge in creating TEVGs is the lack of the ability in creating a multi-layered concentric conduit inside natural ECMs and gels replicating the hierarchical architecture of biological blood vessels more accurately. Attalla et al. reported a new microfluidic nozzle design capable of multi-axial extrusion which can 3D print the vascular graft and pattern bi- and tri-layered hollow channel structures in a fast, simple and low-cost manner. The long-term cell viability and growth without compromising the structural integrity and improved cellular adhesion were achieved by incorporating materials with high mechanical strength (i.e. alginate) and various layers of different cell-friendly materials (such as collagen and fibrin) [68].

8.2.3.5 Simulation of an in vitro and in vivo environment

Several approaches used to simulate in vivo vascularization (layering of SMCs and fibroblasts around microcapillaries and use of chemotaxis, haptotaxis, and mechanotaxis enhancing ECs migration) and creating extensive and perfusable vascular networks in vitro in 3D cell cultures have been discussed in the literature. 3D bioprinting of engineered constructs with cell-laden biomaterials using a simple coaxial-nozzle-based printing for the development of tissue constructs in vitro via one-step gelling gelatin bioink containing different cell types has been reported. The study concluded that a radial distribution of multiple vascular cells can be achieved by using a synthetic GPT bioink combined with a coaxial nozzle printing system for one-step generation of vascular constructs and this technique has a potential for organized 3D vascular graft [69]. A novel rotary 3D bioprinting approach for biofabricating fibrin-based vascular constructs has been recently reported. The researchers incorporated the fibrinogen with gelatin to prepare a new bioink to achieve a desired shear-thinning property and found that the unprintable fibrinogen can be turned into a printable biomaterial by blending heat-treated gelatin with fibrinogen by leveraging the favorable rheological properties of gelatin [70]. Piard et al. [71] have discussed the importance of the distance between cell populations and the effect of cell patterning and suggested that in vitro vascularization strategies can greatly be improved with coculture of HUVECS and MSCs. Their results supported the notion that indirect cell contact prompts the paracrine signals and stimulates the angiogenesis or the process of vasculagenesis is affected by the separation and distance between ECs and MSCs populations via modulation of cell-cell communication. The study showed that HUVECs grown $>400\,\mu m$ apart from MSCs showed characteristics (migration/ proliferation) of an early stage of angiogenesis with an increased upregulation of VEGF, FGF-2, and ITGA3 (integrins) and a smaller fold change in VE-Cadherin and Ang-1 expression while grown $\leq 200\,\mu m$ showed increased Ang-1 and VE-cadherin expression and tighter monolayer formation suggesting stabilization.

8.3 Conclusion and future directions

The role of gene therapy and stem cell therapy has been well established in the field of vascular regeneration. As discussed in this chapter, there is a need to improve the limitations of these therapies for a better outcome. The efficiency and effectiveness of TEVGs have also been demonstrated using animal models, however, the effectiveness of the TEVGs in the clinical applications remains to be elucidated. The results of various studies suggest that there is still a need to improve the techniques of tissue engineering and regenerative medicine for the best outcome. The emphasis is on to carefully monitor the material biocompatibility, printing time, printing conditions, the choice of cell types, different molecular factors, flow rates, nutrient and oxygen diffusion, and cell behavior. The excellent potential of accurately positioning cell-laden constructs via 3D bioprinting technology has revolutionized TE and has aided in preparing an efficient and extensive network of microvessels mimicking the in vivo conditions to support cell viability and neovascularization. The two major merits of 3D printing are automation and high cell density and this technique may further be made more efficient by improving resolution, printing speed, and available materials for 3D bioprinting [56]. Finally, the establishment of blood vessels and its vascularization can be achieved by combinations of 3D bioprinting techniques and 4D printing concepts through patterning proangiogenic factors. This combination may enable the novel solution for implantation of thick constructs [72]. Finally, TEVGs will be of immense help in vascular emergencies and in the cases where we need to replace the vessels but limited by donors. However, for the regenerated vessels or TEVGs to integrate physiologically and functionally with the body, there is a need to understand and monitor various signaling pathways governing the natural vessel physiological and functional aspects. With that notion, if the implanted cells or regenerated endothelial cells and smooth muscle cells took over the normal function as in a natural vessel, the regenerated vessel may mimic but there is a need for deep insight in to the factors regulating the physiological and functional aspects of a natural vessel.

References

[1] Gupta GK, Agrawal T, Rai V, Del Core MG, Hunter 3rd WJ, et al. Vitamin D supplementation reduces intimal hyperplasia and restenosis following coronary intervention in atherosclerotic swine. PLoS One 2016;11: e0156857.
[2] Rai V, Agrawal DK. Role of Vitamin D in cardiovascular diseases. Endocrinol Metab Clin North Am 2017;46:1039−59.
[3] Rao VH, Rai V, Stoupa S, Agrawal DK. Blockade of Ets-1 attenuates epidermal growth factor-dependent collagen loss in human carotid plaque smooth muscle cells. Am J Physiol Heart Circ Physiol 2015;309: H1075−86.
[4] Rao VH, Rai V, Stoupa S, Subramanian S, Agrawal DK. Tumor necrosis factor-alpha regulates triggering receptor expressed on myeloid cells-1-dependent matrix metalloproteinases in the carotid plaques of symptomatic patients with carotid stenosis. Atherosclerosis 2016;248:160−9.
[5] Gupta GK, Agrawal T, Del Core MG, Hunter 3rd WJ, Agrawal DK. Decreased expression of vitamin D receptors in neointimal lesions following coronary artery angioplasty in atherosclerotic swine. PLoS One 2012;7: e42789.
[6] Rai V, Agrawal DK. The role of damage- and pathogen-associated molecular patterns in inflammation-mediated vulnerability of atherosclerotic plaques. Can J Physiol Pharmacol 2017;95:1245−53.
[7] Rai V, Agrawal DK. Pathogenesis of the plaque vulnerability in diabetes mellitus. Mechanisms of vascular defects in diabetes mellitus. Springer; 2017. p. 95−107.
[8] Rai V, Rao VH, Shao Z, Agrawal DK. Dendritic cells expressing triggering receptor expressed on myeloid cells-1 correlate with plaque stability in symptomatic and asymptomatic patients with carotid stenosis. PLoS One 2016;11:e0154802.
[9] Rao VH, Rai V, Stoupa S, Subramanian S, Agrawal DK. Data on TREM-1 activation destabilizing carotid plaques. Data Brief 2016;8:230−4.
[10] Rai V, Jadhav GP, Boosani CS. Role of transcription factors in regulating development and progression of atherosclerosis. Ann Vasc Med Surg 2019;2(1):1007.
[11] Tornai D, Furi I, Shen ZT, Sigalov AB, Coban S, et al. Inhibition of triggering receptor expressed on myeloid cells 1 ameliorates inflammation and macrophage and neutrophil activation in alcoholic liver disease in mice. Hepatol Commun 2019;3:99−115.

[12] Wang H, Gao J, Lu J. Pravastatin improves atherosclerosis in mice with hyperlipidemia by inhibiting TREM-1/DAP12. Eur Rev Med Pharmacol Sci 2018;22:4995—5003.

[13] Derive M, Boufenzer A, Gibot S. Attenuation of responses to endotoxin by the triggering receptor expressed on myeloid cells-1 inhibitor LR12 in nonhuman primate. Anesthesiol: J Am Soc Anesthesiol 2014;120:935—42.

[14] Gibot S, Kolopp-Sarda M-N, Béné M-C, Bollaert P-E, Lozniewski A, et al. A soluble form of the triggering receptor expressed on myeloid cells-1 modulates the inflammatory response in murine sepsis. J Exp Med 2004;200:1419—26.

[15] Chen J, Zhang J, Xu L, Xu C, Chen S, et al. Inhibition of neointimal hyperplasia in the rat carotid artery injury model by a HMGB1 inhibitor. Atherosclerosis 2012;224:332—9.

[16] Musumeci D, Roviello GN, Montesarchio D. An overview on HMGB1 inhibitors as potential therapeutic agents in HMGB1-related pathologies. Pharmacol Therapeut 2014;141:347—57.

[17] Li N, Gu H-F, Xu Z-Q, Hu L, Li H, et al. Chronic unpredictable mild stress promotes atherosclerosis via HMGB1/TLR4-mediated downregulation of PPAR-γ/LXRα/ABCA1 in apoE-/-mice. Front Physiol 2019;10:165.

[18] Goulopoulou S, McCarthy CG, Webb RC. Toll-like receptors in the vascular system: sensing the dangers within. Pharmacol Rev 2016;68:142—67.

[19] Suzuki H, Iso Y. Clinical application of vascular regenerative therapy for peripheral artery disease. BioMed Res Int 2013;2013.

[20] Caputo B, Dani FR, Horne GL, N'Fale S, Diabate A, et al. Comparative analysis of epicuticular lipid profiles of sympatric and allopatric field populations of Anopheles gambiae s.s. molecular forms and An. arabiensis from Burkina Faso (West Africa). Insect Biochem Mol Biol 2007;37:389—98.

[21] Fujita Y, Kawamoto A. Stem cell-based peripheral vascular regeneration. Adv Drug Deliv Rev 2017;120:25—40.

[22] Sahoo S, Klychko E, Thorne T, Misener S, Schultz KM, et al. Exosomes from human CD34(+) stem cells mediate their proangiogenic paracrine activity. Circ Res 2011;109:724—8.

[23] Vacanti JP, Langer R. Tissue engineering: the design and fabrication of living replacement devices for surgical reconstruction and transplantation. Lancet 1999;354:S32—4.

[24] Wang Y, Yin P, Bian G-L, Huang H-Y, Shen H, et al. The combination of stem cells and tissue engineering: an advanced strategy for blood vessels regeneration and vascular disease treatment. Stem Cell Res Ther 2017;8:194.

[25] Kang JM, Yoon JK, Oh SJ, Kim BS, Kim SH. Synergistic therapeutic effect of three-dimensional stem cell clusters and angiopoietin-1 on promoting vascular regeneration in ischemic region. Tissue Eng Part A 2018;24:616—30.

[26] Rai V, Dilisio MF, Dietz NE, Agrawal DK. Recent strategies in cartilage repair: a systemic review of the scaffold development and tissue engineering. J Biomed Mater Res A 2017;105:2343—54.

[27] Kim B-S, Mooney DJ. Development of biocompatible synthetic extracellular matrices for tissue engineering. Trends Biotechnol 1998;16:224—30.

[28] Demirbag B, Huri PY, Kose GT, Buyuksungur A, Hasirci V. Advanced cell therapies with and without scaffolds. Biotechnol J 2011;6:1437—53.

[29] Fortunato T, De Bank P, Pula G. Vascular regenerative surgery: promised land for tissue engineers? Int J Stem Cell Res Transplant 2017;5:268—76.

[30] Avolio E, Alvino VV, Ghorbel MT, Campagnolo P. Perivascular cells and tissue engineering: current applications and untapped potential. Pharmacol Therapeut 2017;171:83—92.

[31] Zhang H, van Olden C, Sweeney D, Martin-Rendon E. Blood vessel repair and regeneration in the ischaemic heart. Open Heart 2014;1:e000016.

[32] Gu W, Hong X, Potter C, Qu A, Xu Q. Mesenchymal stem cells and vascular regeneration. Microcirculation 2017;24:e12324.

[33] Huang NF, Li S. Mesenchymal stem cells for vascular regeneration. Regen Med 2008;3:877—92.

[34] Leeper NJ, Hunter AL, Cooke JP. Stem cell therapy for vascular regeneration: adult, embryonic, and induced pluripotent stem cells. Circulation 2010;122:517—26.

[35] Wang D, Li LK, Dai T, Wang A, Li S. Adult stem cells in vascular remodeling. Theranostics 2018;8:815—29.

[36] Costa-Almeida R, Granja P, Soares R, Guerreiro S. Cellular strategies to promote vascularisation in tissue engineering applications. Eur Cell Mater 2014;28:51—66.

[37] Wang Z, Liu C, Xiao Y, Gu X, Xu Y, et al. Remodeling of a cell-free vascular graft with nanolamellar intima into a neovessel. ACS Nano 2019;13:10576—86.

[38] Akintewe OO, Roberts EG, Rim N-G, Ferguson MA, Wong JY. Design approaches to myocardial and vascular tissue engineering. Annu Rev Biomed Eng 2017;19.

[39] Ji W, Sun Y, Yang F, van den Beucken JJ, Fan M, et al. Bioactive electrospun scaffolds delivering growth factors and genes for tissue engineering applications. Pharm Res 2011;28:1259−72.

[40] Nichol JW, Khademhosseini A. Modular tissue engineering: engineering biological tissues from the bottom up. Soft Matter 2009;5:1312−19.

[41] Rerkasem K, Rothwell PM. Systematic review of randomized controlled trials of patch angioplasty versus primary closure and different types of patch materials during carotid endarterectomy. Asian J Surg 2011;34:32−40.

[42] Muto A, Nishibe T, Dardik H, Dardik A. Patches for carotid artery endarterectomy: current materials and prospects. J Vasc Surg 2009;50:206−13.

[43] Mao AS, Mooney DJ. Regenerative medicine: current therapies and future directions. Proc Natl Acad Sci 2015;112:14452−9.

[44] Cunnane EM, Weinbaum JS, O'Brien FJ, Vorp DA. Future perspectives on the role of stem cells and extracellular vesicles in vascular tissue regeneration. Front Cardiovasc Med 2018;5:86.

[45] Crisan M, Corselli M, Chen WC, Péault B. Perivascular cells for regenerative medicine. J Cell Mol Med 2012;16:2851−60.

[46] Corselli M, Chin CJ, Parekh C, Sahaghian A, Wang W, et al. Perivascular support of human hematopoietic stem/progenitor cells. Blood 2013;121:2891−901.

[47] Fakoya AOJ. New delivery systems of stem cells for vascular regeneration in ischemia. Front Cardiovascular Med 2017;4:7.

[48] Catto V, Farè S, Freddi G, Tanzi MC. Vascular tissue engineering: recent advances in small diameter blood vessel regeneration. ISRN Vasc Med 2014;2014.

[49] de Souza FCB, de Souza RFB, Drouin B, Mantovani D, Moraes ÂM. Comparative study on complexes formed by chitosan and different polyanions: potential of chitosan-pectin biomaterials as scaffolds in tissue engineering. Int J Biol Macromol 2019;132:178−89.

[50] de Souza FCB, de Souza RFB, Drouin B, Popat KC, Mantovani D, et al. Polysaccharide-based tissue-engineered vascular patches. Mater Sci Eng: C 2019;104:109973.

[51] Assmann A, Delfs C, Munakata H, Schiffer F, Horstkötter K, et al. Acceleration of autologous in vivo recellularization of decellularized aortic conduits by fibronectin surface coating. Biomaterials 2013;34:6015−26.

[52] Sarker M, Naghieh S, Sharma N, Chen X. 3D biofabrication of vascular networks for tissue regeneration: a report on recent advances. J Pharm Anal 2018;8:277−96.

[53] Sasmal P, Datta P, Wu Y, Ozbolat IT. 3D bioprinting for modelling vasculature. Microphysiol Syst 2018;2.

[54] Nemeno-Guanzon JG, Lee S, Berg JR, Jo YH, Yeo JE, et al. Trends in tissue engineering for blood vessels. BioMed Res Int 2012;2012.

[55] Chang WG, Niklason LE. A short discourse on vascular tissue engineering. NPJ Regenerat Med 2017;2:1−8.

[56] Jafarkhani M, Salehi Z, Aidun A, Shokrgozar MA. Bioprinting in vascularization strategies. Iran Biomed J 2019;23:9.

[57] Kelm JM, Lorber V, Snedeker JG, Schmidt D, Broggini-Tenzer A, et al. A novel concept for scaffold-free vessel tissue engineering: self-assembly of microtissue building blocks. J Biotechnol 2010;148:46−55.

[58] L'heureux N, Pâquet S, Labbé R, Germain L, Auger FA. A completely biological tissue-engineered human blood vessel. FASEB J 1998;12:47−56.

[59] L'Heureux N, Dusserre N, Konig G, Victor B, Keire P, et al. Human tissue-engineered blood vessels for adult arterial revascularization. Nat Med 2006;12:361.

[60] Norotte C, Marga FS, Niklason LE, Forgacs G. Scaffold-free vascular tissue engineering using bioprinting. Biomaterials 2009;30:5910−17.

[61] Plunkett NA, O'Brien FJ. IV. 3. Bioreactors in tissue engineering; 2010.

[62] Lee J, Sato M, Kim H, Mochida J. Transplantation of scaffold-free spheroids composed of synovium-derived cells and chondrocytes for the treatment of cartilage defects of the knee. Eur Cell Mater 2011;22:90.

[63] Nakamura TM, Morin GB, Chapman KB, Weinrich SL, Andrews WH, et al. Telomerase catalytic subunit homologs from fission yeast and human. Science 1997;277:955−9.

[64] Jiang X-R, Jimenez G, Chang E, Frolkis M, Kusler B, et al. Telomerase expression in human somatic cells does not induce changes associated with a transformed phenotype. Nat Genet 1999;21:111.

[65] Poh M, Boyer M, Solan A, Dahl SL, Pedrotty D, et al. Blood vessels engineered from human cells. Lancet 2005;365:2122−4.

[66] Chiu Y-C, Shen Y-F, Lee AK-X, Lin S-H, Wu Y-C, et al. 3D printing of amino resin-based photosensitive materials on multi-parameter optimization design for vascular engineering applications. Polymers 2019;11:1394.

[67] Shafiq M, Wang L, Zhi D, Zhang Q, Wang K, et al. In situ blood vessel regeneration using neuropeptide substance P-conjugated small-diameter vascular grafts. J Biomed Mater Res Part B: Appl Biomater 2019;107:1669−83.

[68] Attalla R, Puersten E, Jain N, Selvaganapathy PR. 3D bioprinting of heterogeneous bi-and tri-layered hollow channels within gel scaffolds using scalable multi-axial microfluidic extrusion nozzle. Biofabrication 2018;11:015012.

[69] Hong S, Kim JS, Jung B, Won C, Hwang C. Coaxial bioprinting of cell-laden vascular constructs using a gelatin−tyramine bioink. Biomater Sci 2019;7:4578−87.

[70] Freeman S, Ramos R, Chando PA, Zhou L, Reeser K, et al. A bioink blend for rotary 3D bioprinting tissue engineered small-diameter vascular constructs. Acta Biomater 2019;95:152−64.

[71] Piard C, Jeyaram A, Liu Y, Caccamese J, Jay SM, et al. 3D printed HUVECs/MSCs cocultures impact cellular interactions and angiogenesis depending on cell-cell distance. Biomaterials 2019;222:119423.

[72] Miri AK, Khalilpour A, Cecen B, Maharjan S, Shin SR, et al. Multiscale bioprinting of vascularized models. Biomaterials 2019;198:204−16.

Musculoskeletal Regeneration of Tissues

Oral tissue regeneration: Current status and future perspectives

Maji Jose[1], Vrinda Rajagopal[2] and Finosh G. Thankam[3]

[1]Department of Oral Pathology & Microbiology, Yenepoya Dental College, Yenepoya (Deemed to be University), Mangalore, India [2]Department of Biochemistry, University of Kerala, Karyavattom, India [3]Department of Translational Research, Western University of Health Sciences, Pomona, CA, United States

9.1 Introduction

The oral cavity is defined as the space extending from the lips anteriorly to faucial pillars posteriorly and is bounded laterally by the cheeks, superiorly by the palate, and inferiorly by the muscular floor and the tongue. The oral cavity is arbitrarily divided into two parts: the oral cavity proper which is the region medial to the teeth that houses the tongue and the oral vestibule, which is the space that separates the lips and cheeks from the teeth. Functionally, the oral cavity acts as the first part of the digestive tract and also serves as a secondary respiratory conduit. The digestive function begins in the oral cavity and has a significant role in the mastication and swallowing of food. In addition, the oral cavity is important for speech, and acts as a chemosensory organ (taste sensation). As the oral cavity being functionally very important part of human body, even a minor disruption in the function seriously jeopardize the quality of life [1].

9.2 Histology of oral tissues

Structurally, the oral cavity comprises of different types of tissues. The hard tissue components are teeth and supporting bone, while the soft tissue components include oral mucosa, the moist lining of the oral cavity and underlying connective tissue components. Teeth are unique structures composed of crown and root. The crown of the tooth is visible in the oral cavity, while the root is embedded in the supporting bone and is attached to the same through periodontal ligament fibers. Tooth comprises of three mineralized tissues: enamel forms the outer covering of the crown, cementum covers the root and dentin that makes up the major bulk of the tooth. The central part of the tooth has a space that contains soft tissue components, the pulp tissue. Pulp contains the blood vessels that provide nutrition to associated tissues of the teeth and are innervated.

The hard tissue components of the teeth comprise of minerals, primarily calcium and phosphate in the form of hydroxyapatite crystals of varying size with variable amounts of other trace elements and organic constituents. Enamel is distinct from the other hard tissues of teeth owing to its ectodermal origin, higher mineral content (greater than 96%) and the presence of characteristic enamel proteins such as enamelin and amelogenins. Furthermore, it is nonvital, avascular insensitive in nature, inert to any kinds of stimulus. In addition, the formation of enamel is limited to a specific period, till the desired thickness of enamel is formed. Once the enamel formation is completed the ameloblasts, the enamel forming cells are trans-differentiated to perform protective and desmolytic functions, until the tooth erupts into the oral cavity. Hence, the enamel in an erupted tooth devoid of synthetic cells has minimal regeneration potential. In contrast, dentin and cementum are viable tissues possessing significant regenerative capacity. The associated synthetic cells, the odontoblasts continue the formative function as long as the tooth is vital. Dentine—Pulp complex possesses its own protective mechanism to maintain the vitality of the tooth.

An attachment apparatus called the periodontium — comprising of periodontal ligament, lamina propria, which is the connective tissue component of gingiva, cementum and alveolar socket - is responsible for holding the tooth in the jawbone. The principal component of periodontium is periodontal ligament composing of distinct groups of collagen fibers that are attached to the alveolar bone on one end and cementum of tooth on the other end [1]. A healthy periodontium is highly important for the functional integrity of tooth.

The alveolar bone is the part of the upper and lower jawbone, which holds the teeth and is directly contiguous with basal bone, maxilla and mandible. The alveolar bone is structurally similar to buccal and cortical plates with spongy bone in between. The alveolar socket in which the teeth are located is lined by a thin layer of dense compact bone termed as alveolar bone proper. Alveolar bone being a part of attachment apparatus of the teeth, the structural integrity of this bone is highly important to maintain the tooth integrity.

The soft tissue component of the oral cavity includes mucosal membrane which is the protective covering of the oral cavity. Oral mucosa comprises of stratified squamous epithelium that cover and protect the other underlying connective tissue components including lamina propria (the sub epithelial connective tissue) and the submucosa and structures including musculature. The oral cavity is kept moistened by saliva secreted by major and numerous minor salivary glands and reaches the oral cavity through ducts. The health and integrity of all structural components are essential for the functional efficiency of oral environment.

9.3 Oral microbiology in health and diseases

Oral cavity is a complex ecosystem inhabiting more than 700 bacterial species, comprising at least 11 bacterial phyla and 70 genera [2—4]. The bacterial load of oral cavity vary depends on oral hygiene practices of individuals. The number vary from 100,000 to 1 billion bacteria on each tooth surface in individuals depending on the oral care practices. Though the number is alarming, limited species are harmful and cause serious disease whereas the others are beneficial in preventing diseases. Thus, the oral cavity is inhabited by complex multispecies bacterial communities that usually exist in homeostasis with the host immune system. A dysbiotic community, resulting from an imbalance between microbial interactions, results in many disease conditions affecting oral tissues. Occasionally, nonpathogenic oral commensal bacteria cause diseases due to compromised immune responses [5].

Since the oral cavity acts as the entrance for air and food, it is constantly exposed to foreign substances, including bacteria and viruses. Furthermore, the oral environment is influenced by various local and systemic factors. The local factors include dental plaque, tartar, teeth alignment, occlusion, incompatible prosthesis, and unhealthy lifestyle habits including smoking, and consumption of alcohol. Major systemic factors are endocrine disturbances, various stresses, genetic factors and immunological and hematological disturbances. These local and systemic factors adversely affect the oral environment that pave the ways to diverse oral diseases [6].

9.4 Oral diseases, prevalence and management

According to WHO, oral diseases are the most common noncommunicable diseases and affect people throughout their lifetime, causing pain, discomfort, disfigurement and even death. Oral diseases affect half of the world's population according to the Global Burden of Disease Study 2016 and oral diseases that necessitate the possibility of regenerative therapy include dental caries, periodontal diseases, developmental anomalies such as bone defects, and oral cancer.

9.4.1 Dental caries

Dental caries is an irreversible microbial disease involving dental hard tissues characterized by acid dissolution of enamel and dentin resulting in cavitation [7]. It is a widely prevalent disease world-wide and has a definite negative impact on quality of life. The data currently available indicates that untreated cavitated dentine carious lesions in permanent teeth are the most prevalent health condition across the globe, affecting 2.4 billion people, and that untreated cavitated dentine carious lesions in deciduous teeth constituted the 10th most prevalent health condition, affecting 621 million children worldwide. According to Global Oral Health Data Bank, prevalence varies from 49% to 83% across the globe. Though published evidences indicate a worldwide decline in the prevalence of dental caries, it is reflected mainly in high income countries. Still, in low and middle-income countries the dental caries continues to be a major health problem [8–10].

Treatment option presently available for dental caries is debridement of the cavity and reconstructing the lost tooth structure by filling the defects with appropriate restorative materials. Unless treated in early stages the disease process extends to involve the pulp tissue leading to inflammation followed by necrosis, eventually resulting in infection and tooth loss. Once the dental caries proceeds to advance stage, treatment options are either extraction or endodontic (root canal) treatment. This treatment involves removal of entire infected pulp tissue, debride both the pulp chamber and root canal and filling with root canal filling-materials [11]. Though this treatment procedure offers the possibility of retaining natural tooth, an endodontically treated tooth becomes a dead tooth which is brittle and functionally inefficient. Dental pulp regeneration therapy is an alternative procedure which retains the vitality of affected teeth [12].

9.4.2 Periodontal diseases

Periodontal diseases comprising of gingivitis and periodontitis are the diseases involving the supporting structure of the teeth. The gingiva, due to its peculiar relationship with tooth, provide a favorable site for accumulation of food residues, microbial flora, and saliva and thus creating a unique micro niche for the pathogenesis of gingivitis. Gingivitis can be prevented by practicing appropriate oral hygiene measures that clear the food and microbial accumulation. Gingivitis unless treated in time, progresses to the tooth supporting bone and periodontal ligament tissue, resulting in loss of attachment apparatus eventually leading to mobility of teeth and exfoliation. Thus, periodontitis is the major cause of tooth loss leading to edentulism and masticatory dysfunction, thereby affecting nutritional status and quality of life of affected individuals [13].

According to Global Burden of Diseases (GBD) 2010 study and other surveys, severe periodontitis is sixth most common disease globally with the overall prevalence rate of 10.8%, affecting 743 million people aged 15–99 worldwide. The prevalence of severe periodontitis increased with age, with a steep increase between the third and fourth decades of life, reaching peak prevalence at the age of 40 and usually remain stable thereafter [9,11,13]. Periodontitis is preventable, can be easily diagnosed and successfully treated and controlled following appropriate professional care. The primary objective of periodontal treatment is to control the disease and prevent further disease progression and tooth loss. Therefore, periodontal treatment consists of a series of sequential phases of care. After controlling the risk factors, measures are taken to reduce periodontal inflammation through self-motivated oral hygiene procedures combined with professional oral prophylaxis. Subjects who respond appreciably to the first phase of treatment, but present with persistent periodontal pockets are subjected to surgical correction of the anatomical lesions caused by the disease process in order to regain periodontal health. Several treatment modalities have been attempted to regenerate lost periodontal tissue, including condition of root surfaces, and bone grafting. The limitations of these procedures in attaining complete tissue regeneration have led the clinicians to search for new treatment options in regenerative medicine and tissue engineering [14,15].

9.4.3 Craniofacial bone defect/anomalies

Orofacial cleft conditions have been estimated to have a global annual prevalence of 7.94 cases per 10,000 live births with high variations across regions and countries and showing differences between ethnicities [16]. Overall, higher rates have been reported in Asians and American Indians (one in 500 births), and lower rates have been reported in African-derived populations (one in 2500 births) [17]. Females are more prone to develop isolated cleft palate compared to males, at a ratio of 2:1, while a male predominance is noted for cleft lip with or without cleft palate with a ratio of 2:1 [18].

Craniofacial defects involving the bone or soft tissue can result from congenital disorders, trauma or diseases such as cysts or tumors. The most common congenital craniofacial defect is cleft lip with or without cleft palate. This condition is characterized by clefting of lip with maxillary alveolar cleft, oro-nasal communication and partially erupted or unerupted teeth within the cleft. These multiple defects cause functional, aesthetical and psychosocial problems to the affected patients [19]. It is thought to result from multifactorial inheritance, i.e. an interaction between the person's genes and specific environmental factors. Dysfunctions of molecular pathway that guide a precise synchronization and balance of cell adhesion, proliferation, and differentiation, and gene regulation, are responsible for disturbed embryonic development of soft and hard tissues of the orofacial region [19].

Current treatments are early surgery and face reconstruction procedures. The goal of the surgery is to achieve a normal facial appearance as well as the ability to feed, speak, and hear without affecting the ultimate facial appearance of the subject. The treatment approach is multidisciplinary that includes plastic surgeons, maxillofacial surgeons (cleft surgeons) and other specialists. However, functional and anatomic deficits are not fully corrected even after restorative cleft surgery due to suboptimal regeneration of skin, mucosa, and skeletal muscle, causing negative impact of quality of life and self-esteem of affected individuals. The field of tissue engineering and regenerative medicine gives a ray of hope as it offers a promising possibility of restoring the normal anatomy and physiology of tissues [19].

9.5 Oral mucosal lesions

Oral mucosal lesions range from mere alteration in color, variations in surface characteristics, swelling, or loss of integrity of the oral mucosal surface. Multiple etiologic factors contribute to these group of lesions, including microbial infections, local trauma or irritation, systemic diseases, and consumption of tobacco, betel quid, and alcohol [20]. Mostly, these mucosal lesions are benign and require only symptomatic treatment, which interfere with daily quality of life in affected patients through impacts on mastication, swallowing, and speech with symptoms of burning, irritation, and pain. Besides, some lesions may present a significant pathology; particularly oral potentially malignant disorders which may progress into malignancy and oral squamous cell carcinoma [21].

The prevalence of oral mucosal lesions widely varies among different countries and populations and with age, systemic health of the individuals and tobacco habits. Epidemiologic studies, though fewer compared with reports on dental caries or periodontal diseases, provide information on the prevalence of mucosal lesions in global population which range from 4.9% to 64.7%. The overall prevalence reported in a Chinese population was 10.8%, while in Lebanese sample was 61.8%. Amarodi et al. reported a prevalence of 31.7% of oral mucosal lesion in the teenager group [21–23].

The epidemiologic studies have demonstrated diverse prevalence rates in habits related oral mucosal lesions in different populations. Prevalence of these diseases are more in Indian population owing to the increased tobacco consumption. Along with various tobacco related oral mucosal lesions such as nicotinic stomatitis, tobacco pouch keratosis, smoker's melanosis, mild keratosis of the palate, and chewer's mucosa, high prevalence of the oral potentially malignant disorders such as leukoplakia (10.1%), and oral submucous fibrosis (4.7%) also have been reported [24,25].

Current treatment modalities for oral mucosal lesions including inflammatory or ulcerative lesions are use of antiinflammatory analgesics, primarily aimed at a short term and symptomatic relief. Habit discontinuation followed by the administration of antioxidants has been tried in the initial phases of oral potentially malignant disorders. In advanced cases, surgical management is mandatory which causes certain degree of morbidity. Advances in stem cell technology have opened new vistas for the treatment of these lesions. The research in animal models have demonstrated the scope of stem cell therapies in the treatment of oral mucosal lesions such as oral ulcers, mucositis, immune mediated diseases such as oral lichen planus and pemphigus vulgaris and in potentially malignant disorder — Oral submucous fibrosis. More human research trials are warranted to ascertain the role of stem cells in management of oral mucosal diseases and management [26–28].

9.6 Oral cancer

Oral cancer has been considered to be a worldwide public health problem which is the malignancy arising in the squamous epithelium lining the oral cavity which present as red or white patches, nonhealing ulcers or ulcero-proliferative growth involving tongue, lip, gingival, palate, floor of the mouth or buccal mucosa. There is a wide variation in the prevalence of oral cancer in different zones of the world or even within the same countries from the minorities or subpopulations. Oral cancer is the sixth most frequent type of cancer. Various studies from different part of the world report a prevalence rate ranging between 0.15% and 24.8%. The tumor invades deeply into adjacent tissues of the tongue and the floor of the mouth, as well as into the bones. Late diagnosis, high mortality rates and morbidity leading to significant disfigurement are characteristics of the disease worldwide. The mortality and morbidity caused due to disease *per se* and as side effects of different treatment modalities and impart significant adverse physical and psychosocial impact on the patients. Oral cancers have a multifaceted etiology with the most important risk factors being tobacco use and alcohol abuse, which have synergistic effect. Along with these, a variety of suspected risk factors such as chronic irritation, poor oral hygiene, Human Papilloma Virus (HPV) infection, malnutrition has been proposed for the development of oral cancer along with genetic predisposition [29,30]. Two-thirds of the global incidence of oral cancer occurs in low- and middle-income countries; half of those cases are in South Asia [31–33].

Preventive strategies are of at most significance in tackling oral cancer which is targeted to prevent tobacco and alcohol use and promoting the consumption of fruits and vegetables. Early detection through screening and relatively inexpensive treatment is also equally important to avoid oral cancer related mortality and morbidity. However, oral cancer continues to be a major cancer in India, East Asia, Eastern Europe, and parts of South America [32,33]. Surgery and radiotherapy either as single modalities or in combination are the treatment of choice for oral cancer. Functional and cosmetic outcomes is a major concern in this treatment strategy. Despite many advances in treatment modalities, recurrence of the disease and high mortality rate to occur in a significant percentage of patients [34].

9.7 Oro-dental tissue engineering: a modern epoch in tooth management

Tissue loss due to injury, diseases or other developmental disorders is one of the major global medical concerns. Tissue replacement surgeries have proven to be effective in restoring the structure and function of the tissues; however, the availability of viable human tissues and organs pose challenge. As an alternative, there is an emphasis on shifting from a surgical approach to regenerative strategies for the laboratory engineering of functional tissues. Tissue engineering refers to the field of science that drive into the remarkable ability of the human body to regenerate and engineer tissues. It is a multidisciplinary field, with the knowledge of anatomy of the tissue of interest, molecular pathways for neo-tissue formation, versatile biomaterial scaffolding and the knowledge of and engineering. In addition, the tissue engineering strategies are based on the knowledge of medical sciences aiming to translate the scientific essence into clinical arena.

The loss of tooth and the diseases of associated tissue, due to any one of the abovementioned causes, exert physiological and psychological effects on the affected subjects. Epidemiological surveys show that more than half of adult population globally are affected with various degrees of periodontitis [35]. Eke et al. [36] reported that 46% of US adults, representing 64.7 million people had periodontal disease from 2009 to 2012. A remarkable surge in the rate of periodontitis was also observed [36,37]. Periodontitis leads to tooth loss and if left untreated activates the progression of other chronic diseases such as cardiovascular disease, obesity, diabetes, chronic nephritis and cancer. Hence, effective therapeutic interventions are necessary in periodontal diseases.

The pathology of alveolar bone resorption in denture wearers due to tooth loss has been well studied [38]. Also, the oro-dental reconstruction is essential for esthetics as well as maintaining the functionality in such cases [39]. Osseointegration is the direct interaction between a load bearing artificial implant with a live bone. Bothe, Beaton, and Davenport, in 1940, first observed this phenomenon and in 1952, Per-Ingvar Branemark coined the term osseointegration due to the observation of fusion between bone and titanium [40,41]. Since then, the durable and biocompatible titanium has been the gold standard for tooth replacement. However, the osseointegration approach fails to provide the cushioning of the periodontium and cementum tissues which help to modulate the mechanical stress and the friction due to mastication. Hence, the tissue engineering strategies address such issues by engineering the implants for better integration with the associated periodontal tissues [42]. Generally, the oro-dental tissue engineering (ODTE) exploits either synthetic or natural biodegradable scaffolds, to mimic the native extracellular matrix (ECM) to encourage the synthesis of new tissue with the desired structure and function.

The dental structure is unique in its development. In natural tooth development, the oro-dental bones are derived from neural crest and the paraxial mesoderm. The ectomesenchyme derived from the neural crest, is the principal signaling source of tooth development. The dental epithelium differentiates into enamel secreting ameoblasts and adjacent mesenchyme cells differentiate into dentin secreting odontoblasts [43]. Enamel is 90% hydroxyapatite and the hardest tissue in the body. A population of odontoblasts derived from inner enamel epithelium are terminally differentiated. On the other hand, the functional odontoblasts enter quiescent phase following the dentin formation and are programmed to synthesize secondary dentin upon stimuli by injury suggesting the regeneration potential of dentine tissue, unlike the enamel [44]. The dental papilla generates pulp tissue composing of blood vessels, lymph vessels, fibroblasts, nerves and odontoblasts. As the development of tooth progresses, a transient structure called dental follicle is established which gives rise to 3 types of cells; cementoblasts, osteoblasts and fibroblasts. The cementoblasts secrete cementum in the root of the tooth, osteoblasts synthesize alveolar bone around the root and fibroblasts synthesize periodontal ligament tissues attaching the root of the tooth to the alveolar bone [45].

Understanding regarding the development of dental tissue has been utilized to design viable bioengineered tooth and/or associated tissues that are compatible to the native dental tissue. Even before the evolution of the concept of tissue engineering, regeneration of dentin and pulp tissues were studied using calcium hydroxide capping materials [46]. The advancement of knowledge and research in stem cells and of the discovery of biocompatible materials led the transition of dental tissue engineering from concept to reality. For example, the human embryonic stem cells (ESCs) are renewable resource of cells that differentiate into multiple lineages including the dental tissues [47, 48]. However, the teratoma formation and the ethical issues in the usage of ESCs offers challenge for the wide range application of ESCs.

The resident dental stem cells (DSCs), include dental pulp stem cells, were found to differentiate into odontoblast-like cells producing dentin and expressing the neuronal marker, nestin [49,50]. Another population of stem cells from the tooth root apex of human teeth called DSCs of the apical papilla (SCAP) were also differentiates into odontoblasts. SCAP are superior to DSCs regarding the proliferative potential and are promising for cell-based dental root regeneration [51, 52]. Periodontal ligament stem cells (PDLSCs) exhibit limited capacity of regeneration in response to mild injury. In culture conditions, PDLSCs differentiated into cementoblast-like cells, adipocytes and fibroblast; also, when seeded in 3D scaffolds, PDLSCs exhibited higher osteogenic potential [53]. Early attempts for periodontal tissue regeneration were not satisfactory owing to the failure of integration with the components of periodontium, including gingival, periodontal ligament tissue, cementum and alveolar bone. However, the principles of guided tissue regeneration, has merged to be the gold standard for periodontal tissue regeneration [54]. A biocompatible membrane such as polytetrafluoroethylene, polylactide/glycolide or any of the biomimicking collagen biopolymers are employed to cover the periodontal space and the adjacent alveolar bone as the negatively charged collagen membrane triggers tissue regeneration. Cytokines and growth factors such as transforming growth factor-β1(TGF-β), bone morphogenic protein-2 (BMP-2) and several others have been explored for periodontal regeneration (Table 9.1) [55].

Thus, successful attempts at periodontal tissue regeneration are set to improve the current oro-dental implant therapies. Regeneration of whole tooth via tissue engineering

TABLE 9.1 Role of biologic signaling mediators in tissue engineering.

Mediators	Function
TGF	Mediator of inflammation, proliferation of osteoblasts and ECM formation
BMPs	Differentiation of MSCs into osteoblasts and chondroblasts for the formation of bone and cartilage
PDGF	Promotes cell growth and division of mesenchymal cells and
EGF	Regulates cell growth, proliferation and differentiation
IGF	Bone matrix formation and proliferation of cells
VEGF	Stimulates neo-angiogenesis
EMD	Stimulates cell growth, differentiation and angiogenesis

consists of two methods: *in vivo* implantation of the tooth structure grown in vitro from any of the dental stem cells; or implanting polymer scaffolds seeded with in vitro cultured dental progenitor cells [56,57]. Tremendous research outcome has been obtained in these techniques and the following section overviews the present scenario of oro-dental tissue engineering.

9.8 Strategies adapted for periodontal tissue regeneration

In order to address the defect in the oral tissues, the general idea of exploiting the principles of tissue engineering to accelerate the healing by providing enough viable cells and mediators to the target site. These cells facilitate the production of new cells and maintain the structure of the tissue. However, this can be achieved if the cells are stable and are shielded from the pathological signals at the target site. This calls for the need of a mechanical construct as a scaffold that holds the cells and mediators and establishes a proper communication with the surviving tissue to accelerate the healing responses. Appropriate levels of signaling molecules are also required to regulate cell growth, along with instigating the necessary gene expression patterns to convert the seeded cells into the desired tissue phenotype [56]. Finally, regeneration of the defective oro-dental tissue is garnered by neovascularisation and the upregulation of pro-healing mediators and growth factors. In addition, another way of tissue reconstruction is to engineer the tissue structure *ex vivo* and transplanting to the defective tissue *in vivo*. This section focuses on the various tissue engineering success-stories regarding the regeneration of oro-dental tissues.

9.9 ECM regeneration with ECM-based scaffolds

The regeneration of periodontal tissue is a highly integrated process that includes cell-cell and cell-ECM interactions. In normal conditions, structurally and functionally intact suspensory periodontal ligament functions in the proper positioning of teeth. The remodeling events in the alveolar bone facilitate the tooth to fall out; and during adulthood, the bone modeling acts to stabilize the tooth with the help of periodontal ligament. The key to this process is deposition of inorganic calcium (calcification) and phosphate-containing crystals into the ECM of teeth and bone which is known as mineralization [58,59]. The nano-sized apatite crystals, the major functional element taking part in the mineralization process, acts collectively to stabilize as well as harden these collagenous extracellular matrices and serves to counteract compressive forces of mastication [60,61].

The tooth ECM and bone possess permanent collagenous protein phenotypes intermingled with other noncollagenous proteins and inorganic materials that provide necessary stability for the tooth. The scaffolding base of the ECM for bone, dentin and cementum are all rich in type I collagen and 50−70% calcified with carbonate-substituted apatite [3,62]. The ECM assembly during osteogenesis, dentinogenesis and cementogenesis occur through successive, highly ordered events and orchestrated by the tissue forming blast cells; the osteoblasts, odontoblasts and cementoblasts respectively. Basement membrane of tooth consists of collagen IV, laminin (Lama5), perlecan, nidogen/entactin, and other molecules that interact with each other to form the supramolecular structure. The ECM proteins are secreted, modified by enzymes and organized into macromolecular assemblies along with a mineralization front to form the finalized tissue construct. In the ECM, the nano- apatite crystals reside within and between collagen fibrils. They are also present within the lumen of the chondrocyte-, osteoblast- and odontoblast-derived matrix vesicles that are shed from the cells [63,64]. The noncollagenous proteins in teeth are made up of highly acidic phosphoproteins that bind strongly to the mineral phase through negatively charged phosphate groups and regulate apatite crystal growth.

Acellular dermal matrix has been effectively utilized for the regeneration of the oro-dental ECM; however, has been reported to be highly susceptible to shrinkage and scarring [65]. Human amniotic membrane is composed of a single epithelial layer in a thick elastic membrane, a collagen layer and homes ample growth factors. These growth factors possess immense healing potential which made human amniotic membrane to be an ideal choice for the reconstruction of oro-dental ECM [66]. Porcine bilayered collagen matrix comprising the collagen subtype 1 and 3 has been widely used in periodontal plastic surgery procedures and possess the advantage of increased deposition of keratinized tissue, augmentation of soft tissue thickness, minimal surgical time, and increased patency. In addition, such materials exhibited excellent integration with the host tissue with minimal adverse reactions and increased performance of fibroblasts and keratinocytes [67]. Similar observation has been reported for porcine derived acellular dermal matrix as well [68]. Also, the porcine-derived matrix from the submucosa of the small intestine preserves the native ECM structure and architecture which is ideal for the repopulation of fibroblasts, and angiogenesis [69]. The clinical consideration regarding the ECM based scaffolds for the regeneration of oro-dental tissues is lacking in the literature and warrants further investigation.

9.10 Biomaterial scaffolds for oro-dental tissue regeneration

The dental biomaterials are largely associated with the pulp capping, cavity filling and dentine regeneration where the recurrent inflammatory episodes induced by the synthetic materials often results in the failure of the implants. In addition, the permeability of the dentine-like scar tissue paves the way for microbial recontamination followed by secondary inflammation, eventually leading to tissue necrosis [70]. Such limitations are potentially due to the lack of specific temporal and spatial integration over the biological signaling which warrants further research. The current trend of regenerative dentistry focusses on the development of more reliable, effective and translationally worthwhile avenues by integrating elements from restorative/surgery, biomaterials science, cell and molecular biology and medicine [70]. The classical definition for the features for ideal biomaterials for various tissue engineering applications have been reported extensively [2,4,71−74]. This section throws light to the recent advances in polymeric biomaterials-based tissue engineering for oro-dental tissue regeneration.

Alginate, a natural heteropolysaccharide, scaffolds promoted the differentiation of dental pulp stem cells to odontoblast-like cells and facilitated the formation of calcified bodies, and deposition of collagen subtypes and dentine sialoprotein [75]. Stem cell loaded alginate-calcium phosphate paste has also exhibited appreciable outcomes for oro-dental tissue regeneration [76]. Scaffolds based on chitosan, another polysaccharide, has reported to support the growth and differentiation of dental pulp stem cells which has been widely used for oro-dental tissue engineering and drug delivery [77,78]. In addition, Feng et al.; demonstrated that the superior conductivity of chitosan based scaffolds are ideal for the neural differentiation of dental pulp stem cells [79]. Hyaluronic acid based sponges exhibited cell enrichment in the dentine defect suggesting its regenerative potential which possess an added advantage of injectability [80,81]. Sumitha et al. utilized collagen sponges for tooth-tissue engineering and reported the superior performance of collagen scaffolds over polyglycolic acid fiber mesh scaffold [82]. Similarly, the gelatin-based hydrogels have been effectively utilized for the generation of calcified tissue for repairing dentine defects [83]. However, the poor mechanical properties of gelatin encouraged the development of gelatin-based composites for oro-dental applications [84]. The tooth bud cell loaded silk protein hydrogels implanted in rat models revealed appreciable mineralization along with shape and size guided osteodentin regeneration [85].

The synthetic polymers, especially poly lactic acid (PLA) and poly glycolic acid (PGA) possess excellent biomimetic characters ideal for various biomedical applications including dental tissue regeneration. However, the increased degradation rate and the acidic degradation products were reported to induce mild inflammatory responses. Interestingly, the hybrid combinations of PLA and PGA exhibited promising effects for oro-dental tissue regeneration [86]. Takashi et al; developed the injectable polyethylene glycol-maleate-citrate system for the direct capping of pulp tissue which exhibited excellent regenerative potential of the pulp stem cells [87]. The increased mineralization potential exhibited by the poly caprolactone (PCL) scaffolds has attracted its application for oro-dental tissue engineering which supported the growth and odontogenic differentiation of human dental pulp stem cells [88,89]. Poly(propylene fumarate) (PPF) reinforced with calcium phosphate cement composites facilitated the survival of active osteoblasts suggesting its regenerative potential [90]. Biodegradable polyanhydrides have been exploited as a delivery vehicle by linking low molecular weight drugs for treating various oral pathologies [90]. Poly carbonates have been used as reinforcing agents for metallic orthodontic wires which promoted the growth of gingival fibroblasts and supported the regeneration of oro-dental tissue [90]. Poly ethylene glycol (PEG) has been extensively used in regenerative dentistry as coating material, drug delivery vehicle, and tissue engineering scaffold owing to its exceptional biocompatibility [90].

The ceramic biomaterials and bioactive glass are being used for restorative and regenerative dentistry owing to their biocompatibility and similarities with native dental tissues. The ceramic materials lead to the formation of mineralized layer over the implantation site which is biologically equivalent to the native hydroxyapatite (HAP) [70]. HAP and its derivatives including tricalcium phosphate (TCP), and biphasic calcium phosphate (DCP) have been widely tested for bone regeneration and are being extensively used for dental application as well [70]. The ceramic biomaterials have been hailed for their ability to favor osteogenesis, osteoconductivity, bio/immunocompatibility and bone resorption. However, the brittle nature of ceramic biomaterials remains to be a major hurdle for their load bearing functions [91]. The bioactive glass materials offer supreme crystallinity, superior mechanical properties, translucency/opacity and resorbability and have been used in regenerative/restorative dentistry veneers, crowns, and bridges as well as tissue engineering [92,93].

9.11 Stem cell biology updates for oro-dental tissue regeneration

In oro-dental tissue regeneration, introduction of inherently programmable cells, such as stem cells, to the defect site is a simple and minimally invasive strategy. Single cell suspensions or cells suspended in medium can be utilized for delivery. The effectiveness of this approach is limited due to inadequate cell supply, insufficient localization of the cell due to spreading of injected cells to surrounding tissue and low rate of engraftment [94,95]. Also, immunological rejection offers major hurdle for this technique which warrants immunocompatible delivery vehicles [96]. Such delivery vehicles carry the cells and deliver them to a localized area of the defect, without interfering the immune system [97]. The cell delivery can be coupled with tissue engineering strategies which offers the advantage of simultaneous delivery of regenerative signals and therapeutics along with the cells of interest. This section focuses on the overview of stem cell research and stem cell-based tissue engineering strategies adopted for the regeneration of oro-dental tissues.

Among the diverse cell types employed for oro-dental tissue engineering, the stem cells harvested from various sources gained immense attention owing to their self-renewal, multipotency and regenerative potential. The stem cells are of various types depending on the lineage of cell types differentiated from the parent stem cell and their location of origin inside the body [98]. Post-natal stem cells have been isolated from various tissues including bone marrow, blood, neural tissue, adipose, skin, and retina [99–101]. Recent findings identified oral tissues to be a good source for stem cells. The oral cavity derived stem cells include dental pulp stem cells [102], periodontal ligament stem cells [103,104], exfoliated deciduous teeth stem cells [105], and gingival stem cells [106,107]. Dental pulp stem cells are clonogenic, rapidly proliferative and able to differentiate into neurogenic, osteogenic, dentinogenic, and myogenic cell lineages [108]. Periodontal ligament stem cells (PDLSCs) have the potential to generate cementum and PDL-like tissues *in vivo*. Human gingiva derived MSCs are easy to isolate, proliferate fast, homogenous and are not teratogenic [109] which has been a better alternative for bone marrow derived MSCs. Also, the stem cells isolated from the apical dental papilla (SCAP) facilitate the regeneration of pulp/dentin organ and the periodontal ligament (PDL) which exhibited 2- to threefold proliferative potential than those in the pulp in cultures [52,53].

HAP composites loaded with dental pulp stem cells promoted the secretion of dentine like structures in the immunocompromised mouse model [110]. Interestingly, the genetic profiling of dental pulp stem cells revealed considerable difference in the expression of genes between native osteoblast cells suggesting altered mechanisms executed by these cell types [104]. The implantation of stem cells from human exfoliated deciduous teeth (SHEDs) grown on poly-L-lactic acid (PLLA) subcutaneously in mice resulted in their differentiation into odontoblasts. And, the cotransplantation of endothelial cells triggered angiogenesis in the scaffolds suggesting the regeneration potential [111]. The transplantation of periodontal ligament stem cells (PLSCs) in mice resulted in the formation of bone, cementum and periodontal ligament and healed the periodontal lesions in swine model [51,112]. In addition, the PLSCs derived osteoblasts exhibited increased production of calcium and NO revealing a promising strategy for the management of periodontal lesions [113]. Dental follicle stem cells (DFSCs) have been reported to be differentiated into osteoblasts and promoted the formation of cementum and periodontal ligament [111]. Differentiation potential of the stem cells from the dental apical papilla (SCAPs) to multiple lineages including osteoblasts, odontoblasts and adipocytes has been established and has been considered to be more efficient for tooth formation than the above-mentioned dental stem cells [111]. It has been reported that the transplantation of SCAPs along with HAP resulted in the formation of dentine in mice suggesting their translational potential [51]. In addition, the mesenchymal and epithelial stem cells harvested from the dental tissues of various animal models and grown onto various biomaterials scaffolds revealed encouraging outcomes [111].

The continuous advancements in the knowledge of stem cell biology has expanded the horizons of regenerative dentistry and have formed the fertile substrates for several novel translational avenues. During embryogenesis, the development of teeth occurs via the interplay between cranial neural crest-derived mesenchymal stem cells and oral-derived epithelial stem cells which pave the ways for the generation of hard tissue structures [114]. Even though the differentiation and mineralization process have been considered to be the prominent features for oro-dental tissue regeneration, the exact physiological definition for a regenerated dental tissue has not been established yet [115]. The advancements in the stem cell biology, tissue engineering and biotechnology have opened immense avenues to repair and regenerate the damaged dental tissue. The versatile sources of dental derived stem cells and their differentiation potential towards oro-dental tissue lineage have been extensively studied. However, the selection and choice of these stem cells in conjunction with the response of appropriate biomaterials is important for the successful tissue engineering strategies. In addition, the investigations regarding the differentiation of mesenchymal stem cells derived from the sources other than oro-dental tissues are very limited in the literature. Also, the utilization of adult stem cell systems for dental tissue regeneration warrants further research. Moreover, the innervation and neovascularization are equally important for the functional performance of the regenerated dental tissue structures. Tissue engineering strategies focusing on these aspects are still in infancy. Finally, the infection and inflammation associated with the biomaterials following the implantation may aggravate the pathology and impair the regenerative responses. These observations suggest that further improvements and advanced knowledge are warranted for the management of various oro-dental pathologies.

9.12 Summary

Oral cavity comprising of complex biological tissues is a site for multiple disorders that seriously jeopardize quality of life. Dental caries and periodontitis are very common diseases leading to and tooth loss, which is the most common organ failure, affecting a vast majority of the population. Strategies based upon regenerative medicine that facilitates the repair or replacement of damaged teeth may hold promising means to restore physiologically functioning tooth over conventional dental treatment modalities. Advancement in regenerative dentistry is emerging with many breakthrough innovations. Researchers have been successful in establishing regenerative strategies in dentistry for in situ partial pulp regeneration, *de novo* pulp replacement, regeneration or replacement of mineralized tissues such as dentin, enamel, bone, and cementum, and regeneration of the periodontium. Tooth regeneration is certainly a leap forward which could have far-reaching implications for the field of regenerative medicine. In addition to regeneration of tissue lost due to dental diseases, tissue engineering also promises functional restoration of structurally and functionally impaired or damaged tissues due to congenital defects, oral cancer, and traumatic injuries. Though there are reports on successful oral tissue regeneration in laboratory models and experimental animals, the outcome in clinical scenario and long-term success warrants further investigation. The characteristic oral environment inhabited by complex multispecies bacterial communities that usually exist in homeostasis with the host immune system and structural complexity are major hurdles for regenerative techniques. Despite all limitations, regenerative dentistry, offers a wealth of opportunities, paving the way for superior treatment techniques which restores the lost orofacial and dental tissues closer to the natural condition.

References

[1] Dinesh KC, Karuna D. Dysphagia evaluation and management in otolaryngology. Elsevier; 2019.
[2] Chenicheri S, Komeri R. 9 — Integration of dental implants: molecular interplay and microbial transit at tissue—material interface. In: Sharma CP, editor. Biointegration of medical implant materials. 2nd ed. Woodhead Publishing; 2020. p. 221—43.
[3] Chenicheri S, RU, Ramachandran R, et al. Insight into oral biofilm: primary, secondary and residual caries and phyto-challenged solutions. Open Dent J 2017;11:312—33. Available from: https://doi.org/10.2174/1874210601711010312.
[4] Chenicheri S, Thankam FG, Ramachandran R. Chapter 14 — Drug delivery in nano-dimensions: a focus on oro-dental infections. In: Sharma CP, editor. Drug Delivery Nanosystems For Biomedical Applications. Elsevier; 2018. p. 303—31.
[5] Zhang Y, Wang X, Li H, et al. Human oral microbiota and its modulation for oral health. Biomed Pharmacother 2018;99:883—93. Available from: https://doi.org/10.1016/j.biopha.2018.01.146.
[6] Nazir MA. Prevalence of periodontal disease, its association with systemic diseases and prevention. Int J Health Sci (Qassim) 2017;11:72—80.

[7] Pitts NB, Zero DT, Marsh PD, et al. Dental caries. Nat Rev Dis Prim 2017;3:17030. Available from: https://doi.org/10.1038/nrdp.2017.30.

[8] Vos T, Abajobir AA, Abate KH, et al. Global, regional, and national incidence, prevalence, and years lived with disability for 328 diseases and injuries for 195 countries, 1990—2016: a systematic analysis for the Global Burden of Disease Study 2016. Lancet 2017;390:1211—59. Available from: https://doi.org/10.1016/S0140-6736(17)32154-2.

[9] Frencken JE, Sharma P, Stenhouse L, et al. Global epidemiology of dental caries and severe periodontitis — a comprehensive review. J Clin Periodontol 2017;44:S94—105. Available from: https://doi.org/10.1111/jcpe.12677.

[10] Janakiram C, Antony B, Joseph J, Ramanarayanan V. Prevalence of dental caries in india among the WHO Index age groups: a meta-analysis. JCDR 2018;. Available from: https://doi.org/10.7860/JCDR/2018/32669.11956.

[11] Saunders W. Latest concepts in root canal treatment. Br Dent J 2005;198:515—16. Available from: https://doi.org/10.1038/sj.bdj.4812297.

[12] Jung C, Kim S, Sun T, et al. Pulp-dentin regeneration: current approaches and challenges. J Tissue Eng 2019;10. Available from: https://doi.org/10.1177/2041731418819263.

[13] Tonetti MS, Jepsen S, Jin L, Otomo-Corgel J. Impact of the global burden of periodontal diseases on health, nutrition and wellbeing of mankind: a call for global action. J Clin Periodontol 2017;44:456—62. Available from: https://doi.org/10.1111/jcpe.12732.

[14] Hägi TT, Laugisch O, Ivanovic A, Sculean A. Regenerative periodontal therapy. Quintessence Int 2014;45:185—92. Available from: https://doi.org/10.3290/j.qi.a31203.

[15] Lin N-H, Gronthos S, Bartold PM. Stem cells and future periodontal regeneration. Periodontol 2009;51:239—51. Available from: https://doi.org/10.1111/j.1600-0757.2009.00303.x.

[16] IPDTOC Working Group. Prevalence at birth of cleft lip with or without cleft palate: data from the International Perinatal Database of Typical Oral Clefts (IPDTOC). Cleft Palate Craniofac J 2011;48:66—81. Available from: https://doi.org/10.1597/09-217.

[17] Dixon MJ, Marazita ML, Beaty TH, Murray JC. Cleft lip and palate: understanding genetic and environmental influences. Nat Rev Genet 2011;12:167—78. Available from: https://doi.org/10.1038/nrg2933.

[18] Mossey PA, Little J, Munger RG, et al. Cleft lip and palate. Lancet 2009;374:1773—85. Available from: https://doi.org/10.1016/S0140-6736(09)60695-4.

[19] Martín-del-Campo M, Rosales-Ibañez R, Rojo L. Biomaterials for cleft lip and palate regeneration. IJMS 2019;20:2176. Available from: https://doi.org/10.3390/ijms20092176.

[20] Chandroth SV, Venugopal HKV, Puthenveetil S, et al. Prevalence of oral mucosal lesions among fishermen of Kutch coast, Gujarat, India. Int Marit Health 2014;65:192—8. Available from: https://doi.org/10.5603/IMH.2014.0037.

[21] Feng J, Zhou Z, Shen X, et al. Prevalence and distribution of oral mucosal lesions: a cross-sectional study in Shanghai, China. J Oral Pathol Med 2015;44:490—4. Available from: https://doi.org/10.1111/jop.12264.

[22] El Toum S, Cassia A, Bouchi N, Kassab I. Prevalence and distribution of oral mucosal lesions by sex and age categories: a retrospective study of patients attending Lebanese School of Dentistry. Int J Dent 2018;2018:1—6. Available from: https://doi.org/10.1155/2018/4030134.

[23] Amadori F, Bardellini E, Conti G, Majorana A. Oral mucosal lesions in teenagers: a cross-sectional study. Ital J Pediatr 2017;43. Available from: https://doi.org/10.1186/s13052-017-0367-7.

[24] Krishna Priya M, Srinivas P, Devaki T. Evaluation of the prevalence of oral mucosal lesions in a population of eastern coast of South India. J Int Soc Prevent Communit Dent 2018;8:396. Available from: https://doi.org/10.4103/jispcd.JISPCD_207_17.

[25] Aslesh OP, Paul S, Paul L, Jayasree AK. High prevalence of tobacco use and associated oral mucosal lesion among interstate male migrant workers in urban Kerala, India. Iran J Cancer Prev 2015;8. Available from: https://doi.org/10.17795/ijcp-3876.

[26] Abdel Aziz Aly L, El- Menoufy H, Ragae A, et al. Adipose stem cells as alternatives for bone marrow mesenchymal stem cells in oral ulcer healing. Int J Stem Cell 2014;7:167. Available from: https://doi.org/10.15283/ijsc.2014.7.2.167.

[27] Zhang Q, Nguyen AL, Shi S, et al. Three-dimensional spheroid culture of human gingiva-derived mesenchymal stem cells enhances mitigation of chemotherapy-induced oral mucositis. Stem Cell Dev 2012;21:937—47. Available from: https://doi.org/10.1089/scd.2011.0252.

[28] Arora M, Lakhanpal M, Suma G. Stem cell therapy: a novel treatment approach for oral mucosal lesions. J Pharm Bioall Sci 2015;7:2. Available from: https://doi.org/10.4103/0975-7406.149809.

[29] Dhanuthai K, Rojanawatsirivej S, Thosaporn W, et al. Oral cancer: a multicenter study. Med Oral 2017;. Available from: https://doi.org/10.4317/medoral.21999.

[30] Mehanna H, Paleri V, West CML, Nutting C. Head and neck cancer — Part 1: epidemiology, presentation, and preservation. Clin Otolaryngol 2011;36:65—8. Available from: https://doi.org/10.1111/j.1749-4486.2010.02231.x.

[31] Bray F, Ren J-S, Masuyer E, Ferlay J. Global estimates of cancer prevalence for 27 sites in the adult population in 2008. Int J Cancer 2013;132:1133—45. Available from: https://doi.org/10.1002/ijc.27711.

[32] Dhillon PK, Mathur P, Nandakumar A, et al. The burden of cancers and their variations across the states of India: the Global Burden of Disease Study 1990—2016. Lancet Oncol 2018;19:1289—306. Available from: https://doi.org/10.1016/S1470-2045(18)30447-9.

[33] Pavão Spaulonci G, Salgado de Souza R, Gallego Arias Pecorari V, Lauria Dib L. Oral cancer knowledge assessment: newly graduated versus senior dental clinicians. Int J Dent 2018;2018:1—12. Available from: https://doi.org/10.1155/2018/9368918.

[34] Vijayakumar M, Burrah R, Sabitha KS, et al. To operate or not to operate n0 neck in early cancer of the tongue? A prospective study. Indian J Surg Oncol 2011;2(172—175). Available from: https://doi.org/10.1007/s13193-011-0083-5.

[35] Dye BA. Global periodontal disease epidemiology. Periodontol 2000 2012;58:10—25. Available from: https://doi.org/10.1111/j.1600-0757.2011.00413.x 2000.

[36] Eke PI, Dye BA, Wei L, et al. Update on prevalence of periodontitis in adults in the United States: NHANES 2009 to 2012. J Periodontol 2015;86:611—22. Available from: https://doi.org/10.1902/jop.2015.140520.

[37] Carasol M, Llodra JC, Fernández-Meseguer A, et al. Periodontal conditions among employed adults in Spain. J Clin Periodontol 2016;43:548—56. Available from: https://doi.org/10.1111/jcpe.12558.

[38] Tallgren A. The continuing reduction of the residual alveolar ridges in complete denture wearers: a mixed-longitudinal study covering 25 years. J Prosthet Dent 1972;27:120—32. Available from: https://doi.org/10.1016/0022-3913(72)90188-6.

[39] Zaky SH, Cancedda R. Engineering craniofacial structures: facing the challenge. J Dent Res 2009;88:1077—91. Available from: https://doi.org/10.1177/0022034509349926.

[40] Bothe RT, Beaton LE, Davenport HA. Reaction of bone to multiple metallic implants surgery. Gynecol Obstet 1940;71:598—602.

[41] Rudy RJ, Levi PA, Bonacci FJ, et al. Intraosseous anchorage of dental prostheses: an early 20th century contribution. Compend Contin Educ Dent 2008;29:220—2 224, 226-228 passim.

[42] Lin C, Dong Q-S, Wang L, et al. Dental implants with the periodontium: a new approach for the restoration of missing teeth. Med Hypotheses 2009;72:58—61. Available from: https://doi.org/10.1016/j.mehy.2008.08.018.

[43] Mina M, Kollar EJ. The induction of odontogenesis in non-dental mesenchyme combined with early murine mandibular arch epithelium. Arch Oral Biol 1987;32:123—7. Available from: https://doi.org/10.1016/0003-9969(87)90055-0.

[44] Huang X, Xu X, Bringas P, et al. Smad4-Shh-Nfic signaling cascade-mediated epithelial-mesenchymal interaction is crucial in regulating tooth root development. J Bone Min Res 2010;25:1167—78. Available from: https://doi.org/10.1359/jbmr.091103.

[45] Ruch JV. Odontoblast commitment and differentiation. Biochem Cell Biol 1998;76:923—38.

[46] Pontoriero R, Nyman S, Ericsson I, Lindhe J. Guided tissue regeneration in surgically-produced furcation defects. An experimental study in the beagle dog. J Clin Periodontol 1992;19:159—63. Available from: https://doi.org/10.1111/j.1600-051x.1992.tb00632.x.

[47] Thomson JA, Itskovitz-Eldor J, Shapiro SS, et al. Embryonic stem cell lines derived from human blastocysts. Science 1998;282:1145—7. Available from: https://doi.org/10.1126/science.282.5391.1145.

[48] Ma L, Makino Y, Yamaza H, et al. Cryopreserved dental pulp tissues of exfoliated deciduous teeth is a feasible stem cell resource for regenerative medicine. PLoS One 2012;7:e51777. Available from: https://doi.org/10.1371/journal.pone.0051777.

[49] Wang Y, Zhao Y, Jia W, et al. Preliminary study on dental pulp stem cell-mediated pulp regeneration in canine immature permanent teeth. J Endod 2013;39:195—201. Available from: https://doi.org/10.1016/j.joen.2012.10.002.

[50] Orhan EO, Maden M, Sengüven B. Odontoblast-like cell numbers and reparative dentine thickness after direct pulp capping with platelet-rich plasma and enamel matrix derivative: a histomorphometric evaluation. Int Endod J 2012;45:317—25. Available from: https://doi.org/10.1111/j.1365-2591.2011.01977.x.

[51] Sonoyama W, Liu Y, Fang D, et al. Mesenchymal stem cell-mediated functional tooth regeneration in swine. Plos One 2006;1:e79. Available from: https://doi.org/10.1371/journal.pone.0000079.

[52] Abe S, Yamaguchi S, Watanabe A, et al. Hard tissue regeneration capacity of apical pulp derived cells (APDCs) from human tooth with immature apex. Biochem Biophys Res Commun 2008;371:90–3. Available from: https://doi.org/10.1016/j.bbrc.2008.04.016.

[53] Seo B-M, Miura M, Gronthos S, et al. Investigation of multipotent postnatal stem cells from human periodontal ligament. Lancet 2004;364:149–55. Available from: https://doi.org/10.1016/S0140-6736(04)16627-0.

[54] Retzepi M, Donos N. Guided bone regeneration: biological principle and therapeutic applications. Clin Oral Implant Res 2010;21:567–76. Available from: https://doi.org/10.1111/j.1600-0501.2010.01922.x.

[55] Kozlovsky A, Aboodi G, Moses O, et al. Bio-degradation of a resorbable collagen membrane (Bio-Gide) applied in a double-layer technique in rats. Clin Oral Implant Res 2009;20:1116–23. Available from: https://doi.org/10.1111/j.1600-0501.2009.01740.x.

[56] Hu B, Nadiri A, Kuchler-Bopp S, et al. Tissue engineering of tooth crown, root, and periodontium. Tissue Eng 2006;12:2069–75. Available from: https://doi.org/10.1089/ten.2006.12.2069.

[57] Morsczeck C, Götz W, Schierholz J, et al. Isolation of precursor cells (PCs) from human dental follicle of wisdom teeth. Matrix Biol 2005;24:155–65. Available from: https://doi.org/10.1016/j.matbio.2004.12.004.

[58] McKee MD, Hoac B, Addison WN, et al. Extracellular matrix mineralization in periodontal tissues: noncollagenous matrix proteins, enzymes, and relationship to hypophosphatasia and X-linked hypophosphatemia. Periodontol 2000 2013;63:102–22. Available from: https://doi.org/10.1111/prd.12029.

[59] Putnam AJ, Mooney DJ. Tissue engineering using synthetic extracellular matrices. Nat Med 1996;2:824–6. Available from: https://doi.org/10.1038/nm0796-824.

[60] Wagner HD, Weiner S. On the relationship between the microstructure of bone and its mechanical stiffness. J Biomech 1992;25:1311–20. Available from: https://doi.org/10.1016/0021-9290(92)90286-A.

[61] McKee MD, Nakano Y, Masica DL, et al. Enzyme replacement therapy prevents dental defects in a model of hypophosphatasia. J Dent Res 2011;90:470–6. Available from: https://doi.org/10.1177/0022034510393517.

[62] Chaussain-Miller C, Fioretti F, Goldberg M, Menashi S. The role of matrix metalloproteinases (MMPs) in human caries. J Dent Res 2006;85:22–32. Available from: https://doi.org/10.1177/154405910608500104.

[63] Wuthier RE, Lipscomb GF. Matrix vesicles: structure, composition, formation and function in calcification. Front Biosci (Landmark Ed) 2011;16:2812–902. Available from: https://doi.org/10.2741/3887.

[64] Nanci A, Moffatt P. Embryology of craniofacial bones. Mineralized tissues in oral and craniofacial science. John Wiley & Sons, Ltd; 2013. p. 1–11.

[65] Shulman J. Clinical evaluation of an acellular dermal allograft for increasing the zone of attached gingiva. Pract Periodontics Aesthet Dent 1996;8:201–8.

[66] Jain A, Jaiswal GR, Kumathalli K, et al. Comparative evaluation of platelet rich fibrin and dehydrated amniotic membrane for the treatment of gingival recession – a clinical study. J Clin Diagn Res 2017;11:ZC24–8. Available from: https://doi.org/10.7860/JCDR/2017/29599.10362.

[67] Tavelli L, McGuire MK, Zucchelli G, et al. Extracellular matrix-based scaffolding technologies for periodontal and peri-implant soft tissue regeneration. J Periodontol n/a: 10.1002/JPER.19-0351.

[68] Shirakata Y, Sculean A, Shinohara Y, et al. Healing of localized gingival recessions treated with a coronally advanced flap alone or combined with an enamel matrix derivative and a porcine acellular dermal matrix: a preclinical study. Clin Oral Investig 2016;20:1791–800. Available from: https://doi.org/10.1007/s00784-015-1680-4.

[69] Nevins M, Nevins ML, Camelo M, et al. The clinical efficacy of DynaMatrix extracellular membrane in augmenting keratinized tissue. Int J Periodontics Restor Dent 2010;30:151–61.

[70] Moussa DG, Aparicio C. Present and future of tissue engineering scaffolds for dentin-pulp complex regeneration. J Tissue Eng Regener Med 2019;13:58–75. Available from: https://doi.org/10.1002/term.2769.

[71] Thankam FG, Muthu J. Influence of physical and mechanical properties of amphiphilic biosynthetic hydrogels on long-term cell viability. J Mech Behav Biomed Mater 2014;35:111–22. Available from: https://doi.org/10.1016/j.jmbbm.2014.03.010.

[72] Finosh GT, Jayabalan M. Hybrid amphiphilic bimodal hydrogels having mechanical and biological recognition characteristics for cardiac tissue engineering. RSC Adv 2015;5:38183–201. Available from: https://doi.org/10.1039/C5RA04448K.

[73] Finosh GT, Jayabalan M. Regenerative therapy and tissue engineering for the treatment of end-stage cardiac failure: new developments and challenges. Biomatter 2012;2:1–14. Available from: https://doi.org/10.4161/biom.19429.

[74] Gnanaprakasam Thankam F, Muthu J, Sankar V, Kozhiparambil Gopal R. Growth and survival of cells in biosynthetic poly vinyl alcohol–alginate IPN hydrogels for cardiac applications. Colloids Surf B: Biointerfaces 2013;107:137–45. Available from: https://doi.org/10.1016/j.colsurfb.2013.01.069.

[75] Fujiwara S, Kumabe S, Iwai Y. Isolated rat dental pulp cell culture and transplantation with an alginate scaffold. Okajimas Folia Anat Jpn 2006;83:15–24. Available from: https://doi.org/10.2535/ofaj.83.15.

[76] Wang P, Song Y, Weir MD, et al. A self-setting iPSMSC-alginate-calcium phosphate paste for bone tissue engineering. Dent Mater 2016;32:252–63. Available from: https://doi.org/10.1016/j.dental.2015.11.019.

[77] Aguilar A, Zein N, Harmouch E, et al. Application of chitosan in bone and dental engineering. Molecules 2019;24. Available from: https://doi.org/10.3390/molecules24163009.

[78] Bakopoulou A, Georgopoulou A, Grivas I, et al. Dental pulp stem cells in chitosan/gelatin scaffolds for enhanced orofacial bone regeneration. Dent Mater 2019;35:310–27. Available from: https://doi.org/10.1016/j.dental.2018.11.025.

[79] Feng X, Lu X, Huang D, et al. 3D porous chitosan scaffolds suit survival and neural differentiation of dental pulp stem cells. Cell Mol Neurobiol 2014;34:859–70. Available from: https://doi.org/10.1007/s10571-014-0063-8.

[80] Inuyama Y, Kitamura C, Nishihara T, et al. Effects of hyaluronic acid sponge as a scaffold on odontoblastic cell line and amputated dental pulp. J Biomed Mater Res Part B Appl Biomater 2010;92:120–8. Available from: https://doi.org/10.1002/jbm.b.31497.

[81] Park SH, Seo JY, Park JY, et al. An injectable, click-crosslinked, cytomodulin-modified hyaluronic acid hydrogel for cartilage tissue engineering. NPG Asia Mater 2019;11:1–16. Available from: https://doi.org/10.1038/s41427-019-0130-1.

[82] Sumita Y, Honda MJ, Ohara T, et al. Performance of collagen sponge as a 3-D scaffold for tooth-tissue engineering. Biomaterials 2006;27:3238–48. Available from: https://doi.org/10.1016/j.biomaterials.2006.01.055.

[83] Ishimatsu H, Kitamura C, Morotomi T, et al. Formation of dentinal bridge on surface of regenerated dental pulp in dentin defects by controlled release of fibroblast growth factor-2 from gelatin hydrogels. J Endod 2009;35:858–65. Available from: https://doi.org/10.1016/j.joen.2009.03.049.

[84] Chocholata P, Kulda V, Babuska V. Fabrication of scaffolds for bone-tissue regeneration. Materials (Basel) 2019;12. Available from: https://doi.org/10.3390/ma12040568.

[85] Xu W-P, Zhang W, Asrican R, et al. Accurately shaped tooth bud cell-derived mineralized tissue formation on silk scaffolds. Tissue Eng Part A 2008;14:549–57. Available from: https://doi.org/10.1089/tea.2007.0227.

[86] Willerth SM, Sakiyama-Elbert SE. Combining stem cells and biomaterial scaffolds for constructing tissues and cell delivery. Stemjournal 2019;1:1–25. Available from: https://doi.org/10.3233/STJ-180001.

[87] Komabayashi T, Wadajkar A, Santimano S, et al. Preliminary study of light-cured hydrogel for endodontic drug delivery vehicle. J Investig Clin Dent 2016;7:87–92. Available from: https://doi.org/10.1111/jicd.12118.

[88] Kim J-J, Bae W-J, Kim J-M, et al. Mineralized polycaprolactone nanofibrous matrix for odontogenesis of human dental pulp cells. J Biomater Appl 2014;28:1069–78. Available from: https://doi.org/10.1177/0885328213495903.

[89] Qu T, Jing J, Jiang Y, et al. Magnesium-containing nanostructured hybrid scaffolds for enhanced dentin regeneration. Tissue Eng Part A 2014;20:2422–33. Available from: https://doi.org/10.1089/ten.TEA.2013.0741.

[90] Conte R, Di Salle A, Riccitiello F, et al. Biodegradable polymers in dental tissue engineering and regeneration. AIMS Mater Sci 2018;5:1073–101. Available from: https://doi.org/10.3934/matersci.2018.6.1073.

[91] Yuan Z, Nie H, Wang S, et al. Biomaterial selection for tooth regeneration. Tissue Eng Part B Rev 2011;17:373–88. Available from: https://doi.org/10.1089/ten.TEB.2011.0041.

[92] El-Gendy R, Yang XB, Newby PJ, et al. Osteogenic differentiation of human dental pulp stromal cells on 45S5 Bioglass® based scaffolds in vitro and in vivo. Tissue Eng Part A 2013;19:707–15. Available from: https://doi.org/10.1089/ten.TEA.2012.0112.

[93] Misra SK, Ansari T, Mohn D, et al. Effect of nanoparticulate bioactive glass particles on bioactivity and cytocompatibility of poly(3-hydroxybutyrate) composites. J R Soc Interface 2010;7:453–65. Available from: https://doi.org/10.1098/rsif.2009.0255.

[94] Golub EE. Role of matrix vesicles in biomineralization. Biochim Biophys Acta 2009;1790:1592–8. Available from: https://doi.org/10.1016/j.bbagen.2009.09.006.

[95] Langer R. Tissue engineering: a new field and its challenges. Pharm Res 1997;14:840–1. Available from: https://doi.org/10.1023/a:1012131329148.

[96] Ravichandran R, Venugopal JR, Sundarrajan S, et al. Minimally invasive injectable short nanofibers of poly (glycerol sebacate) for cardiac tissue engineering. Nanotechnology 2012;23:385102. Available from: https://doi.org/10.1088/0957-4484/23/38/385102.

[97] Park H, Choi B, Hu J, Lee M. Injectable chitosan hyaluronic acid hydrogels for cartilage tissue engineering. Acta Biomater 2013;9:4779−86. Available from: https://doi.org/10.1016/j.actbio.2012.08.033.

[98] Amini AA, Nair LS. Injectable hydrogels for bone and cartilage repair. Biomed Mater 2012;7:024105. Available from: https://doi.org/10.1088/1748-6041/7/2/024105.

[99] Bianco P, Riminucci M, Gronthos S, Robey PG. Bone marrow stromal stem cells: nature, biology, and potential applications. Stem Cell 2001;19:180−92. Available from: https://doi.org/10.1634/stemcells.19-3-180.

[100] Fuchs E, Segre JA. Stem cells: a new lease on life. Cell 2000;100:143−55. Available from: https://doi.org/10.1016/s0092-8674(00)81691-8.

[101] Paz AG, Maghaireh H, Mangano FG. Stem cells in dentistry: types of intra- and extraoral tissue-derived stem cells and clinical applications. Stem Cell Int 2018;2018:1−14. Available from: https://doi.org/10.1155/2018/4313610.

[102] Bonab MM, Alimoghaddam K, Talebian F, et al. Aging of mesenchymal stem cell in vitro. BMC Cell Biol 2006;7:14. Available from: https://doi.org/10.1186/1471-2121-7-14.

[103] Gronthos S, Brahim J, Li W, et al. Stem cell properties of human dental pulp stem cells. J Dent Res 2002;81:531−5. Available from: https://doi.org/10.1177/154405910208100806.

[104] Gronthos S, Mankani M, Brahim J, et al. Postnatal human dental pulp stem cells (DPSCs) in vitro and in vivo. Proc Natl Acad Sci USA 2000;97:13625−30. Available from: https://doi.org/10.1073/pnas.240309797.

[105] Bartold PM, Shi S, Gronthos S. Stem cells and periodontal regeneration. Periodontol 2006;40:164−72. Available from: https://doi.org/10.1111/j.1600-0757.2005.00139.x 2000.

[106] Huang GT-J, Gronthos S, Shi S. Mesenchymal stem cells derived from dental tissues vs. those from other sources: their biology and role in regenerative medicine. J Dent Res 2009;88:792−806. Available from: https://doi.org/10.1177/0022034509340867.

[107] Miura M, Gronthos S, Zhao M, et al. SHED: stem cells from human exfoliated deciduous teeth. Proc Natl Acad Sci USA 2003;100:5807−12. Available from: https://doi.org/10.1073/pnas.0937635100.

[108] Zhang Q-Z, Su W-R, Shi S-H, et al. Human gingiva-derived mesenchymal stem cells elicit polarization of m2 macrophages and enhance cutaneous wound healing. Stem Cell 2010;28:1856−68. Available from: https://doi.org/10.1002/stem.503.

[109] Mitrano TI, Grob MS, Carrión F, et al. Culture and characterization of mesenchymal stem cells from human gingival tissue. J Periodontol 2010;81:917−25. Available from: https://doi.org/10.1902/jop.2010.090566.

[110] Zhang W, Walboomers XF, van Kuppevelt TH, et al. The performance of human dental pulp stem cells on different three-dimensional scaffold materials. Biomaterials 2006;27:5658−68. Available from: https://doi.org/10.1016/j.biomaterials.2006.07.013.

[111] Lymperi S, Ligoudistianou C, Taraslia V, et al. Dental stem cells and their applications in dental tissue engineering. Open Dent J 2013;7:76−81. Available from: https://doi.org/10.2174/1874210601307010076.

[112] Liu Y, Zheng Y, Ding G, et al. Periodontal ligament stem cell-mediated treatment for periodontitis in miniature swine. Stem Cell 2008;26:1065−73. Available from: https://doi.org/10.1634/stemcells.2007-0734.

[113] Orciani M, Trubiani O, Vignini A, et al. Nitric oxide production during the osteogenic differentiation of human periodontal ligament mesenchymal stem cells. Acta Histochem 2009;111:15−24. Available from: https://doi.org/10.1016/j.acthis.2008.02.005.

[114] Smith MM, Fraser GJ, Mitsiadis TA. Dental lamina as source of odontogenic stem cells: evolutionary origins and developmental control of tooth generation in gnathostomes. J Exp Zool B Mol Dev Evol 2009;312B:260−80. Available from: https://doi.org/10.1002/jez.b.21272.

[115] George A, Eapen A. Dentin phosphophoryn in the matrix activates AKT and mTOR signaling pathway to promote preodontoblast survival and differentiation. Front Physiol 2015;6. Available from: https://doi.org/10.3389/fphys.2015.00221.

Further reading

Antonio N. Ten Cate's oral histology − 9th ed. https://www.elsevier.com/books/ten-cates-oral-histology/nanci/978-0-323−48518-0; [accessed 22.11.19].

10

Regenerative technologies for oral structures

Prachi Hanwatkar[1] and Ajay Kashi[2]

[1]Rochester General Hospital, Rochester, NY, United States [2]Private Practice, Rochester, NY, United States

10.1 Introduction

The oral cavity is the gateway to the digestive system of the human body. It encompasses hard and soft tissues, including muscles, teeth and its supporting structures, basal and alveolar bone including the temporomandibular joint, neurovascular structures and the minor salivary glands. All these structures are critical but are predisposed to physiologic and pathologic changes. For example, it is not rare to find individuals losing their teeth owing to some of the commonly prevalent oral diseases including dental caries and periodontitis. Trauma plays a significant role in tooth loss and damage to adjoining tissues. In addition, malignancy has the potential to affect each of these organs/structures extensively leading to impaired loss of function. People are living longer than ever before in the United States as well as in other parts of the world. Due to this increase in the life expectancy, better and improved technologies to replace damaged/lost body tissues will become important areas of research in biomedicine [1].

Clinical dentistry can often be grouped into two distinct categories when it involves replacing lost tissues. For example, replacing a missing portion of a natural tooth can be accomplished by artificial or a combination (i.e., naturally derived and artificial combinations) of restorative materials to ensure that the tooth is brought back to normal function. However, if the entire tooth/multiple teeth or all teeth are missing the treatment is often aimed towards regeneration as one of the preferred treatment options. This can be accomplished in several ways including surgical techniques. Often conventional treatment performed in dentistry can restore esthetics and function following disease or injury, but these treatments do not promote regeneration of the affected structures [2]. For example, denture therapy would constitute a non-surgical treatment approach whereas implants to support prosthetics would constitute a combination of surgical/non-surgical approaches.

In recent years, rapid advancements in research and technologies have led to new theories for tooth and tissue regeneration [3]. Similarly the management of facial defects has rapidly changed during the past few decades [4,5]. In the case of maxillofacial reconstruction (i.e., in cases of post-carcinoma or trauma), the current techniques available for the reconstruction of extended defects involves the transfer of vascularized osseous structures and osseo-myo cutaneous free flaps in particular as part of oncologic surgery of the head and neck region [4]. The defects can broadly be classified as congenital defects and acquired defects. It is beyond the scope of this chapter to provide more detailed information about these defects [6]. Readers are encouraged to refer to other literature available.

10.1.1 Classification of regenerative therapies

Regenerative therapies can be classified into three categories:

1. Stem cell based therapies — rely on the implantation of previously isolated and expanded mesenchymal stem cells (MSC) which are seeded on scaffolds [2,7].
2. Growth factor mediated therapies (GF-MT) — relies on the ability of scaffolds and implanted growth factors (GF) to attract MSC to the damaged site [2,8].
3. Bioactivity of scaffolds charged (or not) with biomolecules are able to provide adhesion and proliferation of implanted or recruited cells [7,9].

10.2 Embryology of oral structures

Authors have reported that vertebrate organ development, from the initiation to terminal differentiation depends upon inductive interactions typically between the epithelium and the adjacent mesenchyme [10,11]. Furthermore, the structures of the head and neck are derived from the cephalic portion of the neural tube which also give rise to the five pairs of branchial arches [12]. The first branchial arch gives rise to the maxillary and mandibular processes and the mesenchymal portion of the first branchial arch gives rise to the muscles of mastication [12]. Thesleff et al. have reported that the first morphological sign of a tooth is indicated by the local thickening of the dental epithelium [10,11]. Reciprocal and sequential interactions between the epithelium and mesenchyme during early tooth development lead to the formation of root dentin, cementum and periodontal tissues [10,13]. Please refer to Table 10.1. Reproduced with permission from Chapter 1 in Stem cell biology and tissue engineering in dental sciences, edited by Ajaykumar Vishwakarma, Paul Sharpe, Ongtao Shi, Murugan Ramalingam, 2015 [8].

10.3 Regeneration of teeth

Teeth are highly differentiated organs formed by the development of tooth germ tissue located in the jaw [8]. The human tooth is comprised of the outer highly calcified enamel layer followed by an inner softer dentin, pulp tissue and the periodontal ligament around the root portion. The entire odontogenic complex is housed within the alveolar bone of the

TABLE 10.1 List of different cell sources and types of scaffolds used in tissue engineering applications. The field of regenerative medicine has largely focused toward the regeneration, replacement or supporting the functions of organs that are affected. Authors have proposed that by either using a biological approach, classical engineering or combination approach regenerative efforts are being investigated [8].

Cells	Scaffold	Application	Reference
Porcine third impacted tooth bud germ cell	PGA/PLLA and PLGA scaffolds	Fabrication of bioengineered teeth	Young et al. [13]
Rat osteoblasts	PLGA/PEO scaffolds	Cell adhesion studies	Saltzman and Olbricht [15]
Smooth muscle cells	Collagen/PLLA nanofibers	Nanofiber cell interaction studies	He et al. [16]
Dermal fibroblast cells	Gelatin/HA/faujasite porous scaffolds	Wound healing applications	Ninan et al. [17]
Osteoprogenitor cells	Porous bioactive glass scaffolds	Cortical bone defects engineering	Livingston et al. [18]
Mesenchymal stem cells	Silk fibroin scaffolds	Anterior cruciate ligament (ACL) regeneration	Fan et al. [19]
Adipose-derived MSCs	rhBMP-2 and -TCP granules autologous scaffold	Maxillary bone defect	Mesimaki et al. [20]
Bone marrow-derived MSCs	Hydroxyapatite particles	Bone tissue engineering	Meijer et al. [21]
Placenta-derived MSCs	PEG-PCL triblock copolymers nanofiber scaffolds	Osteogenic differentiation potential	Zhang et al. [22]
Human bone marrow derived stem cells	Poly (L-Lactic acid) scaffolds	Cartilate tissue engineering	Izal et al. [23]
Cord blood mesenchymal stem cells	Tissue engineering constructs	Vascularized tissue engineering	Cetrulo [24]
Hematopoietic stem cells	Bioceramics composite scaffolds	Bone tissue engineering	Mishra et al. [25]
Dental pulp stem cells	Phase separated nanofibrous PLLA scaffold	Odontogenic differentiation potential and dentine tissue engineering	[26,28]

jaws that constantly remodels during an individual's lifetime. This can consequently influence the physiologic response of teeth. The average chewing forces transmitted during mastication have been reported to be in the range of 170−630 N with forces increasing from the incisor toward the molars. It is also known that males exert higher masticatory forces when compared to females [29]. Furthermore, teeth are constantly subjected to rapid cyclic temperature changes from their exposure to foods that range from freezing to boiling hot. The textures of food being chewed, the surface area of teeth available (i.e., number of healthy teeth present) and the structural stability (i.e., periodontally stable teeth) are important factors for a healthy and viable tissue. In our attempts to regenerate teeth it is not only important to ensure that these biomechanical and chemical factors are considered but there is also a need to allow for neurophysiological phenomena such as proprioception to be taken in account.

One of the main drawbacks of dental implants is the lack of a periodontal ligament that is present in natural teeth. Therefore, in order to regenerate teeth and their associated structures the odontogenic potential which is the capability of a tissue to induce gene expression in an adjoining tissue to commence tooth genesis is vital. Similarly the capability of a tissue to reciprocate to odontogenic signals and to support tooth genesis which constitutes odontogenic competence is critical [29].

Researchers in the field of tissue engineering have attempted to regenerate dental tissues as well as whole tooth [8,26]. Authors have reported about several regenerative technologies including conventional approaches as well as novel cell manipulation methods. Some of these include but are not limited to [3,8]:

1. Biodegradable Scaffold Method
2. Cell Aggregation Method
3. Three-Dimensional Cell-Manipulation Method: The "Organ Germ Method"
4. Functional Whole-Tooth Replacement Using the Bioengineered Tooth [26].

Enamel has unique properties due to its high mineral content (i.e., hydroxyapatite) thus rendering it extremely hard (i.e., to compressive loads) but at the same time making it brittle (i.e., to tensile loads). These apatite crystals are arranged individually in a parallel fashion [8,30]. While enamel has excellent mechanical properties in some respects at the same time it lacks a dedicated vascular supply thus preventing attempts at delivering biomolecules and other signaling molecules/enzymes to promote their in vivo/in vitro regeneration [31,34]. Researchers have invested millions of dollars, time and resources to date in attempts to regenerate enamel tissue; however this has remained a unique biotechnology challenge [30].

In the case of dental pulp regeneration, the aim would be to replace diseased pulp tissue instead of alloplastically restoring it which is the current approach and standard of clinical care with root canal therapy. Regenerative endodontics is a new and exciting area of science that encompasses two approaches to pulp tissue regeneration, namely cell free and cell-based method Lin and Kahler [35]. Dental pulp stem cells, stem cells from the apical papilla as well as from exfoliated deciduous teeth have been the hallmarks of the cell-based approach [3,8]. Although it is possible to regenerate bone and periodontal tissue to some extent, several challenges have been encountered regarding the regeneration of the pulp tissue [2]. According to authors promising approaches to dental pulp regeneration by cell transplantation may not be economically viable or competitive with current root canal therapies and/or dental implants [36]. From a clinical perspective the goal of pulpal regeneration should be the restoration of tooth vitality and not necessarily the regeneration of missing tissues.

Newer concepts for dental pulpal regeneration are being presented by molecular and cell biologists. One among these approaches has been the topic of cell homing where chemotactic cell homing is utilized to regenerate the pulp tissue [36]. Another emerging technique that is utilized currently includes cell transplantation [36]. Some of the challenges with these newer approaches of regenerative medicine include a lack of clinical data from randomized double-blind trials to adequately demonstrate the efficacy, long-term success and viability. Platelet-Rich Plasma (PRP) and Platelet-Rich Fibrin (PRF) contains high concentrations of transforming growth factor-beta (TGF-ß), vascular endothelial growth factor (VEGF), epithelial growth factor (EGF) and insulin-like growth factor could potentially play a key role in regeneration [8,36].

Neural re-innervation and vascular re-innervation which are commonly used to assess functional regeneration need to be further evaluated as there is a lack of in vitro and in vivo evidence [31]. Additionally, currently there is also a lack of consensus of how to measure functional pulp-dentin tissue regeneration [32]. Please refer to Fig. 10.1—10.3 reproduced by permission from Functional Tooth Regeneration Using a Bioengineered Tooth Unit as a Mature Organ Replacement Regenerative Therapy [37].

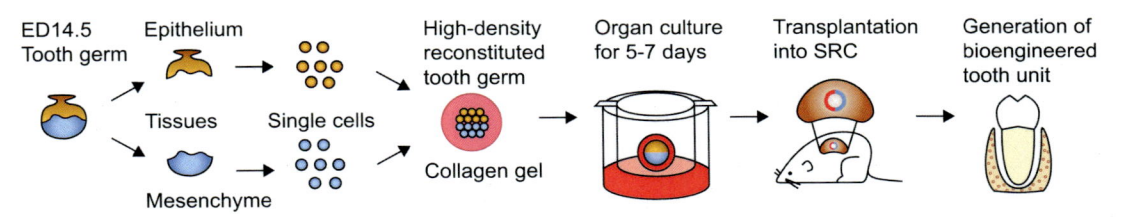

FIGURE 10.1   Generation of a bioengineered tooth unit. A Schematic representation of the generative technology of bioengineered tooth unit (Oshima et al. [37]).

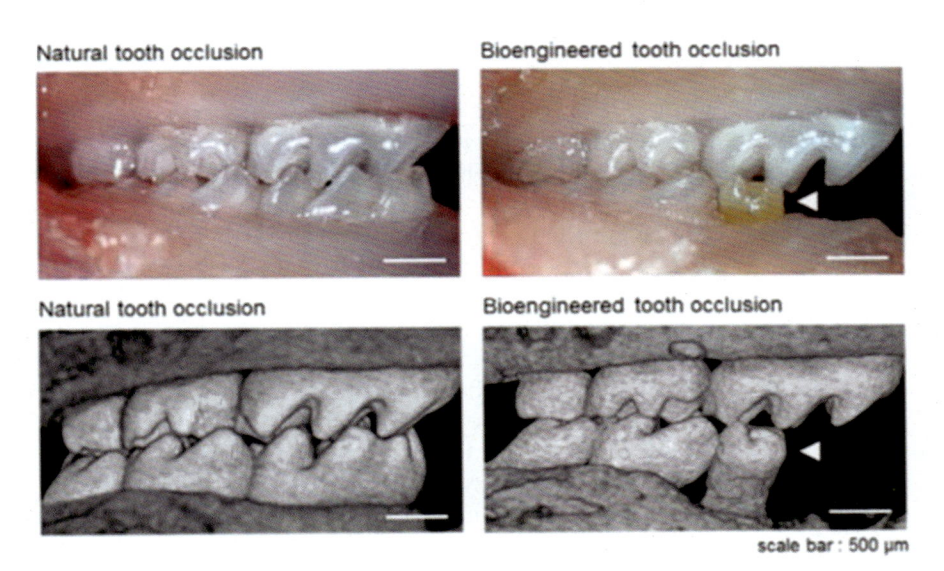

FIGURE 10.2   Engraftment and occlusion of a bioengineered tooth unit in a tooth loss model. Oral photographs (upper) and micro-CT (lower) images showing occlusion of natural (left) and bioengineered teeth (right). Scale bar, 500 mm (Oshima et al. [37]).

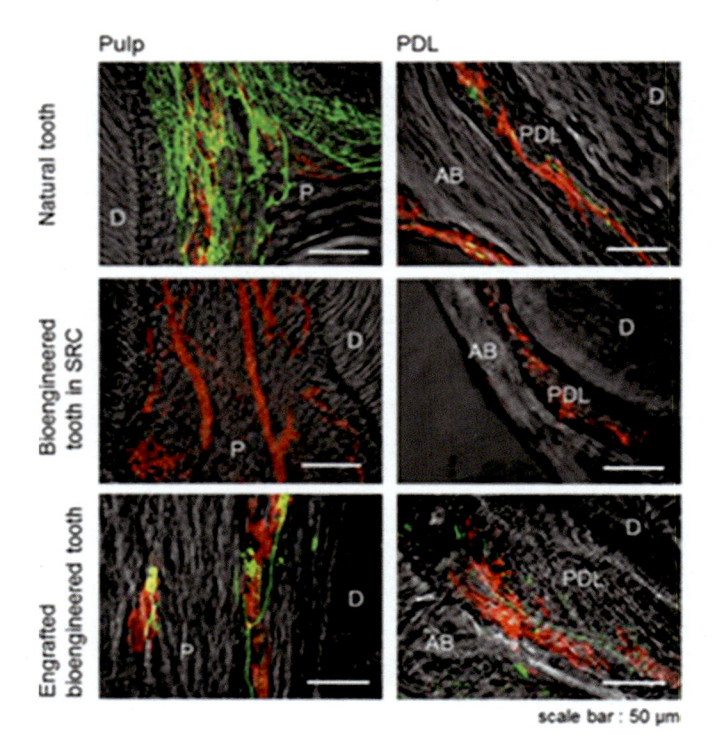

FIGURE 10.3 Experimental tooth movement and pain response to mechanical stress. Nerve fibers and blood vessels in the pulp and PDL of a natural tooth (top), a bioengineered tooth unit in an SRC (middle), and a bioengineered tooth at 40 days after transplantation (bottom) were analyzed immunohistochemically using specific antibodies for neurofilament (NF; green) and von Willebrand Factor (vWF; red). Scale bar, 50 mm. D, dentin; P, pulp; AB, alveolar bone; PDL, periodontal ligament (Oshima et al. [37]).

10.4 Regeneration of muscles/tongue

Orofacial clefts and trauma are conditions that warrant orofacial muscle regeneration. Oral cancer is the third most common cancer in the United States [38,39] and it is often associated with devastating consequences. The most common subsite for oral squamous cell carcinoma involves the tongue [41]. In addition to oral carcinoma, other orofacial trauma often results in defective muscle regeneration and fibrosis [42]. The current surgical approaches to treat some of these conditions involve radical procedures to achieve optimal negative margins and optimal local tumor control [41]. Extensive defects of the maxillofacial complex are often rehabilitated with maxillofacial prosthetic approaches and attempts at regeneration of extensive defects with optimum outcomes are yet to be realized. Some of the negative clinical effects of these previously mentioned conditions include severe compromises in the quality of life as well as significant social and psychological problems [43]. Despite improvements in the treatment modalities over the past couple of decades restoring the clinical function of individuals afflicted with cancerous or traumatic lesions has remained one of the major challenges of modern oral surgery and medicine.

In order to address some of these challenges research efforts have focused toward the introduction of cell and molecular biological approaches to regenerate muscle tissues. Authors have reported that orofacial muscle development is regulated by cranial neural crest cells and is of ectodermal origin with a higher tendency for fibrosis [44]. This is in contrast to the limb muscle tissue which is of mesodermal origin and has a lesser tendency for fibrosis [42]. Growth factor derived scaffolds can be very helpful in guiding the alignment of new myofibers, preventing fibrosis and therefore aiding in the muscle regeneration after surgical repairs of clefts and soft palate defects [43]. Similarly with some specific delivery systems growth factors and cytokines may stimulate proliferation and differentiation of resident satellite cells in the soft palate muscles after surgical repair [43].

10.5 Regeneration of bone

Bone regeneration is a significant challenge and constant efforts are being attempted in order to develop newer and more advanced technologies. Bone defects can vary in size and morphology and typically range from minor to more advanced/extensive defects. The etiologies of such defects are also due to various reasons which is beyond the scope of this chapter. The human maxillary and mandibular bone is comprised of the basal and alveolar processes that support the teeth and their associated structures. A distinction between replacement and regeneration of bone defects would be useful to explain here. In terms of utilizing xenograft-based materials or alloplastic materials for reconstruction of bone defects they can broadly constitute replacement and not necessarily regeneration as they might lack osteoconductive and osteoinductive potential [45]. With the use of autogenous bone or cadaver bone along with growth factors and biologically active components mixed, such combinations may have varying degrees of potential to stimulate the native bone to regenerate. However, the extent of the bone defect often dictates the quality of the bone that is regenerated with smaller defects being more successfully treated when compared to larger defects.

Advanced atrophy or jaw defects often warrant extensive horizontal and vertical bone augmentation in order to enable patients to be restored with implants [4,5]. Autologous bone (i.e., bone harvested from a patient's own jaw) is capable of bone regeneration due to its osteoconductive potential and it also brings with it several advantages including minimal risk for infections, rejection and predictable long-term success. With the advent of stem cell-based therapies there has been some promise toward true regeneration of bone tissue [46]. The human palate has been shown to be of critical importance in the regenerative process due to the cells derived from it [4]. According to researchers the cells from the human palate have been shown to possess specific properties such as increased proliferation rates and are representative of a regenerative phenotype [40]. Grimm et al have reported that multipotent ecto-mesenchymal palatal-derived stem cells (paldSCs) can tolerate low oxygen concentrations and have the potential for osteogenic, chondrogenic, and adipogenic differentiation, depending on the external stimulus by the cytokines [4]. Furthermore the authors mention that paldSCs are capable of differentiating both into the neuronal lineage as well as the osteogenic lineage and can be accessed easily from patients undergoing open flap minimally invasive periodontal surgery [4].

Studies have also demonstrated that autogenous bone blocks when covered with PRF in the anterior maxillary area results in increased bone width and decreased bone resorption [48]. This data may or may not be valid in other regions of the upper/lower jaws and further studies are needed to confirm the same. Also, PRP and PRF have been shown able to promote bone regeneration with histological confirmation [2].

10.6 Regeneration of TMJ

Temporomandibular disorders (TMD) are a collective series of various disorders that can affect the TMJ [49]. Degenerative arthropathy is known to be the most common outcome of long-standing TMD; however, there is yet a lack of understanding about the pathogenesis and progression of TMD [49]. The mandible is the largest of the maxillofacial bones and it articulates with the base of the cranium with a bilateral ginglymoarthroidal joint [50]. It supports several functions including mastication and airway communication. Further it is understood that the mandible is an important component of the pharyngeal arch which plays a role in the development and maintenance of a patent airway [50]. Authors have reported that the TMJ is formed from three separate condensations of mesenchyme and the presence of the disc is necessary for condylar regeneration [50,51].

The lack of consistent success with some of the common treatment approaches has necessitated the development of novel solutions such as those based in tissue engineering [49]. The principal elements of tissue engineering include cells, stimuli, and scaffolds [8,49]. Studies have shown that the mechanical stimuli approximating physiological loading conditions have improved mechanical and biochemical properties in engineered TMJ discs [49]. Bioreactors have also been used for engineering bone in a condylar shape [8]. In spite of these advances, an imminent challenge of surgically accessing the ailing TMJ tissue and integrating the neotissues within the native milieu still exists [49]. Biomechanical testing is critical regardless of any developments in regenerative technologies. The biomechanical behavior of TMJ implants with various screw configurations has been reported in studies, albeit these studies focused towards alloplastic replacement of damaged joint structures [52].

10.7 Regeneration of salivary glands

The human salivary gland consists of epithelial acini, connecting ducts, branching structures, vascular and neuronal networks that function collectively to produce/secrete saliva [53]. Xerostomia is an associated symptom of irreversible salivary gland hypofunction resulting from many systemic diseases including Sjogren's syndrome, granulomatous diseases, graft-versus-host disease, cystic fibrosis, uncontrolled diabetes, human immunodeficiency virus infection, thyroid disease and late-stage liver disease [54,55]. Hyposalivation is also a prominent complication for more than 550,000 patients that are annually diagnosed with head and neck cancer globally [39].

Saliva is essential for digestion, lubrication, oral homeostasis and protection against a variety of microbial and environmental hazards [53]. Reduced salivary flow can lead to difficulty swallowing, loss of taste, rampant caries, painful mucositis, oral fungal infections and taste loss [8,53]. There are reports alluding the beneficial effects of non-Salivary gland, non-epithelial cells and their ability to regenerate salivary glands that are irradiated [53].

Bone Marrow (BM)-derived cells, BM-derived mesenchymal stem cells (MSC), human adipose-derived MSCs, salivary gland derived MSC-like amniotic cells, embryonic stem cells (ESC) and induced-pluripotent stem cells (IPS) have been reported as having varying degrees of efficacy in the regeneration of salivary gland tissues [8,53,54]. Complete functional regeneration requires restoration of all tissue types, and gene therapy alone may not be able to achieve that [47]. Recent studies have revealed the dynamic aspects of submandibular gland (SMG) branching, including cell—cell and cell—matrix interactions, as well as cell migration and proliferation [56].

Fibronection (FN) has also been reported to regulate bud branching and facilitate duct elongation [56]. Fibronectin is a glycoprotein known to assist cellular adhesion, migration, growth, and differentiation [56—58]. It also promotes cleft formation and SMG tissue branching by converting cell—cell adhesions to cell—matrix adhesions in the cleft-specific regions [56,58]. Low concentrations of FN has been shown to induce ductal formation in 3D SMG cell aggregates [56].

FN along with FGF7 (FGF7 is a secreted protein also known as keratinocyte growth factor (KGF)) have synergistic effects [56]. FN can significantly increase bud number and ductal elongation in the cultured epithelial rudiments that are supplemented with FGF7 [56]. Despite significant advances being made during the last decade, definitive therapies for salivary gland hypofunction have not been successful and can be attributed to some of the fundamental challenges that are associated with each approach [53]. For example translating the research findings from animal models to human salivary gland function continue to be a challenge [8,53].

10.8 Microgravity

According to the National Aeronatics and Space Administration (NASA) the acceleration of an object toward the ground caused by gravity alone, near the surface of Earth, is called "normal gravity," or $1\,g$ and the condition of microgravity occurs whenever an object is in free fall. Any object on earth and the larger universe is affected by gravitational force. It has been reported that gravity renders a key mechanical stimulus on our bodies at all times [59]. Microgravity environments are typically encountered only in space or during free fall [60—63]. Simulating microgravity environments have been reported to be useful in studying cell behavior and therapeutic effects following cell transplantation [64—67]. Researchers have reported that at the cellular level, gravity serves as a useful factor for the determination of cell features [64—67]. Similarly researchers have previously reported about the impact of mechanical loading/mechanical stress and their effects on osteogenic differentiation of MSCs [68], osteoblast proliferation [69] and myogenic differentiation [70], cell differentiation into tissues such as the smooth muscle of human airways [71], human myocardium [72], keratocytes [73], and dentin-like tissue [74].

When a comparison study between cultured murine bone marrow stromal cells and control cells under $1\,{}^{x}g$ (1 G) conditions were carried out it was demonstrated that most of the differential gene expressions between cells under the two gravity levels were related to neural development, neuron morphogenesis, and transmission of nerve impulse and synapse [75]. Other researchers have also reported differential gene expression in cell lines [76]. Due to the dearth of substantial high-level evidence pertaining to the effects of microgravity at the cellular levels further funding need to be dedicated towards such studies. Additionally, the effects of microgravity on in vitro bioengineered tissue constructs have not been studied adequately. Initial reports of the effects of microgravity in the areas of cell biology and regenerative medicine have been promising. However future studies including designing safety protocols, clinical trial designs and regulatory practices need to be developed.

10.9 Ethical considerations

The goal of regenerative medicine is to replace/regenerate damaged or degenerated tissue(s) without generating chronic inflammation, unwanted cell growth, rejection or long-term adverse events. In the context of regenerative technologies in healthcare it is important to highlight some of the ethical implications. Some of these include the private banking of cells, safety of cells, informed consent, intellectual property issues, allocation of federal research funds and international laws and regulations. We have previously discussed and reported relevant issues related to most of these aforementioned topics [77].

As the life expectancy of populations in the developed world and the developing world is increasing greater efforts at addressing emerging ethical problems will need to be encouraged. With the wide availability of treatment options worldwide and with the increasing use of the internet patients are finding it easier to access treatments away from their home countries. This can bring with it some unique challenges including the regulation of expensive unproven therapies or treating adverse effects from treatments offered in foreign countries. We have previously reported about the potential risks and pitfalls of availing clinical therapies in foreign countries with lax regulations (data presented at The 7th International Conference on Ethical Issues in Biomedical Engineering, Brooklyn, NY, April 20–21, 2013). In an era where healthcare costs are increasing it is only natural to expect patients to look for cheaper alternatives even if it means crossing international borders. Patients who obtain healthcare in foreign countries and return for after care to their home countries can present with unique challenges including the use of unapproved products, poor infection control protocols, unproven clinical approaches and undesired long-term outcomes. These and other related areas of concern are only expected to increase in the future and they need to be debated further.

There are reports that a large segment of scientific/clinical research is not reproducible with some studies suggesting that scientists are unable to reproduce as much as 70% of previous research [78,79]. Most of this is generally attributed to bad research design and inherent biases [80]. With these low probabilities of reproducibility several important factors need to be discussed including the allocation of resources, treatment efficacy, the placebo and nocebo effects.

References

[1] Kochanek KD, Sherry L, Murphy SL, Xu J, Arias E. Deaths: final data for 2017; 2017.
[2] Chisini LA, Conde MCM, Grazioli G, Martin ASS, Carvalho RV, Sartori LRM, et al. Bone, periodontal and dental pulp regeneration in dentistry: a systematic scoping review. Braz Dental J 2019;30(2):77−95 [pii]. doi: S0103-64402019000200077.

[3] Zhai Q, Dong Z, Wang W, Li B, Jin Y. Dental stem cell and dental tissue regeneration. Front Med 2019;13 (2):152–9. Available from: https://doi.org/10.1007/s11684-018-0628-x.

[4] Grimm WD, Dannan A, Giesenhagen B, Schau I, Varga G, Vukovic MA, et al. Translational research: Palatal-derived ecto-mesenchymal stem cells from human palate: a new hope for alveolar bone and cranio-facial bone reconstruction. Int J Stem Cell 2014;7(1):23–9. Available from: https://doi.org/10.15283/ijsc.2014.7.1.23.

[5] Urban IA, Montero E, Monje A, Sanz-Sánchez I. Effectiveness of vertical ridge augmentation interventions: a systematic review and meta-analysis. J Clin Periodontol 2019;46(S21):319–39. Available from: https://doi.org/10.1111/jcpe.13061.

[6] Gupta AD, Aviral Verma A, Islam JI, Agarwal S. Maxillofacial defects and their classification: a review. Int J Adv Res 2016;4:109–14. Available from: https://doi.org/10.21474/IJAR01.

[7] Conde MC, Chisini LA, Demarco FF, Nör JE, Casagrande L, Tarquinio SB. Stem cell-based pulp tissue engineering: variables enrolled in translation from the bench to the bedside, a systematic review of literature. Int Endod J 2016;543–50. Available from: https://doi.org/10.1111/iej.12489 PubMed - NCBI.

[8] Vishwakarma A, Sharpe P, Shi S, Ramalingam M. *An introduction to stem cell biology and tssue engineering*; 2015.

[9] Demarco FF, Conde MCM, Cavalcanti B, Casagrande L, Sakai VT, Nör JE. Dental pulp tissue engineering. Braz Dental J 2011;22(1):3–13. Available from: https://doi.org/10.1590/S0103-64402011000100001.

[10] Thesleff, I., Vaahtokari, A., & Partanen, A. M. (1995). Regulation of organogenesis. common molecular mechanisms regulating the development of teeth and other organs. Spain: Retrieved from https://www.ncbi.nlm.nih.gov/pubmed/7626420

[11] Young CS, Terada S, Vacanti JP, Honda M, Bartlett JD, Yelick PC. Tissue engineering of complex tooth structures on biodegradable polymer scaffolds. J Dental Res 2002;81(10):695–700. Available from: https://doi.org/10.1177/154405910208101008.

[12] Zohrabian VM, Poon CS, Abrahams JJ. Embryology and anatomy of the jaw and dentition. SemUltrasound CT MRI 2015;36(5):397–406. Available from: https://doi.org/10.1053/j.sult.2015.08.002.

[13] Li J, Parada C, Chai Y. Cellular and molecular mechanisms of tooth root development. Feb 1 2017;144 (3):374–84. Available from: https://www.ncbi.nlm.nih.gov/pubmed/28143844.

[14] Zhang YD, Chen Z, Song YQ, Liu C, Chen YP. Making a tooth: growth factors, transcription factors, and stem cells. Cell Res 2005;15(5):301–16. Available from: https://doi.org/10.1038/sj.cr.7290299.

[15] Saltzman WM, Olbricht WL. Building drug delivery into tissue engineering. Nat Rev Drug Discov 2002;1 (3):177–86. Available from: https://doi.org/10.1038/nrd744.

[16] He W, Ma Z, Yong T, Teo WE, Ramakrishna S. Fabrication of collagen-coated biodegradable polymer nanofiber mesh and its potential for endothelial cells growth. Biomaterials 2005;26(36):7606–15 doi:S0142-9612(05)00414-X [pii].

[17] Ninan N, Muthiah M, Park IK, Elain A, Wong TW, Thomas S, et al. Faujasites incorporated tissue engineering scaffolds for wound healing: In vitro and in vivo analysis. ACS Appl Mater Interfaces 2013;5 (21):11194–206. Available from: https://doi.org/10.1021/am403436y.

[18] Livingston T, Ducheyne P, Garino J. In vivo evaluation of a bioactive scaffold for bone tissue engineering. J Biomed Mater Res 2002;62(1):1–13. Available from: https://doi.org/10.1002/jbm.10157.

[19] Fan H, Liu H, Toh SL, Goh JC. Anterior cruciate ligament regeneration using mesenchymal stem cells and silk scaffold in large animal model. Biomaterials 2009;30(28):4967–77. Available from: https://doi.org/10.1016/j.biomaterials.2009.05.048.

[20] Mesimaki K, Lindroos B, Tornwall J, Mauno J, Lindqvist C, Kontio R, et al. Novel maxillary reconstruction with ectopic bone formation by GMP adipose stem cells. Int J Oral Maxillofac Surg 2009;38(3):201–9. Available from: https://doi.org/10.1016/j.ijom.2009.01.001.

[21] Meijer GJ, de Bruijn JD, Koole R, van Blitterswijk CA. Cell based bone tissue engineering in jaw defects. Biomaterials 2008;29(21):3053–61. Available from: https://doi.org/10.1016/j.biomaterials.2008.03.012.

[22] Zhang D, Tong A, Zhou L, Fang F, Guo G. Osteogenic differentiation of human placenta-derived mesenchymal stem cells (PMSCs) on electrospun nanofiber meshes. Cytotechnology 2012;64(6):701–10. Available from: https://doi.org/10.1007/s10616-012-9450-5.

[23] Izal I, Aranda P, Sanz-Ramos P, Ripalda P, Mora G, Granero-Molto F, et al. Culture of human bone marrow-derived mesenchymal stem cells on of poly(L-lactic acid) scaffolds: potential application for the tissue engineering of cartilage. Knee Surg Sports Traumatol Arthrosc: Off J ESSKA 2013;21(8):1737–50. Available from: https://doi.org/10.1007/s00167-012-2148-6.

[24] Cetrulo CL. Cord-blood mesenchymal stem cells and tissue engineering. Stem Cell Rev 2006;2(2):163−8 doi: SCR:2:2:163 [pii].

[25] Mishra S, Rajyalakshmi A, Balasubramanian K. Compositional dependence of hematopoietic stem cells expansion on bioceramic composite scaffolds for bone tissue engineering. J Biomed Mater Res Part A 2012;100(9):2483−91. Available from: https://doi.org/10.1002/jbm.a.34145.

[26] Oshima M, Tsuji T. Functional tooth regenerative therapy: tooth tissue regeneration and whole-tooth replacement. 2015;102(2):123−36. Available from: https://www.ncbi.nlm.nih.gov/pubmed/25052182https://www.ncbi.nlm.nih.gov/pubmed/25052182

[27] Deng X, Xu M, Li D, Sui G, Hu XY, Yang X. Electrospun PLLA/MWNTs/HA hybrid nanofiber scaffolds and their potential in dental tissue engineering. Key Eng Mater − KEY ENG MAT 2007;330−332:393−6. Available from: https://doi.org/10.4028/www.scientific.net/KEM.330-332.393.

[28] Xu MM, Mei F, Li D, Yang XP, Sui G, Deng XL, et al. Electrospun poly(L-lacticacid)/nano-hydroxyapatite hybrid nanofibers and their potential in dental tissue engineering. Key Eng Mater 2007;330−332:377−80. Available from: https://doi.org/10.4028/www.scientific.net/KEM.330-332.377.

[29] Umesh S, Padma S, Asokan S, Srinivas T. Fiber Bragg grating based bite force measurement [Abstract]. J Biomech 2016;49(13):2877−81 doi:S0021-9290(16)30732-1 [pii].

[30] Mirali P, Thomas GHD. Enamel biomimetics—fiction or future of dentistry. Dallas Tx: Springer Nature; 2019.

[31] Zhang W, Yelick PC. Vital pulp therapy-current progress of dental pulp regeneration and revascularization. 2010;2010:856087−9. Available from: http://dx.doi.org/10.1155/2010/856087

[32] Fawzy El-Sayed KM, Ahmed GM, Abouauf EA, Schwendicke F. Stem/progenitor cell-mediated pulpal tissue regeneration: a systematic review and meta-analysis. Nov 2019;52(11):1573−85. Available from: https://onlinelibrary.wiley.com/doi/abs/10.1111/iej.13177.

[33] Xu R, Zhou Y, Zhang B, Shen J, Gao B, Xu X, et al. Enamel regeneration in making a bioengineered tooth Curr Stem Cell Res Ther 2015;10(5):434Retrieved from. Available from: https://www.ncbi.nlm.nih.gov/pubmed/25741712.

[34] Bartlett JD. Dental enamel development: proteinases and their enamel matrix substrates. ISRN Dent 2013;2013. Available from: https://doi.org/10.1155/2013/684607 684607-24.

[35] Lin LM, Kahler B. A review of regenerative endodontics current protocols and future directions. J Istanb Univ Fac Dent 2017;41−51. Available from: https://doi.org/10.17096/jiufd.53911.

[36] Kim JY, Xin X, Moioli EK, Chung J, Lee CH, Chen M, et al. Regeneration of dental-pulp-like tissue by chemotaxis-induced cell homing. Tissue Eng Part A 2010;16(10):3023−31. Available from: https://doi.org/10.1089/ten.tea.2010.0181.

[37] Oshima M, Mizuno M, Imamura A, Ogawa M, Yasukawa M, Yamazaki H, et al. Functional tooth regeneration using a bioengineered tooth unit as a mature organ replacement regenerative therapy. PLoS One 2011;6(7):e21531. Available from: https://doi.org/10.1371/journal.pone.0021531.

[38] NIDCR. Oral cancer causes, symptoms, diagnosis, treatment. National Institute of Dental and Craniofacial Research; 2018. Retrieved from https://www.nidcr.nih.gov/health-info/oral-cancer/more-info.

[39] Bray F, Ferlay J, Soerjomataram I, Siegel RL, Torre LA, Jemal A. Global cancer statistics 2018: GLOBOCAN estimates of incidence and mortality worldwide for 36 cancers in 185 countries. CA: Cancer J Clin 2018;68(6):394−424. Available from: https://doi.org/10.3322/caac.21492.

[40] Greiner JFW, Hauser S, Widera D, Müller J, Qunneis F, Zander C, et al. Efficient animal-serum free 3D cultivation method for adult human neural crest-derived stem cell therapeutics. 2011 Dec 17;22:403−19. Available from: https://www.ncbi.nlm.nih.gov/pubmed/22179938

[41] Xu Q, Shanti RM, Zhang Q, Cannady SB, O'Malley BW, Le AD. A gingiva-derived mesenchymal stem cell-laden porcine small intestinal submucosa extracellular matrix construct promotes myomucosal regeneration of the tongue. Tissue Eng Part A 2017;23(7−8):31−312. Available from: https://doi.org/10.1089/ten.tea.2016.0342.

[42] Rosero Salazar DH, Carvajal Monroy PL, Wagener FADTG, Von den Hoff JW. Orofacial muscles: Embryonic development and regeneration after injury. J Dental Res 2019;. Available from: https://doi.org/10.1177/0022034519883673.

[43] Carvajal Monroy PL, Grefte S, Kuijpers-Jagtman AM, Wagener FADTG, Von den Hoff JW. Strategies to improve regeneration of the soft palate muscles after cleft palate repair. Tissue Eng Part B: Rev 2012;18(6):468−77. Available from: https://doi.org/10.1089/ten.teb.2012.0049.

[44] Rinon A, Lazar S, Marshall H, Büchmann-Møller S, Neufeld A, Elhanany-Tamir H, et al. Cranial neural crest cells regulate head muscle patterning and differentiation during vertebrate embryogenesis. Development 2007;134(17):3065−75. Available from: https://doi.org/10.1242/dev.002501.

[45] Jepsen S, Schwarz F, Cordaro L, Derks J, Hämmerle CHF, Heitz-Mayfield LJ, et al. Regeneration of alveolar ridge defects. consensus report of group 4 of the 15th European workshop on periodontology on bone regeneration. J Clin Periodontol 2019;46(S21):277−86. Available from: https://doi.org/10.1111/jcpe.13121.

[46] Kaigler D, Avila-Ortiz G, Travan S, Taut AD, Padial-Molina M, Rudek I, et al. Bone engineering of maxillary sinus bone deficiencies using enriched CD90 + stem cell therapy: a randomized clinical trial. J Bone Miner Res: Off J Am Soc Bone Miner Res 2015;30(7):1206−16. Available from: https://doi.org/10.1002/jbmr.2464.

[47] Holmberg KV, Hoffman MP. Anatomy, biogenesis and regeneration of salivary glands [Internet]. Basel, Switzerland: S. Karger AG; 2014. 13 p. (Saliva: Secretion and Functions; vol. 24). Available from: https://www.karger.com/Article/FullText/358776

[48] Al-Hamed FS, Mahri M, Al-Waeli H, Torres J, Badran Z, Tamimi F. Regenerative effect of platelet concentrates in oral and craniofacial regeneration. Front Cardiovasc Med 2019;6:126. Available from: https://doi.org/10.3389/fcvm.2019.00126.

[49] Aryaei A, Vapniarsky N, Hu J, Athanasiou K. Recent tissue engineering advances for the treatment of temporomandibular joint disorders. Curr Osteoporos Rep 2016;14(6):269−79. Available from: https://doi.org/10.1007/s11914-016-0327-y.

[50] Stocum D, Roberts W. Part I: Development and physiology of the temporomandibular joint. Curr Osteoporos Rep 2018;16(4):360−8. Available from: https://doi.org/10.1007/s11914-018-0447-7.

[51] Hayashi H, Fujita T, Shirakura M, Tsuka Y, Fujii E, Terao A, et al. Role of articular disc in condylar regeneration of the mandible. Exp Anim 2014;63(4):395−401. Available from: https://doi.org/10.1538/expanim.63.395.

[52] Roy Chowdhury A, Kashi A, Saha S. Optimization of surgical procedure, screw dimensions and screw hole locations for a temporomandibular joint implant: a finite element study. J Biomech 2011;44(14):2584−7. Available from: https://doi.org/10.1016/j.jbiomech.2011.06.002.

[53] Lombaert I, Movahednia MM, Adine C, Ferreira JN. Concise review: salivary gland regeneration: Therapeutic approaches from stem cells to tissue organoids. Stem Cell 2017;35(1):97−105. Available from: https://doi.org/10.1002/stem.2455.

[54] Mendi A, Ulutürk H, Ataç MS, Yılmaz D. Stem cells for the oromaxillofacial area: could they be a promising source for regeneration in dentistry? Adv Exp Med Biol 2019;1144:101Retrieved from. Available from: https://www.ncbi.nlm.nih.gov/pubmed/30725365.

[55] von Bültzingslöwen I, Sollecito TP, Fox PC, Daniels T, Jonsson R, Lockhart PB, et al. Salivary dysfunction associated with systemic diseases: systematic review and clinical management recommendations. Oral Surg Oral Med Oral Pathol Oral Radiol Endodontol 2007;103:S57.e1−57.e15. Available from: https://doi.org/10.1016/j.tripleo.2006.11.010.

[56] Farahat M, Kazi G, Taketa H, Hara E, Oshima M, et al. Fibronectin induced ductal formation in salivary gland self-organization model. Dev Dyn 2019;248(11). Available from: https://doi.org/10.1002/dvdy.78.

[57] Taketa H, Sathi GA, Farahat M, Rahman KA, Sakai T, Hirano Y, et al. Peptide-modified substrate for modulating gland tissue growth and morphology in vitro. Sci Rep 2015;5(1):11468. Available from: https://doi.org/10.1038/srep11468.

[58] Yang T, Young T. The enhancement of submandibular gland branch formation on chitosan membranes. Biomaterials 2008;29(16). Available from: https://doi.org/10.1016/j.biomaterials.2008.02.014.

[59] Herranz R, Medina FJ. Cell proliferation and plant development under novel altered gravity environments. Plant Biol (Stuttgart, Ger) 2014;16(Suppl 1):23−30. Available from: https://doi.org/10.1111/plb.12103.

[60] Leach CS, Dietlein LF, Pool SL, Nicogossian AE. Medical considerations for extending human presence in space. Acta Astronaut 1990;21(9):659−66. Available from: https://doi.org/10.1016/0094-5765(90)90077-x.

[61] LeBlanc A, Schneider V, Shackelford L, West S, Oganov V, Bakulin A, et al. Bone mineral and lean tissue loss after long duration space flight. J Musculoskelet Neuronal Interact 2000;1(2):157−60.

[62] Vico L, Collet P, Guignandon A, Lafage-Proust MH, Thomas T, Rehaillia M, et al. Effects of long-term microgravity exposure on cancellous and cortical weight-bearing bones of cosmonauts. Lancet (London, England) 2000;355(9215):1607−11 doi:S0140673600022170 [pii].

[63] Yumi K, Louis Y. Simulated microgravity based stem cell cultures enhance their utility for cell-based therapy; 2013. https://doi.org/10.2174/22115501113029990020.

[64] Imura T, Nakagawa K, Kawahara Y, Yuge L. Stem cell culture in microgravity and its application in cell-based therapy. Stem Cell Dev 2018;27(18):1298−302. Available from: https://doi.org/10.1089/scd.2017.0298.

[65] Kim J, Montagne K, Nemoto H, Ushida T, Furukawa KS. Hypergravity down-regulates c-fos gene expression via ROCK/rho-GTP and the PI3K signaling pathway in murine ATDC5 chondroprogenitor cells. PLoS One 2017;12(9):e0185394. Available from: https://doi.org/10.1371/journal.pone.0185394.

[66] Otsuka T, Imura T, Nakagawa K, Shrestha L, Takahashi S, Kawahara Y, et al. Simulated microgravity culture enhances the neuroprotective effects of human cranial bone-derived mesenchymal stem cells in traumatic brain injury. Stem Cell Dev 2018;27(18):1287−97. Available from: https://doi.org/10.1089/scd.2017.0299.

[67] Yuge L, Sasaki A, Kawahara Y, Wu SL, Matsumoto M, Manabe T, et al. Simulated microgravity maintains the undifferentiated state and enhances the neural repair potential of bone marrow stromal cells. Stem Cell Dev 2011;20(5):893−900. Available from: https://doi.org/10.1089/scd.2010.0294.

[68] Ozcivici E, Luu YK, Adler B, Qin YX, Rubin J, Judex S, et al. Mechanical signals as anabolic agents in bone. Nat Rev Rheumatol 2010;6(1):50−9. Available from: https://doi.org/10.1038/nrrheum.2009.239.

[69] Kapur S, Mohan S, Baylink DJ, Lau KH. Fluid shear stress synergizes with insulin-like growth factor-I (IGF-I) on osteoblast proliferation through integrin-dependent activation of IGF-I mitogenic signaling pathway. J Biol Chem 2005;280(20):20163−70 doi:M501460200 [pii].

[70] Ciofani G, Ricotti L, Rigosa J, Menciassi A, Mattoli V, Monici M. Hypergravity effects on myoblast proliferation and differentiation. J Biosci Bioeng 2012;113(2):258−61. Available from: https://doi.org/10.1016/j.jbiosc.2011.09.025.

[71] Asano S, Ito S, Morosawa M, Furuya K, Naruse K, Sokabe M, et al. Cyclic stretch enhances reorientation and differentiation of 3-D culture model of human airway smooth muscle. Biochem Biophys Rep 2018;16:32−8. Available from: https://doi.org/10.1016/j.bbrep.2018.09.003.

[72] Ruan JL, Tulloch NL, Saiget M, Paige SL, Razumova MV, Regnier M, et al. Mechanical stress promotes maturation of human myocardium from pluripotent stem cell-derived progenitors. Stem Cell (Dayton, Ohio) 2015;33(7):2148−57. Available from: https://doi.org/10.1002/stem.2036.

[73] Ferrell N, Cheng J, Miao S, Roy S, Fissell WH. Orbital shear stress regulates differentiation and barrier function of primary renal tubular epithelial cells. ASAIO J (Am Soc Artif Intern Organs: 1992) 2018;64(6):766−72. Available from: https://doi.org/10.1097/MAT.0000000000000723.

[74] Hashmi B, Mammoto T, Weaver J, Ferrante T, Jiang A, Jiang E, et al. Mechanical induction of dentin-like differentiation by adult mouse bone marrow stromal cells using compressive scaffolds. Stem Cell Res 2017;24:55−60 doi:S1873-5061(17)30166-6 [pii].

[75] Monticone M, Liu Y, Pujic N, Cancedda R. Activation of nervous system development genes in bone marrow derived mesenchymal stem cells following spaceflight exposure. J Cell Biochem 2010;111(2):442−52. Available from: https://doi.org/10.1002/jcb.22765.

[76] Mattei C, Alshawaf A, D'Abaco G, Nayagam B, Dottori M. Generation of neural organoids from human embryonic stem cells using the rotary cell culture system: effects of microgravity on neural progenitor cell fate. Stem Cell Dev 2018;27(12):848−57. Available from: https://doi.org/10.1089/scd.2018.0012.

[77] Kashi A, Saha S. Ethical aspects of tissue engineered products. In: Vishwakarma A, Sharpe P, Shi S, Wang X-P, Ramalingam M, editors. Stem cell biology and tissue engineering in dental sciences − Chapter 84. Elsevier; 2013 (in press).

[78] Baker M. Is there a reproducibility crisis? Nature 2016;533(7604):452Retrieved from. Available from: https://search.proquest.com/docview/1792376854.

[79] Baker M, Dolgin E. Reproducibility project yields muddy results Nature 2017;541(7637):269Retrieved from. Available from: https://search.proquest.com/docview/1861002886.

[80] Kak S. On ternary coding and three-valued logic; 2018. Retrieved from https://arxiv.org/abs/1807.06419.

11

State-of-the-art strategies and future interventions in bone and cartilage repair for personalized regenerative therapy

Yogendra Pratap Singh[1], Joseph Christakiran Moses[1], Ashutosh Bandyopadhyay[1], Bibrita Bhar[1], Bhaskar Birru[1], Nandana Bhardwaj[2] and Biman B. Mandal[1,3]

[1]Biomaterial and Tissue Engineering Laboratory, Department of Biosciences and Bioengineering, Indian Institute of Technology Guwahati, Guwahati, India [2]Department of Biotechnology, National Institute of Pharmaceutical Education and Research Guwahati, Guwahati, India [3]Centre for Nanotechnology, Indian Institute of Technology Guwahati, Guwahati, India

11.1 Introduction

Orthopedic surgeries associated with bone, cartilage, and articular joints have been on the rise since the last decade, contributed due to the increase in aging population, sedentary lifestyle, sports/ traumatic injuries, and pathological conditions. With over 3.48 million total knee replacement surgeries and more than 0.5 million total hip replacement surgeries estimated to occur by 2030 in USA alone [1], and this staggering statistics point towards the dire need for clinically viable treatment solutions for orthopedic care and management. The global market and demand for orthopedic grafts and implants segment in regional markets in Asia-Pacific, Europe, USA and rest of the world was estimated to be at US$ 46.5 billion by 2017 from US$ 21.1 billion in 2007, compounded with an annual growth rate of 8.2% between 2007 and 2019 [2]. Though bone has innate ability to self-heal, it is significantly compromised in pathological condition or with age, while cartilage damage affirmatively needs intervention since it is avascular tissue, and any damage worsens the plight even more. Surgical interventions for large volume bone defect still rely on autogenic/allogenic graft, which suffers from donor site morbidity and inadequacy of the required volume of bone graft for transplantation. On the other hand, palliative cartilage repair involves either bone marrow stimulation procedures such as subchondral bone drilling to allow the mesenchymal stem cells to repopulate the damaged hyaline cartilage or matrix-assisted approaches such as autologous chondrocyte implantation (ACI) and chondroplasty [3,4]. These conventional strategies are now wading away for tissue-engineered strategies, which in the last two decades has made some remarkable strides in clinical translation. For instance, there are tissue-engineered grafts already available from the US and European markets such as DeNovo® NT, Arthrex® Biocartilage, Chondrofix®, CeruleauNuProbe® and KineSpring®, which have been used in clinical settings. However, these grafts suffer from shortcomings such as high failure rates (40−70%), which result in poor performance of these grafts, and often a second surgical intervention is necessitated in such conditions.

The current biomaterials used in clinic mostly fall under first or second-generation biomaterials, which are either bio-inert or bioactive in nature. Hence, their use in tissue engineering strategies for clinical use though has been credited as innovative, but the success stories are mostly anecdotal. For instance, the concept of using decellularized bone graft loaded with bone morphogenetic proteins (BMPs), generated vascularized constructs and gained favorable outcomes initially but never bore providence [5]. Similarly, concerns regarding the compliance of implanted chondral grafts looms at large for clinicians, as the survival rate at 2 years is 86%, while at 9 years is 69% [6], and decreases drastically after 10 years with the need for a revision surgery. This has instigated the research towards third-generation biomaterials and tissue engineering strategies, more focussed on addressing the drawbacks of graft compliance, survival rate, and bioresorbability. These third-generation biomaterials utilize multidisciplinary approaches in biomaterial design, fabrication procedures inspired by nature, in vitro tissue maturation strategies inclusive of recombinant engineering, or use of 'cell-free' strategies to improve the chances of clinical success. According to the notion of this chapter, our interpretation of 'third generation' generation biomaterials can be delineated as follows:

1. emphasis on grafts fabricated through rapid prototyping methods, recapitulating the form or physiological function of bone/cartilage/interface;
2. use of biomaterials laden with cell responsive molecules such as exosomes or nanoparticles loaded with biomolecules/exogenous DNA/mRNA for mediating functions relating to cell homing, immunomodulation, tissue maturation or differentiation;
3. use of biomaterials that are responsive to stimuli such as mechanical stresses/magnetic or electric field/temperature or pH changes to bring about changes to cellular phenotype or biomaterial function.

   Outlined in this chapter, we focus on the state-of-the-art strategies now being developed and used in the laboratory and clinical stages specifically for repair and regeneration of bone, cartilage, and osteochondral interface region.

11.2 State-of-the-art strategies for regeneration

11.2.1 3D bioprinting

Bioprinting represents the high throughput automated fabrication of structurally mimicking and biologically functional 3D tissue constructs by blending together specific cells, bioactive molecules, and biomaterials formulating a "Bioink." This fabrication process is generally followed by a maturation cycle [7–9]. Recent advancements in 3D bioprinting have enabled the fabrication of patient-specific scaffolds and tissue constructs that possess both the native structure and function of the targeted tissue. The design of the constructs is typically derived from an Magnetic Resonance Imaging (MRI) or Computed Tomography (CT) scan of the patients, which is then reconstructed by the aid of computer aided design (CAD) software and converted into printable codes (GCODES). These codes are thereafter fed to the bioprinter that contains the necessary bioinks for layer by layer execution of these codes into 3D constructs. The various physico-chemical properties, such as porosity, microarchitecture, mechanical strength, and substrate stiffness can be tuned for better host tissue integration [10–12]. Though many bioprinting methods have developed in the recent times, the spatial layer-by-layer deposition of living cells with biomaterials require special care for maintaining high cellular survival for the long-term performance of the printed constructs [7,13,14]. Inkjet, laser-based, and micro-extrusion driven methods are the most common methods for 3D bioprinting [15–17]. Micro-extrusion-based methods are the most affordable and convenient modality. It involves the extrusion of a cell-laden bioink containing growth factors and other biologics through a small orifice/syringe nozzle. Micro-extrusion can be further classified into many forms but, the most common forms are fused deposition modeling (FDM) and direct ink writing (DIW) modalities. The complex graded structure and the difference in physiology and mechanics of each layer of the bone, cartilage, and osteochondral interface renders a unique opportunity for the application of bioprinting in the fabrication of these tissues. As opposed to conventional fabrication strategies, bioprinting offers mimicry of geometry, pore size, distribution and interconnectivity. These affect the cellular viability, nutrient transport and cellular migration leading to adequate maturation of tissues as well as enhanced ability to integrate with the host tissue. The ability to fabricate seamless multi-layered constructs allows the application of multiple bioinks simultaneously and fabrication of multi-phase constructs with precise control of material deposition and recapitulation of the anisotropy, gradient, and heterogeneity of the constructs.

11.2.1.1 Bioprinting for bone regeneration

Bioprinting strategies of bone can be classified into various approaches such as biomimetic, directed cellular assembly, or reconstruction using building blocks [18]. The simulation of cells and extracellular matrix (ECM) components of the tissue to match the native tissue composition is known as the biomimetic approach. Byambaa et al. [19] demonstrated the bioprinting of a vascular endothelial growth factor (VEGF) gradient using a gelatin methacryloyl (GelMA) ink to mimic the vascularity of bone from the intramedullary cavity towards the cortical sections. Following the principles of developmental biology for the self-assembly of micro- and macro-bodies by virtue of signaling molecules, to form biomimetic tissues is known as directed cellular assembly. Guided placement of small building blocks of a tissue in a layer-by-layer fashion leads to the fabrication of mature tissue in reconstructive fabrication approach [20].

Biomimetic approach to bone bioprinting involves the extraction of native bone ECM and blending it with suitable additives to form optimal bioinks that can be employed in printing durable and mechanically robust constructs [21,22]. The decellularization protocols often involve the use of harsh solvents or detergents [23], and involves batch variations in the constructs. Moreover, the low viscosity of decellularized extracellular matrix (dECM) based bioinks makes them mechanically weak, which leads to the rise in the usage of blends. Incorporation of a rigid and porous additive [24–26] or a thermoplastic polymer framework such as polycaprolactone (PCL) [27], PCL:demineralized bone matrix (DBM) [28], PCL encapsulated with Ca/Polyphosphate microparticles [29], PCL conjugated with thermosensitive chitosan [30], alginate-polyvinyl alcohol blended with hydroxyapatite (HA) [31] or PCL blended with Pluronic F127, gelatin and fibrinogen [32] can help in fabrication of biomimetic and robust constructs for long-term culture durability. Additionally, naturally available structural proteins such as silk fibroin have been blended with polymers such as gelatin, crosslinked using various chemical agents such as EDC (1-Ethyl-3-(3-dimethylaminopropyl)-carbodiimide)−NHS(N-hydroxysuccinimide esters) and have demonstrated promising results for bone regeneration [33].

Restoration of vascularity is necessary for the survival of regenerated bone tissue and improving integration with the host tissue. The most facile approach to the fabrication of such constructs involves the encapsulation of growth factors for angiogenic differentiation alongside osteogenic differentiation. Kolesky et al. [34,35] biofabricated hybrid constructs with multiple cells and controlled release of growth factors for improved regeneration potential. Moreover, pre-vascularized bone tissue has been integrated with the host vasculature to accelerate bone regeneration [36]. The addition of chemical additives such as bioactive borate glass in PCL constructs has shown extensively vascularized bone formation [37]. Additionally, unique direct-writing methods have been used to form pre-vascularized physiologically relevant bone tissue constructs using a blend of GelMA, sodium alginate, and 4-arm poly(-ethylene glycol)-tetra-acrylate (PEGTA) with coculture of human mesenchymal stem cells (hMSCs) and human umbilical vein endothelial cells (HUVECs) [19].

A comparatively under-explored approach is the addition of hypoxic components to trigger ossification and angiogenic factor secretion leading to a better-engineered bone [38,39]. Multi-material printing is gaining momentum due to its ability to mimic the heterogeneity of tissues and organs. Recently, vascularized bone was biofabricated using fused deposition modeling (FDM) printed polylactic acid (PLA), and stereolithography (SLA) printed GelMA containing BMP and VEGF [40]. This results in the slow release of growth factors and the promotion of better-vascularized bone formation. In another approach, inkjet printing of hMSCs with photocross-linkable polyethylene glycol-GelMA (PEG-GelMA) demonstrated enhanced osteogenic differentiation as well as considerably high compressive moduli [41].

11.2.1.2 Bioprinting for cartilage regeneration

3D bioprinting of cartilage tissues has been a widely explored area, and there is a multitude of techniques that have exploited for the fabrication of cartilages. SLA based fabrication of cartilage regeneration scaffolds has been reported with various photocurable polymeric materials such as trimethylene carbonate (TMC)/trimethylolpropane (TMP), poly(trimethylene carbonate) (PTMC), polyacrylamide hydrogel, PEG hydrogel and poly-propylene fumarate (PPF). Though the constructs fabricated by these polymers show excellent cytocompatibility [42], they have insufficient mechanical strength [43–45] and are unable to match the internal architecture of the natural articular cartilage. Moreover, the use of photo-initiator and photopolymerizable materials that possess higher hardness than natural articular cartilage limits the application of SLA technology [43]. Similarly, there have been reports of SLS (selective laser sintering) based PCL scaffolds for the regeneration of cartilage [46–48]. These scaffolds possessed good hydrophilicity, cytocompatibility and induction of articular cartilage formation when integrated with bioactive materials such as hyaluronic acid/PCL microspheres [46]. FDM of acrylonitrile butadiene styrene (ABS) and polylactic acid (PLA), PCL, and polyglycolic acid (PGLA) [49–51] have been reported for the fabrication of highly porous, mechanically durable and biomimetic scaffolds in conjunction with collagen and tricalcium phosphate (TCP) for articular cartilage regeneration. Liquid deposition modeling is a technique similar to FDM and has been reported for fabrication of biphasic and multi-material fabrication of cartilage tissue. Biodegradable polyurethane (PU) [52], poly(lactide-co-glycolide) (PLGA) with collagen [53,54], hydroxyapatite (HA) [54], and dECM [55], manganese with TCP [56], have been reported to enhance regeneration of cartilaginous and subchondral bone for a more wholesome approach to regeneration of the articular defect sites. Apart from these acellular techniques that require the seeding of cells externally, various cell-laden methods have been adapted for articular cartilage bioprinting. Chondrocytes from different sources have been used in conjunction with bioactive and biocompatible materials for the formation of articular cartilage tissues [57–61]. Live cell-laden constructs were fabricated by Sun et al. [62] consisting of human adipose-derived stem cells (hASCs) and (poly-D,L-lactic acid/polyethylene glycol)/HA matrix using SLA method. MSC laden constructs with GelMA and polyethylene glycol diacrylate (PEGDA) as base biomaterial with transforming growth factor (TGF)-β1-embedded nanospheres were fabricated by Zhu et al. [63] using SLA technique. These growth factor laden nanospheres improved chondrogenic differentiation. In another study, Xu et al. used a combination of inkjet and electrospinning system for fabricating PCL/fibroin–collagen constructs laden with chondrocytes from rabbit to form functional cartilage [64].

Recently, silk fibroin based bioprinting of cartilaginous tissue has seen a spur owing to the biocompatible and chondro-supportive traits exhibited by these biopolymers [65]. With our expertise in this domain, we have developed a few silk-based bioinks. In one approach, we presented a cross-linker free silk-based bioink exhibiting excellent print fidelity (Fig. 11.1I), which enabled in bioprinting of chondrocytes (Fig. 11.1II) maintaining their chondrogenic phenotype over 2 weeks in culture as indicated by the cardinal cartilage genes and extracellular matrix upregulation [66]. In another approach, we opted for an acellular approach for printing meniscal tissues much akin to the patient's needs by recovering CT scans, recapitulating the intricate architecture of the native meniscus tissue as shown in Fig. 11.1(III) [67]. The robust, biomimetic meniscal analogues were able to withstand physiologically relevant compressive loads without deformation for

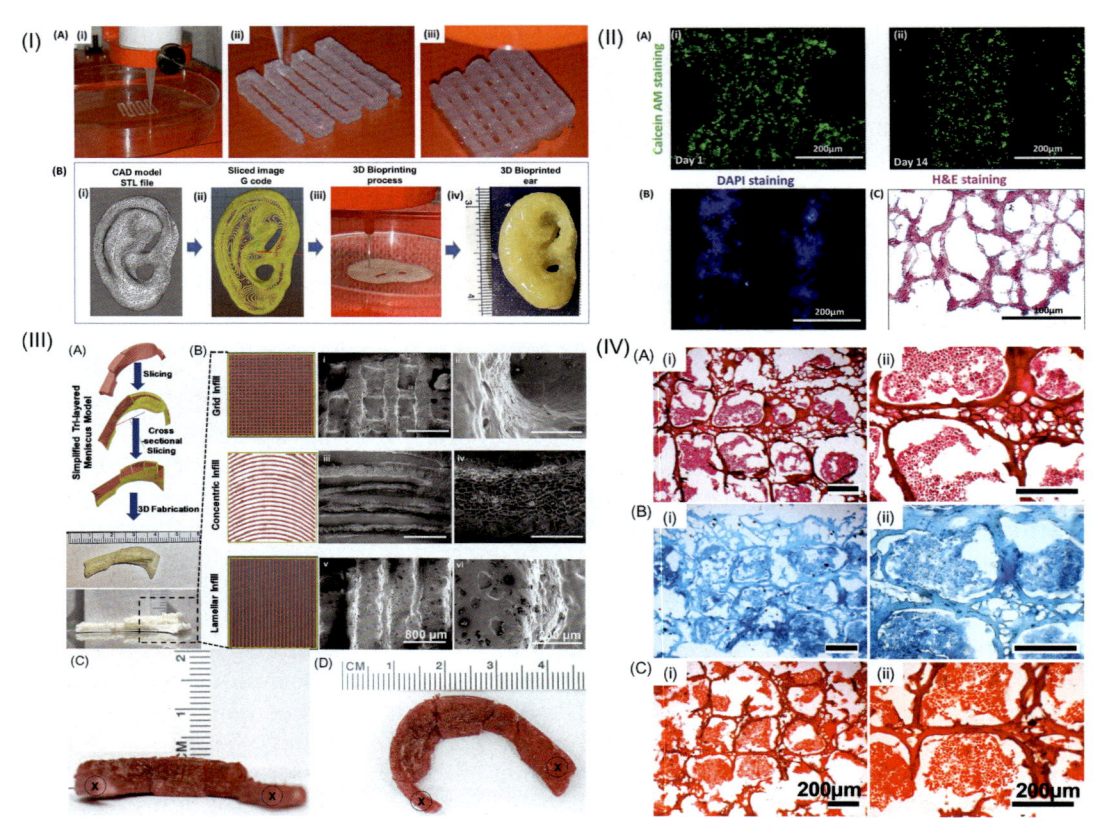

FIGURE 11.1 (I) Print fidelity of the crosslinker free silk-based bioink for cartilage 3D bioprinting showing (A) printing of grid-like structure (i) ink extrusion through the nozzle, (ii) initial layers, and (iii) formed grid-like pattern; (B) fabrication of anatomical human ear using (i) CAD model and generated STL file, (ii) generated G code, (iii) printing process, and (iv) printed ear structure [66]. (II) Biological evaluation of cell-laden bioprinted constructs using crosslinker free silk-based bioink. (A) Calcein AM-stained images of the chondrocyte-laden bioprinted construct after (i) 1 day and (ii) 14 days. Histological assessment of in vitro-cultured bioprinted constructs post 14 days of culture (B) DAPI-labeled blue cells and (C) H&E-stained section of the construct [66]. (III) Acellular tri-layered patient-specific 3D printed knee meniscus scaffolds using silk-based bioink. (A) Schematic of tri-layered meniscus model sectioned and 3D printed to show the three different layers. (B) Representative field emission scanning electron microscopy images for grid (i,ii), concentric (iii,iv) and lamellar (v,vi) infill layers. (C) Side and (D) top views of a tri-layered meniscus scaffold. The X marks indicate suturing points [67]. (IV) Meniscus-specific ECM secretion of fibrochondrocyte cells seeded on 3D printed meniscus scaffolds shown using representative histological sections matured for 21 days in vitro. (A) H&E stained sections showing the distribution and alignment of cells. (B) Alcian Blue stained sections, showing the deposition of sGAG by the fibrochondrocytes. (C) Picrosirus Red-stained sections, showing the deposition of collagen by fibrochondrocytes. A(i), B(i) and C(i) are lower-magnification micrographs. A(ii), B(ii) and C(ii) are magnified micrographs focusing within the region. Scale bar: 200 μm [67]. *(I and II) Copyright 2019, Reproduced with permission from American Chemical Society. (III and IV) Reproduced with permission from IOP Publishing, 2019.*

200 cycles and even supported the growth and maturation of porcine meniscal derived fibro-chondrocytes (Fig. 11.1IV) assuring its fruitful translation as off-the-shelf grafts for meniscus repair.

11.2.1.3 Bioprinting for osteochondral regeneration

Mimicking the osteochondral interface tissue involves precise gradient controlled deposition of osteogenic and chondral counterparts. Facile fabrication of such bi-/tri-phasic interfacial tissue constructs can be achieved by the virtue of 3D bioprinting. Holmes et al. [68] developed mechanically resilient biphasic PLA scaffolds that contained surface modification for enhanced cytocompatibility. These scaffolds possessed mechanical properties that matched the cartilage and subchondral bone. Fedorovich et al. [61] bioplotted osteochondral tissues with osteogenic and chondrogenic progenitors in an alginate scaffold. This resulted in distinct regions of ECM deposited within the scaffolds. Holmes et al. [68] have contributed to printing a line of 3D printed filamentous scaffolds with biphasic geometry and appropriate mechanical strength for the promotion of niche-specific stem cell differentiation. Nowicki et al. [69] reported the use of tuneable nanohydroxyapatite (nHA) based pore gradient scaffolds with anisotropic pores for mechanically resilient and biologically improved osteochondral scaffolds. Though hydrogel-based constructs of alginate [70], collagen [71], and silk fibroin/gelatin [72] have been reported for cartilage and osteochondral repair, they lack the mechanical stability to support in vivo applications. Hence, polymer blends with better load-bearing capacity have seen a upsurge in their application for osteochondral repair. Hong et al. [73] formulated a tough poly (ethylene glycol) (PEG)/sodium alginate blend with high fracture toughness for 3D bioprinting of osteochondral tissues. Similarly, Zhu et al. [74] and Yang et al. [75] fabricated high strength hydrogel based bioinks for the development of constructs with high mechanical strength. Though high in strength, these hydrogels require a cumbersome addition of bulking agents for biofabrication. Moreover, any photo-initiator and photo-crosslinking using UV irradiation result in the loss of cellular viability. These shortcomings need to be addressed by direct printability and controlled swelling/shrinking materials. Gao et al. [58] have fabricated using a 3D printed biohybrid hydrogel, which has excellent tensile and compressive strength. This bioink possesses shear-thinning that allows precise deposition into complex 3D architectures. The encapsulation of TGF-β1 and β-TCP in the scaffold promoted the differentiation of hBMSCs regionally and formation of osteochondral interfacial tissues. Similarly, Du et al. [46] reported an SLS fabricated multilayer gradient scaffold using PCL and HA/PCL for osteochondral defect repair. Lithium calcium silicate ($Li_2Ca4Si_4O_{13}$, $L_2C_4S_4$) [76] scaffolds have been utilized as a wholesome approach for the regeneration of both chondral and subchondral bone generating an approach for osteochondral repair (Fig. 11.2I). In addition, as discussed previously, many bioprinted constructs have been fabricated for regeneration of both articular cartilage and osteochondral regeneration. Bestowing cell-instructive traits to bioinks is becoming a major driving factor in the new age of bioprinting era, where researchers are trying to shun the use of biologicals/growth factors due to its unintentional drawbacks. In these lines, we have developed multifunctional silk-based bioinks which by the virtue of their physico-chemical nature, help in preferentially preserving the chondrogenic phenotype and osteogenic phenotype in stem cell -laden silk bioinks [77]. The hierarchically biomimetic nature of the osteochondral constructs thus bioprinted (Fig. 11.2II) maintained cellular crosstalk due to the permeable gel network. Herein bone phase exhibited osteo-inductive traits while chondral phase enabled maintenance of the chondrogenic phenotype, during the 2 weeks culture period in vitro (Fig. 11.2III).

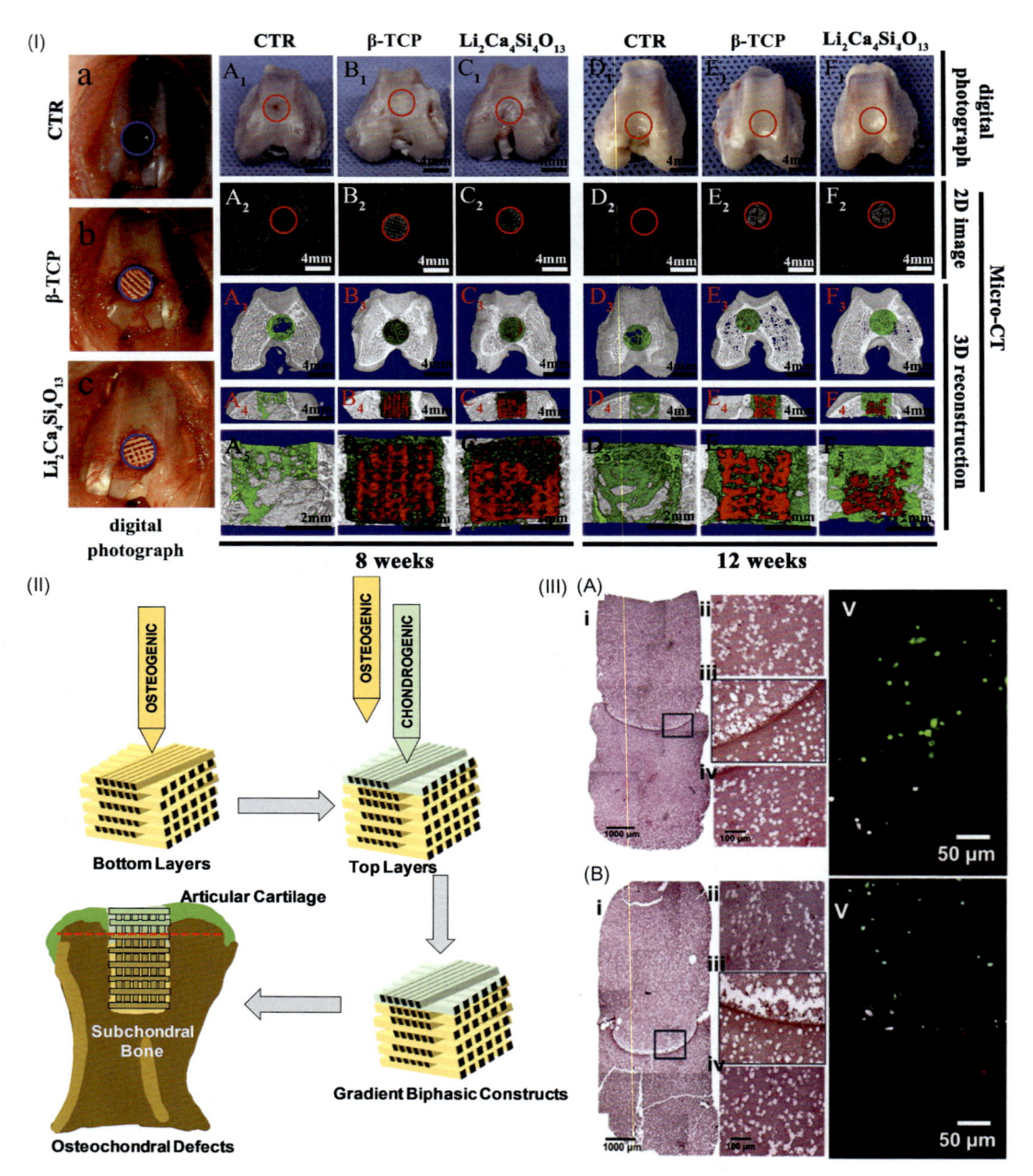

FIGURE 11.2 (I) (a−c) Photographs of defects created in the three experimental groups taken during surgery. Photographs and micro-CT images of defects after 8 weeks and 12 weeks post-surgery. A1−A5 and D1−D5 represent blank control scaffolds, B1−B5 and E1−E5 represent pure β-TCP scaffolds, and C1−C5 and F1−F5 represent Li2Ca4Si4O13 scaffolds after 8 and 12 weeks of surgery respectively. A2−F2 show 2D projection images, A3−F3 show transverse view of 3D reconstruction images of the three experimental groups and A4−F4 show sagittal view of 3D reconstruction images of the three experimental groups at week 8 and 12, respectively. A5−F5 are high-magnification images of A4−F4. The off-white color, green color, and red color stand for primary bone, new bone and scaffold, respectively [76]. (II) Scheme demonstrating bioprinting of biphasic osteochondral constructs using osteogenic and chondrogenic bioinks for repair of interfacial osteochondral defects. (III) (A) ITE1 and (B) ITE2 optimized bioinks for bioprinting osteochondral tissues. (i) Stitched H&E stained images of the entire construct, (ii) osteo phase, (iii) osteochondral interface, (iv) chondral phase, (v) immunohistochemistry for chondrogenic marker aggrecan (green) and osteogenic marker collagen-I (red) at the interface region [77]. *(I) Reproduced with permission from Elsevier Co. 2019. (III) Reproduced with permission from Elsevier Co. 2019.*

11.2.2 Gene therapy

The gene itself when used as a therapeutic agent and introduced into the patient-specific cells or tissues to treat the diseases is called gene therapy. The robust usage of therapeutic molecules and their direct administration has impeded the safety of the therapy. Thus, gene therapy is a promising alternative to conventional therapeutic molecule-based treatment. The efficacy of gene therapy is determined by gene insertion and delivery modes. The viral and non-viral modes of gene delivery for a variety of tissue engineering applications have been studied, in which a variety of strategies have been invented for sustainable controlled delivery to the targeted site. Gene therapy is being evaluated in in vivo and ex vivo levels and the different approaches have been represented in Fig. 11.3A. In general, in vivo approach considers the gene delivery to the targeted site through viral or non-viral vectors with anticipation of cells to be transfected, which can be termed as in situ transfection. In ex vivo approach, autologous cells are harvested and transfected externally. Subsequently, transfected cells are administered for therapeutic applications. Gene activated matrix has attracted the attention to adopt gene therapy for tissue engineering applications, wherein the ECM can be used as a medium to enhance transfection efficiency and achieve localized controlled delivery.

Numerous viruses such as adeno, lenti, and retro viruses have been widely investigated for their potential ability in gene transfection. The recombinant technology has boosted the viral-mediated gene transfection approach to improve the efficiency of transfection in various cells. Bio/synthetic molecules including peptides, cationic polymers, nanomaterials are employed for non-viral gene delivery applications. Gene therapy for bone, cartilage, and osteochondral interface has been discussed in the following sections.

11.2.2.1 Gene therapy for bone regeneration

The genes involved in encoding bone morphogenic proteins (BMPs) have been introduced into the viral vectors such as adeno, retrovirus for bone regeneration. The retroviral vector MFG was used to transfect the BMP-4 gene into rat marrow stromal cells, and it induced higher secretion of BMP-4, which enhanced bone regeneration, which was confirmed with ectopic bone formation in in vivo studies [78]. The osteogenic potential of hMSCs on silk fibroin biomaterials using adenovirus vector-mediated human BMP-2 transduction has been confirmed by the enhanced bone regeneration with higher expression of bone sialoprotein (BSP) and osteopontin (OP) [79]. Non-viral gene delivery has addressed the issues of viral gene delivery including immunogenicity and occurrence of mutagenesis. The cationic polymer polyethylenimine (PEI) has been employed for BMP-2 transduction of bone marrow stromal cells, wherein cells expressed higher amounts for BMP-2, which in turn promoted bone formation in vitro and in vivo [80]. The recombinant human BMP-2 (rhBMP-2) loaded collagen sponge carriers for bone regeneration have been approved as INFUSE bone graft by US Food & Drugs Administration (FDA). However, ectopic bone formation is the limitation of this bone graft, for which covalent or non-covalent carrier molecules for systemic controlled delivery is an area of ongoing research for overcoming the clinical challenges. Plasmid DNAs of BMP-2 and VEGF embedded gene activated demineralized bone have been used in clinical trials on ulnar pseudarthrosis patients in the year 2017. They showed promising results in terms of callus formation and reduction of pain after implantation [81]. VEGF gene activated collagen-hydroxyapatite scaffold has

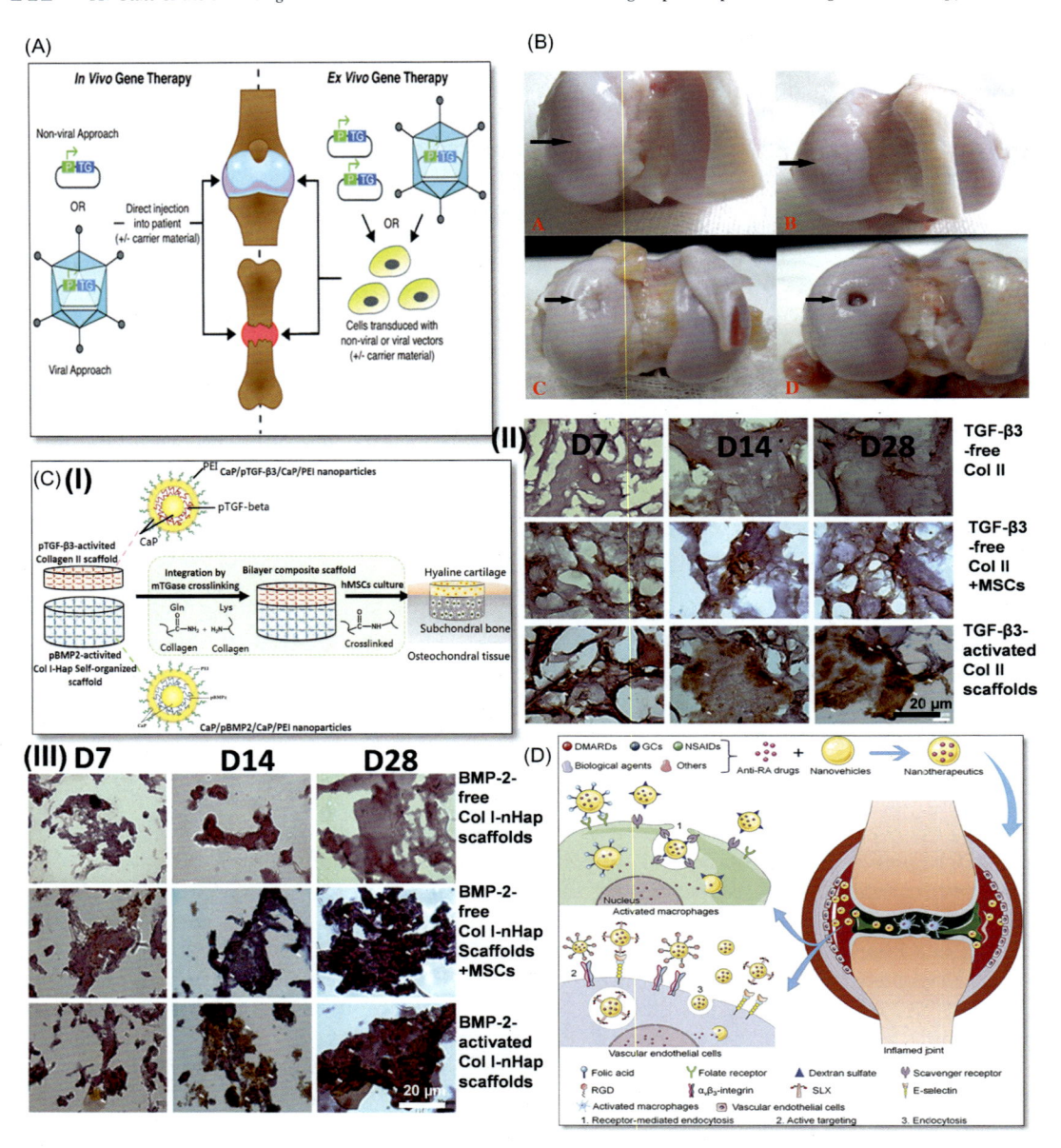

FIGURE 11.3 Gene therapy for bone, cartilage, and osteochondral tissue engineering. (A) Different gene therapy strategies [93]. (B) The cartilage regeneration evaluation after 16 weeks of surgery [91]. (C) Gene activated scaffold for osteochondral tissue engineering. (I) Scheme for the development of Col-II/Col I-nHap gene activated scaffolds. (II) The collagen II immunostaining for free and gene activated scaffolds to evaluate hyaline cartilage formation. (III) The osteocalcin immunostaining of BMP-2 for free and gene activated scaffolds to evaluate bone matrix deposition [94]. (D) Schematic representation of nanoparticle assisted therapeutic agents delivery for management of rheumatoid arthritis [95]. *(A) Copyright 2018, Reproduced with permission from Elsevier. (B) Copyright 2012, Reproduced with permission from Elsevier. (C) Copyright 2017, Reproduced with permission from Elsevier. (D) Copyright 2017, Reproduced with permission from Elsevier.*

been investigated for the treatment of large bone defects in the clinical trials and has been confirmed for the removal of non-unions [82]. The transfection efficiency is dependent on the strong pDNA binding ability of the biomaterial, which in turn, dictates the bone healing mechanism. Henceforth, current research is being carried out in developing appropriate biomaterial scaffolds for higher cell transfection to make clinically relevant gene activated scaffolds for bedside applications.

11.2.2.2 Gene therapy for cartilage repair

Numerous studies with varied combinations of genetic material, vector, and scaffold have been evaluated for cartilage regeneration. TGF-β1 is an essential growth factor for cartilage regeneration, maintenance, and repair. TissueGene-C clinical trials have been recently completed, validating its role in cartilage regeneration, where transduced chondrocytes using retroviral vectors induces overexpression of TGF-β1 [83]. The recombinant retrovirus used to transduce bone marrow aspirates with chondrogenic factors (SOX-9, TGF-β, and IGF-1), resulting in elevated secretion of proteins that promote cartilage regeneration [84]. The clinical trials on Invossa™ for osteoarthritis has been conducted in the US and South Korea. It has been approved as a cell-mediated gene therapy named Invossa™ by South Korea Ministry of Food and Drug Safety [85,86]. However, the clinical trials on Invossa™ therapy are still ongoing in the United States. Non-viral gene delivery for cartilage regeneration research is vastly proceeding, and yet these findings have not reached to bedside applications. A variety of cationic polymers have been studied for transduction of autologous cells with chondro-reparative genes for developing an efficient, affordable treatment for cartilage defects. SOX are the common cartilage transcription factor that plays a vital role in the downregulation of endochondral ossification while RunX2 inhibits chondrocyte hypertrophy. Thus, a plethora of research studies have been conducted on the combinatorial use of non-viral gene delivery of TGF-β1 and SOX-9 for cartilage regeneration. PLGA microspheres have been chosen as agent for the SOX-9 transfection of MSCs to enhance the chondrogenic gene expression and improve cartilage ECM formation [87]. Further research on cartilage regeneration has been aimed at inhibition of hypertrophy and enhancing chondrogenic differentiation. For instance, the combination of chondrogenic inductors such as TGF-β1, FGF-2, and BMP-2, and hypertrophy inhibitors SOX-9, parathyroid hormone-related protein (PTHrP) has been proposed for developing translational approaches in stable cartilage repair [88,89].

11.2.2.3 Gene therapy for osteochondral interface regeneration

Research inclined towards osteochondral tissue engineering has greatly attracted attention in the past two decades. Various scaffolds have been developed for osteochondral interface regeneration. Recently, silk scaffolds maintaining the hierarchical anatomical structure of native tissue are a potential candidate, which was preliminarily confirmed with in vivo studies in animal models [90]. Gene therapy alone and in combination with scaffolds also has been successfully validated for repair of interface defects. Insulin growth factor-1 (IGF-1) gene using plasmid vector (Fig. 11.3B) has been used to transfect bone marrow MSCs, and gene-modified scaffolds have been shown to successfully fill the knee osteochondral defects (OCD) [91]. In another study, IGF-1 & FGF-2 have been used in combination, in alginate scaffolds for in vitro and ex vivo regeneration. Gene modified articular chondrocytes have been used for stimulating osteochondral regeneration [92]. Gene activated scaffolds for induction of MSCs for osteochondral tissue regeneration has been studied extensively. Collagen-II/collagen-I-nano-hydroxyapatite (nHap) bilayer composite scaffolds in different combinations including gene free and gene activated scaffolds were evaluated for osteochondral tissue regeneration and the scheme for the development of composite scaffolds is given in Fig. 11.3CI. This study confirmed that TGF-β3 activated Col-II scaffold and BMP-2 activated Col-I-nHap scaffold resulted in the induction of bone and cartilage differentiation in MSCs. The histology results confirmed that gene activated matrix showed higher hyaline cartilage regeneration and bone matrix deposition (Fig. 11.3CII–III). The subchondral bone and cartilage regeneration ability and bilayered native structure confirmed that gene activated scaffolds possess the potential for osteochondral tissue regeneration. In another study, the influence of hIGF-1 gene on regeneration of osteochondral defect of knee has been studied. Complete regeneration at the defect site and neo tissue formation was achieved with localized gene transfection of hIGF-1 gene (Fig. 11.3CIV). The sustained release of plasmid DNAs in a spatio-temporal manner within the composite scaffolds has shown enhanced subchondral bone and cartilage regeneration. Invossa™ tissue gene therapy was approved in South Korea for osteoarthritis which involves the use of human chondrocytes transfected with TGF-β3 for the treatment of degenerative osteoarthritis. Spherox-spheroids of human autologous chondrocytes and Invossa™ have been recently approved in Europe for osteoarthritis treatment.

11.2.3 Nanotherapy

Nanotherapy in regenerative medicine is the current trending research. Tissue-engineering research has also shifted towards identifying novel therapeutic approaches using nanoparticles for tissue regeneration. Nanoparticles have been widely used for drug, gene, and growth factor delivery. Nanoparticle embedded scaffolds have been proven to improvise or tune mechanical strength, cell adhesion, and biocompatibility. In addition to this, nanoparticles for stem cell labeling and targeting specific sites for guiding cells to a specific location for therapeutic intervention, is the emerging research in regenerative medicine. The surface topography modifications play a vital role in facilitating the microenvironment suitable in enhancing the biological and physico-chemical interaction of engineered graft with the native in vivo environment. Given this, materials science research has aimed to provide novel material combinations to expedite bone regeneration and integration. Herein, the native physical structure is an important factor to be considered in developing composites for regeneration along with the improvement in biocompatibility. Composite mechanical strength and degradation phenomenon are intended to modulate material properties according to the specific requirements of tissue; in particular, the mechanical strength of the composite should be higher for a bone implant. In order to attempt this, metal nanoparticle, nano-hydroxyapatite has been widely used. Notable research has been reported for the development of nanocomposites, nanoparticle embedded scaffolds, functionalized biomaterials with nanoparticles for bone, cartilage, and interface tissue engineering. In the following sections, the clinical advances of nanotherapy for bone, cartilage, and the osteochondral interface regeneration are briefly discussed.

11.2.3.1 *Nanotherapy for bone regeneration*

Nanoparticle-based bone graft substitutes have been developed for bone regeneration owing to their ability to mimic native bone tissue structure, cell adhesive properties, and higher surface area. Moreover, the use of nanoparticles aids in mimicking the hierarchical bone structure composed of collagen and other proteins arranged in 100 nm size or less. Thus, the quest for developing nanocomposites for bone regeneration has been augmented throughout the world. However, inorganic nanoparticles used in allo bone grafts has been limited to calcium phosphate and hyaluronic acid (HA) [96]. Calcium phosphate-based synthetic bone graft, i.e. Vitoss (Stryker), is one of the first approved grafts under the nanomedicine category. This bone graft possesses a structure that mimics native tissue, and clinical trials have proved that their superior cell adhesive properties boost bone regeneration. Ostim (Heraseus Kuzler), OsSatura (IsoTIs Orthobiologics), NanOss (Rti Surgical), and EquivaBone (Zimmer Biomet) are HA-based bone grafts. All of the above grafts except EquivaBone graft possess cell adhesion properties. All HA-based synthetic bone grafts possess the native structure of bone. Various scaffolds have been developed with the combination of HA for bone regeneration studies. The extensive literature survey on HA composite scaffolds has declared that HA is more feasible to mimic the native bone structure and greatly improves osseo-integration. HA used in microscale for bone graft has been compared with nanoscale approaches, and reports show that microscale addition does not mimic natural bone properties owing to slow degradation and brittleness. In addition to synthetic bone grafts, nanoparticles have been widely demonstrated for delivery of growth factors BMP-2 and BMP-7 for bone regeneration. These growth factors are FDA approved and well explored for sustained delivery. In particular, plasmid DNA, miRNA delivery for transduction of growth factors into cells for overexpression of proteins for bone regeneration, is mostly targeted for translational research. Chitosan-PEI nanoparticles have found application in BMP-t transfection for upregulating osteogenic differentiation [97].

11.2.3.2 *Nanotherapy for cartilage repair*

Hydrogels have been investigated extensively for cartilage tissue engineering. The lower stiffness and high water content are the major limitations of hydrogels. The native cartilage structure composed of collagen and proteoglycans is known to be arranged in the nanoscale range. Thus, nanocomposite hydrogels have emerged as plausible alternatives for recapitulation of cartilage ECM for regeneration applications. The physical and chemical crosslinking approaches have been employed for the development of nanocomposite hydrogels. Several types of nanoparticles including organic, inorganic, metallic, and polymeric nanoparticles have been used in the polymeric network to form nanocomposite hydrogels.

11.2.3.3 *Nanotherapy for osteochondral tissue repair*

For osteochondral tissue engineering, nanoparticles have been widely studied to develop gene, and cell-based therapies. The targeted delivery of anti-rheumatoid arthritis (RA) agents, which are encapsulated in nanocarriers for active or passive targeting is a very recent approach that has been developed for treating osteochondral degeneration (Fig. 11.3D). Liposomal encapsulation of methotrexate (MTX) contributes to its sustained delivery while down-regulating the expression of pro-inflammatory cytokines such as IL-

1β, IL-6, and prostaglandin-E-2 (PGE2). Various nano-vehicles have been developed for localized targeted delivery of RA treatment agents. Though clinically approved treatment have not been found for RA due to numerous factors, including frequent administration, severe systemic side effects, and tolerance from long-lasting administration as well as high costs. However, nanomaterial-based approaches are a promising avenue towards developing composite scaffolds for interface tissue engineering and some of the scaffolds have already been translated to bedside applications. The multilayer nanocomposite scaffold, MaioRegen, composed of a superficial layer of deantigenated type I equine collage, intermediate layer of magnesium-enriched hydroxyapatite (Mg-HA) and collagen and bottom layer of Mg-HA has been found to mimic the native osteochondral interface. Cell-free versions of MaioRegen have been taken to clinical trial in eight patients and they have shown good osteointegration and osteoconduction in seven patients. These nanocomposite scaffolds have successfully filled the cartilage defect with improved osteointegration, which was confirmed 18 months after implantation [98].

11.2.4 Exosome based therapeutic strategies

Mesenchymal stromal/stem cell (MSC) derived secretome has been under the scope of many researchers for the development of cell-free therapeutic approaches for a multitude of tissues. Apart from possessing multi-lineage potential, MSCs possess paracrine secretion properties that form a complex network of factors that confer stability to the cellular micro-environment and regenerative response amplification [99]. MSC based therapies have shown that they release biologically active factors such as cytokines, chemokines, growth factors and miRNAs dubbing them as the local drug stores for the site of injury [100,101]. The secretome is a repertoire of factors secreted by the MSCs and has been documented to produce effects on angiogenic, proliferative, differentiation, immuno-modulating, wound healing, osteo-regenerative and kidney, and cardiac regenerative processes [102–107]. MSCs are also documented to possess paracrine anti-inflammatory effects by releasing cytokines such as IL-10, and TNFα-stimulated gene-6 (TSG-6) [108]. Moreover, angiogenesis and neo-vessel formation potential due to the secretion of VEGF-A and Ang1/Ang2 by MSCs has also been documented [108]. The recent focus of secretome based therapies has been extracellular vesicles (EVs), which can be sub-divided based on their size, density, surface markers, and origin into any of the three categories, namely: apoptotic bodies, microparticles, and exosomes. Exosomes are the fundamentally researched EVs owing to their ability to achieve high-throughput interaction with the target cells and their property to selectively modify cell signaling [109]. EVs have properties similar to liposomes while having the gratuity of being naturally assembled as targeted complex biological entities. In vivo studies have been reported suggesting the positive effects of EV administration in a variety of diseases and conditions, but very few have been translated clinically [110]. Secretome and EV based therapy can be used to produce niches that can help in the regeneration of the various target tissues. The most common delivery methods of EVs in murine animal experiments have been intravenously (i.v.) or intraperitoneally (i.p.) in a repeated or single dose [111,112]. Animal studies have suggested that EVs are readily cleared from circulation upon injection throughout the body [113] and their biodistribution is influenced by their cell of origin [114]. To date, there is no evidence that EVs administered for regeneration preferentially home to their action site that reduces the effectivity of EVs. The availability of the secretome/EV needs to be, hence, confined to the areas where the therapy is to be executed therefore producing optimal results.

Exosomes from MSCs are being studied for their therapeutic application in osteochondral repair and regeneration in recent times. Zhang et al. showed (Fig. 11.4I) that exosomes derived from MSCs could repair and regenerate large osteochondral defects in rat models. These exosomal vesicles were injected intrarticularly and proved the efficacy of regeneration as compared to no-treatment controls [115]. Results have demonstrated enhanced proliferation, migration, and matrix synthesis by chondrocytes [118]. Another report by Hassan et al. states the efficacy of Wharton's Jelly derived secretome for cartilage-specific gene upregulation and thereby shows the potential of such components for cartilage/osteochondral regeneration [119]. Researchers have found EV from various sources such as monocytes activated with lipopolysaccharide (LPS) [120], platelet-lysate derived [121], MSCs [122–127], and hiPSC [128] promote osteogenic differentiation and pose a viable solution for osteochondral regeneration studies. In an interesting study, Kawai et al. [129] reported the secretome derived from bone marrow MSCs enhanced the osteogenesis and angiogenesis for enhanced periodontal tissue regeneration. In another report by Xie et al. [116] demonstrated pro-angiogenic and pro-bone regeneration potential of EV decorated decalcified bone matrix scaffolds (Fig. 11.4II–III). Xu et al. [117] demonstrated the consolidatory and regenerative effects of fetal MSC derived secretome on bone in distracted osteogenesis therapy (Fig. 11.4IV). Zhu et al. [130] have studied and shown the efficacy of secretome derived from iPSCs and synovial MSCs in mitigating osteoarthritis.

11.2.5 Smart biomaterial technologies/tissue engineering strategies

The use of smart biomaterials in tissue engineering has been recently spurred due to the advantages it garners over the traditional biomaterials in use clinically. The term *"smart"* could be appropriated to any biomaterial which could (i) trigger a cellular response innately, by virtue of the physical or chemical cues presented by the biomaterial; (ii) possess material responsiveness to adapt based on the microenvironment (pH, ionic or temperature changes); (iii) able to be modulated through external stimuli (magnetic or electric) to achieve mechano-transduction or release biomolecules/drugs in a controlled or "on-demand" manner. The use of such "cell instructive" or intelligent biomaterials as scaffolds for tissue engineering application has seen to greatly exploit the therapeutic effect of cells to a greater extent in comparison to traditional 'non-responsive' scaffolding biomaterials. These smart biomaterials have opened up avenues in mechano-transduction, immunomodulation, developing responsive peptides (degradable by matrix metalloproteinases, MMPs), and tissue-on-chip models to study the organ physiology or disease modeling. Though most of the currently commercialized biomaterials marketed by companies for bone/cartilage applications are well-established and acquainted by regulatory authorities [131], but the use of these smart biomaterials are guaranteed improvement in treatment strategies and hence hold promise in clinical translation.

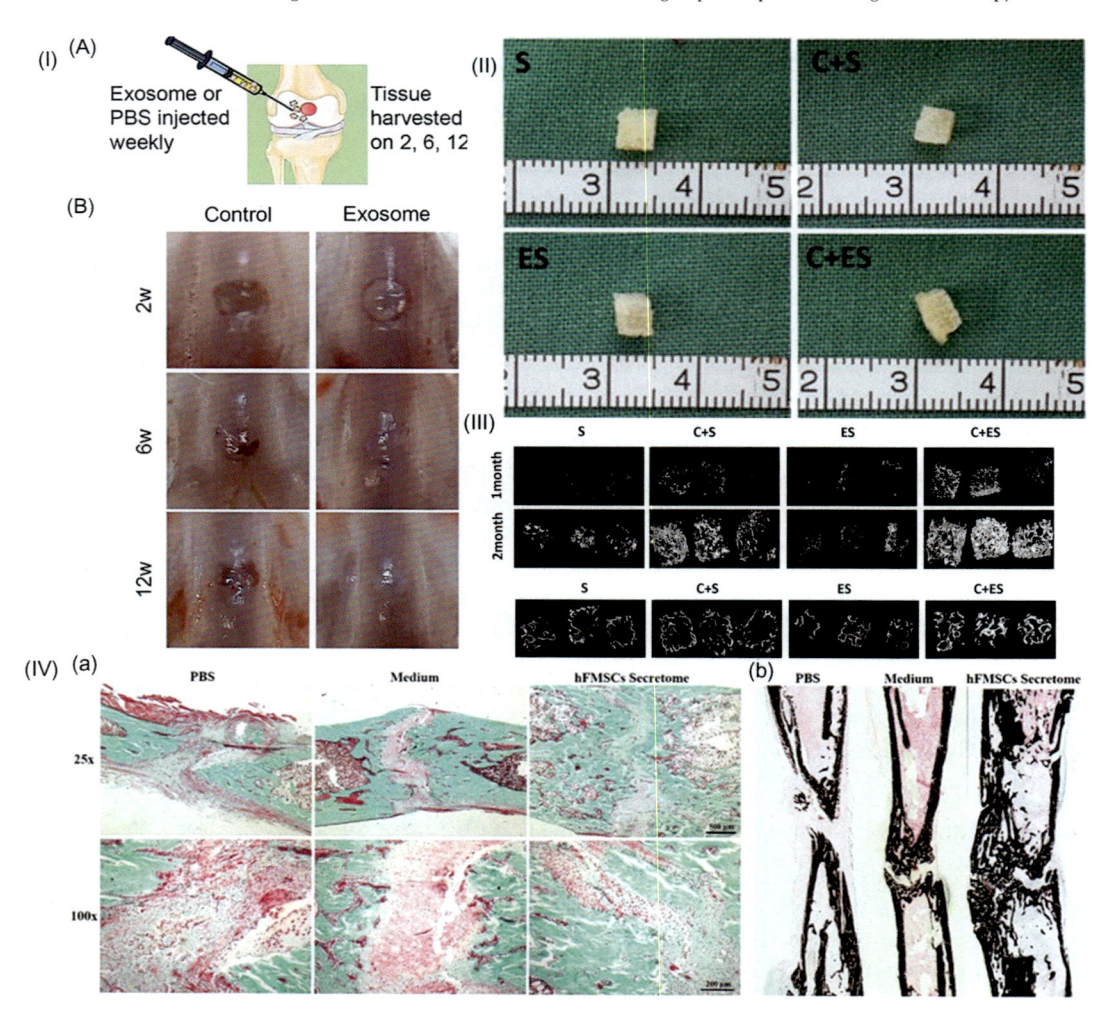

FIGURE 11.4   (I) (A) Scheme showing the evaluation of cartilage repair by exosomes isolated from MSC. (B) Macroscopic evaluation of cartilage repair in animals with PBS control and exosome injection [115]. (II) Macroscopic views of the grafts after 2 months of implantation. S represents unmodified scaffolds, C + S represents scaffolds coated with fibronectin, ES represents the EV modified scaffolds and C + ES represent scaffolds with primed osteogenic MSC and EVs. (III) Reconstructed micro-CT images of the grafts in each group at 1 and 2 months [116]. (IV) Histological analysis shows new callus formation post secretome intervention. (a) Representative Trichrome Goldner stained sections that show better callus formation in the experimental secretome group as compared to the other controls. (b) Von Kossa stained sections show consolidation of new bone in the secretome treated group after 6 weeks [117]. *(I) Reproduced with permission from Elsevier Co. 2018. (II and III) Reproduced under CC BY License from Springer Nature Publishing, 2017. (IV) Reproduced under CC BY License from Springer Nature Publishing, 2016.*

11.2.5.1 Smart biomaterials in bone tissue engineering

Large volume bone defects are often plagued by concerns relating to donor site morbidity, the inadequacy of allogenic/autogenic graft for transplant, the improper onset of vascularization in conventional bone grafts leading to failure of grafts. Addressing these critical issues, we had developed resorbable silk-bioactive glass composites, which were attributed with osteoinductive and proangiogenic traits [132]. This was achieved through the innate physical and chemical cues of the composite, where copper was doped into the bioactive glass to trigger hypoxia-inducible factor (HIF-1α) to activate a hypoxic mimic that facilitated the migration and survival of endothelial cells through CXC4/SDF-1 signaling (Fig. 11.5AI,II). While the silk scaffolding and amorphous bioactive glass are resorbable, the composites helped in complete restoration of rabbit bone defect with periosteal regeneration with no remnants of scaffolds observable after 3 months (Fig. 11.5AIII). In similar notions, we also looked into the synergistic effects of mediating osteogenesis and angiogenesis by doping silicon/zinc doped brushite cement and silk scaffolding, which also indicated similar impact [133]. These approaches enable in triggering cell responses without the use of any additional growth factors (such as BMPs; vascular endothelial growth factor, VEGF), thus proving to be an alternative for conventionally used strategies [134,135]. Another approach now commonly employed is the use of oxygen-generating biomaterials such as sodium percarbonate, calcium peroxide, magnesium peroxide, where they decompose to hydrogen and oxygen, thus negating hypoxic core generally encountered in implants for large volume defects [136]. However, the burst release of oxygen tends to create hyperoxia, leading to elevated reactive oxygen species (ROS) hampering tissue regeneration. Addressing this, more recently, hollow polycaprolactone (PCL) microparticles loaded with oxygen generator, perfluorooctane enabled the survival and osteogenic differentiation of human periosteal derived mesenchymal stem cells under hypoxic condition in vitro while showing enhanced vascularization potential in mandibular defects created in miniature pigs [137].

Bone's proven piezoelectric properties have been harnessed to mediate regeneration through the use of piezoelectric materials [138]. In these lines, dopamine coated ferroelectric BaTiO3 nanoparticles were solution cast in poly(vinylidene fluoridetrifluoroethylene) P(VDF-TrFE) to form piezoelectric membranes, and its potential to mediate bone regeneration was assessed [139]. By tuning the percentage of BaTiO3 content (5% w/v), the composites were able to maintain a polarization potential of -76 mV stable over 22 days in vitro and in vivo along the poling direction of the electric field. The electric field encouraged osteogenic differentiation of rat bone marrow mesenchymal stem cells in vitro and helped in faster regeneration of rat calvarial bone defects. The surface of the implant mediates its performance and compliance in vivo. In the case of metal bone implants, which have become indispensable part of orthopedic surgeries as fixtures are constantly plagued by implant loosening, fibrous tissue formation and infection at implant site which greatly diminish the survival post-implantation. Addressing this, innovative strategies have been adopted to engineer the metal implant's surface with smart coatings [140] which are (i) non-adhesive to prevent bacterial biofilm formation [141]; (ii) to enhance the osteointegration by using peptides or ECM moieties which harbor cell-binding motifs such as RGD

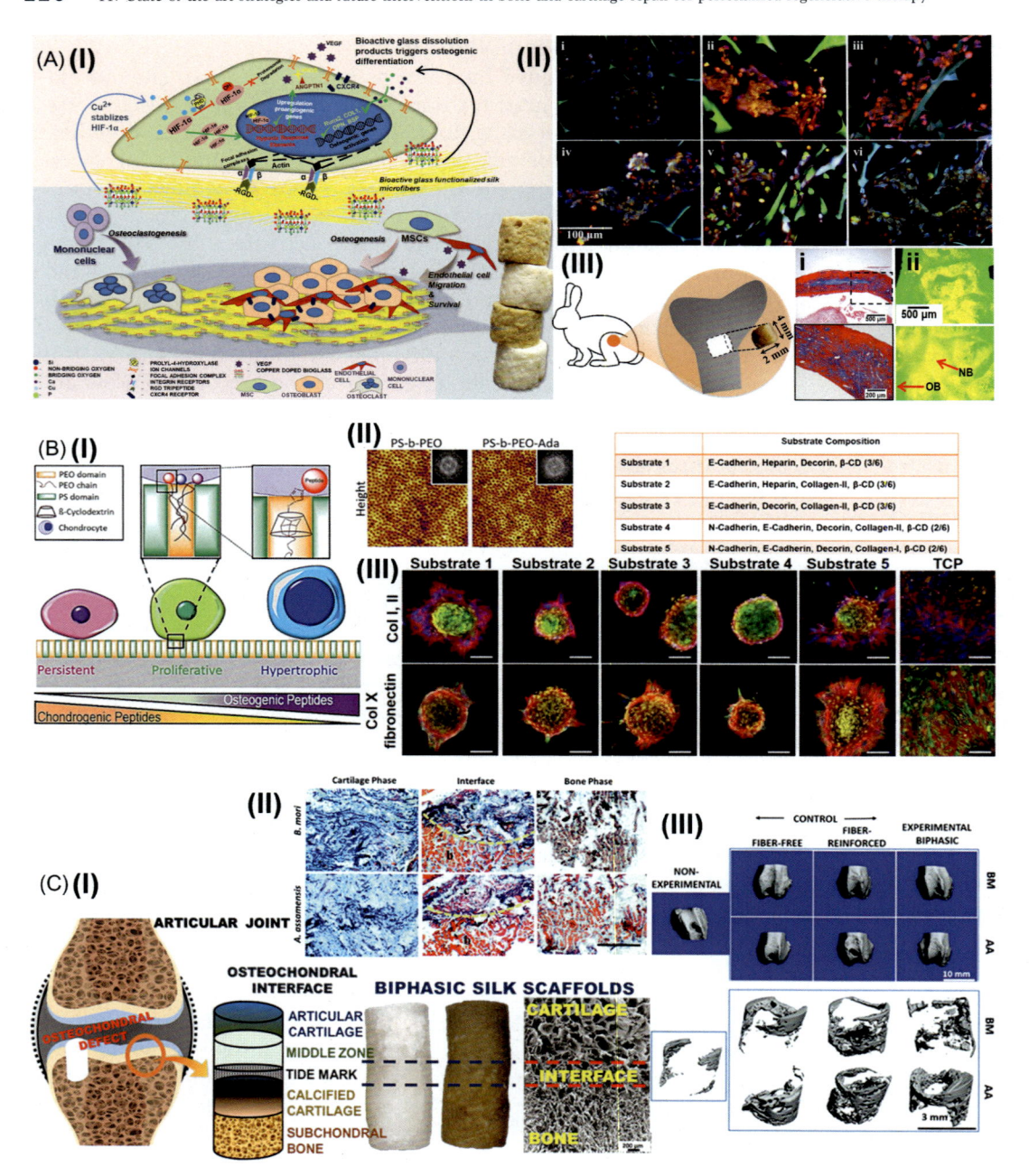

FIGURE 11.5   Smart biomaterial fabrication/ tissue engineering strategies for (A) targeting large volume bone defects with impetus on vascularization, osteoinductiveness and resorbability, (I) scheme representing the activation of cell signaling pathways relating to hypoxia response element, (II) in vitro histological evaluation showing forming of primitive vascular network within (copper doped) *Antheraea assama* fiber reinforced scaffolds (vi)

(arginine—glycine—aspartate) [142]; and (iii) compartmentalized core—shell nanoparticle systems which could house hydrophobic (osteogenic biomolecule) and hydrophilic (bactericidal drug) to have multifunctional effect at the implant surface site [143]. In these lines, a common gut microbe (*Lactococcus lactis*) was recombinantly engineered to express BMP-2 (in the secretory form) or fibronectin (presented on the cell surface) in either inducible manner (through nisin addition) or constitutively [144]. Since bacteria can easily form biofilm on a wide variety of surfaces (ceramics, metal implants or even collagen sponges), the authors exploited this feature to form a dynamic biofilm which could be triggered to express BMP-2 to mediate osteogenic differentiation or enable stem cell adhesion (to enhance the osteointegration) by expression of fibronectin. Thus, such smart approaches, which tune the material interface to address critical shortcomings related to material implant survival are key for paving way for next generation intuitive biomaterials possibly improving the clinical outcomes in orthopedic care.

11.2.5.2 Smart biomaterials in cartilage tissue engineering

One of the utmost challenges in cartilage tissue engineering is the maintenance of chondrogenic phenotype of isolated donor chondrocytes or chondrogenically differentiated stem cells for culture and expansion in vitro prior to transplantation in cell-based therapies. Hyaline cartilage is composed of chondrocytes in three phenotypic forms, namely persistent, transient, and hypertrophic which secrete ECM that compositionally differ along the cartilage gradient which gives it mechanical resilience. In order to recapitulate the ECM microenvironment and differentiate mesenchymal stem cells (MSCs) to a very definitive chondrogenic lineage (persistent, transient or hypertrophic) (Fig. 11.5BI), nano-patterned surfaces (polystyrene-b-polyethylene oxide di-block copolymer with a adamantyl group (PS-b-PEO-Ada) to conjugate any biopolymer moiety) presenting different biopolymer moieties such as E-cadherin, N-cadherin, collagen-I, collagen-II and decorin, were studied (Fig. 11.5BII) [145]. Interestingly, these engineered surfaces helped MSCs attachment, directing site-specific chondrogenic differentiation (Fig. 11.5BIII), thus proving to be a useful method to expand differentiating chondrocytes in vitro for cell therapies without the use of growth factors, but purely based on the synergistic presentation of nano-patterned cues and cell-binding motifs. Though cellular approaches are doubted for their therapeutic efficacy, acellular approaches have been the norm for cartilage therapies, with agarose or agarose based composites being set as the gold standards

collagen-I osteoblasts stained red, angiogenic marker von Willebrand factor (vWF) stained green, (III) restoration of volumetric defect in rabbits with (i) complete degradation of scaffolds at 3 months with periosteal regeneration noticed, which is confirmed by (ii) new bone (NB) formation within old bone (OB) region defect margins using fluorochrome labeling study [132]; (B) expansion of chondrocytes in vitro for cellular therapies on (I) nanoplatforms which harbor cartilage (II) inspired peptides to definitively drive the (III) chondrogenic differentiation and maturation [145]; (C) fabrication of biomimetic seamless osteochondral interfaces using silk scaffolding strategy (I), which helped in regeneration of osteochondral defects in rabbits as seen through the (II) histological examination for alizarin red (calcium deposits) and alcian blue (for glycosaminoglycan deposition) and (III) micro-computed tomography analysis for detection of mineralized region in subchondral bone [90]. *(A) Copyright 2018, Reproduced with permission from Wiley. (B) Copyright 2019, Reproduced with permission from Elsevier. (C) Copyright 2018, Reproduced with permission from The Royal Society of Chemistry.*

for cartilage repair [146]. However, the articular cartilage experiences mechanical stresses in different force magnitudes, length scales and utilizing this to deliver biomolecules 'on-demand' based on mechano-sensitivity have been recently reported [147]. These PLGA based mechanically activated microcapsules (MAMCs) with a core-shell morphology were fabricated in a microfluidic system with varying thickness to diameter (t/D) ratios. Thus, when subjected to compression loads ranging between 0 and 1 N, these MAMCs exhibited mechano-activation profile dependent on t/D ratio, releasing the encapsulated cargo (transforming growth factor beta-3, TGF-β3) in an applied load-dependent manner. Such mechanical stimuli instigated on-demand release platforms shows great potential in acellular cartilage repair strategies, to help home MSCs from the subchondral bone and differentiate chondrocytes to a very definitive chondrocyte lineage.

Mesenchyme condensation and subsequent ECM deposition postulate the cartilage maturation during developmental stages. Taking incentives from such developmental stages, researchers have devised strategies to condense MSCs to form aggregates leading them to chondrocyte clusters with the use of a magnetic field [148] in the absence of any scaffold. Recently, the same approach was ensued for cell seeding and condensation in the presence of a scaffold (pullan/dextran based) [149]. Superparamagnetic 8 nm maghemite (γ-Fe$_2$O$_3$) nanoparticle stabilized by negatively charged citrate ligand were used to label MSCs and by applying a magnetic field of 30 mT/mm scaffold surface area, the authors were able to seed, condense the MSCs within the scaffold while successfully achieving chondrocyte differentiation within the scaffold under dynamic culture conditions for 21 days. Another approach now gaining importance is the use of in situ gelling injectable hydrogel systems in cartilage repair owing to their ease of deployment at the defect site in a minimally invasive manner and complete coverage of irregular defect margins [4150]. The crosslinking mechanism determines how fast the gelation occurs or the degree of crosslinking, which correlates to its stability or degradation in vivo and also offers the feasibility to develop pH or temperature-responsive gels. In this regard, a click chemistry-based crosslinking of hyaluronic acid hydrogel (tetrazine modified hyaluronic acid and transcyclooctene modified hyaluronic acid) encapsulating cytomodulin-2 (a chondrogenic differentiation factor) for effective delivery and chondrogenic differentiation of human periodontal ligament stem cells [151]. Another interesting prospect in the injectable systems is the use of sutureless, bioadhesive biopolymers to be deployed at the chondral defect site for harboring the implanted scaffold or hydrogel for cartilage repair. Adhesive properties of a well-explored glycosaminoglycan, chondroitin sulfate was improved by modifying it by catechol groups (DOPA), which showed potential alternatives for sutureless cartilage repair strategies [152]. Bhardwaj et al. have recently reported the development of a 3D co-culture model using articular chondrocytes (ACs) and adipose-derived human mesenchymal stem cells (ADhMSCs) for modulation of chondrogenesis and hypertrophy. The results revealed synergistic interactions between cells as indicated by the interaction index value ranging from 2 to 3. This modulation is effected by the 3D microenvironment induced by physicochemical and biological properties of silk scaffolds, synergistic interactions between cells, and paracrine signaling in the co-culture system [153].

11.2.5.3 Smart biomaterials in osteochondral interface tissue engineering

Recapitulating the osteochondral defect characterized by its anisotropic gradient is one of the challenges in osteochondral interfacial tissue engineering [154]. In our endeavors we had devised two unique approaches, wherein the bulk modulus and chemical modulus in biphasic silk scaffolding systems to spatiotemporally guide the maturation of chondrocytes and osteoblasts preferentially in seeded osteochondral constructs have been showcased. In our first approach, we had fabricated biphasic bioactive glass/silk electrospun mats, which resembled the hierarchical complexity of native osteochondral interface with microfibrous bone phase (constituted by electrospun bioactive glass) and nanofibrous cartilage phase (constituted by electrospun silk). These composite membranes by its differential wettability, bulk modulus, and chemical cues mediated, maturation of osteoblasts and chondrocytes preferentially without the addition of any growth factors [3]. Following suit to this, we tried another approach to devise a seamless way of fabricating only silk scaffolding system (Fig. 11.5CI) to support chondrocyte survival in a spongy cartilage phase and osteoblast maturation in silk fiber-reinforced bony phase, in a biphasic composite scaffold [90]. These seamless biphasic silk constructs addressed critical problems relating to layer delamination often noticed in osteochondral grafts and also helped in regeneration of rabbit osteochondral defects with well-developed cartilage phase characterized by ECM deposition (Fig. 11.5CII), mineralized subchondral bone (with mineral density of 600–700 mg hydroxyapatite per cm^3) post 8 weeks after implantation (Fig. 11.5CIII). More recently, researchers have utilized the notion of buoyancy-driven gradients to fabricate an inhomogeneous native-like osteochondral interface [155]. In this work, the authors utilized agarose with different densities to form gradients which can house cargoes (liposomes, nanoparticles, extracellular vesicles), by a facile approach to inject denser to lighter agarose using electronic auto-pipette, which clarified within 10 s post-injection. The researchers also corroborated that the same principle could be applied for other polymers also, where they made density different gradients using heparin methacryloyl and gelatin methacryloyl (photo-crosslinked post density separation) with a BMP-2 gradient which guided the mineralization along the direction of BMP-2 gradient in 28-days culture in vitro using hMSCs.

11.3 Disease models

11.3.1 In vitro and in vivo models

One of the emerging approaches of tissue engineering is exploring techniques to treat malignancies by developing disease models that mimic the clinical manifestation to develop effective modes of diagnosis, therapeutic intervention, and strategies of prevention. Bone is commonly affected by a vast majority of metastatic cancers, which interact with the physical microenvironment of bone tissue and destroy the tissue. The recapitulation of the 3D microenvironment needs the presence of all the cellular components involved and appropriate substrate to maintain tissue architecture, essential for cellular communication. 3D tissue-like structure based on co-culture of human mesenchymal stem cells with improved osteogenic capacity (OEhMSCs) and a bone–tumor cell line (MOSJ-Dkk1) attached to ECM coated micro-carrier beads have been developed to recapitulate the mechanism of osteoinhibition [156] (Fig. 11.6A). The tissue-engineered human bone, grown from hMSCs has been used as a niche to culture various cancer cells or spheroids such as metastatic prostate cancer cells [157], breast cancer cells [157], and Ewing's sarcoma spheroids [158] to study cross-talk between crucial bone microenvironment and cancer cells. Osteosarcoma model has been designed using 3D hydrogel-based co-culture system of adipose-derived stem cells (ADSCs) and primary osteosarcoma cells (SaOS-2) to investigate the protumor effect of ADSCs. Metastatic effect of osteosarcoma has been successfully recapitulated by the introduction of SaOS-2 into the humanized bone microenvironment, which led to spontaneous lung metastasis in mice model [159]. Besides investigations on bone–tumor cross talks, in vitro biomimetic models for osteoporosis have been generated by co-culturing of osteoblasts and osteoclasts on a substrate such as silk-hydroxyapatite films for the assessment of therapeutics [160]. The conventional bone defect models such as calvarial bone defects, large-bone segmental defects are made in animal models by introducing mechanical injuries using trephine bur, and drill or saw, respectively [161]. Removal of ovary or ovariectomy in female animal models also leads to osteoporosis as estrogen withdrawal stimulates the differentiation of myeloid precursor cells into osteoclasts which leads to substantial increase in bone turnover [162,163].

In cartilage tissue engineering, chondrocytes play significant role which maintains the architecture of ECM in cartilage, but they show low proliferative ability and lose their phenotype in monolayer-culture conditions [164]. Therefore, 3D microenvironment is essential to recapitulate physio-pathological conditions. Galuzzi et al. have developed various tissue scaffolds to mimic cartilaginous pathology by pellet culture of chondrocytes, decellularized ECM of cartilage, and two alginate-based scaffold (microcarriers and beads) [165]. In vitro cartilage explants have been utilized to generate damaged cartilage models by providing inflammatory cytokines (IL-1β and TNF-α) as media supplements and the application of injurious compression [166]. Age-related changes in cartilage remodeling has been investigated by the development of human osteoarthritis (OA) model, co-culturing chondrocytes and synoviocytes, isolated from patients [167]. The stimulatory effect of cartilage fragments in progression of osteoarthritis has been studied using a co-culture system of macrophages and chondrocytes [168]. Interactive tri-cultured platform of rheumatoid arthritis has been used to study the effect of cytokines release from lipopolysaccharide (LPS)-activated macrophages on chondrocytes and synovial cells [169]. An osteochondral-synovial membrane explant based co-culture system was exposed to mechanical and chemical injury which led to cartilage degeneration and inflammatory reaction similar to naturally occurring OA [170] (Fig. 11.6B). 3D printed platforms have been designed with accessible individual compartments to develop osteochondral microsystem which has a cartilage/bone biphasic structure containing functional interface and compatible to drug assessment [171].

Naturally occurring or post-traumatic osteochondral anomalies are studied in animal models including mouse [172], rat [173], rabbit [174], guinea pig [175], ovine [176], dog [177], pig [178], goat [179], sheep [179,180], and horse [181]. Commonly used surgically-induced models are developed by anterior cruciate ligament transection (most common) [182], medial meniscal tear, partial or total meniscectomy [179,183] (Fig. 11.6C), and ovariectomy [175,176,184]. Several studies have been performed to induce OA chemically in animal models by intra-articular injections of monosodium iodoacetate [173,185] and

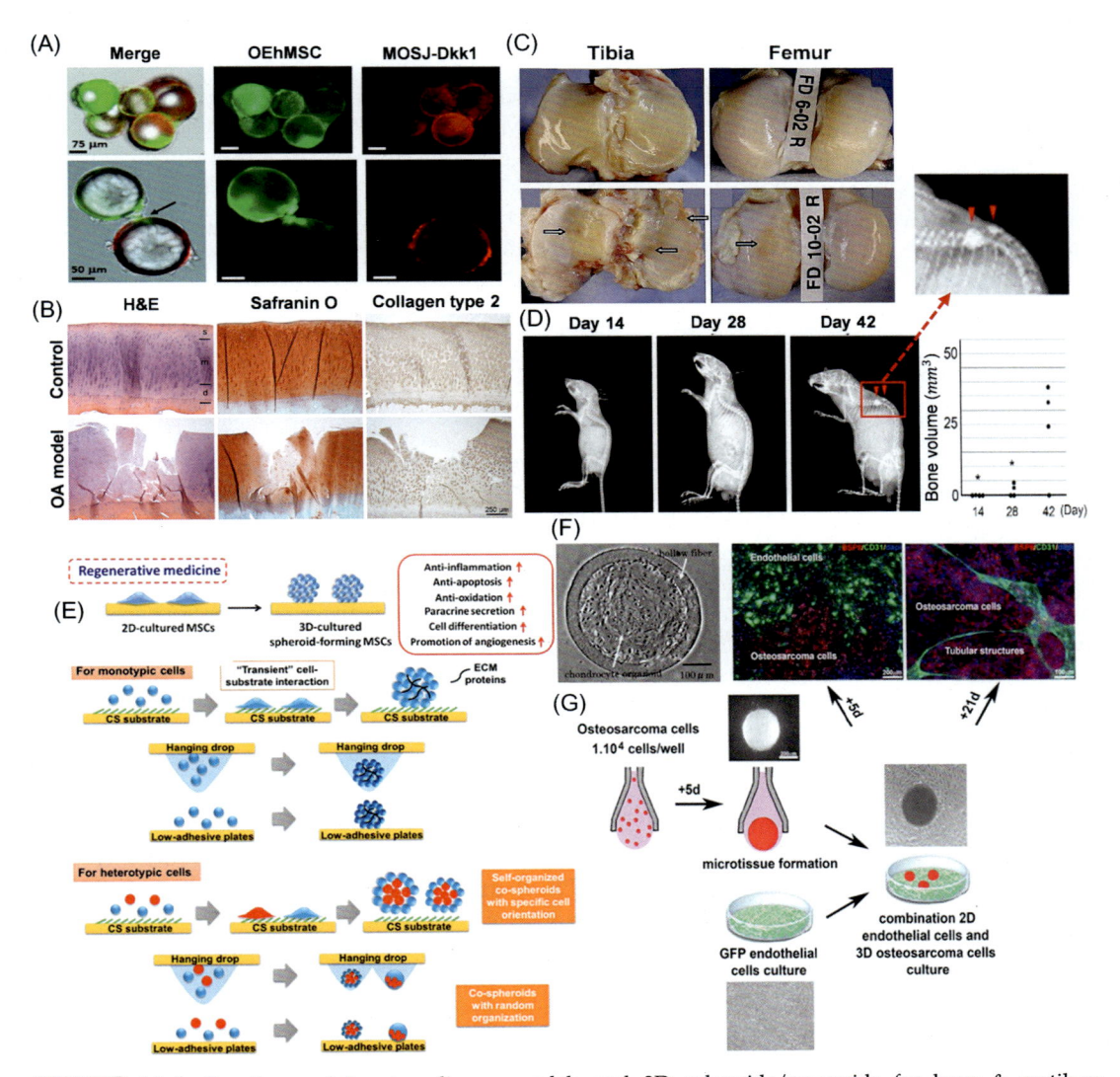

FIGURE 11.6 In vitro and in vivo disease models and 3D spheroids/organoids for bone & cartilage defects. (A) Microscopic images of co-cultures of GFP-labeled OEhMSCs (center), and RFP-labeled MOSJ-Dkk1 cells(right), seeded on collagen I-coated beads, with merged images (left) [156]. (B) Histochemical analysis (Hematoxylin and Eosin staining, Safranin O staining, and immunostaining for collagen type-2) of OA model in adult horses after 3rd week of implantation [170]. (C) Postoperative macroscopic images of meniscectomy-induced OA in sheep model. Cartilage pathology has been compared between normal tibial or femoral condyles (above) and defected condyles (below) [179]. (D) X-ray images of mice transplanted with hiPSCs of patients with FGFR3 chondrodysplasia after 14, 28 and 42 days of transplantation. Ossification is indicated by red arrowheads. Micro CT analysis of total bone volume was plotted above [172]. (E) Different methods for the formation of monotypic and heterotypic spheroids using substrates [187]. (F) Histological analysis of organoid cultured in a hollow fiber using rat chondrocyte for 28 days, showing the presence of secreted ECM [188]. (G) 3D spheroids of osteosarcoma cells were cultured on monoculture of HUVEC. Immunofluorescence images are showing the formation of tubule-like structure formed by HUVECs after 21 days, indicating the successful formation of vascularization [189]. *(A) Copyright 2018, Reproduced with permission from Nature Springer under CC BY License. (B) Copyright 2019, Reproduced with permission from PLoS One under CC BY License. (C) Copyright 2010, Reproduced with permission from Elsevier. (D) Copyright 2018, Reproduced with permission from Elsevier. (E) Copyright 2019, Reproduced with permission from MDPI under CC BY License. (F) Copyright 2008, Reproduced with permission from Elsevier. (G) Copyright 2017, Reproduced with permission from Elsevier.*

collagenase [186]. Chondral injury was created surgically in pig model, and efficacy of MSCs collected from human umbilical cord (HUMSC) for treatment was evaluated [178]. Patient-specific FGFR3 chondrodysplasia has been generated using human induced pluripotent stem cells (hiPSCs), collected from patients' cartilages, which was subcutaneously transplanted 4 weeks old immunodeficient SCID mice to study disease pathology and drug assessment [172] (Fig. 11.6D). Such humanized models may improve our level of understanding of cartilage pathology and joint dysfunction to provide insight into its prospective treatments.

11.3.2 3D spheroid model

In scaffold-based techniques, there are many specifications, such as the proper distribution of cells throughout the scaffold, biocompatibility, and stability of material that have to be critically monitored. To elude these problems, the scaffold-free culture or spheroid culture methods have attracted attention in recent trends of bone-tumor interaction. The self-organized organoid models have been developed by various techniques (Fig. 11.6E) such as hanging drop methods, low-adhesive methods, spinner flasks or stirred-tank cultures, and rotating cell culture bioreactor. The organoid technology has prevailed to be superior as compared to 2D culture system for high-throughput screening of therapeutics. Tissue organoid culture (Fig. 11.6F) has been shown to secrete ECM to reconstitute the tissue-like structure for regeneration [188]. Besides regenerative tissue-like culture, spheroid culture techniques are being used to study pathological processes, for example, 3D spheroids of osteosarcoma MG-63 cells were formed using the hanging drop technique and cultured on a monolayer of human umbilical vein endothelial cells (HUVEC) to mimic tumor angiogenesis in 3D culture. The study has shown 3D spheroids allowed expression of ECM proteins such as osteocalcin, osteopontin, and BSPII and angiogenic factors such as VEGF, CXCR4, and ICAM1 for the formation of microtissue [189] (Fig. 11.6G). To study the metastatic effect of breast cancer, the osteoid matrix was created using murine osteoblasts in 3D bioreactor, and the introduction of breast cancer cells led to degradation of the matrix, which suggests the negative effect of the breast cancer cells on bone–matrix mineralization in diseased state [190]. For cartilage malignancy, the spheroid-based culture chondrosarcoma model using SW1353 cells has been developed to identify salinomycin as a promising drug for the treatment [191]. Similarly, human chondrosarcoma spheroid model has shown to have a high hypoxic core, rich in glycosaminoglycans (GAGs) and VEGF excretion and investigated for drug screening using hypoxia-activated prodrug TH-302 and doxorubicin [192]. Not only drugs, 3D organoids to assess patient-specific responses to chemotherapy may also become a future direction for personalized therapy. Furthermore, the spheroid model can be introduced to gene silencing, genome editing, and transcriptome modulations using small interfering RNA (siRNA) or CRISPR-Cas9, to develop gene-based therapeutics [193]. Although the exact in vivo system cannot be recreated using 3D organoid culture, new disease models are being developed and validated with time, significantly tending to accurate recapitulation of complex systems to establish robust preclinical platforms for rapid assessment of cancer therapeutics.

11.3.3 Immune response and Immunomodulation

Inflammatory milieu at wound site may determine the success of tissue repair process. Innate and acquired immune cells from surrounding tissue and peripheral blood immediately invade at the fractured site [194]. The ruptured blood vessels cause hematoma, which leads to the initial inflammatory phase of healing, characterized by hypoxic environment due to the anaerobic energy turnover, low pH, low oxygen levels, and high lactate concentrations. At first, neutrophil infiltration occurs which attracts macrophages to the fractured site [195]. Both the cells secrete pro-inflammatory cytokines, clean dead cells and debrides and promote the migration of MSCs towards injured site [196]. The fracture hematoma is reorganized, and the deposition of fibrin thrombus gets initiated [197]. Infiltrated macrophages acquire pro-inflammatory M1 phenotype, characterized by the production of increased levels of pro-inflammatory cytokines including tumor necrosis factor-α (TNFα), interferon-γ (IFN-γ), and interleukin-6 (IL-6) to mediate pathogen-resistance, high production of reactive oxygen and nitrogen intermediates and activation of adaptive immune response through T cell responses [198]. Regulatory T cells (Tregs) stimulate bone formation by secretion of IL-4 and the downregulation of TNF-α and IFN-γ. CD4 + T cells prevail in hypoxic environment as they can adapt well to energy insufficiencies. CD8 + T cells have a negative effect on bone regeneration by secretion of pro-inflammatory cytokines, such as TNF-α and IFN-γ [199]. The MSCs can differentiate into osteogenic lineage for intramembranous ossification at the edges of the fractured site and chondrogenic lineage for the synthesis of a cartilaginous matrix of soft callus [200]. MSCs act as immunosuppressive which aids to resolve the inflammation at injury site [201] and promotes angiogenesis for further enhancing the fracture healing cascade [202]. Due to the change of surrounding microenvironment, the macrophages predominantly acquire an anti-inflammatory and angiogenic M2 phenotype, which secrete immunomodulatory cytokines including transforming growth factor β (TGF-β) and IL-10 and also enhance angiogenesis through VEGF secretion. TGF-β induces chondrogenic differentiation of MSCs to form the soft, cartilaginous callus [198]. Macrophages can be polarized into any type by exposing them to specific microenvironments [203] (Fig. 11.7A) to modulate the immune reaction for the acceptance of tissue-engineered graft. The bone marrow MSC-based bioengineered cartilaginous tissue can promote M2 polarization and inhibit T cell proliferation by G^0 cell cycle arrest [193]. The viral vector-mediated gene delivery systems had been explored for intra-articular expression of immunomodulatory cytokines for treatment of age-related pathology osteoarthritis [204,205]. Exosomes derived from MSCs have been utilized for cartilage repair by rapid cellular proliferation and infiltration, increased matrix synthesis and infiltration of regenerative M2 macrophages [206]. Lee et al. has reported immunomodulatory property of human fetal cartilage progenitor cells (hFCPCs) in co-culture system with Tregs [207]. Besides cells, the choice of biomaterials used to make the construct is also important for the modulation process. Implantation of biomaterials may induce inflammation in the affected area. Surface properties such as hydrophilicity, charge, roughness, and porosity of the biomaterials are needed to be considered to develop a construct for the immunomodulation process. Sustained and sequential release of immunomodulatory chemicals, proteins, and cytokines from tissue-engineered [208] (Fig. 11.7B), hydrogel [209] and 3D printed constructs [210] (Fig. 11.7C) have been demonstrated to promote M2 polarization and enhanced

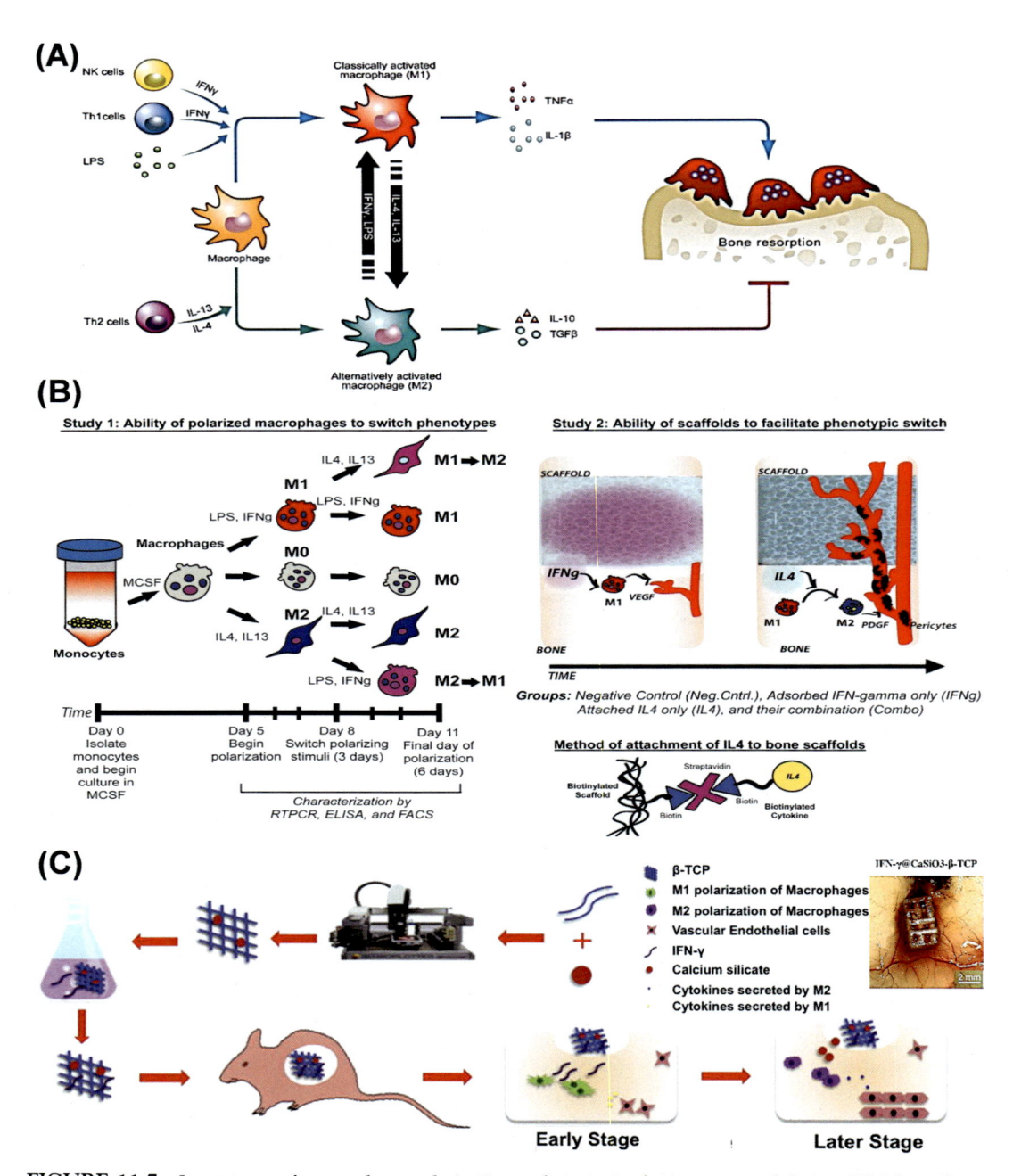

FIGURE 11.7 Importance of macrophage polarization and strategies for immunomodulation. (A) Macrophages can be activated and polarized into M1 and M2 phenotypes based on the presence of respective cytokines, respectively. M1 macrophages can induce bone resorption by secreting TNFα and IL-1β. M2 macrophages can inhibit bone resorption by secreting IL-10 and TGFβ [203]. (B) Graphical representation of experiment which involves switching of phenotypes of polarized macrophages in the presence of M1-or M2-polarizing stimuli. Study 2 is showing the scaffold-based sequential release of immunomodulatory cytokines to induce the M1-to-M2 transition of macrophages [208]. (C) 3D printed construct is made calcium silicate/β-tricalcium phosphate (CaSiO3-β-TCP) designed to polarize M1 and M2 macrophages by sequential release of IFN-γ and Si, respectively. Subcutaneous implantation of the construct in mice showing angiogenesis after 4 weeks [210]. *(A) Copyright 2017, Reproduced with permission from Elsevier. (B) Copyright 2015, Reproduced with permission from Elsevier. (C) Copyright 2018, Reproduced with permission from Elsevier.*

vascularization. Therefore, the controlled delivery of immunomodulators has potential to improve the regenerative strategies and a promising field for further investigation.

11.4 Graft substitutes

Graft substitutes are increasingly being used as surgical methods to enhance tissue regeneration in orthopedic procedures. A graft substitute when transplanted from one site to another of the same individual is known as an autograft, and when transferred between two genetically different individuals of the same species is called allograft. When transplantation of the graft is between members of different species, it is known as xenograft. Clinically, autografts are considered as the "gold standard" owing to their histocompatibility and their chances to provoke immune reactions are minimal; however, their usage is hindered mainly by donor site morbidity. On the other hand, the use of allografts is restricted due to a possible immune reaction elicited by the host in response to the foreign tissue of the allograft, along with the possibility for disease transmission [211]. As a result of these drawbacks, the demand and development of new orthobiologic materials to help in the treatment of orthopedic defects is growing.

Bone grafts are bone-like materials derived from sources such as living donor, deceased organ donors, or fabricated artificially. They are commonly used in clinical settings to augment bone repair and regeneration towards healing, strengthening, and improving the bone function during disease or injury. The ideal bone grafting material is expected to have several important properties, namely biocompatibility, bio-absorbability, durability, vascularity, cost-effective, and ease of availability [212]. Additionally, the graft should have the three crucial characteristics of osteogenesis, osteoinduction, and osteoconduction. Bone autografts serve as the gold standard as graft substitutes as they are immuno-compatible and non-immunogenic, and they possess the required properties of a bone graft material such as osteoinduction through growth factors, osteogenesis through progenitor cells and osteoconduction through the 3D porous matrix. However, autografts are expensive procedures, and they may result in significant donor site morbidity, deformity, and scarring [211]. The use of allografts is the second most suitable option for orthopedic surgeons. Allogenic tissue is likely to be histocompatible, but it has reduced osteoinductive properties as they are devoid of any cellular constituent. Additionally, the allografts are associated with higher risks of immune reactions and the possibility of infection transmission [213]. To avoid the risk of transmission of any infection or disease, the allografts are treated using methods such as tissue-freezing, and freeze-drying, that destroys most of the osteogenic cells thus relieving it of any osteoinductive capabilities; followed by treatment with sterilizing agents such as gamma radiation, ethylene oxide, and electron-beam radiation [214]. This processing is the reason that most allograft tissues are devoid of its osteoinductive capacity, leaving it to serve mostly as an osteoconductive scaffold in bone grafting. A xenograft, having to use the tissue from other species but humans, is scarcely used as bone substitutes. For its use, these xeno-materials are often freeze-dried or demineralized or deproteinized and usually prepared as a calcified matrix. Coral based xenografts are used as "coral derived granules" (CDG) and are mainly composed of calcium carbonate, while the natural human bone is composed primarily of HA along with calcium phosphate and carbonate [215]. Thus, the CDGs are processed to be transformed industrially into HA via a hydrothermal process, forming a non-resorbable xenograft that could be enhanced using other molecules and growth factors. Interpore and Pro-osteon (Interpore International, Inc., Irvine, CA) are some of the clinically accessible coral-based products [216].

Synthetic materials to be used for developing grafting materials should have the properties of the native bone (Table 11.1). A large number of synthetic material alternatives with various mechanisms of action have emerged, which could be used for grafting applications. Ceramics for instance, are osteoconductive and are composed of calcium phosphate (Ca-P). Ceramics possess the specific phase composition and porous structural features that make it ideal for interacting with signaling factors and ECM of the host system, that aids in creating a local niche for neo-bone formation [217]. Grafts of composites, on the other hand, combines osteoconductive matrix and bioactive agents for providing osteoinductive as well as osteogenic potential [218]. Polymers are another class of materials used to develop grafts due to their better bone-matrix interference and its biomimetic ability (Table 11.2) [218].

The field of bone tissue engineering (BTE) is another potential technique for developing graft substitutes. BTE focuses on alternative treatment options with a focus to eliminate the limitations of the current clinically used treatments such as graft rejection, donor site morbidity, restricted availability, and disease transmission. The approach uses a biocompatible scaffold that is biomimetic to bone ECM, osteogenic or progenitor cells, signaling molecules/factors, and sufficient vascularization. Several biomaterials are used as matrices or scaffolding materials for BTE, such as silk fibroin, collagen, hyaluronic acid, alginate, etc. [4]. Demineralized bone matrix (DBM) is an important biomaterial used as a scaffold in BTE for the development of bone graft substitutes. DBM is commonly produced from allograft bone sources through decalcification or an acid extraction method retaining the most of the collagen proteins. Post-processing, its osteoconductive and osteoinductive properties are retained but not the osteogenic property [219]. Its ease and higher availability, along with reduced immune response due to processing and makes it a sought out for material. Some commercially available DBM bone graft substitutes are Grafton™, Opteform (Exactech), Osteofil (Medtronic), Dynagraft (SeaSpine).

As opposed to bone, there are fewer literature reports of the development of cartilage and osteochondral grafts. However, over the past decade, osteochondral allograft transplantation have gained attention. This is mainly due to the advancements in the storage techniques that have demonstrated enhanced chondrocyte viability for longer time intervals, thereby increasing the graft availability [220]. Single-step autologous minced cartilage is another procedure that is used increasingly for filling cartilage and osteochondral defects and is currently under investigation with some success [221]. DeNovo® NT Graft is an FDA-listed off-the-shelf human allograft consisting of juvenile cartilage tissue and used by the surgeons for early interventions in the repair and restoration of damaged cartilage tissue. The allograft is composed of pieces of human hyaline cartilage and implanted in a one-step procedure with a fibrin glue adhesive. Studies are being conducted to evaluate long-term outcomes, including pain relief and improvement of function, for both knee and ankle cartilage repair [222,223]. Similarly, RevaFlex™ (ISTO Technologies) is a scaffold-free tissue-engineered juvenile cartilage graft intended for the treatment of

TABLE 11.1 List of few commercially available synthetic materials based grafts and their applications.

Name of the product	Material	Properties	Applications
Osteograf	Ceramic	Osteoconductive, limited osteoinductive when mixed with bone marrow	Bone void filler
NovaBone	Bioactive glass	Osteoconductive, limited osteoinductive when mixed with bone marrow	Filling surgical or traumatic bone gaps
Osteosat	Surgical grade calcium phosphate	Osteoconductive and bioresorbable	Hip and knee joint repair
Calceon 6	Calcium sulfate	Osteoconductive and bioresorbable	Bone void filler; provides strength
Norian	Monocalcium phosphate, tricalcium phosphate, and calcium carbonate	Good compressive strength	Skull bone defect; injectable paste, craniofacial reconstructions
Hard tissue-replacement (HTR)	Poly methyl methacrylate (PMMA)	Good strength, durable, and surface osteoconductive	Craniofacial reconstruction
Alpha BSM	Calcium phosphate cement	Good compressive strength	Dental application for bone and cartilage defects
Mimix	Synthetic hydroxyapatite tetra-tricalcium phosphate	Good compressive strength	Cranial defects
ELIZ (Kyeron)	Composed of (40%) β-tricalcium phosphate and of (60%) hydroxyapatite	Ultrahigh porosity, biocompatible, and osteoconductive	It has been successfully implanted in more than 1200 patients without any side-effects
OSIQ (Kyeron)	Fully synthetic ultrapure nano-hydroxyapatite	Ready to use, injectable, and biodegradable	Filling or reconstruction of small and medium bone defects
AXOZ QS (Kyeron)	Resorbable phosphocalcic compounds and a polymer	Injectable and fully resorbs	Supports bone growth
COLLAPAT II (Symatese)	Composed of a collagen structure in which ceramised hydroxyapatite granules are dispersed	Strong hemostatic power, completely resorbable in a few weeks, and osteoconductive	Induces bone substance replacement in maxillofacial surgery and odontostomatology
CopiOs (Zimmer Biomet) Bone Void Filler	Calcium phosphate, dibasic (DICAL), and highly purified Type I bovine collagen	DICAL provides significantly more calcium and phosphate ions at equilibrium than either β-TCP or HA	CopiOs paste acts as an osteoconductive scaffold for the growth of new bone

Adapted from Dahiya UR, Mishra S, Bano S. Application of bone substitutes and its future prospective in regenerative medicine. In: Biomaterial-supported tissue reconstruction or regeneration. IntechOpen; 2019.

TABLE 11.2 List of some commercially available polymer-based graft materials and their applications.

Name of the product	Material	Properties	Applications
Cortoss	Polymer system with reinforcing particle bioactive glass	Forms biological interface	Augmentation of screws in osteoporotic bone (hip, spine, etc.)
Open porosity polylactic acid polymer (OPLA)	Polylactic acid	Osteoconductive and bioresorbable	Articular cartilage regeneration
Collagraft	Mixture of tricalcium phosphate, bovine collagen, and hydroxyapatite	Bioresorbable and osteoconductive	Use for the treatment of long bone fracture and void filling
DynaGraft	Demineralized bone matrix	Heat sensitive copolymer, injectable gel, limited osteo-induction	Dental bone graft substitute
MedPor	Porous polyethylene	Higher porosity	Orbital reconstruction and facial contouring
Collapro/matrix	Human collagen in lyophilized strip	Lack of immunogenic property	Use in development
Healos	Hyaluronic acid-coated collagen sponge	Osseo-inductive property	Replacement of autograft/autograft extender for spinal fusion
Immix	PGA/PLA polymer to be produced in chip, flex forms	Provides structural support	Bone graft extender
OsteoScaf (Bonetec)	Macroporous poly(lactide-*co*-glycolide)/calcium phosphate (PLGA/ CaP) foam matrices	Fully resorbable, osteoconductive, and mechanically robust	Heal tissue defects

Adapted from Dahiya UR, Mishra S, Bano S. Application of bone substitutes and its future prospective in regenerative medicine. In: Biomaterial-supported tissue reconstruction or regeneration. IntechOpen; 2019.

articular cartilage lesions [224]. Other cartilage and osteochondral matrices under investigation include the Cartilage Autograft Implantation System (CAIS, Johnson and Johnson), and BioCartilage® (Arthrex, Naples, Florida).

11.5 Clinical status

Many scaffolds/constructs have been designed, developed, and explored in bone, cartilage, and osteochondral tissue engineering. Of those, only a few have been progressed to clinical trials [225]. In the case of articular cartilage, the focal and degenerative lesions significantly decrease the quality of patient's life. In spite of the progress in several therapies, including surgical treatment, a definitive treatment strategy towards cartilage repair is not yet known [88]. The regeneration and formation of fibrocartilage instead of the desired hyaline cartilage of the neo-tissue is a significant challenge for clinicians. There is a constant requirement of new and improved strategies with the best cost-effective treatment for cartilage regeneration, with high quality randomised clinical trials. Biomimetic ceramics and composites have recently garnered success in bone regeneration therapies in patients [226,227]. A recent comprehensive review by Oryan et al. reports the status of the various commercially available bone graft substitutes and their clinical applications [228], and states that autografts remain as the gold standard for bone regeneration by functioning better to the tissue engineering scaffolds. Therefore, much understanding and advancements are required to fabricate clinically relevant constructs that mimic the natural bone. Similarly, in the case of osteochondral tissue, several multilayer scaffolds and composite constructs have been explored with a focus on the specific anatomical details of each tissue layer. These constructs are investigated comprehensively in osteoarthritic patients for repair and regeneration of osteochondral defects, with some of them achieving limited success and commercially available in the market [225].

The preclinical and clinical trials help us to understand where we stand and create the correct strategies towards better and efficient tissue repair with improved cost-benefit ratios. Towards clinical translation of grafts, most of the pre-clinical studies are either ongoing or are waiting to undergo clinical trials, with the results from several current trials of bone, cartilage and osteochondral grafts showing promising results for future clinical applications. Table 11.3 lists the studies and their status for the clinical trials of bone, cartilage and osteochondral grafts/constructs extracted from ClinicalTrials.gov.

TABLE 11.3 List of ongoing/completed clinical trials for cartilage, bone, and osteochondral tissues from ClinicalTrails.gov.

NCT number/ status	Study title	Sponsor	Interventions/study type/phase
NCT04000659 Not yet recruiting	Episealer® Knee System IDE Clinical Study	Episurf Medical Inc.	Device: Episealer Knee System Procedure: Microfracture Interventional NA
NCT00314236 Completed	Trial Comparing BST-CarGel and Microfracture in Repair of Articular Cartilage Lesions in the Knee	Piramal Healthcare Canada Ltd	Device: BST-CarGel with Microfracture Procedure: Microfracture without BST-CarGel Interventional NA
NCT03545269 Completed	Study to Assess the Efficacy and Safety of Treatment of Articular Cartilage Lesions With CartiLife®	Biosolution Co., Ltd	Drug: CartiLife® Procedure: Microfracture Interventional Phase 2
NCT00989794 Active, not recruiting	Study to Evaluate the Safety and Performance of Treatment of Articular Cartilage Lesions Located on the Femoral Condyle With gelrinC	Regentis Biomaterials	Device: GelrinC Interventional Phase 1, Phase 2

(Continued)

TABLE 11.3   (Continued)

NCT number/ status	Study title	Sponsor	Interventions/study type/phase
NCT01791062 Completed	Safety and Efficacy Study of HYTOP® in the Treatment of Focal Chondral Defects	TRB Chemedica AG	Device: HYTOP® Interventional NA
NCT02391506 Completed	Safety and Effectiveness of Cartiva Implant in the Treatment of First CMC Joint Osteoarthritis (GRIP)	Cartiva, Inc.	Device: Cartiva Interventional NA
NCT02732873 Not yet recruiting	Porous Tissue Regenerative Silk Scaffold for Human Meniscal Cartilage Repair (REKREATE)	Orthox Limited	Device: FibroFix Meniscal Scaffold Interventional NA
NCT01221441 Completed	Study of TG-C in Patients With Grade 3 Degenerative Joint Disease of the Knee	KolonTissueGene, Inc.	Biological: TissueGene-C Drug: Normal Saline Interventional Phase 2
NCT02609074 Completed	Pilot Clinical Trial of CPC/rhBMP-2 Microffolds as Bone Substitute for Bone Regeneration	East China University of Science and Technology	Procedure: Minimally invasive internal fixation surgeries Device: CPC/rhBMP-2 micro-scaffolds Device: CPC paste Interventional Phase 4
NCT00841152 Unknown	Comparison of Bioactive Glass and Beta-Tricalcium Phosphate as Bone Graft Substitute (BAGvsTCP)	Turku University Hospital	Device: Bioactive glass Device: Beta-tricalcium phosphate (ChronOs) Procedure: Autograft Procedure: Allograft (frozen femoral head) Observational
NCT00147823 Completed	Comparing Synthetic Bone Alone Versus Synthetic Bone With Bone Marrow in Bone Lesions	State University of New York — Upstate Medical University	Device: Vitoss Alone Device: Vitoss with Bone Marrow Aspirate Interventional NA
NCT03024008 Recruiting	Enhancement of Bone Regeneration and Healing in the Extremities by the Use of Autologous BonoFill-II	BonusBio Group Ltd	Biological: BonoFill-II Interventional NA
NCT03884790 Recruiting	Pre-market Study to Evaluate Safety and Performance of GreenBone Implant (Long Bone Study)	GreenBone Ortho S. r.l.	Device: surgical repair of long bone defects Interventional NA
NCT04018287 Recruiting	Circulating miRNAs and Bone Microstructure in Adults With Hypophosphatasia	Medical University of Vienna	Other: HR-pQCT scans, BMD measurements, bone specific circulating microRNAs (miRNAs) Observational
NCT01878084 Unknown	Bioactive Glass (Sol–gel) for Alveolar Bone Regeneration After Surgical Extraction	Alexandria University	Other: bioactive glass (sol–gel) Other: empty extraction socket Interventional Phase 1, Phase 2
NCT03103295 Unknown	3D Tissue Engineered Bone Equivalent for Treatment of Traumatic Bone Defects (3D-TEBE)	A.A. Partners, LLC	Biological: 3D-Tissue Engineered Bone Equivalent Interventional Phase 1, Phase 2
NCT03916328		MU-JHU CARE	Drug: B/F/TAF

(Continued)

TABLE 11.3   (Continued)

NCT number/ status	Study title	Sponsor	Interventions/study type/phase
Not yet recruiting	BONE: STAR (Switching to TAF Based Anti-Retroviral Therapy) Study (BONE: STAR)		Drug: TDF/3TC/EFV or DTG or NVP Other: DMPA Interventional Phase 4
NCT03941028 Recruiting	Clinical Effects of Large Segmental Bone Defects With 3D Printed Titanium Implant	Peking University Third Hospital	Device: 3D printed titanium implant Interventional NA
NCT01012921 Completed	Comparison of a PEG Membrane and a Collagen Membrane for the Treatment of Bone Dehiscence Defects at Bone Level Implants	Institut Straumann AG	Device: barrier membrane Device: MembraGel Other: Bio-Gide® membrane Interventional NA
NCT02613663 Completed	Immediate Implant Using "Nanobone" Versus "Autogenous Bone" for Treatment of Patients With Unrestorable Single Tooth	Cairo University	Procedure: Immediate implants with nanobone graft Procedure: immediate implants with autogenous bone graft Interventional NA
NCT00206791 Unknown	Osteoconduction Potential of an Injectable Calcium Phosphate in Orthopedic Surgery in Fillings of Osseous Defects	Biomatlante	Device: MBCP-Gel (tm) Interventional NA
NCT03978962 Recruiting	Performance and Safety of the Resorbable Collagen Membrane "Ez Cure"	Biomatlante ATLANSTAT	Device: Guided Tissue Regeneration Observational
NCT01815658 Unknown	"Bio-logical" Carbon Fiber Intramedullary Nail Biomechanical Test and Preliminary Clinical Results	Hillel Yaffe Medical Center	Device: Carbon fiber intramedullary nail (PEEK-CF) Observational NA
NCT02293031 Unknown	Gene-activated Matrix for Bone Tissue Repair in Maxillofacial Surgery	NextGen Company Limited	Device: Gene-activated matrix "Nucleostim" Interventional NA
NCT03166917 Recruiting	Clinical Application of Personal Designed 3D Printing Implants in Bone Defect Restoration	fang guofang Shenzhen Hospital of Southern Medical University	Procedure: 3D printing implant Procedure: autogenous bone grafting Interventional NA
NCT00872066 Completed	A Study to Assess the Long-term Performance of SmartSet® HV and SmartSet® GHV Bone Cements in Primary Total Hip Replacement	DePuy International	Device: SmartSet® HV bone cement Device: SmartSet® GHV bone cement Interventional Phase 4
NCT03601130	Assessment of HydroxyColl Bone Graft Substitute in High Tibial Osteotomy Wedge Grafting. (HColl_HTO)	SurgaColl Technologies Limited	Device: HydroxyColl Bone Graft Substitute Device: DePuy Synthes TomoFix Medial High Tibial Plate (MHT) fixation (Standard of Care) Interventional NA
NCT03185286 Recruiting	3D-Printed Personalized Metal Implant in Surgical Treatment of Ankle Bone Defects	Southwest Hospital, China	Device: 3D-printed personalized metal implant Interventional NA

(Continued)

TABLE 11.3 (Continued)

NCT number/ status	Study title	Sponsor	Interventions/study type/phase
NCT02748343 Recruiting	The Clinical Therapeutic Effects and Safety of Tissue-engineered Bone	Xijing Hospital	Device: allograft bone Device: tissue-engineered bone Interventional Phase 1, Phase 2
NCT03735199 Recruiting	3D Printed Scaffold Device for Ridge Preservation After Tooth Extraction	National Dental Centre, Singapore	Device: PCL-TCP scaffold Device: Geistlich Bio-Gide collagen membrane Interventional NA
NCT03608280 Not yet recruiting	Efficiency of 3D-printed Implant Versus Autograft for Orbital Reconstruction (TOR-3D)	Hospices Civils de Lyon	Procedure: Bone autograft Procedure: Orbital reconstruction by 3D-printed porous titanium implant Interventional NA
NCT02423629 Active, not recruiting	Agili-C™ Implant Performance Evaluation in the Repair of Cartilage and Osteochondral Defects	Cartiheal (2009) Ltd	Device: Agili-C™ implantation procedure Interventional NA
NCT02430558 Unknown	Second Line Treatment of Knee Osteochondral Lesion With Treated Osteochondral Graft	TBF Genie Tissulaire	Biological: OD-PHOENIX Interventional Phase 1, Phase 2
NCT01477008 Active, not recruiting	BiPhasic Cartilage Repair Implant (BiCRI) IDE Clinical Trial — Taiwan	BioGend Therapeutics Co., Ltd	Device: BiPhasic Cartilage Repair Implant Procedure: Marrow Stimulation Interventional NA
NCT03696394 Recruiting	A Study to Evaluate the Efficacy of BioCartilage® Micronized Cartilage Matrix in Microfracture Treatment of Osteochondral Defects	St. Paul's Hospital, Canada	Device: BioCartilage® Micronized Cartilage Matrix Interventional NA
NCT01209390 Terminated	A Prospective, Post-marketing Registry on the Use of ChondroMimetic for the Repair of Osteochondral Defects (OMCM)	TiGenixn.v.	Device: Chondromimetic Observational
NCT03036878 Recruiting	ReNu™ Marrow Stimulation Augmentation	NuTech Medical, Inc.	Other: ReNu Interventional NA
NCT03347877 Enrolling by invitation	Autologous Osteo-periosteal Cylinder Graft Transplantation for Hepple V Osteochondral Lesions of the Talus	Peking University Third Hospital	Procedure: autologous osteo-periosteal cylinder graft transplantation Procedure: autologous osteochondral graft transplantation Device: autologous osteo-periosteal cylinder graft Interventional NA

https://clinicaltrials.gov/, accessed 26.10.19.

ClinicalTrials.gov is a web-based database maintained by the U.S. National Library of Medicine, which offers information on the status of more than 320,000 research studies/ clinical trials from over 200 countries. The database offers the patients, health care professionals, researchers, and the public with data of these clinical studies for an extensive variety of conditions and diseases.

11.6 Conclusion and future perspectives

Skeletal defects and/or diseases related to bone, cartilage, and osteochondral tissues represent a pertinent issue as it leads to a massive burden on the healthcare sector. Various treatment methods starting with autologous cell therapy, prosthetic joints, microfracture, transplantation, and others, were attempted but associated with challenges in terms of quality and performance of the repaired tissue. Nonetheless, these methods do not provide complete recovery, which further manifested the need for the exploration of advanced technologies for repair and regeneration of skeletal tissues (bone and cartilage). Herein, we have discussed in detail the innovative state of art strategies for bone and cartilage repair. The current treatment methods for bone and cartilage repair include advanced tissue engineering approaches in terms of designing smart biomaterials, nanotherapy, gene therapy, 3D bioprinting, exosomes, grafts, in vitro, and in vivo disease models etc.

Tissue engineering-based approach involves cells, biomaterials and different stimuli such as growth factors, mechanical stimuli for the repair of different diseased or damaged tissues. A variety of biomaterials starting from natural to synthetic have been explored for accelerated bone and cartilage repair and regeneration, which provides prerequisite biochemical and biophysical cues for the repair process. Furthermore, smart biomaterials and 3D printed matrices provided more advancement in repair process by providing biomimetic matrices more close to native tissues in terms of architecture and functionality. Despite of the successful role of different advanced and smart biomaterials for cartilage and bone repair, few biomaterials such as collagen and gelatin demonstrated adaptive immune response in host and indicated granuloma formation and necrosis postimplantation. Therefore, the definition of ideal biomaterial/scaffold is still unclear and needs much more exploration in future as far as enhanced repair is concerned. To tackle this challenge, the utilization of natural tissue ECM came into the forefront using decellularization technology for mimicking biological and biomechanical properties. Both physical and chemical methods can do decellularization of different tissues with pros and cons associated with each method. Currently, a combination of the physical and chemical method has been utilized for skeletal tissue repair. Although, much progress has been made using decellularized matrices in in vitro and in vivo studies; however, further assessment should be done in the future to evaluate the immune reactions, systemic effects induced by degradation products, etc. Therefore, decellularized matrices should be evaluated extensively for host immune response before using it clinically.

In skeletal tissue repair, the pivotal requirement is a combination of biochemical and biophysical cues, which is provided by scaffolds. However, certain additional features need to be incorporated in designing of scaffolds particularly for bone repair, which could promote both osteogenesis and angiogenesis. Unlike bone, cartilage tissue repair does not require angiogenesis owing to its avascular nature. There is a correlation of osteogenesis and angiogenesis for bone regeneration and both the processes work in a coordinated manner during the repair process. Therefore, the development of the dual functioning scaffolds which could promote functional vascularization and osteogenesis is much required in bone regeneration. These scaffolds can be fabricated by using either natural or synthetic biomaterials with the aim to retain the functionality for promoting osteogenesis coupled with angiogenesis. Earlier reports have also confirmed the correlation between these two processes through an indication of the existence of cross-talk between bone progenitor cells and endothelial cells. The dual-functioning scaffolds may be designed by incorporating single or dual bioactive factors for promoting both osteoinduction and angioinduction. The combined approach to fabricate dual functioning scaffolds using more than bioactive factors are more explored and has gained tremendous success; however, it leads to an increase in complexity in designing the scaffolds with the increase of constituents. Therefore, the focus is more towards the development of dual functioning scaffolds using a single bioactive factor, which can induce both osteogenesis and angiogenesis. Furthermore, osteogenesis coupled angiogenesis synergy could be obtained by using a combination of physical and biochemical stimuli for functional bone formation.

Furthermore, gene therapy-based strategies have also demonstrated huge improvements in currently available treatments for skeletal diseases. Various gene therapy-based products are currently in clinical trials for repair and regeneration of bone and cartilage repair. Recently, gene therapy product Invossa™ (from Korea, 2017) has got approval by the FDA for the treatment of moderate knee osteoarthritis using TGF-β1, which further signifies humongous growth in this rapidly developing field. Many mono and combinatorial gene therapy targets and approaches have been evaluated in different (small and large) animal models and indicated exciting results in treatment of joint disease. Taken together, mono and combinatorial gene therapies with a regulated expression of different factors within the disease milieu may improve the therapeutic efficacy in future development. In addition, mitigation of negative side effects could be achieved. Therefore, the development of these achievable therapeutics strategies would accelerate the translation of preclinical success into the clinical platform.

Recently, extracellular vesicles such as exosomes (40–150 nm size) have been projected as promising intermediate cell mediator for various functions related to repair process. Exosomes are secreted by various cell sources, but mesenchymal stem cells secreted exosomes have shown much success in skeletal tissue repair. Exosomes are a reservoir of various molecules such as proteins, mRNAs, and miRNA, which can modulate various physiological processes of repair. The natural exosomes have shown superiority over synthetic carrier agents such as liposomes or synthetic nanoparticles in terms of stability, toxicity, aneuploidy occurrence, and risk of immunological rejection. Owing to these advantages, exosomes demonstrated high potential for clinical trials. Despite the promising results, there still remain some problems with usage of exosomes. The most important shortcomings include the choice of suitable cell sources for obtaining exosomes as it should secrete large quantities and can be scalable for regeneration purposes. In addition, special storage condition would be required for exosomes as it contains various molecules. Therefore, in the future, if exosomes are loaded with suitable biological molecules/drugs would be utilized for clinical trials and provides an alternative to common treatments available for cartilage and bone repair.

The recent innovation of high throughput bioprinting technology revolutionized the potential of biomedical sciences arena for creating scaffolds/matrices for tissue and organ transplantation, regenerative medicine and drug screening field. Various 3D bioprinted bone and cartilage constructs have been developed for repair and regeneration and even few translated to clinical trials. Despite great advancement, the choice of suitable bioinks and designing of complex tissue fabrication still remains a challenge. The foremost challenge is the maintenance of the cell viability in bioinks and its protection from damage encountered during different types of printing process. Therefore, quality-control and standardization of bioprinting methods would be required to look into in the future. Recently, bioprinting has moved to 4D from 3D bioprinting, which employs 'smart' 3D constructs which can be tuned into different shapes and functions in response to external stimuli such as ultraviolet light, heat, energy sources, current, and pressure. Despite the current success of 3D bioprinting in bone and cartilage repair process, considerable research would be required for the fabrication of constructs for full organ transplantation and development of biocompatible tissue grafts. Overall, 3D bioprinted bone and cartilage constructs would be utilized for personalized medicine in the near future through optimization of different printing parameters. Taken together, currently, available different innovative approaches hold great promise for the development of biomimetic bone and cartilage constructs in the near future. However, much research and optimization would be required to make it for patient-specific therapeutics.

Acknowledgment

B.B.M. and N.B. are thankful for the generous funding support from Department of Biotechnology (DBT) and Department of Science and Technology (DST), Government of India. Y.P.S., J.C.M., A.B., and B.B. are grateful to the Ministry of Human Resource Development (MHRD), Government of India, for fellowship.

References

[1] Cheng A, et al. Advances in porous scaffold design for bone and cartilage tissue engineering and regeneration. Tissue Eng Part B: Rev 2019;25(1):14−29.

[2] Bhattacharjee P, et al. Chapter 12 − Silk-based matrices for bone tissue engineering applications. In: Grumezescu AM, editor. Nanostructures for the engineering of cells, tissues and organs. William Andrew Publishing; 2018. p. 439−72.

[3] M JC, et al. Mimicking hierarchical complexity of the osteochondral interface using electrospun silk−bioactive glass composites. ACS Appl Mater & Interfaces 2017;9(9):8000−13.

[4] Singh YP, et al. Injectable hydrogels: a new paradigm for osteochondral tissue engineering. J Mater Chem B 2018;6(35):5499−529.

[5] Crist BD, Leach JK, Lee MA. Is tissue engineering helping orthopaedic care in trauma? J Orthopaedic Trauma 2019;33:S12−19.

[6] Mei XY, et al. Fresh osteochondral allograft transplantation for treatment of large cartilage defects of the femoral head: a minimum two-year follow-up study of twenty-two patients. J Arthroplasty 2018;33(7):2050−6.

[7] Groll J, et al. Biofabrication: reappraising the definition of an evolving field. Biofabrication 2016;8(1):013001.

[8] Moroni L, et al. Biofabrication: a guide to technology and terminology. Trends Biotechnol 2018;36(4):384−402.

[9] Moroni L, et al. Biofabrication strategies for 3D in vitro models and regenerative medicine. Nat Rev Mater 2018;3(5):21.

[10] Hedayati R, et al. Isolated and modulated effects of topology and material type on the mechanical properties of additively manufactured porous biomaterials. J Mech Behav Biomed Mater 2018;79:254—63.

[11] Liu X, et al. Relationship between osseointegration and superelastic biomechanics in porous NiTi scaffolds. Biomaterials 2011;32(2):330—8.

[12] Malda J, et al. The effect of PEGT/PBT scaffold architecture on the composition of tissue engineered cartilage. Biomaterials 2005;26(1):63—72.

[13] Murphy SV, Atala A. 3D bioprinting of tissues and organs. Nat Biotechnol 2014;32(8):773.

[14] Ozbolat IT, Hospodiuk M. Current advances and future perspectives in extrusion-based bioprinting. Biomaterials 2016;76:321—43.

[15] Cui X, et al. Synergistic action of fibroblast growth factor-2 and transforming growth factor-beta1 enhances bioprinted human neocartilage formation. Biotechnol Bioeng 2012;109(9):2357—68.

[16] Schuurman W, et al. Gelatin-methacrylamide hydrogels as potential biomaterials for fabrication of tissue-engineered cartilage constructs. Macromol Biosci 2013;13(5):551—61.

[17] Visser J, et al. Biofabrication of multi-material anatomically shaped tissue constructs. Biofabrication 2013;5 (3):035007.

[18] Burke M, Carter BM, Perriman AW. Bioprinting: uncovering the utility layer-by-layer. J 3D Print Med 2017;1 (3):165—79.

[19] Byambaa B, et al. Bioprinted osteogenic and vasculogenic patterns for engineering 3D bone tissue. Adv Healthc Mater 2017;6(16):1700015.

[20] Daly AC, et al. 3D bioprinting of developmentally inspired templates for whole bone organ engineering. Adv Healthc Mater 2016;5(18):2353—62.

[21] Jang J, et al. Tailoring mechanical properties of decellularized extracellular matrix bioink by vitamin B2-induced photo-crosslinking. Acta Biomater 2016;33:88—95.

[22] Pati F, et al. Printing three-dimensional tissue analogues with decellularized extracellular matrix bioink. Nat Commun 2014;5:3935.

[23] Kim BS, et al. Decellularized extracellular matrix: a step towards the next generation source for bioink manufacturing. Biofabrication 2017;9(3):034104.

[24] Jariwala SH, et al. 3D printing of personalized artificial bone scaffolds. 3D Print Addit Manuf 2015;2 (2):56—64.

[25] Levato R, et al. Biofabrication of tissue constructs by 3D bioprinting of cell-laden microcarriers. Biofabrication 2014;6(3):035020.

[26] Tan YJ, et al. Hybrid microscaffold-based 3D bioprinting of multi-cellular constructs with high compressive strength: a new biofabrication strategy. Sci Rep 2016;6:39140.

[27] Kim WJ, Yun H-S, Kim GH. An innovative cell-laden α-TCP/collagen scaffold fabricated using a two-step printing process for potential application in regenerating hard tissues. Sci Rep 2017;7(1):3181.

[28] Nyberg E, et al. Comparison of 3D-printed poly-ε-caprolactone scaffolds functionalized with tricalcium phosphate, hydroxyapatite, bio-oss, or decellularized bone matrix. Tissue Eng Part A 2017;23(11—12):503—14.

[29] Neufurth M, et al. Engineering a morphogenetically active hydrogel for bioprinting of bioartificial tissue derived from human osteoblast-like SaOS-2 cells. Biomaterials 2014;35(31):8810—19.

[30] Dong L, et al. 3D-printed poly (ε-caprolactone) scaffold integrated with cell-laden chitosan hydrogels for bone tissue engineering. Sci Rep 2017;7(1):13412.

[31] Bendtsen ST, Quinnell SP, Wei M. Development of a novel alginate-polyvinyl alcohol-hydroxyapatite hydrogel for 3D bioprinting bone tissue engineered scaffolds. J Biomed Mater Res Part A 2017;105(5):1457—68.

[32] Kang H-W, et al. A 3D bioprinting system to produce human-scale tissue constructs with structural integrity. Nat Biotechnol 2016;34(3):312.

[33] Midha S, et al. Advances in three-dimensional bioprinting of bone: progress and challenges. J Tissue Eng Regener Med 2019;13(6):925—45.

[34] Kolesky DB, et al. Three-dimensional bioprinting of thick vascularized tissues. Proc Natl Acad Sci 2016;113 (12):3179—84.

[35] Kolesky DB, et al. 3D bioprinting of vascularized, heterogeneous cell-laden tissue constructs. Adv Mater 2014;26(19):3124—30.

[36] Park JY, et al. 3D printing technology to control BMP-2 and VEGF delivery spatially and temporally to promote large-volume bone regeneration. J Mater Chem B 2015;3(27):5415—25.

[37] Murphy C, et al. 3D bioprinting of stem cells and polymer/bioactive glass composite scaffolds for bone tissue engineering. Int J Bioprint 2017;3(1):53—63.

[38] Kuss MA, et al. Short-term hypoxic preconditioning promotes prevascularization in 3D bioprinted bone constructs with stromal vascular fraction derived cells. RSC Adv 2017;7(47):29312—20.

[39] Rankin EB, Giaccia AJ, Schipani E. A central role for hypoxic signaling in cartilage, bone, and hematopoiesis. Curr Osteoporos Rep 2011;9(2):46—52.

[40] Cui H, et al. Hierarchical fabrication of engineered vascularized bone biphasic constructs via dual 3D bioprinting: integrating regional bioactive factors into architectural design. Adv Healthc Mater 2016;5(17):2174—81.

[41] Gao G, et al. Improved properties of bone and cartilage tissue from 3D inkjet-bioprinted human mesenchymal stem cells by simultaneous deposition and photocrosslinking in PEG-GelMA. Biotechnol Lett 2015;37 (11):2349—55.

[42] Lee S-J, et al. Application of microstereolithography in the development of three-dimensional cartilage regeneration scaffolds. Biomed Microdevices 2008;10(2):233—41.

[43] Elomaa L, et al. Preparation of poly (ε-caprolactone)-based tissue engineering scaffolds by stereolithography. Acta Biomater 2011;7(11):3850—6.

[44] Schüller-Ravoo S, et al. Flexible and elastic scaffolds for cartilage tissue engineering prepared by stereolithography using poly (trimethylene carbonate)-b ased resins. Macromol Biosci 2013;13(12):1711—19.

[45] Zhai C, et al. Repair of articular osteochondral defects using an integrated and biomimetic trilayered scaffold. Tissue Eng: Part A 2018;24(21—22):1680—92.

[46] Du Y, et al. Selective laser sintering scaffold with hierarchical architecture and gradient composition for osteochondral repair in rabbits. Biomaterials 2017;137:37—48.

[47] Chen C-H, Chen J-P, Lee M-Y. Effects of gelatin modification on rapid prototyping PCL scaffolds for cartilage engineering. J Mech Med Biol 2011;11(05):993—1002.

[48] Lee M-Y, et al. Laser sintered porous polycaprolactone scaffolds loaded with hyaluronic acid and gelatin-grafted thermoresponsive hydrogel for cartilage tissue engineering. Bio-med Mater Eng 2013;23(6):533—43.

[49] Yen H-J, et al. Evaluation of chondrocyte growth in the highly porous scaffolds made by fused deposition manufacturing (FDM) filled with type II collagen. Biomed Microdevices 2009;11(3):615—24.

[50] Schumann D. Design of bioactive, multiphasic PCL/collagen type I and type II-PCL-TCP/collagen composite scaffolds for functional tissue engineering of osteochondral repair tissue by using electrospinning and FDM techniques. Methods Mol Biol 2007;140:101—24.

[51] Rosenzweig D, et al. 3D-printed ABS and PLA scaffolds for cartilage and nucleus pulposus tissue regeneration. Int J Mol Sci 2015;16(7):15118—35.

[52] Hung KC, Tseng CS, Hsu Sh. Synthesis and 3D printing of biodegradable polyurethane elastomer by a water-based process for cartilage tissue engineering applications. Adv Healthc Mater 2014;3(10):1578—87.

[53] Zhang T, et al. Biomimetic design and fabrication of multilayered osteochondral scaffolds by low-temperature deposition manufacturing and thermal-induced phase-separation techniques. Biofabrication 2017;9(2):025021.

[54] Huang J, et al. Preparation and biocompatibility of diphasic magnetic nanocomposite scaffold. Mater Sci Eng: C 2018;87:70—7.

[55] Xu Y, et al. Construction of bionic tissue engineering cartilage scaffold based on three-dimensional printing and oriented frozen technology. J Biomed Mater Res: Part A 2018;106(6):1664—76.

[56] Deng C, et al. 3D printing of bilineage constructive biomaterials for bone and cartilage regeneration. Adv Funct Mater 2017;27(36):1703117.

[57] Park JY, et al. A comparative study on collagen type I and hyaluronic acid dependent cell behavior for osteochondral tissue bioprinting. Biofabrication 2014;6(3):035004.

[58] Gao F, et al. Direct 3D printing of high strength biohybrid gradient hydrogel scaffolds for efficient repair of osteochondral defect. Adv Funct Mater 2018;28(13):1706644.

[59] Liu Y, et al. In vitro engineering of human ear-shaped cartilage assisted with CAD/CAM technology. Biomaterials 2010;31(8):2176—83.

[60] Kundu J, et al. An additive manufacturing-based PCL—alginate—chondrocyte bioprinted scaffold for cartilage tissue engineering. J Tissue Eng Regener Med 2015;9(11):1286—97.

[61] Fedorovich NE, et al. Biofabrication of osteochondral tissue equivalents by printing topologically defined, cell-laden hydrogel scaffolds. Tissue Eng Part C: Methods 2011;18(1):33—44.

[62] Sun AX, et al. Projection stereolithographic fabrication of human adipose stem cell-incorporated biodegradable scaffolds for cartilage tissue engineering. Front Bioeng Biotechnol 2015;3:115.

[63] Zhu W, et al. 3D bioprinting mesenchymal stem cell-laden construct with core—shell nanospheres for cartilage tissue engineering. Nanotechnology 2018;29(18):185101.

[64] Xu T, et al. Hybrid printing of mechanically and biologically improved constructs for cartilage tissue engineering applications. Biofabrication 2012;5(1):015001.

[65] Chawla S, et al. Silk-based bioinks for 3D bioprinting. Adv Healthc Mater 2018;7(8):1701204.

[66] Singh YP, Bandyopadhyay A, Mandal BB. 3D bioprinting using cross-linker-free silk—gelatin bioink for cartilage tissue engineering. ACS Appl Mater & Interfaces 2019;11(37):33684—96.

[67] Bandyopadhyay A, Mandal BB. A three-dimensional printed silk-based biomimetic tri-layered meniscus for potential patient-specific implantation. Biofabrication 2019;12(1):015003.

[68] Holmes B, et al. Development of novel three-dimensional printed scaffolds for osteochondral regeneration. Tissue Eng: Part A 2014;21(1—2):403—15.

[69] Nowicki MA, et al. 3D printing of novel osteochondral scaffolds with graded microstructure. Nanotechnology 2016;27(41):414001.

[70] Markstedt K, et al. 3D bioprinting human chondrocytes with nanocellulose—alginate bioink for cartilage tissue engineering applications. Biomacromolecules 2015;16(5):1489—96.

[71] Rhee S, et al. 3D bioprinting of spatially heterogeneous collagen constructs for cartilage tissue engineering. ACS Biomater Sci Eng 2016;2(10):1800—5.

[72] Das S, et al. Bioprintable, cell-laden silk fibroin—gelatin hydrogel supporting multilineage differentiation of stem cells for fabrication of three-dimensional tissue constructs. Acta Biomater 2015;11:233—46.

[73] Hong S, et al. 3D printing of highly stretchable and tough hydrogels into complex, cellularized structures. Adv Mater 2015;27(27):4035—40.

[74] Zhu F, et al. 3D-printed ultratough hydrogel structures with titin-like domains. ACS Appl Mater Interfaces 2017;9(13):11363—7.

[75] Yang F, Tadepalli V, Wiley BJ. 3D printing of a double network hydrogel with a compression strength and elastic modulus greater than those of cartilage. ACS Biomater Sci Eng 2017;3(5):863—9.

[76] Chen L, et al. 3D printing of a lithium-calcium-silicate crystal bioscaffold with dual bioactivities for osteochondral interface reconstruction. Biomaterials 2019;196:138—50.

[77] Moses JC, Saha T, Mandal BB. Chondroprotective and osteogenic effects of silk-based bioinks in developing 3D bioprinted osteochondral interface. Bioprinting 2020;e00067.

[78] Peng H, et al. Development of an MFG-based retroviral vector system for secretion of high levels of functionally active human BMP4. Mol Ther 2001;4(2):95—104.

[79] Meinel L, et al. Osteogenesis by human mesenchymal stem cells cultured on silk biomaterials: comparison of adenovirus mediated gene transfer and protein delivery of BMP-2. Biomaterials 2006;27(28):4993—5002.

[80] Lü K, et al. BMP-2 gene modified canine bMSCs promote ectopic bone formation mediated by a nonviral PEI derivative. Ann Biomed Eng 2011;39(6):1829—39.

[81] Masgutov R, et al. Use of gene-activated demineralized bone allograft in the therapy of ulnar pseudarthrosis. case report. BioNanoScience 2017;7(1):194—8.

[82] Bozo I, et al. World's first clinical case of gene-activated bone substitute application. Case Rep Dent 2016;2016.

[83] Zhang W, et al. Current research on pharmacologic and regenerative therapies for osteoarthritis. Bone Res 2016;4:15040.

[84] Frisch J, et al. rAAV-mediated overexpression of sox9, TGF-β and IGF-I in minipig bone marrow aspirates to enhance the chondrogenic processes for cartilage repair. Gene Therapy 2016;23(3):247.

[85] Evans CH, Ghivizzani SC, Robbins PD. Arthritis gene therapy is becoming a reality. Nat Rev Rheumatol 2018;14(7):381.

[86] Kim M-K, et al. A multicenter, double-blind, phase III clinical trial to evaluate the efficacy and safety of a cell and gene therapy in knee osteoarthritis patients. Hum Gene Ther Clin Dev 2018;29(1):48—59.

[87] Hattori T, et al. SOX9 is a major negative regulator of cartilage vascularization, bone marrow formation and endochondral ossification. Development 2010;137(6):901—11.

[88] Mollon B, et al. The clinical status of cartilage tissue regeneration in humans. Osteoarthr Cartil 2013;21 (12):1824—33.

[89] Gonzalez-Fernandez T, Kelly DJ, O'Brien FJ. Controlled non-viral gene delivery in cartilage and bone repair: current strategies and future directions. Adv Therapeutics 2018;1(7):1800038.

[90] Singh YP, et al. Hierarchically structured seamless silk scaffolds for osteochondral interface tissue engineering. J Mater Chem B 2018;6(36):5671–88.

[91] Leng P, et al. Reconstruct large osteochondral defects of the knee with hIGF-1 gene enhanced Mosaicplasty. Knee 2012;19(6):804–11.

[92] Orth P, et al. Transplanted articular chondrocytes co-overexpressing IGF-I and FGF-2 stimulate cartilage repair in vivo. Knee Surg Sports Traumatol Arthrosc 2011;19(12):2119–30.

[93] Grol MW, Lee BH. Gene therapy for repair and regeneration of bone and cartilage. Curr Oppharmacol 2018;40:59–66.

[94] Lee Y-H, et al. Enzyme-crosslinked gene-activated matrix for the induction of mesenchymal stem cells in osteochondral tissue regeneration. Acta Biomater 2017;63:210–26.

[95] Yang M, et al. Nanotherapeutics relieve rheumatoid arthritis. J Controlled Release 2017;252:108–24.

[96] Luo Y, Wu C, Chang J. Inorganic nanomaterials for bone tissue engineering. In: Biomedical nanomaterials; 2016. p. 246.

[97] Zhao L, et al. Effective delivery of bone morphogenetic protein 2 gene using chitosan–polyethylenimine nanoparticle to promote bone formation. RSC Adv 2016;6(41):34081–9.

[98] Mathis DT, et al. Good clinical results but moderate osseointegration and defect filling of a cell-free multi-layered nano-composite scaffold for treatment of osteochondral lesions of the knee. Knee Surg Sports Traumatol Arthrosc 2018;26(4):1273–80.

[99] Park CW, et al. Cytokine secretion profiling of human mesenchymal stem cells by antibody array. Int J Stem Cell 2009;2(1):59.

[100] Caplan AI, Correa D. The MSC: an injury drugstore. Cell Stem Cell 2011;9(1):11–15.

[101] Gnecchi M, et al. Paracrine mechanisms of mesenchymal stem cells in tissue repair. Mesenchymal Stem Cells. Springer; 2016. p. 123–46.

[102] Maguire G. Stem cell therapy without the cells. Commun Integr Biol 2013;6(6):e26631.

[103] Se H, Baber M, Caplan A. Cytokine expression by human marrow-derived mesenchymal progenitor cells in vitro: effects of dexamethasone and IL-1 alpha. J Cell Physiol 1996;166:585–92.

[104] Patschan D, Plotkin M, Goligorsky MS. Therapeutic use of stem and endothelial progenitor cells in acute renal injury: ca ira. Curr Oppharmacol 2006;6(2):176–83.

[105] Togel F, et al. Administered mesenchymal stem cells protect against ischemic acute renal failure through differentiation-independent mechanisms. Am J Physiol – Renal Physiol 2005;289(1):F31–42.

[106] Gnecchi M, et al. Evidence supporting paracrine hypothesis for Akt-modified mesenchymal stem cell-mediated cardiac protection and functional improvement. FASEB J 2006;20(6):661–9.

[107] Timmers L, et al. Reduction of myocardial infarct size by human mesenchymal stem cell conditioned medium. Stem Cell Res 2008;1(2):129–37.

[108] Khubutiya MS, et al. Paracrine mechanisms of proliferative, anti-apoptotic and anti-inflammatory effects of mesenchymal stromal cells in models of acute organ injury. Cytotherapy 2014;16(5):579–85.

[109] Kourembanas S. Exosomes: vehicles of intercellular signaling, biomarkers, and vectors of cell therapy. Annu Rev Physiol 2015;77:13–27.

[110] Silva AM, et al. Extracellular vesicles: immunomodulatory messengers in the context of tissue repair/regeneration. Eur J Pharm Sci 2017;98:86–95.

[111] Benameur T, et al. Microparticles carrying Sonic hedgehog favor neovascularization through the activation of nitric oxide pathway in mice. PLoS One 2010;5(9):e12688.

[112] Arslan F, et al. Mesenchymal stem cell-derived exosomes increase ATP levels, decrease oxidative stress and activate PI3K/Akt pathway to enhance myocardial viability and prevent adverse remodeling after myocardial ischemia/reperfusion injury. Stem Cell Res 2013;10(3):301–12.

[113] Bala S, et al. Biodistribution and function of extracellular miRNA-155 in mice. Sci Rep 2015;5:10721.

[114] Saunderson SC, et al. CD169 mediates the capture of exosomes in spleen and lymph node. Blood 2014;123(2):208–16.

[115] Zhang S, et al. MSC exosomes mediate cartilage repair by enhancing proliferation, attenuating apoptosis and modulating immune reactivity. Biomaterials 2018;156:16–27.

[116] Xie H, et al. Extracellular vesicle-functionalized decalcified bone matrix scaffolds with enhanced pro-angiogenic and pro-bone regeneration activities. Sci Rep 2017;7:45622.

[117] Xu J, et al. Human fetal mesenchymal stem cell secretome enhances bone consolidation in distraction osteo-genesis. Stem Cell Res Ther 2016;7(1):134.

[118] Zhang S, et al. Human mesenchymal stem cell-derived exosomes promote orderly cartilage regeneration in an immunocompetent rat osteochondral defect model. Cytotherapy 2016;18(6):S13.

[119] Famian MH, Saheb SM, Montaseri A. Conditioned medium of Wharton's jelly derived stem cells can enhance the cartilage specific genes expression by chondrocytes in monolayer and mass culture systems. Adv Pharm Bull 2017;7(1):123.

[120] Ekström K, et al. Monocyte exosomes stimulate the osteogenic gene expression of mesenchymal stem cells. PLoS One 2013;8(9):e75227.

[121] Torreggiani E, et al. Exosomes: novel effectors of human platelet lysate activity. Eur Cell Mater 2014;28:137−51.

[122] Qin Y, et al. Bone marrow stromal/stem cell-derived extracellular vesicles regulate osteoblast activity and differentiation in vitro and promote bone regeneration in vivo. Sci Rep 2016;6:21961.

[123] Furuta T, et al. Mesenchymal stem cell-derived exosomes promote fracture healing in a mouse model. Stem Cell Transl Med 2016;5(12):1620−30.

[124] Martins M, et al. Extracellular vesicles derived from osteogenically induced human bone marrow mesen-chymal stem cells can modulate lineage commitment. Stem Cell Rep 2016;6(3):284−91.

[125] Liu X, et al. Exosomes secreted from human-induced pluripotent stem cell-derived mesenchymal stem cells prevent osteonecrosis of the femoral head by promoting angiogenesis. Int J Biol Sci 2017;13(2):232.

[126] Katagiri W, et al. Clinical study of bone regeneration by conditioned medium from mesenchymal stem cells after maxillary sinus floor elevation. Implant Dent 2017;26(4):607−12.

[127] Infante A, Rodríguez CI. Secretome analysis of in vitro aged human mesenchymal stem cells reveals IGFBP7 as a putative factor for promoting osteogenesis. Sci Rep 2018;8(1):4632.

[128] Qi X, et al. Exosomes secreted by human-induced pluripotent stem cell-derived mesenchymal stem cells repair critical-sized bone defects through enhanced angiogenesis and osteogenesis in osteoporotic rats. Int J Biol Sci 2016;12(7):836.

[129] Kawai T, et al. Secretomes from bone marrow-derived mesenchymal stromal cells enhance periodontal tis-sue regeneration. Cytotherapy 2015;17(4):369−81.

[130] Zhu Y, et al. Comparison of exosomes secreted by induced pluripotent stem cell-derived mesenchymal stem cells and synovial membrane-derived mesenchymal stem cells for the treatment of osteoarthritis. Stem Cell Res Ther 2017;8(1):64.

[131] Ratner BD. Biomaterials: been there, done that, and evolving into the future. Annu Rev Biomed Eng 2019;21:171−91.

[132] Moses JC, Nandi SK, Mandal BB. Multifunctional cell instructive silk-bioactive glass composite reinforced scaffolds toward osteoinductive, proangiogenic, and resorbable bone grafts. Adv Healthc Mater 2018;7 (10):1701418.

[133] Moses JC, et al. Synergistic effects of silicon/zinc doped brushite and silk scaffolding in augmenting the osteogenic and angiogenic potential of composite biomimetic bone grafts. ACS Biomater Sci Eng 2019;5 (3):1462−75.

[134] Kempen DH, et al. Effect of local sequential VEGF and BMP-2 delivery on ectopic and orthotopic bone regeneration. Biomaterials 2009;30(14):2816−25.

[135] Zhang S, et al. Accelerated bone regenerative efficiency by regulating sequential release of BMP-2 and VEGF and synergism with sulfated chitosan. ACS Biomater Sci Eng 2019;5(4):1944−55.

[136] Ashammakhi N, et al. Advances in controlled oxygen generating biomaterials for tissue engineering and regenerative therapy. Biomacromolecules 2019;.

[137] Kim HY, et al. Oxygen-releasing microparticles for cell survival and differentiation ability under hypoxia for effective bone regeneration. Biomacromolecules 2019;20(2):1087−97.

[138] Tandon B, Blaker JJ, Cartmell SH. Piezoelectric materials as stimulatory biomedical materials and scaffolds for bone repair. Acta Biomater 2018;73:1−20.

[139] Zhang X, et al. Nanocomposite membranes enhance bone regeneration through restoring physiological elec-tric microenvironment. ACS Nano 2016;10(8):7279−86.

[140] Mas-Moruno C, Su B, Dalby MJ. Multifunctional coatings and nanotopographies: toward cell instructive and antibacterial implants. Adv Healthc Mater 2019;8(1):1801103.

[141] Buxadera-Palomero J, et al. Biofunctional polyethylene glycol coatings on titanium: an in vitro-based comparison of functionalization methods. Colloids Surf B: Biointerfaces 2017;152:367—75.

[142] Llopis-Hernández V, et al. Material-driven fibronectin assembly for high-efficiency presentation of growth factors. Sci Adv 2016;2(8):e1600188.

[143] Rocas P, et al. Installing Multifunctionality on Titanium with RGD-Decorated Polyurethane-Polyurea Roxithromycin Loaded Nanoparticles: Toward New Osseointegrative Therapies. Adv Healthc Mater 2015;4 (13):1956—60.

[144] Hay JJ, et al. Bacteria-based materials for stem cell engineering . Adv Mater 2018;30(43):1804310.

[145] Camarero-Espinosa S, Cooper-White JJ. Combinatorial presentation of cartilage-inspired peptides on nano-patterned surfaces enables directed differentiation of human mesenchymal stem cells towards distinct articular chondrogenic phenotypes. Biomaterials 2019;210:105—15.

[146] Singh YP, Bhardwaj N, Mandal BB. Potential of agarose/silk fibroin blended hydrogel for in vitro cartilage tissue engineering. ACS Appl Mater Interfaces 2016;8(33):21236—49.

[147] Mohanraj B, et al. Mechanically activated microcapsules for "On-Demand" drug delivery in dynamically loaded musculoskeletal tissues. Adv Funct Mater 2019;29(15):1807909.

[148] Fayol D, et al. Use of magnetic forces to promote stem cell aggregation during differentiation, and cartilage tissue modeling. Adv Mater 2013;25(18):2611—16.

[149] Luciani N, et al. Successful chondrogenesis within scaffolds, using magnetic stem cell confinement and bioreactor maturation. Acta Biomater 2016;37:101—10.

[150] Bhunia BK, Mandal BB. Exploring gelation and physicochemical behavior of in situ bioresponsive silk hydrogels for disc degeneration therapy. ACS Biomater Sci & Eng 2019;5(2):870—86.

[151] Park SH, et al. An injectable, click-crosslinked, cytomodulin-modified hyaluronic acid hydrogel for cartilage tissue engineering. NPG Asia Mater 2019;11(1):30.

[152] Zhu W, Iqbal J, Wang D-A. A DOPA-functionalized chondroitin sulfate-based adhesive hydrogel as a promising multi-functional bioadhesive. J Mater Chem B 2019;7(10):1741—52.

[153] Bhardwaj N, Singh YP, Mandal BB. Silk fibroin scaffold-based 3D co-culture model for modulation of chondrogenesis without hypertrophy via reciprocal cross-talk and paracrine signaling. ACS Biomater Sci Eng 2019;5(10):5240—54.

[154] Ansari S, Khorshidi S, Karkhaneh A. Engineering of gradient osteochondral tissue: from nature to lab. Acta Biomater 2019;87:41—54.

[155] Li C, et al. Buoyancy-driven gradients for biomaterial fabrication and tissue engineering. Adv Mater 2019;31 (17):1900291.

[156] McNeill EP, et al. Three-dimensional in vitro modeling of malignant bone disease recapitulates experimentally accessible mechanisms of osteoinhibition. Cell Death Dis 2018;9(12):1161.

[157] Thibaudeau L, et al. A tissue-engineered humanized xenograft model of human breast cancer metastasis to bone. Dis Model Mech 2014;7(2):299—309.

[158] Villasante A, Marturano-Kruik A, Vunjak-Novakovic G. Bioengineered human tumor within a bone niche. Biomaterials 2014;35(22):5785—94.

[159] Wagner F, et al. A validated preclinical animal model for primary bone tumor research. J Bone Jt Surg 2016;98(11):916—25.

[160] Hayden RS, Vollrath M, Kaplan DL. Effects of clodronate and alendronate on osteoclast and osteoblast co-cultures on silk-hydroxyapatite films. Acta Biomater 2014;10(1):486—93.

[161] McGovern JA, Griffin M, Hutmacher DW. Animal models for bone tissue engineering and modelling disease. Dis Model Mech 2018;11(4):dmm033084.

[162] Sun Y, et al. Cajaninstilbene acid inhibits osteoporosis through suppressing osteoclast formation and RANKL-induced signaling pathways. J Cell Physiol 2019;234(7):11792—804.

[163] Li K, et al. Tiliroside is a new potential therapeutic drug for osteoporosis in mice. J Cell Physiol 2019;234 (9):16263—74.

[164] Phull A-R, et al. Applications of chondrocyte-based cartilage engineering: an overview. Biomed Res Int 2016;2016:1—17.

[165] Galuzzi M, et al. Human engineered cartilage and decellularized matrix as an alternative to animal osteoarthritis model. Polymers 2018;10(7) 738-738.

[166] Frank E, et al. Dose-dependent chondroprotective effects of triamcinolone acetonide on inflamed and injured cartilage using an in vitro model. Osteoarthr Cartil 2019;27 S176-S176.

[167] Stellavato A, et al. Novel hybrid gels made of high and low molecular weight hyaluronic acid induce proliferation and reduce inflammation in an osteoarthritis in vitro model based on human synoviocytes and chondrocytes. Biomed Res Int 2019;2019:1—13.

[168] Hamasaki M, et al. A novel cartilage fragments stimulation model revealed that macrophage inflammatory response causes an upregulation of catabolic factors of chondrocytes in vitro. Cartilage 2019; 1947603519828426.

[169] Peck Y, et al. Establishment of an in vitro three-dimensional model for cartilage damage in rheumatoid arthritis. J Tissue Eng Regener Med 2018;12(1):e237—49.

[170] Haltmayer E, et al. Co-culture of osteochondral explants and synovial membrane as in vitro model for osteoarthritis. PLoS One 2019;14(4) e0214709-e0214709.

[171] Lozito TP, et al. Three-dimensional osteochondral microtissue to model pathogenesis of osteoarthritis. Stem Cell Res Ther 2013;4(Suppl 1) S6-S6.

[172] Kimura T, et al. Proposal of patient-specific growth plate cartilage xenograft model for FGFR3 chondrodysplasia. Osteoarthr Cartil 2018;26(11):1551—61.

[173] Takahashi I, et al. Induction of osteoarthritis by injecting monosodium iodoacetate into the patellofemoral joint of an experimental rat model. PLoS One 2018;13(4) e0196625-e0196625.

[174] Arzi B, et al. A proposed model of naturally occurring osteoarthritis in the domestic rabbit. Lab Anim 2012;41(1):20—5.

[175] Dai G, et al. The validity of osteoarthritis model induced by bilateral ovariectomy in guinea pig. J Huazhong Univ Sci Technol 2006;26(6):716—19.

[176] Kreipke TC, et al. Alterations in trabecular bone microarchitecture in the ovine spine and distal femur following ovariectomy. J Biomech 2014;47(8):1918—21.

[177] Mrosek EH, et al. Subchondral bone trauma causes cartilage matrix degeneration: an immunohistochemical analysis in a canine model. Osteoarthr Cartil 2006;14(2):171—8.

[178] Wu K-C, et al. Transplanting human umbilical cord mesenchymal stem cells and hyaluronate hydrogel repairs cartilage of osteoarthritis in the minipig model. Ci ji yi xue za zhi = Tzu-chi Med J 2019;31(1):11—19.

[179] Little CB, et al. The OARSI histopathology initiative — recommendations for histological assessments of osteoarthritis in sheep and goats. Osteoarthr Cartil 2010;18:S80—92.

[180] Burger C, et al. The sheep as a knee osteoarthritis model: early cartilage changes after meniscus injury and repair. Laboratory Anim 2007;41(4):420—31.

[181] McIlwraith CW, et al. The OARSI histopathology initiative — recommendations for histological assessments of osteoarthritis in the horse. Osteoarthr Cartil 2010;18:S93—105.

[182] McCoy AM. Animal models of osteoarthritis. Veterinary Pathol 2015;52(5):803—18.

[183] McDermott ID, Amis AA. The consequences of meniscectomy. J Bone Jt Surg Br 2006;88-B(12):1549—56.

[184] Oestergaard S, et al. Effects of ovariectomy and estrogen therapy on type II collagen degradation and structural integrity of articular cartilage in rats: Implications of the time of initiation. Arthritis Rheumat 2006;54(8):2441—51.

[185] Miyamoto S, et al. Intra-articular injection of mono-iodoacetate induces osteoarthritis of the hip in rats. BMC Musculoskelet Disord 2016;17(1) 132-132.

[186] Khatab S, et al. Mesenchymal stem cell secretome reduces pain and prevents cartilage damage in a murine osteoarthritis model. Eur Cell Mater 2018;36:218—30.

[187] Han H-W, Asano S, Hsu S-H. Cellular spheroids of mesenchymal stem cells and their perspectives in future healthcare. Appl Sci 2019;9(4):627.

[188] Irie Y, et al. Development of articular cartilage grafts using organoid formation techniques. Transplant Proc 2008;40(2):631—3.

[189] Chaddad H, et al. Combining 2D angiogenesis and 3D osteosarcoma microtissues to improve vascularization. Exp Cell Res 2017;360(2):138—45.

[190] Krishnan V, Vogler EA, Mastro AM. Three-dimensional in vitro model to study osteobiology and osteopathology. J Cell Biochem 2015;116(12):2715—23.

[191] Perut F, et al. Spheroid-based 3D cell cultures identify salinomycin as a promising drug for the treatment of chondrosarcoma. J Orthopaed Res 2018;36(8):2305—12.

[192] Voissiere A, et al. Development and characterization of a human three-dimensional chondrosarcoma culture for in vitro drug testing. PLoS One 2017;12(7) e0181340-e0181340.

[193] Ding J, et al. Bone marrow mesenchymal stem cell-based engineered cartilage ameliorates polyglycolic acid/polylactic acid scaffold-induced inflammation through M2 polarization of macrophages in a pig model. Stem Cells Transl Med 2016;5(8):1079—89.

[194] Claes L, Recknagel S, Ignatius A. Fracture healing under healthy and inflammatory conditions. Nat Rev Rheumatol 2012;8(3):133—43.

[195] Schell H, et al. The haematoma and its role in bone healing. J Exp Orthopaed 2017;4(1) 5-5.

[196] Einhorn TA, Gerstenfeld LC. Fracture healing: mechanisms and interventions. Nat Rev Rheumatol 2015;11 (1):45—54.

[197] Marsell R, Einhorn TA. The biology of fracture healing. Injury 2011;42(6):551—5.

[198] Schlundt C, et al. Macrophages in bone fracture healing: their essential role in endochondral ossification. Bone 2018;106:78—89.

[199] Reinke S, et al. Terminally differentiated CD8(+) T cells negatively affect bone regeneration in humans. Sci Transl Med 2013;5(177):177ra36.

[200] Thompson EM, et al. Recapitulating endochondral ossification: a promising route to in vivo bone regeneration. J Tissue Eng Regen Med 2015;9(8):889—902.

[201] Nemeth K, et al. Bone marrow stromal cells use TGF-beta to suppress allergic responses in a mouse model of ragweed-induced asthma. Proc Natl Acad Sci USA 2010;107(12):5652—7.

[202] Perez JR, et al. Tissue engineering and cell-based therapies for fractures and bone defects. Front Bioeng Biotechnol 2018;6:105.

[203] Gu Q, Yang H, Shi Q. Macrophages and bone inflammation. J Orthopaed Transl 2017;10:86—93.

[204] Zhang X, Mao Z, Yu C. Suppression of early experimental osteoarthritis by gene transfer of interleukin-1 receptor antagonist and interleukin-10. J Orthopaed Res 2004;22(4):742—50.

[205] Frisbie DD, et al. Treatment of experimental equine osteoarthritis by in vivo delivery of the equine interleukin-1 receptor antagonist gene. Gene Ther 2002;9(1):12—20.

[206] Zhang W, et al. Cell-free therapy based on adipose tissue stem cell-derived exosomes promotes wound healing via the PI3K/Akt signaling pathway. Exp Cell Res 2018;370(2):333—42.

[207] Lee SJ, et al. Immunophenotype and immune-modulatory activities of human fetal cartilage-derived progenitor cells. Cell Transplant 2019;28(7):932—42.

[208] Spiller KL, et al. Sequential delivery of immunomodulatory cytokines to facilitate the M1-to-M2 transition of macrophages and enhance vascularization of bone scaffolds. Biomaterials 2015;37:194—207.

[209] Chen J, et al. Macrophage phenotype switch by sequential action of immunomodulatory cytokines from hydrogel layers on titania nanotubes. Colloids Surf B: Biointerfaces 2018;163:336—45.

[210] Li T, et al. 3D-printed IFN-γ-loading calcium silicate-β-tricalcium phosphate scaffold sequentially activates M1 and M2 polarization of macrophages to promote vascularization of tissue engineering bone. Acta Biomater 2018;71:96—107.

[211] Amini AR, Laurencin CT, Nukavarapu SP. Bone tissue engineering: recent advances and challenges. Crit Rev Biomed Eng 2012;40(5):363—408.

[212] Dahiya UR, Mishra S, Bano S. Application of bone substitutes and its future prospective in regenerative medicine. In: Biomaterial-supported tissue reconstruction or regeneration. IntechOpen; 2019.

[213] Delloye C, et al. Bone allografts: what they can offer and what they cannot. J Bone Joint Surg Br 2007;89 (5):574—80.

[214] Singh R, Singh D, Singh A. Radiation sterilization of tissue allografts: a review. World J Radiol 2016;8 (4):355.

[215] Souyris Fo, et al. Coral, a new biomedical material. Experimental and first clinical investigations on Madreporaria. J Maxillofac Surg 1985;13(2):64—9.

[216] Bostrom MP, Seigerman DA. The clinical use of allografts, demineralized bone matrices, synthetic bone graft substitutes and osteoinductive growth factors: a survey study. Hss J 2005;1(1):9—18.

[217] Tang Z, et al. The material and biological characteristics of osteoinductive calcium phosphate ceramics. Regener Biomater 2017;5(1):43—59.

[218] Campana V, et al. Bone substitutes in orthopaedic surgery: from basic science to clinical practice. J Mater Science: Mater Med 2014;25(10):2445—61.

[219] Wang W, Yeung KW. Bone grafts and biomaterials substitutes for bone defect repair: a review. Bioact Mater 2017;2(4):224—47.

[220] Garrity JT, et al. Improved osteochondral allograft preservation using serum-free media at body temperature. Am J Sports Med 2012;40(11):2542—8.

[221] Massen FK, et al. One-step autologous minced cartilage procedure for the treatment of knee joint chondral and osteochondral lesions: a series of 27 patients with 2-year follow-up. Orthopaed J Sports Med 2019;7(6) 2325967119853773.

[222] Ryan PM, et al. Comparative outcomes for the treatment of articular cartilage lesions in the ankle with a denovo NT natural tissue graft: open versus arthroscopic treatment. Orthopaed J Sports Med 2018;6(12) 2325967118812710.

[223] Coetzee JC, et al. Treatment of osteochondral lesions of the talus with particulated juvenile cartilage. Foot Ankle Int 2013;34(9):1205—11.

[224] Huang BJ, Hu JC, Athanasiou KA. Cell-based tissue engineering strategies used in the clinical repair of articular cartilage. Biomaterials 2016;98:1—22.

[225] Tamaddon M, et al. Osteochondral tissue repair in osteoarthritic joints: clinical challenges and opportunities in tissue engineering. Bio-Design Manuf 2018;1(2):101—14.

[226] Marcacci M, et al. Stem cells associated with macroporous bioceramics for long bone repair: 6-to 7-year outcome of a pilot clinical study. Tissue Eng: Part A 2007;13(5):947—55.

[227] Cengiz IF, et al. Orthopaedic regenerative tissue engineering en route to the holy grail: disequilibrium between the demand and the supply in the operating room. J Exp Orthop 2018;5(1):14.

[228] Oryan A, et al. Bone regenerative medicine: classic options, novel strategies, and future directions. J Orthopaed Surg Res 2014;9(1):18.

12

Muscle tissue engineering — A materials perspective

John P. Bradford, Gerardo Hernandez-Moreno and Vinoy Thomas

Polymers & Healthcare Materials/Devices, Department of Materials Science & Engineering University of Alabama at Birmingham (UAB), Birmingham, AL, United States

12.1 Introduction

The term "Tissue", as it pertains to human anatomy, derives from cells sharing rudimentary origin and morphological features that are arranged in patterned array to achieve specific physiological functions. The four types of tissues are connective, epithelial, muscle, and nervous tissue. When tissue is damaged, should proper healing not occur, it could have detrimental consequences as serious as organ failure or death [1]. For thousands of years tissue repair and or replacement has been a priority and generically falls into a couple of categories. The oldest technique is tissue grafting, where tissue from a one's own body, with no independent blood supply, is placed in the location of the damaged tissue. Tissue grafting was used for thousands of years, but often failed due to lack of bioactivity, no bio-integration, generation of an immune response, infections, and the general need for additional surgeries [2]. From this lack, came a need to be able to stimulate tissue growth within the geography of impairment or injury. In order to generate cells with the necessary morphology and functionality to create tissue that fully replaces loss tissue in aspects such as mechanical prowess and functionality; while simultaneously, there being lack of immune rejection, faster cell production, acclimation at site of injury, and no secondary surgeries as to be more cost efficient also lowering risk of infection [3–8]. From this need tissue engineering was birthed. The field of tissue engineering had its modern start in 1991, with the first papers covering fundamental questions related to biomaterials and their potential for use in vivo [9]. Knowing that cell geography, relative to the body, causes generation of specific cells, one specific cell are myocytes. Myoblasts are the precursor cells to myocytes that fuse together forming myofibers or muscle fibers [10,11]. These fibers make up the muscle tissue. Muscle tissues are temperamental and receive constant stimulus to cause both voluntary and involuntary contractions. They fall under three categories.

12.1.1 Biocompatibility: tissue specificity

Muscle tissue specificity is an important factor when considering engineering and design decisions. Variance ranges from simple constructs such as the shape and morphology all the way to vascularity difference. These differences could be categorized into more simple parameters: vascular composition, ECM composition, cellular morphology, and cellular arrangement/order.

12.1.1.1 Cardiac muscle

Cardiac muscle tissue forms the muscle surrounding the heart. With the function of the muscle being to cause the mechanical motion of pumping blood throughout the rest of the body, unlike skeletal muscles, the movement is involuntary as to sustain life. Unique to the cells of the cardiac tissue, the cardiomyocytes, they contract in a rhythmic pattern that is independent to any external stimulus, and engineering cardiac related muscle tissues come with unique challenges. The presence of cardiomyocytes is not the sole source of variance as previously discussed. Heart tissue contains various types of cells aside from cardiomyocytes; including, but not limited to endothelial cells, smooth muscle cells, and fibroblasts. In fact, cardiomyocytes may occupy ∼75% of the total volume, but still only make up only <40% of the overall cell count [12]. When culturing cardiomyocytes, like most cells, by varying the conditions, size, the proliferation rates, and cell viability varies as shown in Fig. 12.1.

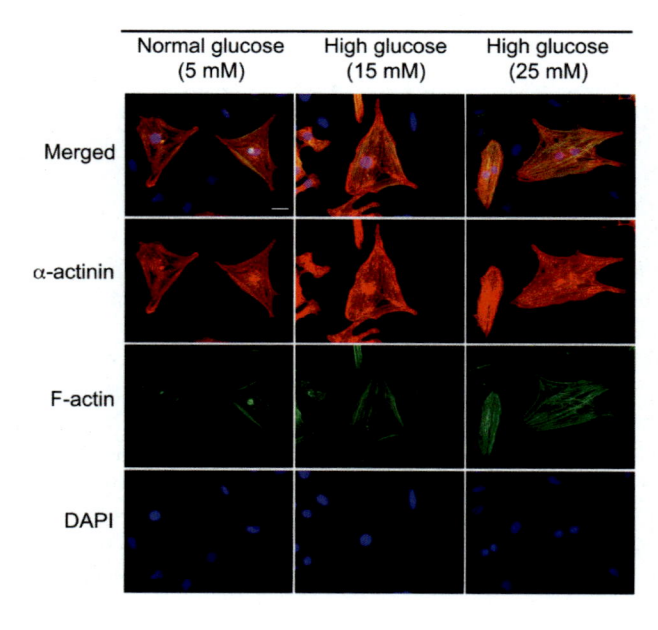

FIGURE 12.1 Cardiomyocytes identified through various stains, and exposed to different amounts of glucose over 24 h. *Adapted and reproduced with permission from Morishima M, Horikawa K, Funaki M. Cardiomyocytes cultured on mechanically compliant substrates, but not on conventional culture devices, exhibit prominent mitochondrial dysfunction due to reactive oxygen species and insulin resistance under high glucose. PLoS One 2018;13(8):e0201891 [13].*

Cardiac muscle tissue's mechanical properties are a factor to consider when developing, scaffolding materials. The importance is related to mechanical biocompatibility between interfacing materials, an issue is exacerbated when selecting higher modulus materials in the case of osteo-implants [14]. Heart tissue works at maximum stresses of 18–44 mN/mm^2 and is important when selecting target material properties in the potential hydrogel polymers for a device such as graft, scaffold or heart patch [12].

The mechanical environment can also have a great impact on the overall development of cardiac cells and tissues [10,11]. Life threatening complications related to the heart can often be due to structural remodeling that results from drastic or often chronic changes to the mechanical environment. New demands from the surrounding tissue are what results in these changes in the mechanical environment. Such demands lead to the structural changes needed in order to compensate for the lack of mechanical output ability. This is often caused by sudden excessive fibrin deposition in the ventricular walls [15]. Mechanical changes can also introduce integrin mediated ECM lead changes in the smooth muscle cells. In turn, ECM proteins are produced that can alter gene expression. This tissue remodeling can lead to serious medical issues such as myocardial infarction [15,16]. As later discussed, cardiac muscle engineering offers some solutions to some of the current complications associated with cardiovascular health.

12.1.1.2 Smooth muscle

Smooth muscle tissues (SMMT) lack striation and is present generically in the lining of organ systems where the primitive function, like other muscles, is to contract, and hold several essential responsibilities from swallowing, digesting food, to the heart beating. The autonomous nervous system is responsible for signaling of involuntary contraction where the stimuli can be hormones, chemical agents, autocrine agents amongst others. More commonly amongst SMMT, is that the contraction will utilize a cross bridge cycling between the actin and myosin proteins. Contraction, in every instance, will require energy and occurs as a result of the phosphorylation of myosin by the cleavage of ATP [17]. This exothermic reaction causes the release of energy, in the form of heat, that results in the interaction between actin and myosin by which ensues contraction of the muscle. Fig. 12.2 is an example of cells that based on expression of specific genes will develop into either smooth or cardiac tissue. This figure specifically displays images from confocal microscopy of cardiac and smooth tissue where the progenitor cells were expressed through different genes.

12.1.1.3 Skeletal muscle

Skeletal muscle tissues (SMT), like cardiac tissue, is striated and attached to bone causing voluntary movement. This voluntary movement requires the energy of ATP and degradation of which, to ADP, is an exothermic process which consequently generates heat. As a result, the locomotion attributed to SMT plays a role in thermal homeostasis of the body, and because all movement by these tissues is voluntary, it is dependent on the nervous system to operate unlike both smooth and cardiac tissue. Each SMT is composed of three layers of connective tissue that causes organization of the muscle fibers. The epimysium, the outer layer, separates the muscle from other structures and causes contractions of the muscle. Individual muscles fibers that are bundled together with in the epimysium which forms the middle layer called the perimysium. The perimysium receives the stimulus of the action potential from the nervous system causing the muscle to contract. Acetylcholine is released by presynaptic axons

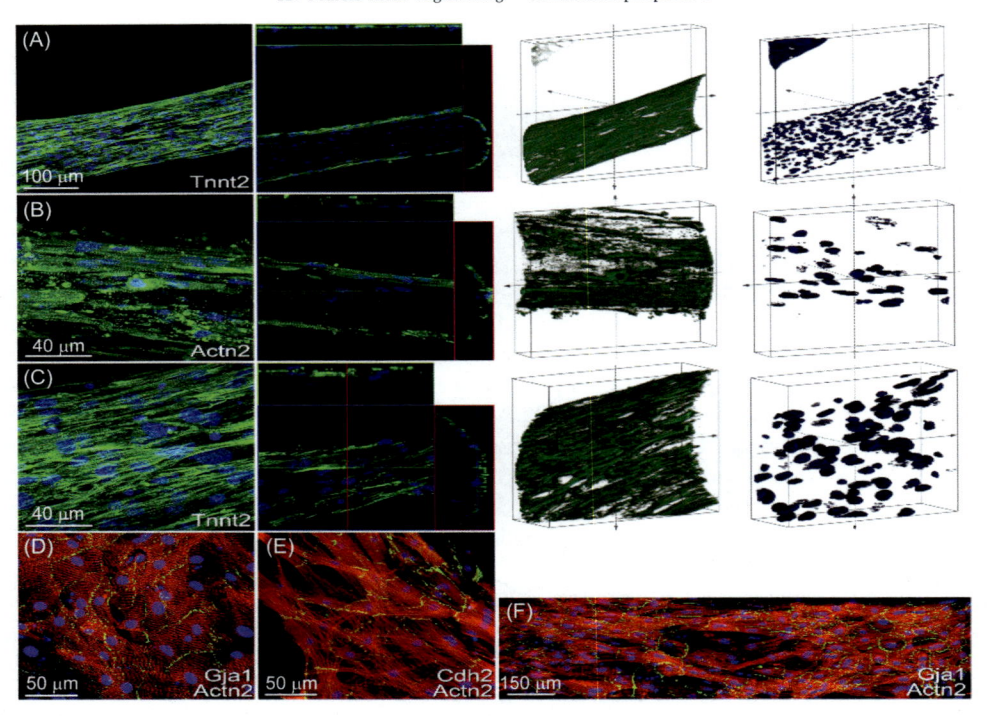

FIGURE 12.2 (A) Within the tissue, cardiac induced pluripotent stem cells where a high-density network of fibers compacts against the hydrogel matrix. (B—C) Cytoskeleton of the cardiomyocytes are stained to determine the which gene is being expressed. (D—F). Cells, with in the tissue, are stained against antibodies showing expressions of either the Cdh2 or Gja1 genes. *Adapted and reproduced with permission from Christoforou N, et al. Induced pluripotent stem cell-derived cardiac progenitors differentiate to cardiomyocytes and form biosynthetic tissues. PLoS One 2013;8 (6):e65963 [18].*

into the synaptic clefts. It binds to the Acetylcholine receptors of the post junctional folds which causes the depolarizing of the cell. This depolarization triggers the action potential that runs up through the perimysium causing the contraction [19]. Inside the perimysium is a mixture of fibers that making up the innermost tissue layer called the endomysial. The endomysial is around the muscle cell and is where the extracellular matrix (ECM) of the muscle resides containing nutrients, nerves, myofibers, blood vessels, etc. [20]. Fig. 12.3A and B below displays the cross-sectional area of the muscle layers.

SMT generally suffer injury through either direct trauma (such as lacerations, strains, etc.) or indirect trauma (disease). Regardless of the type of injury, muscle regeneration occurs through first necrosis or degeneration of the of myofibers, which is tasked to the local proteases. Inflammation from this degradation of the myofibers occurs afterward which causes a hematoma formation. The area of inflammation is flooded by blood vessels, activated macrophages, and T lymphocytes all from the ECM. Cytokines are released into the inflamed area stimulating the activation of satellites cells that proliferate forming new myofibers [22]. This physiological process is followed, at varying time rates, for majority of SMT. Muscle tissue

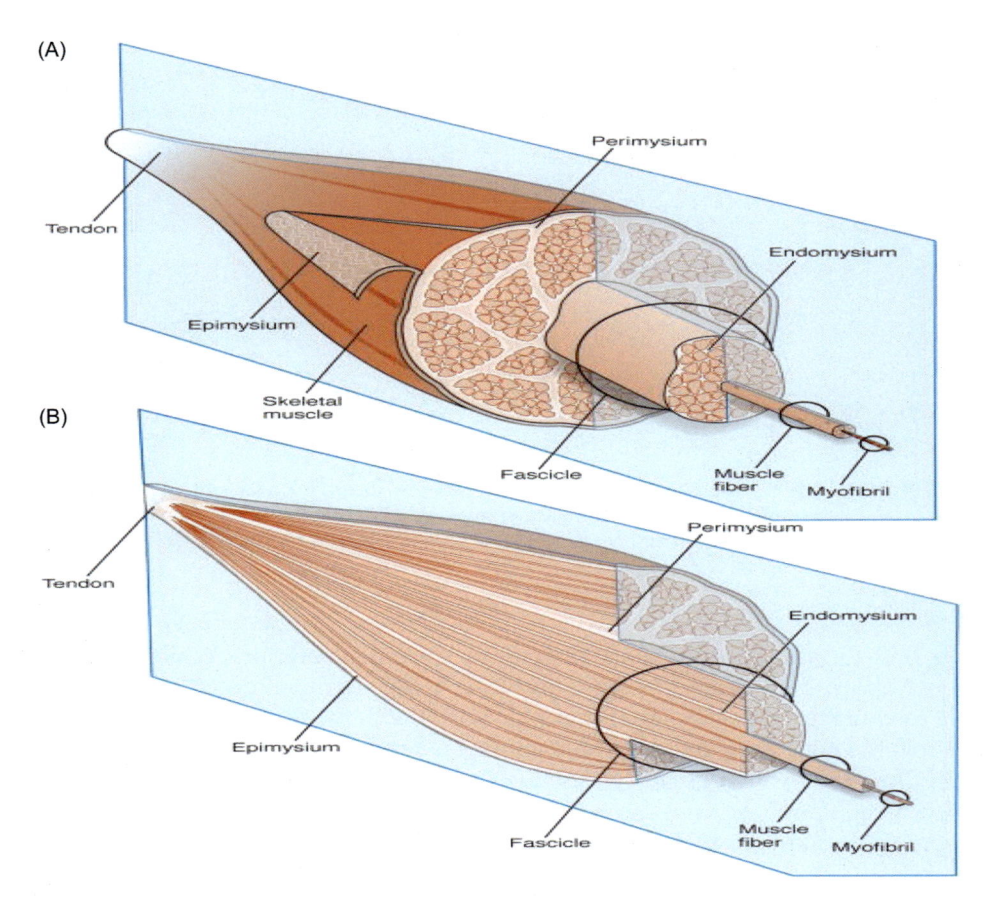

FIGURE 12.3 (A) Full make-up of the SMT ECM consisting of all three layers of the ECM. (B) Cross section area extending from the perimysium going through the tendon. *Adapted and reproduced with permission from Gillies AR, Lieber RL. Structure and function of the skeletal muscle extracellular matrix. Muscle Nerve 2011;44(3):318–31 [21].*

engineering targets different portions of the recovery cycle in order to enhance physiological healing capacities, minimization of detrimental effects during healing, and quantitative amounts of restructuring of skeletal muscle tissue. While much progress has occurred, there is still much that remains unknown. In 2015 it was Hayshai group that confirmed the link between the p21 cytokine and inhibited skeletal muscle tissue growth. They observed that the muscular synthesis gene, after the muscle was damaged, was expressed at a level significantly lower in the muscles tissue of p21 knock out mouse than the control group [23]. Fig. 12.4 below shows the histological difference in the two groups. The P21 knockout mice group do not show a near complete recovery of the tissue until 28 days unlike in the control group, i.e. wildtype mouse signified as WT in Fig. 12.4, which show complete healing after only 14 days.

While this mouse study confirmed the delay as a result of the p21 gene the reasoning as to why was not able to be confirmed in this study. Krawetz group did additional mice

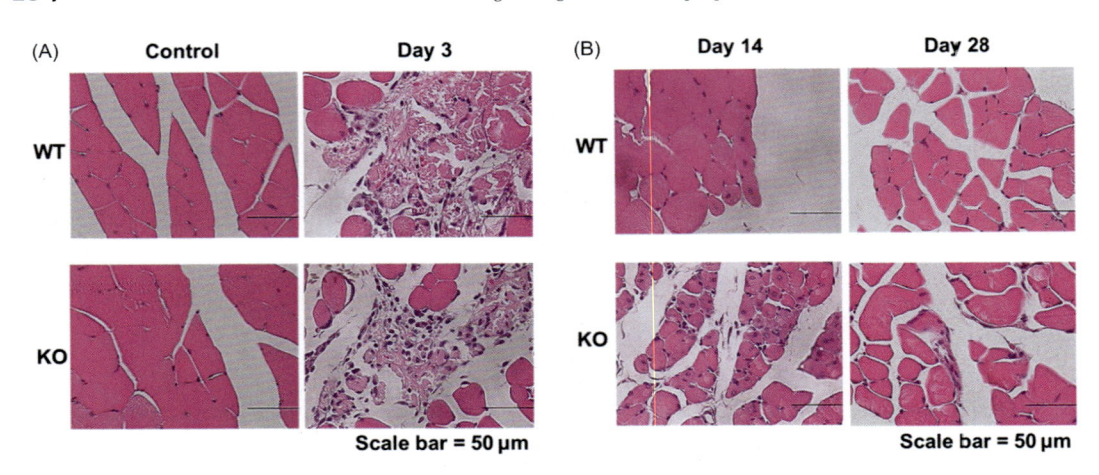

FIGURE 12.4 (A) Displays the histological difference between the p21 Knock out mouse and WT with through Day 3. (B) Distinction in the Knock out mouse sample and WT sample from day 14 and day 28. *Adapted and reproduced with permission from Chinzei N, et al. P21 deficiency delays regeneration of skeletal muscular tissue. PLoSOne 2015;10(5):e0125765 [23].*

studies to find the influence of the p21 gene in bone regeneration. While in their study they did not observe the osteoclasts or osteoblasts being detrimented by the p21 gene, and were able to confirm enhanced fracture repair with the gene, they were unable to confirm the exact role the gene performed in the bone regeneration [24]. P21 is a cytokine gene that is famously known for its regulatory function in the cell cycle and more popularly known for the down regulation of the gene being associated with cancer, so it is astonishing that there is some yet unknown role of p21 in both bone and muscle tissue repair.

Tissue related physiological differences are not the only source of variability in their response to stimuli. Variance of the mechanical environment can up or down-regulate targeted biochemical and physiological mechanisms. An example would be the hypothesis proposed by Bruusgaard group that when the protein small muscle protein x linked (SMPX) is upregulated as a result of the SMT undergoing stretch, that this causes the expression of several transcription factors which inevitably causes a fusion of specialized myoblasts. Larger myofibers would result from this fusion if this hypothesis held. What was found was when the muscle was subjected to an increased load, which caused the upregulation of SMPX, it showed no difference in either fiber length or fiber composition as shown in Fig. 12.5 below.

These results contradict the hypothesis that SMPX has some gene regulating role; regardless, demonstrates that with a change in mechanical environment causes at a biological level.

12.2 Bio-interfacing materials

The field of material science, and by extension, biomaterials is not new. The field has grown and now matured due to significant needs that arise from things like bodily trauma, old age, and abnormal tissue growth in patients over the years [15,26—34]. Such needs include

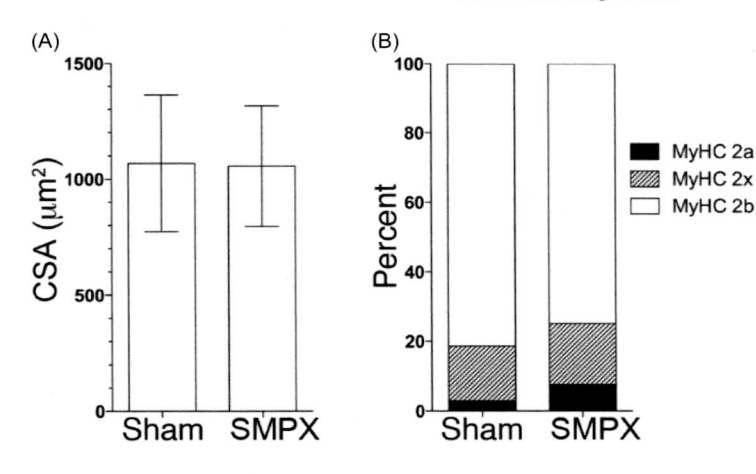

FIGURE 12.5 (A) Lack of variance between the cross-sectional area of the muscle fiber demonstrated as a result of the upregulation of SMPX showed as mean values with sd. (B) Shows the compositions of fibers expressed in both samples. *Adapted and reproduced with permission from Eftestøl E, et al. Overexpression of SMPX in adult skeletal muscle does not change skeletal muscle fiber type or size. PLoS One 2014;9(6):e99232 [25].*

prosthetics, bone/joint implants, internal/external fixation implants, etc. [27,35−39]. On the other hand, biomedical treatment also had to address internal and much less traumatic pathologies which also requires an appreciation for biomaterial development that can expand on the efforts and knowledge gained from earlier work in areas such as prosthetics.

Understanding of physiological markers like heartbeat, blood pressure, and temperature sparked a push for more sensitive and efficient devices to not only collect data from known markers but also to discover new ones. Much of this work has dealt with detecting biochemical signals, which then become extraordinarily useful in a clinical setting when running various types of pathology related assays. This then requires modern medical devices to have great specificity and precision to observe and measure many types of physiological markers. A majority of the advancements made in biomedical devices have primarily relied on the increased mechanistic understanding of cellular physiological processes. These advancements have the potential to develop sensitive materials that can detect, stimulate, and interact with targeted biological signals. Scientific understanding has become more granular with respect to organ system communication, in particular muscle tissue due to its ability to interface with the nervous system in order to execute complex mechanical commands given by the central nervous system (CNS).

12.2.1 Biocompatible polymers

The use of polymers in biomedical applications has pushed the field of biomedical engineering forward. Polymers' inherent viscoelastic properties make them great candidates for tissue engineering. This is primarily due to the fact that these materials allow for greater control in terms of design. Polymeric materials are typically synthesized; however, the materials used have also involved natural or naturally derived polymers such as collagen and melanin.

12.2.1.1 Biological materials

Biomaterials and general material design often take inspiration from the natural world. Bioinspiration is rather successful and continues to improve on various technology sectors [32,40]. Often the most effective and efficient approach is the biomimetic one. Therefore, it is to no one's surprise that there are abundant natural polymeric materials that already hold great value in the tissue engineering field [41—43]. Naturally derived biomaterials have significant advantages over synthetic materials. They are more likely be inherently biocompatible and hold bioactive characteristics. The biocompatibility can be observed in areas such as cell adhesion, migration, and proliferation. This is often the case in situations where a material is a native component in the environment of the cell model used for study [44—48]. Examples commonly used in the field include, but are not limited to: collagen, gelatin, fibrin, and silks.

Collagen is a well-known biologic material that is widely accepted as a great candidate for regenerative approaches. This is mostly due to it being a significant component of the musculo-skeletal ECM [28,49—59]. Therefore, collagen (Type-I) is often used to design scaffold that can mimic the physiological environment found in the native ECM. Some approaches have involved electrospinning composite polymer solutions that contain collagen along with a different polymer (i.e. PCL) to form more manufacturable biomaterials [60]. Other methods make use of a 3-dimensional design in which collagen forms the base structure which is then impregnated with a human fibrin gel in order to promote vascularization [54]. A common material used in tissue engineering is gelatin which a polymer derived via hydrolytic degradation of collagen [61]. This form of collagen can be a better candidate when designing hydrogels while still maintaining many of collagen's biocompatibility features [61—70].

Fibrin is biocompatible natural material abundant in the musculo-skeletal system similarly to collagen [11,15,54,71,72]. Fibrin deposition, especially in cardiac tissue, is often reported in the literature as a characteristic to avoid. This is due to it being associated with myocardial infarction from cardiac ischemias [15]. However, fibrin is often used to engineer scaffolds in conjunction with other polymers, natural or synthetic, to create favorable environments similar to those found in the physiology [11,15,54].

More novel materials include other materials not found in the human body such as silks which are often used as more secondary-like components in many composite materials which will be explored further along in this chapter [73]. Silk is a good biocompatible candidate for conductive polymeric composites that require biomimetic mechanical properties for use in muscle tissue engineering [73—76].

12.2.1.2 Synthetic polymers

Use of synthetic polymers has revolutionized many industries since the introduction of more primitive forms such as trademarked Bakelite. Over time, many new polymeric materials have been discovered and subsequently synthesized. A good example of the advantages would be the use of Ultra-High Molecular Weight Polyethylene (UHMWPE) as a bearing material in total joint arthroplasty [77,78]. In this case, UHMWPE's resistance to wear and semi-crystalline properties help prevent erosion of the femoral head and hip socket of the joint implant. UHMWPE is a form of polyethylene (PE) with significantly high crystallinity along with density. An example of this can be seen below. Fig. 12.6 displays and compares several different prostheses placed into different animal models.

UHMWPE typically has a molecular weight of $\sim$3.1 million g/mol while PE is $\sim$1 million g/mol [80]. PE demonstrates the advantages of synthetic polymers such as property

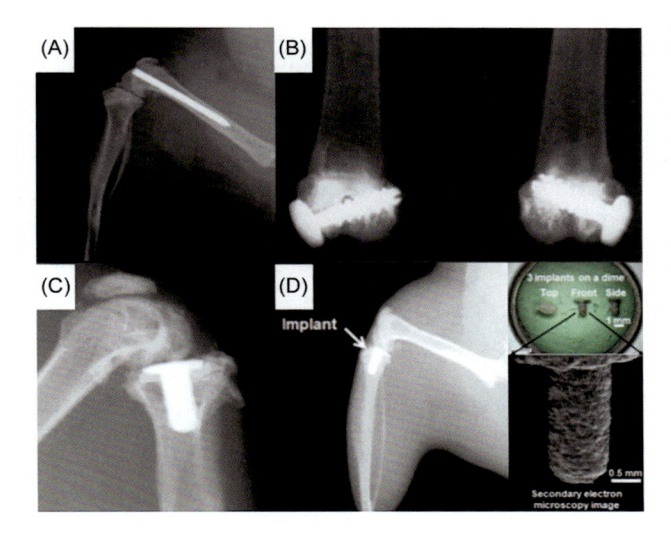

FIGURE 12.6 (A) Kirsherner wire that has been embedded within the femur of the mouse. (B) A stainless-steel nail and an UHMWPE washer used as prostheses in a rabbit model. (C) A 3D printed Ti alloy utilized as a prosthesis in a mouse model. (D) 3D printed prostheses in a mouse model. *Adapted and reproduced with permission from Jie K, et al. Prosthesis design of animal models of periprosthetic joint infection following total knee arthroplasty: a systematic review. PLoS One 2019;14(10):e0223402 [79].*

variability, scalable production, low cost, and allow for more post-processing options [27]. Inherent chain-like properties of polymers allow one to design polymers with varying properties by modifying the polymer chain length.

In the case of PE, lower polymer chains (relative to UHMWPE) produce low density polyethylene (LDPE) which can be used for more mundane uses such as food wrapping film, grocery bags, or other more products requiring more viscoelastic properties. However, due to the inert properties of PE its uses are limited in application to biomedical engineering. PE can be used in the sterilizing wraps or film-type products. Furthermore, as previously mentioned the UHMWPE high polymer chains occur only with highly linear chains are synthesized and subsequently align themselves in a crystalline manner. This orientation of the polymer chains produces a very hard semi-crystalline material that is used in cases where a hard and inert material with some viscoelastic properties is required.

PE, like many other synthetic polymers, allows for the design of varying types of polymeric properties by combining them with new substrates to form novel materials [15,77,78,80–87]. PE often serves as a base material from which many derivatives have been formed over the years. This is primarily done via the addition or modification of functional groups along the polymer chain. As a result, many varieties with varying properties have been produced. Synthetic polymers such conductive polypyrrole, ECM mimicking hydrogels, polymeric nanoparticles, and nanofibers have been developed for muscle tissue engineering materials.

12.2.2 Electrically conductive polymers

Muscle tissue engineering has typically been limited to mechanical and biochemical aspects of the physiology. Often time, research has focused on the cellular physiological characteristics of tissues. However, over time, the field has set its sights on the more macro-perspective components of tissues and their respective organ systems. Tissue specificity, being a key factor in development, has pushed muscle tissue characteristics, such as conductivity, to be the focus many research groups [88]. Although there already exist ways to detect and stimulate the nervous system and by extension muscle tissue, the need to interact in more complex and novel ways has pushed research towards seeking new materials. Common conducting polymers such as polypyrrole (PPy), polyaniline (PANi), polythiophene (PT), and poly (3,4-ethylenedioxythio-phene) (PEDOT) have shown to have translational applications in muscle tissue engineering [89,90]. There are some limitations involved with the use and development of conductive polymers. A significant challenge involves electronic stability over large time scales, particularly for implant development. Conductive polymers can be converted from an oxidized to a neutral form when exposed to air or physiological conditions. The oxidized form being conductive while the neutral form is nonconductive can lead to "poor integration" with the host tissue [91].

Polypyrrole (PPy) has been used in many applications, and various groups' work suggests it is an optimal candidate for muscle tissue engineering applications. PPy is considered biocompatible and has been used for cardiac tissue applications. The polymer can be prepared in conjunction with a base material such as poly(epsilon-caprolactone)/gelatin to form polypyrrole/poly(epsilon-caprolactone)/gelatin. Then using a manufacturing method called electrospinning it can be fabricated into a muscle tissue construct for cardiac cell transplantation [92]. Other groups have incorporated PPy into nanofibers as a way to incorporate electroconductivity in a material [92]. In a related application, Luo group made PPy/chitosan composite mixing PPy nanoparticles with a chitosan solution making a conductive scaffold for nerve regeneration. When electrical stimulation was applied to the scaffold there was an enhanced healing of the axonal. Microscopy of the composite scaffold is shown in Fig. 12.7.

Other groups have incorporated PPy into nanofibers as a way to incorporate electroconductivity in a material. Groups have used the PPy in combination with polyesters to form "interpenetrating networks" throughout silk extracted from *B. mori* silkworms. Potential broad application mentioned by the group includes biomedical application capable of detecting endogenous electrical signals [73].

Biomedical applications continuously push the boundary of what type of materials can be developed to fill the various needs due to tissue specificity. The polymer can used to develop injectable hydrogels using collagen thus improving biocompatibility [52]. Other groups have used it to coat the surface of tissue scaffolds to allow for the use of electrical stimulation (ES), which is then used to promote differentiation of smooth muscle cells to adipose cells and eventually a robust tissue construct [94]. Some groups' research focuses more on the 3D aspects of design. Therefore, PPy has been used in the development of biomechanically active scaffold; the addition of the conductive polymer increases biocompatibility and its prospects as viable technologies [95,96].

Polyaniline (PANi) is a frequently used electroconductive polymer and has been used in several fields outside of tissue engineering. Troitsky fabricated a Langmuir Blodgett (LB) film with PANi to increase conductivity and microscopy of such is shown in Fig. 12.8.

It can be synthesized by combining the monomer (aniline) with an oxidizing agent such as ammonium persulfate (APS) [98]. Its chemical mechanism of reversible acid/base "doping/dedoping" allows for one to control various properties solubility, electroconductivity, optical activity, volume, and solubility [99].

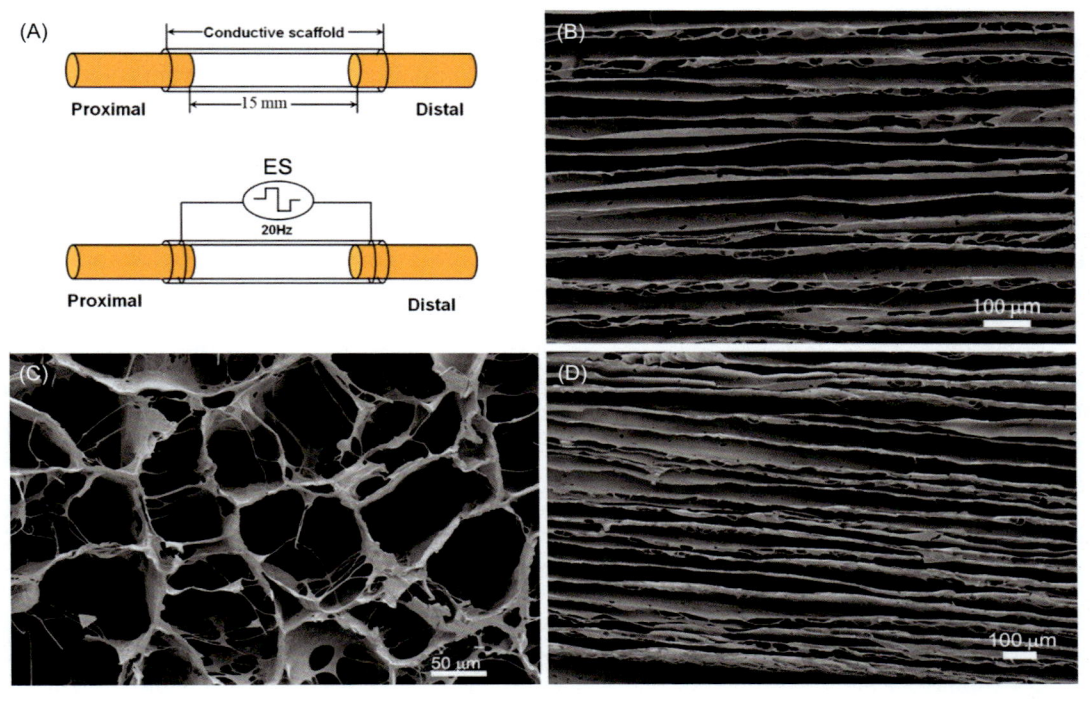

FIGURE 12.7 (A) The combination of the electrical stimulation with the conductive scaffold. (B) Microstructural microscopy of the chitosan/PPy scaffold. (C) The cross section of the scaffold. (D). A longitudinal view of the scaffold. *Adapted and reproduced with permission from Huang J, et al. Electrical stimulation to conductive scaffold promotes axonal regeneration and remyelination in a rat model of large nerve defect. PLoS One 2012;7(6):e39526 [93].*

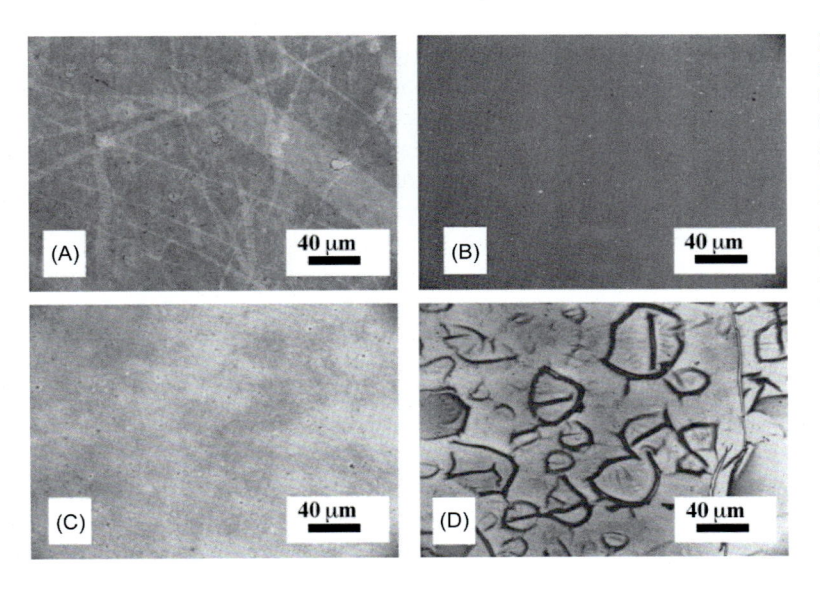

FIGURE 12.8 (A–D) LB films layered with PANi and treated under several different conditions. *Adapted and reproduced with permission from Troitsky VI, Berzina TS, Fontana MP. Langmuir–Blodgett assemblies with patterned conductive polyaniline layers. Mater Sci Eng: C 2002;22 (2):239–44 [97].*

This material has been applied to a wide variety of use-cases [75,100]. It has demonstrated its effectiveness as copolymer material for electrospun scaffolds. A Taiwanese group approached the use of PANi as a way to administer conductive properties to an already characterized material, poly (ε-caprolactone) (PCL) [70,101,102]. The group used an innovative process to produce highly aligned fibers to mimic skeletal muscle tissue orientation [103]. Mechanical function is very significant and should be kept in mind when developing muscle constructs along with conductivity [104]. Other groups continued using PANi in a hybrid form with other polymers. Another group used PANi to develop an electroconductive myocardial patch using a combination of PANi and poly (lactic-co-glycolic acid) (PLGA) [105]. Being that conductivity is the attractive property of PANi, groups have also combined with other materials known for their electrical conductivity. An example from a group in Iran involved developing graphene oxide nanosheets that were then coated with an electrospun composite polymers solution of PANi and polyacrylonitrile (PAN) and tested for biocompatibility using murine skeletal muscle model [106]. Others seek to optimize the conductive polymers such as PANi by doping it with phytic acid during processing portion post-synthesis [91].

A polymer such as polythiophene (PT) shows good promise for future applications. It is also an electroconductive polymeric material that exhibits very similar properties to the previously mentioned materials [90]. PT has been synthesized several ways such as, "nickel-mediated cross-coupling reactions, palladium-catalyzed cross-coupling reactions, oxidative polymerization electrochemical polymerization, and biocatalyzed polymerization". However, a more biocompatible synthesis process like oxidative polymerization using $FeCl_3$ is more appropriate. This process minimizes cytotoxicity issues when used for developing biocompatible devices [107].

12.2.3 Hydrogels and composite approaches

Hydrogels materials can be defined as natural or synthetic in origin that can absorb and retain water within their complex polymer networks synthetically [71,108–111]. These crosslinked networks create a gel phase using physical (reversible) or chemical (permanent) networks. Hydrogel development has taken a lead role in the research of muscle tissue constructs. Hydrogels have the capability of serving as a 3D scaffold, more specifically as a synthetic ECM, but have flexibility as far as mechanical properties according to application [110]. This flexibility can be exploited by different methodologies of synthesis, gelation, and variation of materials employed. Cell arrangement and structure all occur in 3D space. Research has shifted to address that issue and scaffolds now address this via the development of hydrogels that can arrange various cell types in deliberate orientations. This is critically important in the muscle tissue seeing as the fiber orientation dictates many of the mechanical properties of the organ [103]. The manufacturing of the scaffolds using hydrogels provides the cells the mechanical base required to function as their respective in vivo conditions [89]. There a variety of synthetic polymers that are used and synthesis with the intention of forming polymeric hydrogels.

Some approaches involve composite and hybrid materials have been attempted by different groups. Polymers such as such as thiol-2-hydroxyethyl methacrylate (thiol-HEMA) and HEMA at varying ratios have been poured in to a "close-packed lattice of poly (methyl methacrylate) (PMMA) microparticles" in order to form lattice spaces for the cellular deposition [112]. The combined thiol-HEMA/HEMA is then crosslinked by adding tetra (ethylene glycole) dimethacrylate (TEGDMA) and photopolymerization using UV radiation in conjunction with AIBN (photoinitiator). After crosslinking PMMA is removed by immersing the scaffold in dichloromethane (DCM). The scaffold can then be immersed in a potassium tetrachloroaurate (KAuCl$_4$). This tedious process was executed as an alternative to the PPy or PANi. The researchers noted that a limitation of these materials is their failure to mimic physiological moduli. However, the focus on engineering a scaffold with conductivity in mind is important to foster a functional environment for cardiac muscle cells [112].

PANi's use extends to muscle excitation and stimulation purposes. The material has been manufactured as a gelatin-PANi composite doped with camphorsulfonic acid using electrospinning techniques. Subsequent nanofibers formed have been used to culture C2C12 myoblast cells. This approach has allowed for greater enhancement of cellular maturation [76]. Similar electrospinning examples have been tried in the cardiomyocyte space. Another example involved the manufacturing of electrospun of a PANi and poly (L-lactic acid) (PLA) composite material that was then used to produce nanofibers that can serve as bioactuators. This then provided the H9c2 cardiomyoblast cells cultured with the material better cell viability as well promotion of cellular differentiation [113]. Overall, use of hydrogels gives more 3-dimensional control during the design and development process in tissue engineering and is greatly enhanced in muscle tissue engineering with electroconductive materials [114,115].

12.2.4 Biocompatible nanomaterials

Various types of materials have been used in the past to detect musculoskeletal stimuli, many of which were metal due to high electroconductivity. However, as engineering requirements become more robust and complex design issues arise over chronic use such as toxicity, local immune response, and probe corrosion an alternative material must be sought for the electrical conductivity property [116]. Many materials have been developed in recent years that address these issues and also lay the groundwork for more sensitive and specific biosensing capabilities, especially in the case of electroconductivity. More recent types of materials include carbon nanofiber, carbon nanotubes polymeric hydrogels [47,48]. Many groups have approached these materials from a composite perspective. Nanomaterials can be developed for use as "dopants" that can alter the properties of other biocompatible materials [84,87,117−121].

12.2.4.1 Nanotubes and nanofibers

Carbon nanotubes (CNTs) are novel materials that have many favorable properties that are necessary for biomedical applications [122]. CNT's benefit from carbon's stability in bonding with itself which translate to overall increased strength and high electrical conductivity. Electrical conductivity is typically seen in graphene, the most stable form of carbon, along the graphene plane, but that conductivity is limited to the current running parallel to the graphene plane [123]. CNTs on the other hand are graphene sheets wrapped cylindrically around an axis. Therefore, graphene's electroconductivity limitation ceases to apply [90,124]. Their overall properties and bio-interactions can vary depending on their size and type. Orientation can affect other aspects such as electromagnetic properties like their refraction index. Single-walled CNTs (SWNT) tend to share a length of ~1–3 nm in diameter while multi-walled CNT's share lengths between ~10 and 100 nm. These differences are responsible for single-walled CNTs propensity to form bundles when compared to multi-walled CNTs. Therefore, CNTs are not entirely favored biocompatibly due to toxicity issues which can result from aggregation [124]. Groups have noted asbestos-like qualities such as accumulation in lung tissue that often-times leads to pulmonary pathologies typically related to fibrosis and in chronic cases, carcinogenicity due to pulmonary gene damage [125]. However, it is important to note that some groups have observed effects to the contrary [126]. CNTs can have better practical use when combined with other materials to form nanofibers [87,117,127]. This combination mitigates issues that arise when using CNTs in a standalone manner [125].

In order to mitigate in vivo toxicity from CNTs, groups have discovered composite materials in which multi-walled carbon nanotubes (diameters 40–90 nm, length 10–20 μm) have been used as a method of doping polymeric materials such as poly (octamethylene maleate (anhydride) 1,2,4-butanetricarboxylate) (124 polymer) in order to dramatically increase their electrical conductivity. The 124 polymer-CNT materials using PEG-dimethyl ether porogen were mixed with 124 prepolymer in order to better mold the solution into shapes and then they were then set using photo-crosslinking specifically with UV radiation. This approach allows one to engineer conductive materials requiring high conductivity and high elasticity typically found in elastomeric and hydrogel materials [87].

Zhou et al group designed a cardiac tissue construct targeting cardiac repair which incorporated SWNTs into a gelatin hydrogel scaffold crosslinked with glutaraldehyde. The design of the construct allowed for a 3D structure that, combined with the conductive properties from the SWNTs, created a favorable microenvironment for cardiomyocytes. The incorporation of the SWNTs demonstrated stronger contractile and electrical properties. In turn, the use of such constructs provides better support and integration in infarct myocardium for better remodeling during treatment and healing [84].

Similar groups have composed nanofiber yarn networks (NFY-Net) made up of polycaprolactone, silk fibroin, and carbon nanotubes within a hydrogel shell composed of a photocurable methacrylated gelatin (GelMA) [128,129]. In the case of the group Wu Y. et al., the use of an interwoven network of aligned fibers provided similar anisotropic properties typically found in cardiac muscle tissue (or most elongated fibrous tissues). Unidirectional characteristics of cardiac muscle tissue is mimicked with the design of unidirectional conductive fibers. The interwoven pattern mimics the more complex environment established by the ECM that provides crucial support for the surrounding cells [128]. Other groups such as Liu et al. has focused on "tuning" conductivity by varying the concentration of CNTs in the scaffold. They also designed and adjusted the structure of the scaffold by incorporating the use of highly aligned fibers to better mimic myocardial fibers. These changes provide a more biomimetic structural environment that encourages the favorable morphological and physiological changes [130].

The use of nanofibers is important when designing hydrogels, particularly those developed for the muscle tissue constructs. Nanofibers can be prepared using many types of polymers by using post processing methods such as electrospinning.

12.3 Engineering approaches (scaffolds)

12.3.1 Cardiac muscle tissue engineering scaffolding

As aforementioned, cardiovascular health is one of the leading causes of extreme health detriment that results in consequences that can lead to heart failure, and if not corrected consequences are as extreme as death [131]. It is the leading cause of death of the United States, and while a healthy lifestyle can help deter some of the complications, many indulge in a lifestyle of poor health decisions that causes strain to the cardiovascular system resulting in conditions that lead to both cell and tissue necrosis [46]. While drugs offer a pathway in assisting this widespread dilemma, they do not offer a comprehensive solution [132]. Very early on in the journey to differentiate the problems associated with heart disease, thousands of drugs including several beta blockers and ace inhibitors were synthesize, where the hope that the pure quantity of new drugs created would help solve cardiovascular failure or at the minimal be able to help regulate human health in spite of it; however, only a miniscule percentage of these drugs were ever approved and just as small of a percentage showed to be significantly beneficial to the cause [133]. Tissue engineering offers an additional method to aid pharmaceutical solution making ground where medicine lacks in the form of scaffolding. Several groups have taken different approaches to this innovative solution for cardiovascular health. Parker's group observed the deformation of the culture of cardiac cells as cardiomyocytes shorten. They utilized traction force microscopy in order to measure the tractional forces of the cardiac tissues by printing polyacrylamide gels with mechanical stiffness physiologically relative to human myocardium. Fig. 12.9 below shows a diagram illustrating their hypothesis that cardiomyocytes on softer gels would require more energy in the form of ATP in order to contract.

They hypothesized this because on softer gels the cardiomyocytes must perform the assembly of the cytoskeleton in order to contract, unlike Gels that are closer to or above

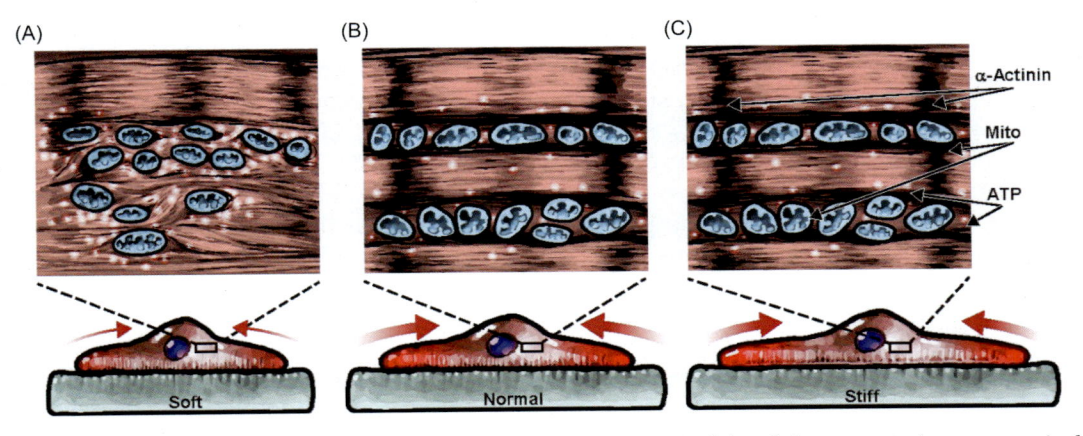

FIGURE 12.9 (A–C) Illustration soft, normal, and stiff gels (respectively) and the amount of energy required for cardiomyocyte contraction by in accordance to differing quantity of ATP. *Adapted and reproduced with permission from Pasqualini FS, et al. Traction force microscopy of engineered cardiac tissues. PLoS One 2018;13(3):e0194706 [134].*

physiological stiffness measurements. Contrary to this hypothesis, they found that cardio-myocytes placed on stiffer gels had decreased contractile action for the amount of energy used, while those placed on softer gels had the most amount of ATP available for the least amount of work. The highest energy efficiency of the cells were those placed on gels closest to physiological mechanical conditions [134].

Bhaarathy created a copolymer Poly L-lactic acid co polycaprolactone (PLCL) silk fibroin and aloe vera blend biocomposite nanofibrous scaffold in order to mediated myocardium tissue death [132]. The scaffold was made through electrospinning the blend solution making a porous electrospun mat. By immunostaining the cardiac cells on the mat, Bhaarathy observed increased differentiation of cardiac cells in comparison to other fibers. With several of the necessary biological cues and the electrospun mat demonstrated necessary mechanical properties which makes it a promising step in scaffolding for cardiac muscle tissue engineering [132].

12.3.2 Skeletal muscle engineering scaffolding

Skeletal muscles have a remarkable ability to repair and regenerate after injury. The ECM within the tissues, engulfing the cells, allow for growth factors, enzymes, proteins, etc. to start the process of necrosis and then the regeneration of new muscle tissue, which allows for muscle volume to stay consistent. Degenerative genetic diseases, such as muscular dystrophia, and severe injuries can force an external response to happen for complete repair. Earliest forms began in the form of grafting thousands of years ago. Today issues ensue such as biocompatibility, wear of material, high cost, risk of infections, etc. [135]. The use of both 2D and 3D polymeric scaffolds may allow for a higher efficiency of repair. When more than 20% of the volume of the muscle is loss, as a result of either disease or injury, the muscle generally can partially repair, but there is an inability to repair the muscle completely and formation of scar tissue follows consequently [56,136]. The end result is the volume of muscle tissue lost is not regained and serves to lower one's quality of life. The use of polymeric scaffolding within tissue engineering is an alternative solution that has an ample amount of potential. Walter's group approached the problem of VML by utilizing an in vivo approach. His group transplanted ECM with stem cells embedded to a mouse with a fractured tibialis anterior versus the control which was the ECM without stem cells to identify whether there was a difference in the functional recovery of the muscle [137]. In their study, the muscle fiber generation was observed for 6 months for both groups and saw that fragments of the damaged tissue remain local to the point of injury throughout the entire healing period for both groups. They believed that this was a result of reinjury to the remaining tissue causing putative atrophy to some of the remaining tissue. While they were not able to explain it, they concluded that, in this study, ECM with stem cells does not promote muscle fiber regeneration; however, this biological ECM does promote a deposition of collagen within the defect areas. This increases the rate of recovery to the mechanical environment, allowing more force to be placed on the area of injury faster [137]. Another example can be taken from Mooney's group who in a study used several variations of scaffolding to key in on what are some optimal conditions in promoting muscle skeletal muscle growth after injury [138]. Using a porous alginate scaffold, they used several groups including: scaffolds implanted with variations of different growth factors, just the alginate scaffold, the alginate scaffold with the growth factors and satellite cells, all in vitro. What resulted was that the scaffold delivering the combination of the growth factors and satellite cells showed rapid growth in the beginning of the healing process. All samples showed that the muscle did heal within the 6-month time span, but there was a significant increase in the number of cells per fiber observed with the scaffold that had only the growth factors. Also, the scaffolds delivering both the cells and growth factors, along with the scaffold with just the growth factors, saw significant increase in the blood vessel density local to the muscle injury at 6 weeks compared to the blank alginate scaffold. Additionally, scaffolds with the growth factors and cells and the scaffold with just the growth factors saw substantial decrease in fibrosis of the muscle. This shows that both of these scaffolds promote muscle regeneration effectively [138].

12.3.3 Cell models for in vitro muscle tissue engineering

Cell studies in the field of tissue engineering has been a key part of its advancement over the years [139]. Compared to an in vivo model cell models may seem limited in scope but, they offer a great opportunity for scientists to mimic the target environment of their material [46]. Animal models are arguably more robust than cell models; however, in pursuit of the "three R's" (Replacement, Reduction, & Refinement) in research, many studies have devised novel ways to incorporate primary cultures for use in dynamic environments that resemble that of an in vivo model [139]. This approach allows for studies to incorporate human cells along with whole tissue when testing and developing novel constructs rather than relying on data derived from a different species [140]. Successful biomaterial discovery entails a great deal of testing biocompatibility with the target tissue. Initially, much of this testing requires the use of a proper cell model that closest resembles that of target. Currently many studies make us of induced pluripotent stem cells (iPSC) which are very versatile and allow for great range of testing parameters for biomaterials to be later used in vivo (Table 12.1) [141].

12.4 Summary

Muscle tissue engineering employs the use of various novel materials as well as sophisticated manufacturing processes to accomplish engineering and design goals. When engineering biomaterials or biomedical devices assessment of overall biocompatibility becomes a significant priority. However, such an assessment can be easily misunderstood and often oversimplified. Biocompatibility correctly requires immune response assessments regardless of the local physiology. It is easy to conclude that the most significant consideration of biomedical design and engineering is that the end result not harm the end user [29,152,153]. Nevertheless, as device applications become more robust, complexity arises and careful consideration is given to areas such as, mechanical, physical, chemical, manufacturing, and environmental properties. In fact, most (if not all) other engineering disciplines hold these properties to be essential factors when designing, validating, and iterating [152]. Therefore, it is no surprise that similar approaches are on the forefront of novel and robust biomaterials exploration.

TABLE 12.1 List of commonly used cell models in tissue engineering studies.

Cell model	Species	Culture type	Cell type	Tissue type	Application	Reference
C2C12	Murine	Immortal	Myoblast	Muscle	Used to cell differentiation, scaffold related immunocytochemistry	[25,49,76,86,96,102,104,114,142−145]
H9C2	Rat	Immortal	Myoblast	Myocardium	Fibrotic muscle tissue scaffolds (i.e. collagen, fibrin)	[49]
MC3T3-E1	Murine	Immortal	Preosteoblast	Calvaria	Fibrotic muscle tissue scaffolds (i.e. collagen, fibrin)	[49,50]
hDPSC	Human	Primary	Multipotent Stem Cell	Dental pulp	Testing Cell differentiation and biomineralization in HA scaffolds	[36]
Cardiomyocytes	Rat	Primary	Cardiomyocyte	Neonatal primary culture	Mitochondrial dysfunction as a function of mechanical biocompatibility	[13,87,105,128,132]
hMSC	Human	Primary	Mesenchymal stem cell	Primary bone marrow aspirate	Cell differentiation in biomimetic scaffolds	[52,69,82,121,146,147]
MDSC	Murine	Primary	Primary muscle derived stem cells	Skeletal Muscle	Testing muscle differentiation in conductive nanofibrous hydrogel	[106,148]
hBMSC	Human	Primary	Stromal cells	Bone marrow	Used to generally characterize PLA scaffolds	[35]
L929	Murine	Immortal	Fibroblasts	Subcutaneous adipose tissue	Testing myogenic differentiation in electroactive scaffold	[60,114]

HL-1	Murine	Immortal	Cardiomyocytes	Atrial cardiomyocyte tumor	Testing biocompatibility of collagen conductive scaffold	[51]
MCEC	Murine	Immortal	Cardiac endothelial cell	Cardiac tissue	Testing mechanical microenvironment via general cell biocompatibility	[31]
hADSC	Human	Primary	Stem cells	Adipose tissue	Testing cell differentiation in fibrotic scaffolds	[54,129]
NIH3T3	Murine	Immortal	Fibroblast	Embryonic cells	Testing general biocompatibility	[96]
iPSC	Murine	Primary	Induced pluripotent stem cells	Cardiac tissue	Used to design synthetic tissues	[18]
HUVECS	Human	Immortal	Endothelial cell	Umbilical vein	Tested cell functionality and overall biocompatibility	[101,149−151]
HFF	Human	Immortal	Fibroblast	Neonatal foreskin	Used to test microenvironment in patterned scaffold	[10]

Tissue specificity is a crucial component of biocompatibility assessment. Tissue type varies key components of physiology such as cell types, ECM environment (biochemical, biophysical, etc.), and general mechanical properties [63,64,152–154]. This focuses engineering constraints on selecting biomaterials with a wide range of chemical and mechanical properties such as polymeric materials [29,153]. These types of materials make it possible to tune various final properties of device design to better interface with the surrounding tissue. This variability is especially important in the field of muscle tissue engineering. Human muscle tissue can exhibit dramatic ranges of movement that can place great demands on implanted devices. This has been serious issue in the overlapping field of bone tissue engineering. Most of the forces experienced by bone are generated via the muscle portion of the musculoskeletal system. Therefore, it is paramount to carefully consider similar mechanical requirements in muscle tissue [66].

Biomaterials do suffer from limitations similar to many other types of materials. These limitations are magnified when they are put in dynamic environments like those found in human physiology [152]. Other engineering fields have historically responded by discovering novel composites that combine many favorable properties of two or more materials into a combined material that responds to the new engineering constraints. This approach is profoundly useful in the field of muscle tissue engineering [90,152,155,156]. Some groups have designed composites manufactured by combining two or more polymeric materials to generate one homogenous or fibrotic combination [157]. Others have generated anisotropic materials that incorporate the use of a polymeric matrix with a different material dispersed throughout [119,155]. The more recent approaches include nanoparticles as dopants that give more established biomaterials novel and greatly useful properties such as electric/thermal conductivity. These dopants can include other nanoparticle polymers, gold nanoparticles, carbon nanotubes, and boron nitride [89,90,109,120,158].

The field of tissue engineering continues to yield novelty. Many new approaches have been interdisciplinary in nature and continue to push the boundary by incorporating a robust physiological awareness. This in turn will produce, not only more complex devices but also more efficient ones.

12.5 Future perspective

Scaffolding, electrospinning, and additives manufacturing all provides partial solutions within the field of muscle tissue engineering, and as many options have been explored there are just as many that provide viable options in the future of muscle tissue engineering. Previous work in tissue engineering where a so called wet-lay process could be a potential additional solution. By having the ability to incorporate either degradable or nondegradable fibers into the composite, amongst variances of the hydrogels, it allows for tunable mechanical and biodegradation properties that with facile are adjusted by varying fiber quantity, fiber length, and crystallinity giving the necessary flexibility for different muscles throughout the body [31,159]. Additive manufacturing is one method of creating 3D scaffolds while allowing for a variety of geometries to be made. This makes it convenient for whatever muscle that is being regenerated; in addition, porosity of the scaffold printed can be varied, allowing for cell attachment and invasion through the scaffold to be easily manipulated, but balancing as to not compromise the mechanical properties [8,35]. Post processing of scaffolds that have been previously explored that have shown to be effective in changing the surface chemistry is low temperature 'cold plasma' processing. This technique, utilized on a scaffold, can manipulate surface roughness, wettability, electrical conductivity, and cellular attachment all of which are variables that can be used in muscle tissue engineering [160,161].

References

[1] Greenhalgh DG, Warden GD. Wound care models. In: Souba WW, Wilmore DW, editors. Surgical research. San Diego: Academic Press; 2001. p. 379–91.

[2] Hollister SJ. Porous scaffold design for tissue engineering. Nat Mater 2005;4(7):518–24.

[3] Ikada Y. Challenges in tissue engineering. J R Soc Interface 2006;3(10):589–601.

[4] Kariduraganavar MY, Kittur AA, Kamble RR. Polymer synthesis and processing. In: Kumbar SG, Laurencin CT, Deng M, editors. Natural and synthetic biomedical polymers. Oxford: Elsevier; 2014. p. 1–31.

[5] Khan F, Tanaka M, Ahmad SR. Fabrication of polymeric biomaterials: a strategy for tissue engineering and medical devices. J Mater Chem B 2015;3(42):8224–49.

[6] Maurus PB, Kaeding CC. Bioabsorbable implant material review. Oper Tech Sports Med 2004;12(3):158–60.

[7] Nair LS, Laurencin CT. Biodegradable polymers as biomaterials. Prog Polym Sci 2007;32(8-9):762–98.

[8] Stratton S, et al. Bioactive polymeric scaffolds for tissue engineering. Bioact Mater 2016;1(2):93–108.

[9] Vacanti C. The history of tissue engineering. J Cell Mol Med 2006;1(3):569–76.

[10] Massia SP, Hubbell JA. An RGD spacing of 440 nm is sufficient for integrin alpha V V beta 3-mediated fibroblast spreading and 140 nm for focal contact and stress fiber formation. J Cell Biol 1991;114(5):1089–100.

[11] Matsumoto T, et al. Three-dimensional cell and tissue patterning in a strained fibrin gel system. PLoS One 2007;2(11):e1211.

[12] Ma SP, Vunjak-Novakovic G. Tissue-engineering for the study of cardiac biomechanics. J Biomech Eng 2016;138(2):021010.

[13] Morishima M, Horikawa K, Funaki M. Cardiomyocytes cultured on mechanically compliant substrates, but not on conventional culture devices, exhibit prominent mitochondrial dysfunction due to reactive oxygen species and insulin resistance under high glucose. PLoS One 2018;13(8):e0201891.

[14] Heary RF, et al. Elastic modulus in the selection of interbody implants. J Spine Surg 2017;3(2):163–7.

[15] Mihalko E, et al. Targeted treatment of ischemic and fibrotic complications of myocardial infarction using a dual-delivery microgel therapeutic. ACS Nano 2018;12(8):7826–37.

[16] Peyton SR, Putnam AJ. Extracellular matrix rigidity governs smooth muscle cell motility in a biphasic fashion. J Cell Physiol 2005;204(1):198–209.

[17] Webb RC. Smooth muscle contraction and relaxation. Adv Physiol Educ 2003;27(1-4):201–6.

[18] Christoforou N, et al. Induced pluripotent stem cell-derived cardiac progenitors differentiate to cardiomyocytes and form biosynthetic tissues. PLoS One 2013;8(6):e65963.

[19] Huard J, Li Y, Fu FH. Muscle injuries and repair: current trends in research. J Bone Jt Surg Am 2002;84(5):822–32.

[20] Pessina P, et al. Novel and optimized strategies for inducing fibrosis in vivo: focus on Duchenne muscular dystrophy. Skelet Muscle 2014;4(1):7.

[21] Gillies AR, Lieber RL. Structure and function of the skeletal muscle extracellular matrix. Muscle Nerve 2011;44(3):318–31.

[22] Uezumi A, Ikemoto-Uezumi M, Tsuchida K. Roles of nonmyogenic mesenchymal progenitors in pathogenesis and regeneration of skeletal muscle. Front Physiol 2014;5(68):68.

[23] Chinzei N, et al. P21 deficiency delays regeneration of skeletal muscular tissue. PLoSOne 2015;10(5):e0125765.

[24] Premnath P, et al. p21(-/-) mice exhibit enhanced bone regeneration after injury. BMC Musculoskelet Disord 2017;18(1):435.

[25] Eftestøl E, et al. Overexpression of SMPX in adult skeletal muscle does not change skeletal muscle fiber type or size. PLoS One 2014;9(6):e99232.

[26] El-Sherbiny IM, Yacoub MH. Hydrogel scaffolds for tissue engineering: progress and challenges. Glob Cardiol Sci Pract 2013;2013(3):316–42.

[27] Middleton JC, Tipton AJ. Synthetic biodegradable polymers as orthopedic devices. Biomaterials 2000;21 (23):2335–46.

[28] Lutolf MP, et al. Repair of bone defects using synthetic mimetics of collagenous extracellular matrices. Nat Biotechnol 2003;21(5):513–18.

[29] He W, Benson R. Polymeric biomaterials. In: Kutz M, editor. Applied plastics engineering handbook. William Andrew Publishing; 2017. p. 145–64.

[30] Liu M, et al. Effect of age on biomaterial-mediated in situ bone tissue regeneration. Acta Biomater 2018;78:329–40.

[31] Wood AT, et al. Wet-laid soy fiber reinforced hydrogel scaffold: fabrication, mechano-morphological and cell studies. Mater Sci Eng C: Mater Biol Appl 2016;63:308–16.

[32] Xue J, et al. Bioinspired multifunctional biomaterials with hierarchical microstructure for wound dressing. Acta Biomater 2019.

[33] Zhang Q, et al. Advanced biomaterials for repairing and reconstruction of mandibular defects. Mater Sci Eng C: Mater Biol Appl 2019;103:109858.

[34] Vasconcelos DP, et al. The inflammasome in host response to biomaterials: Bridging inflammation and tissue regeneration. Acta Biomater 2019;83:1–12.

[35] Gremare A, et al. Characterization of printed PLA scaffolds for bone tissue engineering. J Biomed Mater Res Part A 2018;106(4):887–94.

[36] Sancilio S, et al. Alginate/hydroxyapatite-based nanocomposite scaffolds for bone tissue engineering improve dental pulp biomineralization and differentiation. Stem Cell Int 2018;2018:9643721.

[37] Ahmed MH, et al. Characteristics and applications of titanium oxide as a biomaterial for medical implants. In: Davim JP, editor. The design and manufacture of medical devices. Woodhead Publishing; 2012. p. 1–57.

[38] Li J, et al. Materials evolution of bone plates for internal fixation of bone fractures, a review. J Mater Sci Technol 2019;.

[39] De Meurechy N, Braem A, Mommaerts MY. Biomaterials in temporomandibular joint replacement: current status and future perspectives-a narrative review. Int J Oral Maxillofac Surg 2018;47(4):518–33.

[40] Muller R, et al. Biodiversifying bioinspiration. Bioinspir Biomim 2018;13(5):053001.

[41] Thomas V, Dean D, Vohra Y. Nanostructured biomaterials for regenerative medicine. Curr Nanosci 2006;2:155–77.

[42] Beldjilali-Labro M, et al. Biomaterials in tendon and skeletal muscle tissue engineering: current trends and challenges. Materials (Basel) 2018;11(7):1116.

[43] Lee H, et al. A novel decellularized skeletal muscle-derived ECM scaffolding system for in situ muscle regeneration. Methods 2019.

[44] Youssef AM, et al. Rational design and electrical study of conducting bionanocomposites hydrogel based on chitosan and silver nanoparticles. Int J Biol Macromol 2019.

[45] Kwee BJ, Mooney DJ. Biomaterials for skeletal muscle tissue engineering. Curr Opin Biotechnol 2017;47:16–22.

[46] Chaudhuri R, et al. Biomaterials and cells for cardiac tissue engineering: current choices. Mater Sci Eng C: Mater Biol Appl 2017;79:950–7.

[47] Reis LA, et al. Biomaterials in myocardial tissue engineering. J Tissue Eng Regen Med 2016;10(1):11–28.

[48] Qazi TH, et al. Biomaterials based strategies for skeletal muscle tissue engineering: existing technologies and future trends. Biomaterials 2015;53:502–21.

[49] Kim W, Kim G. A functional bioink and its application in myoblast alignment and differentiation. Chem Eng J 2019;366:150–62.

[50] Kim M, Choe Y, Kim G. Injectable hierarchical micro/nanofibrous collagen-based scaffolds. Chem Eng J 2019;365:220–30.

[51] Sherrell PC, et al. Rational design of a conductive collagen heart patch. Macromol Biosci 2017;17(7).

[52] Ketabat F, et al. Injectable conductive collagen/alginate/polypyrrole hydrogels as a biocompatible system for biomedical applications. J Biomater Sci Polym Ed 2017;28(8):794–805.

[53] Bertram U, et al. Vascular tissue engineering: effects of integrating collagen into a PCL based nanofiber material. Biomed Res Int 2017;2017:9616939.

[54] Chan EC, et al. Three dimensional collagen scaffold promotes intrinsic vascularisation for tissue engineering applications. PLoS One 2016;11(2):e0149799.

[55] Tulloch NL, et al. Growth of engineered human myocardium with mechanical loading and vascular coculture. Circ Res 2011;109(1):47–59.

[56] Kin S, et al. Regeneration of skeletal muscle using in situ tissue engineering on an acellular collagen sponge scaffold in a rabbit model. ASAIO J 2007;53(4):506—13.

[57] Capito RM, Spector M. Collagen scaffolds for nonviral IGF-1 gene delivery in articular cartilage tissue engineering. Gene Ther 2007;14(9):721—32.

[58] Brown B, et al. The basement membrane component of biologic scaffolds derived from extracellular matrix. Tissue Eng 2006;12(3):519—26.

[59] O'Brien FJ, et al. The effect of pore size on cell adhesion in collagen-GAG scaffolds. Biomaterials 2005;26(4):433—41.

[60] Ghosal K, et al. Structural and surface compatibility study of modified electrospun poly(epsilon-caprolactone) (PCL) composites for skin tissue engineering. AAPS PharmSciTech 2017;18(1):72—81.

[61] Yang B, et al. Development of electrically conductive double-network hydrogels via one-step facile strategy for cardiac tissue engineering. Adv Healthc Mater 2016;5(4):474—88.

[62] Yue K, et al. Synthesis, properties, and biomedical applications of gelatin methacryloyl (GelMA) hydrogels. Biomaterials 2015;73:254—71.

[63] Thomas V, Zhang X, Vohra YK. A biomimetic tubular scaffold with spatially designed nanofibers of protein/PDS bio-blends. Biotechnol Bioeng 2009;104(5):1025—33.

[64] Thomas V, et al. Functionally graded electrospun scaffolds with tunable mechanical properties for vascular tissue regeneration. Biomed Mater 2007;2(4):224—32.

[65] Van Vlierberghe S. Crosslinking strategies for porous gelatin scaffolds. J Mater Sci 2016;51(9):4349—57.

[66] Venkatesan J, et al. Alginate composites for bone tissue engineering: a review. Int J Biol Macromol 2015;72:269—81.

[67] Hajiabbas M, et al. Chitosan-gelatin sheets as scaffolds for muscle tissue engineering. Artif Cell Nanomed Biotechnol 2015;43(2):124—32.

[68] Tonsomboon K, Oyen ML. Composite electrospun gelatin fiber-alginate gel scaffolds for mechanically robust tissue engineered cornea. J Mech Behav Biomed Mater 2013;21:185—94.

[69] Shin SR, et al. Carbon-nanotube-embedded hydrogel sheets for engineering cardiac constructs and bioactuators. ACS Nano 2013;7(3):2369—80.

[70] Alvarez-Perez MA, et al. Influence of gelatin cues in PCL electrospun membranes on nerve outgrowth. Biomacromolecules 2010;11(9):2238—46.

[71] Mironi-Harpaz I, et al. Photopolymerization of cell-encapsulating hydrogels: crosslinking efficiency versus cytotoxicity. Acta Biomater 2012;8(5):1838—48.

[72] Taylor SJ, McDonald 3rd JW, Sakiyama-Elbert SE. Controlled release of neurotrophin-3 from fibrin gels for spinal cord injury. J Control Rel 2004;98(2):281—94.

[73] Hardy JG, et al. Into the groove: instructive silk-polypyrrole films with topographical guidance cues direct DRG neurite outgrowth. J Biomater Sci Polym Ed 2015;26(17):1327—42.

[74] Elvin CM, et al. Synthesis and properties of crosslinked recombinant pro-resilin. Nature 2005;437(7061):999—1002.

[75] Kamalesh S, et al. Biocompatibility of electroactive polymers in tissues. J Biomed Mater Res 2000;52(3):467—78.

[76] Ostrovidov S, et al. Gelatin-polyaniline composite nanofibers enhanced excitation-contraction coupling system maturation in myotubes. ACS Appl Mater Interfaces 2017;9(49):42444—58.

[77] Kurtz SM, et al. Advances in the processing, sterilization, and crosslinking of ultra-high molecular weight polyethylene for total joint arthroplasty. Biomaterials 1999;20(18):1659—88.

[78] Grindy SC, et al. Delivery of bupivacaine from UHMWPE and its implications for managing pain after joint arthroplasty. Acta Biomater 2019;93:63—73.

[79] Jie K, et al. Prosthesis design of animal models of periprosthetic joint infection following total knee arthroplasty: a systematic review. PLoS One 2019;14(10):e0223402.

[80] Bracco P, et al. Ultra-high molecular weight polyethylene: influence of the chemical, physical and mechanical properties on the wear behavior. A review. Materials (Basel) 2017;10(7).

[81] Wang K, et al. Synthesis of a novel anti-freezing, non-drying antibacterial hydrogel dressing by one-pot method. Chem Eng J 2019;372:216—25.

[82] Moffat KL, et al. Composite cellularized structures created from an interpenetrating polymer network hydrogel reinforced by a 3D woven scaffold. Macromol Biosci 2018;18(10):e1800140.

[83] Smith AST, et al. Micro- and nano-patterned conductive graphene-PEG hybrid scaffolds for cardiac tissue engineering. Chem Commun (Camb) 2017;53(53):7412—15.

[84] Zhou J, et al. Engineering the heart: evaluation of conductive nanomaterials for improving implant integration and cardiac function. Sci Rep 2014;4:3733.

[85] Kim IL, Mauck RL, Burdick JA. Hydrogel design for cartilage tissue engineering: a case study with hyaluronic acid. Biomaterials 2011;32(34):8771—82.

[86] Park J, et al. Micropatterned conductive hydrogels as multifunctional muscle-mimicking biomaterials: graphene-incorporated hydrogels directly patterned with femtosecond laser ablation. Acta Biomater 2019;97:141—53.

[87] Ahadian S, et al. Moldable elastomeric polyester-carbon nanotube scaffolds for cardiac tissue engineering. Acta Biomater 2017;52:81—91.

[88] del Carmen Ortuño-Costela, M., M. García-López, V. Cerrada, and M.E. Gallardo, iPSCs: a powerful tool for skeletal muscle tissue engineering. J Cell Mol Med, 2019;23(6):3784—3794.

[89] Balint R, Cassidy NJ, Cartmell SH. Conductive polymers: towards a smart biomaterial for tissue engineering. Acta Biomater 2014;10(6):2341—53.

[90] Guo B, Ma PX. Conducting polymers for tissue engineering. Biomacromolecules 2018;19(6):1764—82.

[91] Mawad D, et al. A conducting polymer with enhanced electronic stability applied in cardiac models. Sci Adv 2016;2(11):e1601007.

[92] Kai D, et al. Polypyrrole-contained electrospun conductive nanofibrous membranes for cardiac tissue engineering. J Biomed Mater Res A 2011;99(3):376—85.

[93] Huang J, et al. Electrical stimulation to conductive scaffold promotes axonal regeneration and remyelination in a rat model of large nerve defect. PLoS One 2012;7(6):e39526.

[94] Bjorninen M, et al. Electrically stimulated adipose stem cells on polypyrrole-coated scaffolds for smooth muscle tissue engineering. Ann Biomed Eng 2017;45(4):1015—26.

[95] Gelmi A, et al. Direct mechanical stimulation of stem cells: a beating electromechanically active scaffold for cardiac tissue engineering. Adv Healthc Mater 2016;5(12):1471—80.

[96] Vishnoi T, Kumar A. Conducting cryogel scaffold as a potential biomaterial for cell stimulation and proliferation. J Mater Sci Mater Med 2013;24(2):447—59.

[97] Troitsky VI, Berzina TS, Fontana MP. Langmuir—Blodgett assemblies with patterned conductive polyaniline layers. Mater Sci Eng: C 2002;22(2):239—44.

[98] Yoon S-B, Yoon E-H, Kim K-B. Electrochemical properties of leucoemeraldine, emeraldine, and pernigraniline forms of polyaniline/multi-wall carbon nanotube nanocomposites for supercapacitor applications. J Power Sources 2011;196(24):10791—7.

[99] Huang J, et al. Polyaniline nanofibers: facile synthesis and chemical sensors. J Am Chem Soc 2003;125 (2):314—15.

[100] Zhou G, et al. High-strength single-walled carbon nanotube/permalloy nanoparticle/poly(vinyl alcohol) multifunctional nanocomposite fiber. ACS Nano 2015;9(11):11414—21.

[101] Ku SH, Park CB. Human endothelial cell growth on mussel-inspired nanofiber scaffold for vascular tissue engineering. Biomaterials 2010;31(36):9431—7.

[102] Chen MC, Sun YC, Chen YH. Electrically conductive nanofibers with highly oriented structures and their potential application in skeletal muscle tissue engineering. Acta Biomater 2013;9(3):5562—72.

[103] Chetan DK. Experimental evaluation of fiber orientation based material properties of skeletal muscle in tension. Mol Cell Biomech 2014;11(2):113—28.

[104] Song J, et al. The construction of three-dimensional composite fibrous macrostructures with nanotextures for biomedical applications. Biofabrication 2016;8(3):035009.

[105] Hsiao CW, et al. Electrical coupling of isolated cardiomyocyte clusters grown on aligned conductive nanofibrous meshes for their synchronized beating. Biomaterials 2013;34(4):1063—72.

[106] Mahmoudifard M, et al. The different fate of satellite cells on conductive composite electrospun nanofibers with graphene and graphene oxide nanosheets. Biomed Mater 2016;11(2):025006.

[107] Wang F, et al. Synthesis and characterization of water-soluble polythiophene derivatives for cell imaging. Sci Rep 2015;5:7617.

[108] Ahmed EM. Hydrogel: preparation, characterization, and applications: a review. J Adv Res 2015;6 (2):105—21.

[109] Castilho M, et al. Mechanical behavior of a soft hydrogel reinforced with three-dimensional printed microfibre scaffolds. Sci Rep 2018;8(1):1245.

[110] Drury JL, Mooney DJ. Hydrogels for tissue engineering: scaffold design variables and applications. Biomaterials 2003;24(24):4337—51.

[111] Illeperuma WRK, et al. Fiber-reinforced tough hydrogels. Extreme Mech Lett 2014;1:90−6.

[112] You JO, et al. Nanoengineering the heart: conductive scaffolds enhance connexin 43 expression. Nano Lett 2011;11(9):3643−8.

[113] Wang L, et al. Electrospun conductive nanofibrous scaffolds for engineering cardiac tissue and 3D bioactuators. Acta Biomater 2017;59:68−81.

[114] Zhang M, Guo B. Electroactive 3D scaffolds based on silk fibroin and water-borne polyaniline for skeletal muscle tissue engineering. Macromol Biosci 2017;17(9).

[115] Humpolicek P, et al. Polyaniline cryogels: biocompatibility of novel conducting macroporous material. Sci Rep 2018;8(1):135.

[116] Geddes LA, Roeder R. Criteria for the selection of materials for implanted electrodes. Ann Biomed Eng 2003;31(7):879−90.

[117] Meng X, et al. Novel injectable biomimetic hydrogels with carbon nanofibers and self assembled rosette nanotubes for myocardial applications. J Biomed Mater Res A 2013;101(4):1095−102.

[118] Stout DA, et al. Growth characteristics of different heart cells on novel nanopatch substrate during electrical stimulation. Biomed Mater Eng 2014;24(6):2101−7.

[119] Asiri AM, et al. Greater cardiomyocyte density on aligned compared with random carbon nanofibers in polymer composites. Int J Nanomed 2014;9:5533−9.

[120] Merlo A, et al. Boron nitride nanomaterials: biocompatibility and bio-applications. Biomater Sci 2018;6 (9):2298−311.

[121] Ravichandran R, et al. Gold nanoparticle loaded hybrid nanofibers for cardiogenic differentiation of stem cells for infarcted myocardium regeneration. Macromol Biosci 2014;14(4):515−25.

[122] Eatemadi A, et al. Carbon nanotubes: properties, synthesis, purification, and medical applications. Nanoscale Res Lett 2014;9(1):393.

[123] Brajesh Kumar K. Carbon nanotube: properties and applications. SpringerBriefs Appl Sci Technol 2015;17−37 (9788132220466).

[124] Sajid MI, et al. Carbon nanotubes from synthesis to in vivo biomedical applications. Int J Pharm 2016;501(1-2):278−99.

[125] Kobayashi N, Izumi H, Morimoto Y. Review of toxicity studies of carbon nanotubes. J Occup Health 2017;59(5):394−407.

[126] Foldvari M, Bagonluri M. Carbon nanotubes as functional excipients for nanomedicines: II. Drug delivery and biocompatibility issues. Nanomedicine 2008;4(3):183−200.

[127] Gorain B, et al. Carbon nanotube scaffolds as emerging nanoplatform for myocardial tissue regeneration: a review of recent developments and therapeutic implications. Biomed Pharmacother 2018;104:496−508.

[128] Wu Y, et al. Interwoven aligned conductive nanofiber yarn/hydrogel composite scaffolds for engineered 3D cardiac anisotropy. ACS Nano 2017;11(6):5646−59.

[129] Wu S, et al. Fabrication of aligned nanofiber polymer yarn networks for anisotropic soft tissue scaffolds. ACS Appl Mater Interfaces 2016;8(26):16950−60.

[130] Liu Y, et al. Tuning the conductivity and inner structure of electrospun fibers to promote cardiomyocyte elongation and synchronous beating. Mater Sci Eng C: Mater Biol Appl 2016;69:865−74.

[131] Caspi O, et al. Tissue engineering of vascularized cardiac muscle from human embryonic stem cells. Circ Res 2007;100(2):263−72.

[132] Bhaarathy V, et al. Biologically improved nanofibrous scaffolds for cardiac tissue engineering. Mater Sci Eng C: Mater Biol Appl 2014;44:268−77.

[133] Packer M. The impossible task of developing a new treatment for heart failure. J Card Fail 2002;8(4):193−6.

[134] Pasqualini FS, et al. Traction force microscopy of engineered cardiac tissues. PLoS One 2018;13(3):e0194706.

[135] Cesare G. 10 - Muscle tissue engineering. In: Nukavarapu SP, Freeman JW, Laurencin CT, editors. Regenerative engineering ofof musculoskeletal tissues and interfaces. Woodhead Publishing; 2015. p. 239−68.

[136] Turner NJ, Badylak SF. Regeneration of skeletal muscle. Cell Tissue Res 2012;347(3):759−74.

[137] Corona BT, et al. The promotion of a functional fibrosis in skeletal muscle with volumetric muscle loss injury following the transplantation of muscle-ECM. Biomaterials 2013;34(13):3324−35.

[138] Borselli C, et al. The role of multifunctional delivery scaffold in the ability of cultured myoblasts to promote muscle regeneration. Biomaterials 2011;32(34):8905−14.

[139] Jana S, Levengood SK, Zhang M. Anisotropic materials for skeletal-muscle-tissue engineering. Adv Mater 2016;28(48):10588−612.

[140] MacArthur Clark J. The 3Rs in research: a contemporary approach to replacement, reduction and refinement. Br J Nutr 2018;120(s1):S1—7.

[141] Budhwani KI, et al. Nanofiber and stem cell enabled biomimetic systems and regenerative medicine. J Nanosci Nanotechnol 2016;16(9):8923—34.

[142] Carleton MM, Sefton MV. Injectable and degradable methacrylic acid hydrogel alters macrophage response in skeletal muscle. Biomaterials 2019;223:119477.

[143] Dong R, et al. Biocompatible elastic conductive films significantly enhanced myogenic differentiation of myoblast for skeletal muscle regeneration. Biomacromolecules 2017;18(9):2808—19.

[144] Jana S, et al. Effect of nano- and micro-scale topological features on alignment of muscle cells and commitment of myogenic differentiation. Biofabrication 2014;6(3):035012.

[145] Greco F, et al. Microwrinkled conducting polymer interface for anisotropic multicellular alignment. ACS Appl Mater Interfaces 2013;5(3):573—84.

[146] Fenn SL, Oldinski RA. Visible light crosslinking of methacrylated hyaluronan hydrogels for injectable tissue repair. J Biomed Mater Res B: Appl Biomater 2016;104(6):1229—36.

[147] Jeon O, Alsberg E. Photofunctionalization of alginate hydrogels to promote adhesion and proliferation of human mesenchymal stem cells. Tissue Eng Part A 2013;19(11-12):1424—32.

[148] Hosseinzadeh S, et al. Microfluidic system for synthesis of nanofibrous conductive hydrogel and muscle differentiation. J Biomater Appl 2018;32(7):853—61.

[149] Andukuri A, et al. A hybrid biomimetic nanomatrix composed of electrospun polycaprolactone and bioactive peptide amphiphiles for cardiovascular implants. Acta Biomater 2011;7(1):225—33.

[150] Nichol JW, et al. Cell-laden microengineered gelatin methacrylate hydrogels. Biomaterials 2010;31(21):5536—44.

[151] Stevens KR, et al. Physiological function and transplantation of scaffold-free and vascularized human cardiac muscle tissue. Proc Natl Acad Sci USA 2009;106(39):16568—73.

[152] Williams DF. Specifications for innovative, enabling biomaterials based on the principles of biocompatibility mechanisms. Front Bioeng Biotechnol 2019;7:255.

[153] Poole-Warren LA, Patton AJ. Introduction to biomedical polymers and biocompatibility. In: Poole-Warren L, Martens P, Green R, editors. Biosynthetic polymers for medical applications. Woodhead Publishing; 2016. p. 3—31.

[154] Caló E, Khutoryanskiy VV. Biomedical applications of hydrogels: a review of patents and commercial products. Eur Polym J 2015;65:252—67.

[155] Park J, et al. Micropatterned conductive hydrogels as multifunctional muscle-mimicking biomaterials: graphene-incorporated hydrogels directly patterned with femtosecond laser ablation; 2019 (1878-7568 (Electronic)).

[156] Dixit K, Sinha N. Compressive strength enhancement of carbon nanotube reinforced 13-93B1 bioactive glass scaffolds. J Nanosci Nanotechnol 2019;19(5):2738—46.

[157] Qiu K, Netravali AN. Bacterial cellulose-based membrane-like biodegradable composites using cross-linked and noncross-linked polyvinyl alcohol. J Mater Sci 2012;47(16):6066—75.

[158] Migliaccio L, et al. Evidence of unprecedented high electronic conductivity in mammalian pigment based eumelanin thin films after thermal annealing in vacuum. Front Chem 2019;7:162.

[159] Wood AT, et al. Fiber length and concentration: synergistic effect on mechanical and cellular response in wet-laid poly(lactic acid) fibrous scaffolds. J Biomed Mater Res B: Appl Biomater 2019;107(2):332—41.

[160] Liu R, et al. Low-temperature plasma treatment-assisted layer-by-layer self-assembly for the modification of nanofibrous mats. J Colloid Interface Sci 2019;540:535—43.

[161] Wang M, et al. Cold atmospheric plasma (CAP) surface nanomodified 3D printed polylactic acid (PLA) scaffolds for bone regeneration. Acta Biomater 2016;46:256—65.

SECTION 4

Regenerative Neuroscience

CHAPTER

13

Recent developments and new potentials for neuroregeneration

Sreekanth Sreekumaran[1], Anitha Radhakrishnan[2] and Sanju P. Joy[3]

[1]Department of Biochemistry, University of Kerala, Thiruvananthapuram, India [2]Research and Development Department, Pharmaceutical Corporation (Indian Medicine), Thrissur, India [3]Neurology, National Institute of Mental Health and Neuro-Sciences, Bangalore, India

Damage/degeneration of neuronal tissues results in significant health complications globally, which has a severe impact on the socio-economic status of the sufferers. Adverse neuronal conditions arise from multiple risk factors such as traumatic injury, cardiovascular effects, cancers in the central nervous system (CNS), genetic and epigenetic alterations, and environmental factors. According to the recent report by the Global Burden of Disease, neurological disorders account for the second major cause of death (9.0 million) due to noncommunicable diseases [1]. Generally, the neuronal loss, impaired glial cell function, and neural signaling limit functional restoration of brain function, following brain injuries [2]. The disabilities arising from brain injury due to trauma or stroke, spinal cord injury, degenerative conditions including Parkinson's disease, Huntington's disease, amyotrophic lateral sclerosis, Alzheimer's disease, and multiple sclerosis are increasing at an alarming rate. World Health Organization (WHO) estimates an annual global incidence of 40−80 cases of spinal cord injury per million population. CNS damage causes dysregulated intercellular signaling, cell death, disrupted axonal regeneration, and demyelination that affects the structure and function of the brain or spinal cord [3]. Approximately 40−50 million people currently have Alzheimer's disease and other dementias worldwide [4]. Stroke affects around 15 million people each year worldwide and is one of the main conditions that lead to vascular or Alzheimer's style dementia [5,6]. The conventional treatment strategies for neurodegenerative diseases or neuronal injuries deal with alleviating the symptoms; however, they have limited success in improving the clinical condition. There has been an urge to develop novel and effective strategies for the management of neurodegenerative diseases as WHO predicts neurodegenerative conditions to be the second-most prevalent cause of death after cardiovascular diseases in 20 years [7].

The restoration of neuronal function by regenerative strategies emerges as a promising approach for addressing the long-term disabilities arising from neurological disorders and nerve damage. Neural tissue engineering approaches aim to integrate a favorable microenvironment for the complex intercellular communication in the nervous tissues and restore neuronal functions. Also, the stem cell-based replacement therapies using embryonic stem cells and multipotent stem cells have been reported. Studies have shown that the phenotypic stability and long term viability of these cells are unpredictable, owing to the unique microenvironment of injured brain tissue, which hurdles the regenerative response [8]. Cellular reprogramming technologies such as the generation of induced Pluripotent Stem Cells (iPSCs) and the transdifferentiation of somatic cells to multiple lineages holds excellent potential for neuronal tissue repair and regeneration [9]. Interestingly, the emergence of nanotechnology offers reliable applications of nanostructured materials with tunable physicochemical properties for the delivery of cells or molecules with neuro-regenerative potential [10]. Moreover, the novel approach of utilizing exosomes that encapsulate and transfer microRNAs to modify cellular signaling and promote regeneration is a promising strategy in neuronal engineering [11]. This chapter briefly discusses the current strategies being employed for the repair and the regeneration of neuronal tissues.

13.1 Nervous system, neurodegeneration and regeneration — a nutshell

The central nervous system (CNS) is constituted by the brain and spinal cord whereas the neuronal network that connects the CNS to the organ systems forms the peripheral nervous system (PNS). Neurons and glial cells are the major cell types in the neural tissue. Neurons perform the unique functions of the nervous system by processing and transmitting sensory and motor responses. Approximately, 33—66% of brain mass is constituted by glial cells [12] and 50% of the brain tissue is populated with the star-shaped cells called astrocytes which are the most abundant among the glial cells that have specific roles in maintaining neural homeostasis and supporting neuronal function [13]. The robust immune machinery in the brain is maintained and regulated by microglia. Approximately, 10% of cells in the brain are microglia with differential distribution in various parts of CNS. Relatively more microglial distribution is seen in the hippocampus, basal ganglia, substantia nigra, and olfactory encephalon compared to the cerebellum and brain stem [14]. At the resting stage, microglial cells are highly dynamic and are constantly engaged in immune surveillance of its microenvironment. Following pathological insults, structural and functional changes occur in microglia exhibiting the neurotrophic or neurotoxic effects depending on the severity of injury [15]. Furthermore, the oligodendrocytes are the glial cells responsible for the insulation of neurons. The astrocytes and microglia facilitate the differentiation of oligodendrocytes precursor cells to myelinating oligodendrocytes and is directed by neuronal signals [16]. In general, the myelin sheath exists as a structural extension of the glial plasma membrane forming an insulating sheath around the axons. The coordinated responses initiated by astroglia, microglia and NG2 (polydendrocytes) cells function to reinstate the neuronal homeostasis following neurological pathologies. The increased proliferation of NG2 glial cells is reported during acute brain injury, neuronal defects and neurodegenerative conditions [17]. These are oligodendrocytes progenitor cells uniquely expressing the chondroitin sulfate proteoglycan. Myelin dysfunction in multiple sclerosis is associated with the presence of autoantibodies to NG2 cells. Functional impairment of NG2 cells during neurological disorders explains their significance as mediators of neuronal homeostatic mechanisms. Immunomodulatory responses of NG2 cells are inevitable for preventing neuronal loss and degeneration during pathological conditions [18]. The superficial role of astrocytes in integrating the intricate processes of the nervous system shows that the astrogliosis is fundamental to the understanding of neuropathologic conditions. Studies implicate the beneficial role of astrogliosis in tissue repair and neuronal adaptivity to environmental stress and other insults [19]. Neuronal dysfunction in psychiatric disorders is related to abnormal synaptic connectivity arising from the impaired glutamatergic transmission, which is primarily mediated by astrocytes. It has been reported that astrocytes secrete glypican-6 and glypican-4 that mediate the strengthening of glutamatergic synapses [20]. Brain-derived neurotrophic factor (BDNF), released from microglial cells have been demonstrated to regulate synapses formation and has functional roles in modulating synaptic plasticity [21].

Amyloid plaques and protein aggregates manifested in Alzheimer's disease (AD) trigger the proliferation of pathological phenotype of astrocytes and results in synaptic degeneration. Interaction of reactive astrocytes and activated microglia with $A\beta$ plaques in the AD brain has been reported to have differential effects on neuronal integrity. Neuroinflammatory effects produced by the astroglial response and abnormal neuronal excitability due to downregulation of glutamine synthetase and glutamate transporter-1 in pathological astrocytes, collectively contribute to synaptic dysfunction and neuronal loss [22]. Synapse loss in AD has been correlated with microglia-mediated complement activation and phagocytosis. The process has been identified as a possible target for inhibiting early neuronal loss in AD [23]. Also, the defective neurogenesis in the adult hippocampus has been linked to the progression of AD.

Parkinson's disease (PD) is characterized by the depletion of dopaminergic neurons in the substantia nigra. Abnormal aggregation of alpha-synuclein in presynaptic neuron terminals contributes to the pathogenesis of PD. Impaired dopamine metabolism occurs due to the presence of mutated alpha-synuclein, owing to its regulatory role in the activity of tyrosine hydroxylase, the rate-limiting enzyme in dopamine synthesis [24]. Adult neurogenesis occurs in the subventricular zone (SVZ) of the lateral ventricles of the cerebral cortex and the subgranular zone (SGZ) of the dentate gyrus in the hippocampus of the brain. Albright et al. studied the role of nesting-secreting neuroprogenitor cells in the regeneration of dopaminergic neurons and they identified these cells as a potential source for stem cell therapy for PD [25]. The neuroepithelial stem cell protein or nestin has emerged as an important therapeutic target as reactive astrocytes have been reported to express nestin and other related protein markers of neural stem cells [26]. In addition, the microglia stimulation by aberrant expression of mutated α-synuclein has been demonstrated in many studies and the matrix metalloproteinases (MMPs) released from activated microglia triggers neuroinflammatory responses leading to loss of dopaminergic neurons [27]. Moreover, the involvement of classic complement activation by microglial cells has been shown to eliminate degenerating neurons. Neurogenic stimuli initiated by microglia activated complement cascade can be exploited for regeneration of lost neurons in neurodegenerative conditions [28].

Traumatic brain injury affects the functional integrity of the nervous system and is associated with long term neuronal defects and disruptions in the signaling sequences within the neural network. The damage has been reported to be substantially high in the hippocampal region [29]. The granule cells in the dentate gyrus function to regenerate the neuronal population at the injured site; however, they are subjective to the specific conditions of the tissue microenvironment. Additionally, the therapeutic role of fibroblast growth factor (FGF-2) in CNS disorders has been well documented. Increased expression of FGF-2 stimulates the proliferation of neuroprogenitor cells in the hippocampal region and contributes to neuronal recovery in the post−traumatic brain [30].

Inherent mechanisms of tissue repair and cellular regeneration in the brain are achieved through neuroprogenitor cells distributed in specific locations. Understanding the phenotypes of neuronal cell types from different classes of neural progenitor cells (NPCs) during development is essential for deriving novel regenerative approaches in the management of neurodegenerative diseases [31]. Neurogenesis involves the differentiation of NPCs to mature and functional neuron and is not only confined to the embryonic stage but also extends through the postnatal period to adulthood. NPCs possess excellent self-renewal capacity and can differentiate to functional neurons, astrocytes and oligodendrocytes. Studies have demonstrated the ability of embryonic stem cells or induced pluripotent stem cells to differentiate and generate NPCs in vitro. The primary NPCs or embryonic neural progenitor cells proliferating in the ventricular zone are termed as "Radial Glia" (RG) which arise by the differentiation of neuroepithelial cells associated with the neural tube [32]. The NPCs undergo mitosis and the daughter cells migrate to the subventricular zone to form 'intermediate progenitor cells'(IP cells). The IP cells subsequently divide to produce multiple neurons [33]. Noctor et al. have demonstrated that the radial glial cells from the ventricular zone are superior in translocation and are the possible precursors of astrocytes. Microglia are derived from the hematopoietic stem cells in the yolk sac during early embryogenesis [34]. The NPCs in adult neurogenesis are termed as Type 1 NPCs that are RG like cells and they divide to generate Type 2 cells, which are the precursors of neurons. The differentiation and proliferation of the adult NPCs in the dentate gyrus are associated with the development of cognitive behaviors [35]. Also, the adult NPCs in the subventricular region give rise to glial cells.

Neuronal cell death occurs by necrosis and apoptosis following a pathology or injury. Also, the normal aging process is associated with neuronal death in specific regions of the brain as evident from the decreased adult hippocampal neurogenesis in the aging brain [36]. Apoptotic cell death is associated with the developmental stages of neural precursor cells [37]. Post-mitotic neurons have acquired redundant mechanisms to resist apoptosis, which are inevitable for maintaining the integrity and long term survival of neural tissue whereas these mechanisms are impaired during severe degenerative conditions [38]. Experiments have shown that inhibition of Bax translocation to mitochondria prevents apoptosis in mature neurons and the overexpressed mutant superoxide dismutase gene trigger the apoptotic pathway [39]. In addition, the accumulation of protein aggregates in neurodegenerative conditions disrupts the neuronal homeostasis resulting in the activation of apoptotic pathways. In addition, the energy insufficiency created by cerebral ischemia induces necrotic cell death in neurons.

Despite permanent impairments following traumatic brain injury or stroke, the revival of neuronal cells occurs gradually owing to resident neural stem cells. Neuroplasticity is another contributing factor to neuronal recovery post injury. It refers to the systematic remodeling of neural network in response to various stimuli and results in the functional optimization of the CNS leading to significant impacts on physiology, biochemistry and the regenerative response of injured brain tissues. Although the inherent regenerative potential of neurons is stimulated post injury, the responses to the molecular cues from the pathological microenvironment vary with different types of neurons. The understanding regarding the molecular biology of neurodegeneration and regenerative responses provides a strong background for devising reparative strategies for the management of CNS disorders; however, warrants further research.

13.2 Strategies for neural regeneration and repair

The CNS possesses several endogenous mechanisms to repair or regenerate tissue, but the limited ability of such mechanisms needs interventions to restore neuronal function after injury or disease. Biomaterial assisted tissue engineering approaches has the potential to repair or regenerate damaged neuronal tissue. The 3D architecture of such materials and its ability to deliver growth factors and cells locally enhance regeneration of deteriorating or injured tissue. With the development of nanostructured materials and nanomedicines, novel opportunities arise for the successful treatment strategies for neurodegenerative disorders and CNS injury. The inherent capacity of neuronal cells to deliver regenerative molecules through exosomes has also been adopted in neuronal engineering approaches to achieve efficient delivery of therapeutic molecules to accelerate the regenerative process. Advanced technologies such as three-dimensional bioprinting are promising strategies in developing 3D neuronal tissue constructs for studying pathological conditions, drug screening, and aiding repair and regeneration. This section deals with some of the recent strategies for engineering neurons for therapeutic purposes.

13.2.1 Regenerative biomaterials

The human body is endowed with inherent mechanisms to repair and regenerate tissues/organs in response to injury or disease. The field of tissue engineering attempts to exploit these innate regenerative responses of human tissues by integrating the principles of materials science, engineering and cell biology. A combination of specific cell types, growth factors and three-dimensional matrices converge to simulate an extracellular matrix (ECM) like microenvironment for tissue regeneration and restitution. Scaffold assisted regenerative approaches have documented immense applications in repairing tissues with limited self-regenerative potential as that of neuronal cells [40]. Nerve guide conduits are neural scaffolds employed for bridging gaps in injured neurons and for aiding axonal regeneration for the management of peripheral nerve injury [41]. Also, vascularization plays a vital role in replenishing blood, oxygen and nutrients to regenerating neurons in CNS and PNS [42]. The establishment of neovascularization was observed when implanting neural stem cells on VEGF-releasing PLGA microparticles in an experimental model of stroke cavity [43]. Vascularized neuronal tissue models have been developed by encapsulating neural stem cells and endothelial cells in PEGDA based self-healing glucose-sensitive hydrogel [44]. Neural progenitor cells derived from mesenchymal stem cells seeded on collagen sponges in combination with basic fibroblast growth factor (bFGF) incorporated microspheres were successfully tested in rat stroke model for functional neuronal recovery [45].

Biomimetic scaffolds have been utilized for delivery of neurotrophic factors in spinal cord regeneration as endogenous neuronal regrowth; however, it is hampered by the hostile microenvironment generated by tissue damage [46]. Brain-derived neurotrophic factor is a key neurotrophin that regulates synaptic plasticity, memory, nerve-survival and differentiation of neuronal cells [47]. The primary neuron growth and maturation were observed in response to the controlled release of BDNF from micro-pillared polycaprolactone (MP-PCL) based 3D scaffold. The study demonstrated slow and sustained release of BDNF *in vitro* from MP-PCL indicating the suitability of biomaterial for long-term neurotrophin release after surgical implantation [48]. Interestingly, dual growth factor delivery approach for the release of nerve growth factor (NGF) and glial cell line-derived neurotrophic factor (GDNF) has been demonstrated using bicomponent nanofibrous poly(D, L-lactic acid) (PDLLA) and poly(lactic-*co*-glycolic acid) (PLGA) Liu et al. [49]. These recent findings reflect the advancements in neuronal tissue engineering and guided neuronal tissue regeneration.

The advent of emerging neural engineering methods as a novel therapeutic approach to engineer the complex microenvironment of the nervous system can make significant contributions to alleviate the burden of neurological disease and injury. Self-assembling peptides (SAPs) assemble themselves to form nano-structures that are further organized to form supramolecular 3D architecture suitable for tissue engineering [50]. Reprogramming and functional maturation of neurons from iPSCs have been successfully established using 3D *in vitro* cell culture model fabricated using self-assembled peptide nanofibrous scaffolds (SAPNS) based on the RADA16-I peptide [51]. Synthetic SAPs based *in vitro* 3D model of densely cultured human neuronal stem cells (hNSCs), hNSC-HYDROSAP, was shown to enhance the neurodegeneration in an animal model of spinal cord injury (SCI). The approach succeeded in achieving significant improvement in axon regeneration and motor function recovery after spinal cord injury demonstrating a novel *in vitro* 3D neural network model for a promising regenerative therapy [52]. Despite the increasing wealth of knowledge in neuronal regeneration, the translational avenues for tissue engineering based approaches have not been achieved yet which warrants further investigation.

13.2.2 Neuroregenerative nanomedicine

The blood-brain barrier (BBB) with its tightly interconnected network of endothelial cells (ECs), pericytes and astrocytes, functions as a neurovascular unit to maintain the delicate microenvironment of the brain. The physiological barrier established by the BBB and blood-cerebrospinal fluid barrier (BCSF) restricts the entry of drugs and metabolites into the nervous system. Nanoparticles based drug delivery strategies facilitate controlled drug release and have relatively greater stability, specificity, aiding the transport of both hydrophilic as well as hydrophobic substrates [53]. However, the permeability of BBB varies with the changes in brain microenvironment following injuries, pathology and/or tumor progression. Xinmei et al. have reported the delivery of microRNA-1 (miR-1) in transferrin targeted nanoparticle system to glioblastoma derived stem cell sphere cultures [54]. Owing to the upregulation of transferrin expression on glioblastoma cells, it has emerged as one of the potential targets for drug delivery applications as the drug is targeted through transferrin receptor mediated endocytosis. Significant uptake of transferrin peptide targeted gold nanoparticles incorporated with photodynamic drugs was demonstrated in tumor affected cells [55].

PEG-PCL nanoparticles conjugated with a transferrin mimetic peptide (Lf-NP) developed for the delivery of a neuroprotective peptide NAP, manifested improved hippocampal function in mice models of AD [56]. Copolymeric nanoparticles from polyethylene glycol and polylactic acid loaded with prostaglandin E1 (Nano PGE$_1$) was shown to improve motor neuron function with enhanced angiogenesis and neuronal recovery as observed in rats affected with spinal cord injury (SCI) [57]. Nanomedicine based strategies for addressing the insufficiency of L-DOPA in Parkinson's disease have been reported by various researchers. In a study, poly(lactic-coglycolic acid) (PLGA) nanoparticles were conjugated with L-DOPA for improved permeability through BBB. The particles administered to rat models of Parkinsonism significantly modified PD related pathologies and the rat brains showed better motor neuron activity compared to the control animals [58]. Neuronal growth factor (NGF) and other neurotrophic factors such as glial cell line-derived neurotrophic factor (GDNF) have integral roles in endogenous regenerative mechanisms of the brain. In order to facilitate the permeability of these growth factors through the BBB, nanoparticle-mediated delivery has been successfully demonstrated. Poly (butyl cyanoacrylate) nanoparticles (PBCA) nanoparticles were utilized for the delivery of NGF for the functional recovery of rat models of both PD and amnesia. Significant levels of NGF were found in the brains of the treated animals and a concomitant improvement in the symptoms associated with the degenerative conditions was demonstrated [59]. These studies suggest that the BBB has been tamed for the applications of nanomedicines which offer a pleasant hope to millions of sufferers.

13.2.3 Exosomes assisted neuroregeneration

Exosome mediated delivery of therapeutic molecules has emerged as a novel regenerative strategy to ameliorate tissue injury and degeneration. Exosomes are multivesicular bodies that mediate cell-cell, and cell-ECM communication and regenerative signaling. These nanosized extracellular vesicles loaded with biomolecules such as protein, lipids and RNAs engage in intercellular communication and have been implicated in the modulation of cell differentiation and tissue remodeling [60]. Neurogenesis and neurite remodeling were demonstrated in stroke-affected rats by intravenous injection of MSC derived exosomes [61]. Therapeutic efficacy of exosomes from Adipose derived stem cells (ADSC-exo) has been documented in animal models of Amyotrophic lateral sclerosis (ALS) alleviating the progression of the disease [62]. 3D cultured exosomes (3D-exo) exhibited to improve the memory and cognitive defects in animal models of AD. Also, the exosomes have been isolated from 3D cultures of human umbilical cord derived mesenchymal stem cells (hUMSCs) and characterized for the miRNAs and protein contents to investigate their potential effects on preventing Aβ formation and improving cognitive functions in AD pathology cell model and APP/PS1 transgenic mice respectively [63]. The neuronal release of miRNA containing exosomes was shown to induce and modulate the expression of glutamate transporter in astrocytes, leading to the regulation of synaptic activation [64]. Intranasal administration of MSC derived exosomes loaded with PTEN-siRNA significantly restored neuronal functions in rats with complete spinal cord injury [65].

Owing to their increased permeability through blood brain barrier (BBB), exosomes have emerged as novel tools for the targeted delivery and transport of drugs and other neurotrophic elements in different brain pathologies. Intravenous administration of c (RGDyK)-conjugated curcumin loaded exosomes (cRGD-Exo-cur) was documented to target lesion region of the ischemic brain and effectively suppress inflammatory response and apoptosis [66]. Gold nanoparticle labeled MSC-exo (GNP-labeled MSC-exo) were administrated intranasally in various animal models of brain pathologies including neurovascular (ischemic stroke), neurodegenerative (Parkinson's and Alzheimer's disease), and neuropsychiatric (autism spectrum) disorders to demonstrate the specific ability of the exosome vehicle to migrate and accumulate in diseased regions [67]. These reports suggest the potential application of exosomes in the next-generation neuroregenerative strategies.

13.2.4 3D bioprinting

3D bioprinting is an additive manufacturing process that emerged as a potential regenerative strategy for tissue engineering which relies on computer-assisted manufacture that precisely aligns the cells to form three-dimensional constructs that simulate native microenvironment imparting specific biological function. 3D bioprinting has immense application in the fabrication of tissue models or organoids, replacement of damaged tissue and drug development [68]. Bio-inks, which is the essential component of 3D bioprinting, consist of biomaterial that are specially designed to encapsulate cells or to incorporate growth factors. In the scaffold-free printing methods, 3D tissue constructs are printed using bio-inks that consist of biomaterial and cells, and the scaffold-free approach deposit tissue spheroids in desired patterns to grow into large functional tissue structures [69]. Advanced neural tissue engineering approach relies on bioprinting to fabricate complex 3D neuronal tissue models with specific cell types and environmental cues to understand pathological aspects and regenerative mechanisms of the nervous system to device-specific therapeutic strategies [70]. In vitro generation of 3D brain-like structure has been successfully demonstrated by bioprinting technique using layer by layer printing approach of primary cortical neurons encapsulated in RGD peptide functionalized gellan gum to facilitate the survival and networking of neuronal cells in a complex 3D microenvironment [71]. 3D printed electro conductive scaffold using polyethylene glycol (PEGDA) incorporated with multi-walled carbon nanotubes (MWCNTs) has been documented to exhibit neuronal regenerative potential by enhancing NE-4C neuronal stem cell proliferation and subsequent differentiation [72]. In a scaffold-free approach, the nerve constructs were fabricated from 3D printing spheroids of mesenchymal stem cells from human gingival tissue (GMSCs) which promoted repair and regeneration of facial nerve following the injury in rat model [73]. Similarly, the neuro regenerative therapeutic potential of 3D printed collagen-chitosan implant was documented by improved repair response of nerve fiber as well as the functional recovery in rat model of spinal cord injury [74]. Nerve guide conduits (NGCs) are promising alternatives for nerve graft autografts where the 3D printed porous composite scaffold using polycaprolactone (PCL) and Poly (acrylic acid) was fabricated to form NGCs which displayed immense regenerative potential of the conduits using rat adrenal medulla derived PC12 cells [75]. Even though a handful of reports are being published every year, the field of neuronal 3D bioprinting is still in infancy. The understanding of the basic biology/histomorphometry and regenerative signaling pathways are inevitable for the fabrication of translationally relevant and functionally viable bio-constructs for neuronal regeneration using 3D bioprinting approach.

13.3 Future perspectives

The success of regenerative strategies for the management of neurological disorders largely relies on the effective modulation of the complex microenvironment of the nervous system. The multifactorial aspects of neuronal repair and recovery demand a combinatorial approach integrating the effective delivery of neuro-therapeutics and the implementation of regenerative strategies depending on the type of injury or disease. Induction of a regenerative microenvironment in the injured or diseased brain tissue requires the synergistic interplay of signaling pathways and specific neuronal cell types which is complex and warrants further research. Computational methods for developing predictive models to optimize tissue biomaterial interactions and possible outcomes enable the designing of more efficient regenerative protocols; however, such approaches are still in infancy. The revival of neuronal plasticity is essential for proper behavioral as well as motor responses in post-recovery conditions which require a vivid understanding of the cell biology and signaling. Modulation of neuronal network connectivity and synaptic plasticity are inevitable for the normal functioning of the CNS and tissue engineering approaches focusing on such aspects are yet to be accomplished. The future inputs in the regenerative protocols should inculcate specific topographical or chemical cues for directed tissue genesis and improvement of cognitive functions of the recovered neuronal tissue. The future implications of conjugating electrical stimulation strategies with cell-based therapies could augment the success of current regenerative therapies. Clinical translation of the developed technologies for neuronal regeneration is still in infancy and the scientific world is striving to translate the findings to the therapeutic arena. While the scope of regenerative medicine and stem cell therapies holds great promise for the management of debilitating neurological disorders, further advancements in the field of regenerative neurology warrant focus on its extrapolation to successful clinical trials and integration into routine therapeutic systems.

References

[1] Feigin VL, Nichols E, Alam T, Bannick MS, Beghi E, Blake N, et al. Global, regional, and national burden of neurological disorders, 1990−2016: a systematic analysis for the Global Burden of Disease Study 2016. Lancet Neurol 2019;18(5):459−80. Available from: https://doi.org/10.1016/S1474-4422(18)30499-X.
[2] Struzyna LA, Katiyar K, Cullen DK. Living scaffolds for neuroregeneration. Curr Opsolid State Mater Sci 2014;18(6):308−18. Available from: https://doi.org/10.1016/j.cossms.2014.07.004.

[3] Egawa N, Lok J, Washida K, Arai K. Mechanisms of axonal damage and repair after central nervous system injury. Transl Stroke Res 2017;8(1):14−21. Available from: https://doi.org/10.1007/s12975-016-0495-1.

[4] Nichols E, Szoeke CEI, Vollset SE, Abbasi N, Abd-Allah F, Abdela J, et al. Global, regional, and national burden of Alzheimer's disease and other dementias, 1990−2016: a systematic analysis for the Global Burden of Disease Study 2016. Lancet Neurol 2019;18(1):88−106. Available from: https://doi.org/10.1016/S1474-4422 (18)30403-4.

[5] Savva GM, Stephan BCM. Epidemiological studies of the effect of stroke on incident dementia: a systematic review. Stroke 2010;41(1). Available from: https://doi.org/10.1161/STROKEAHA.109.559880.

[6] Vijayan M, Reddy PH. Stroke, vascular dementia, and Alzheimer's disease: molecular links. J Alzheimer'S Dis 2016;54(2):427−43. Available from: https://doi.org/10.3233/JAD-160527.

[7] Durães F, Pinto M, Sousa E. Old drugs as new treatments for neurodegenerative diseases. Pharmaceuticals 2018;11(2):1−21. Available from: https://doi.org/10.3390/ph11020044.

[8] Kim SU, de Vellis J. Stem cell-based cell therapy in neurological diseases: a review. J Neurosci Res 2009;87 (10):2183−200. Available from: https://doi.org/10.1002/jnr.22054.

[9] Fang L, Wazan LE, Tan C, Nguyen T, Hung SSC, Hewitt AW, et al. Potentials of cellular reprogramming as a novel strategy for neuroregeneration Front Cell Neurosci 2018;12:1−10November. Available from: https://doi.org/10.3389/fncel.2018.00460.

[10] Teleanu RI, Gherasim O, Gherasim TG, Grumezescu V, Grumezescu AM, Teleanu DM. Nanomaterial-based approaches for neural regeneration. Pharmaceutics 2019;11(6):1−22. Available from: https://doi.org/10.3390/pharmaceutics11060266.

[11] Reza-Zaldivar EE, Hernández-Sapiéns MA, Minjarez B, Gutiérrez-Mercado YK, Márquez-Aguirre AL, Canales-Aguirre AA. Potential effects of MSC-derived exosomes in neuroplasticity in Alzheimer's disease Front Cell Neurosci 2018;12:1−16September. Available from: https://doi.org/10.3389/fncel.2018.00317.

[12] Jäkel S, Dimou L. Glial cells and their function in the adult brain: a journey through the history of their ablation Front Cell Neurosci 2017;Frontiers Research Foundation. Available from: https://doi.org/10.3389/fncel.2017.00024.

[13] Xu JP, Zhao J, Li S. Roles of NG2 glial cells in diseases of the central nervous system Neurosci Bull 2011; Springer. Available from: https://doi.org/10.1007/s12264-011-1838-2.

[14] Alibhai JD, Diack AB, Manson JC. Unravelling the glial response in the pathogenesis of Alzheimer's disease. FASEB J 2018;32(11):5766−77. Available from: https://doi.org/10.1096/fj.201801360R.

[15] Nimmerjahn A, Kirchhoff F, Helmchen F. Neuroscience: resting microglial cells are highly dynamic surveillants of brain parenchyma in vivo. Science 2005;308(5726):1314−18. Available from: https://doi.org/10.1126/science.1110647.

[16] Simons M, Nave KA. Oligodendrocytes: myelination and axonal support Cold Spring Harb Perspect Biol 2016;Cold Spring Harbor Laboratory Press. Available from: https://doi.org/10.1101/cshperspect.a020479.

[17] Kang SH, Fukaya M, Yang JK, Rothstein JD, Bergles DE. NG2 + CNS glial progenitors remain committed to the oligodendrocyte lineage in postnatal life and following neurodegeneration. Neuron 2010;68(4):668−81. Available from: https://doi.org/10.1016/j.neuron.2010.09.009.

[18] Nakano M, Tamura Y, Yamato M, Kume S, Eguchi A, Takata K, et al. NG2 glial cells regulate neuroimmunological responses to maintain neuronal function and survival. Sci Rep 2017;7(1):1−15. Available from: https://doi.org/10.1038/srep42041.

[19] Sofroniew MV. Astrogliosis. Cold Spring Harb Perspect Biol 2015;7(2):a020420. Available from: https://doi.org/10.1101/cshperspect.a020420.

[20] Allen NJ, Bennett ML, Foo LC, Wang GX, Chakraborty C, Smith SJ, et al. Astrocyte glypicans 4 and 6 promote formation of excitatory synapses via GluA1 AMPA receptors. Nature 2012;486(7403):410−14. Available from: https://doi.org/10.1038/nature11059.

[21] Parkhurst CN, Yang G, Ninan I, Savas JN, Yates JR, Lafaille JJ, et al. Microglia promote learning-dependent synapse formation through brain-derived neurotrophic factor. Cell 2013;155(7):1596−609. Available from: https://doi.org/10.1016/j.cell.2013.11.030.

[22] Zumkehr J, Rodriguez-Ortiz CJ, Cheng D, Kieu Z, Wai T, Hawkins C, et al. Ceftriaxone ameliorates tau pathology and cognitive decline via restoration of glial glutamate transporter in a mouse model of Alzheimer's disease. Neurobiol Aging 2015;36(7):2260−71. Available from: https://doi.org/10.1016/j.neurobiolaging.2015.04.005.

[23] Shi Q, Chowdhury S, Ma R, Le KX, Hong S, Caldarone BJ, et al. Complement C3 deficiency protects against neurodegeneration in aged plaque-rich APP/PS1 mice. Sci Transl Med 2017;9(392). Available from: https://doi.org/10.1126/scitranslmed.aaf6295.

[24] Peng XM, Tehranian R, Dietrich P, Stefanis L, Perez RG. α-Synuclein activation of protein phosphatase 2A reduces tyrosine hydroxylase phosphorylation in dopaminergic cells. J Cell Sci 2005;118(15):3523−30. Available from: https://doi.org/10.1242/jcs.02481.

[25] Albright JE, Stojkovska I, Rahman AA, Brown CJ, Morrison BE. Nestin-positive/SOX2-negative cells mediate adult neurogenesis of nigral dopaminergic neurons in mice Neurosci Lett 2016;615:50−4February. Available from: https://doi.org/10.1016/j.neulet.2016.01.019.

[26] Choudhury GR, Ding S. Reactive astrocytes and therapeutic potential in focal ischemic stroke Neurobiol Dis 2016;Academic Press Inc. Available from: https://doi.org/10.1016/j.nbd.2015.05.003.

[27] Lee E-J, Woo M-S, Moon P-G, Baek M-C, Choi I-Y, Kim W-K, et al. α-Synuclein activates microglia by inducing the expressions of matrix metalloproteinases and the subsequent activation of protease-activated receptor-1. J Immunol 2010;185(1):615−23. Available from: https://doi.org/10.4049/jimmunol.0903480.

[28] Ekdahl CT. Microglial activation-tuning and pruning adult neurogenesis Front Pharmacol 2012;3Mar. Available from: https://doi.org/10.3389/fphar.2012.00041.

[29] Ngwenya LB, Danzer SC. Impact of traumatic brain injury on neurogenesis Front Neurosci 2019;13:1014Jan. Available from: https://doi.org/10.3389/fnins.2018.01014.

[30] Yoshimura S, Takagi Y, Harada J, Teramoto T, Thomas SS, Waeber C, et al. FGF-2 regulation of neurogenesis in adult hippocampus after brain injury. Proc Natl Acad Sci USA 2001;98(10):5874−9. Available from: https://doi.org/10.1073/pnas.101034998.

[31] Sailor KA, Ming GL, Song H. Neurogenesis as a potential therapeutic strategy for neurodegenerative diseases Expert Opin Biol Ther 2006;Taylor & Francis. Available from: https://doi.org/10.1517/14712598.6.9.879.

[32] Morest DK, Silver J. Precursors of neurons, neuroglia, and ependymal cells in the CNS: what are they? Where are they from? How do they get where they are going? GLIA 2003;43(1):6−18. Available from: https://doi.org/10.1002/glia.10238.

[33] Martínez-Cerdeño V, Noctor SC. Neural progenitor cell terminology Front Neuroanatomy 2018;Frontiers Media S.A. Available from: https://doi.org/10.3389/fnana.2018.00104.

[34] Ginhoux F, Lim S, Hoeffel G, Low D, Huber T. Origin and differentiation of microglia. Front Cell Neurosci 2013;. Available from: https://doi.org/10.3389/fncel.2013.00045.

[35] Song H, Daniel AB, Bond AM, Ming G l. Radial glial cells in the adult dentate gyrus: what are they and where do they come from? F1000Research 2018;Faculty of 1000 Ltd. Available from: https://doi.org/10.12688/f1000research.12684.1.

[36] Boldrini M, Camille AF, Tartt AN, Simeon LR, Pavlova I, Poposka V, et al. Human hippocampal neurogenesis persists throughout aging Cell Stem Cell 2018;22(4):589−99e5. Available from: https://doi.org/10.1016/j.stem.2018.03.015.

[37] Oppenheim RW. Cell death during development of the nervous system. Annu Rev Neurosci 1991;14(1):453−501. Available from: https://doi.org/10.1146/annurev.ne.14.030191.002321.

[38] Kole AJ, Annis RP, Deshmukh M. Mature neurons: equipped for survival Cell Death & Dis 2013;Nature Publishing Group. Available from: https://doi.org/10.1038/cddis.2013.220.

[39] Guégan C, Vila M, Rosoklija G, Hays AP, Przedborski S. Recruitment of the mitochondrial-dependent apoptotic pathway in amyotrophic lateral sclerosis. J Neurosci − − − − − −: Off J Soc Neuroscience 2001;21(17):6569−76. Available from: http://www.ncbi.nlm.nih.gov/pubmed/11517246.

[40] Kornev VA, Grebenik EA, Solovieva AB, Dmitriev RI, Timashev PS. Hydrogel-assisted neuroregeneration approaches towards brain injury therapy: a state-of-the-art review Computational Struct Biotechnol J 2018; Elsevier B.V. Available from: https://doi.org/10.1016/j.csbj.2018.10.011.

[41] Muheremu A, Ao Q. Past, present, and future of nerve conduits in the treatment of peripheral nerve injury Biomed Res Int 2015;2015. Available from: https://doi.org/10.1155/2015/237507.

[42] Liu Y, Hsu SH. Biomaterials and neural regeneration. Neural Regener Res 2020;15(7):1243−4. Available from: https://doi.org/10.4103/1673-5374.272573.

[43] Bible E, Qutachi O, Chau DYS, Alexander MR, Shakesheff KM, Modo M. Neo-vascularization of the stroke cavity by implantation of human neural stem cells on VEGF-releasing PLGA microparticles. Biomaterials 2012;33(30):7435−46. Available from: https://doi.org/10.1016/j.biomaterials.2012.06.085.

[44] Tseng TC, Hsieh FY, Theato P, Wei Y, Hsu S h. Glucose-sensitive self-healing hydrogel as sacrificial materials to fabricate vascularized constructs. Biomaterials 2017;133(July):20—8. Available from: https://doi.org/10.1016/j.biomaterials.2017.04.008.

[45] Matsuse D, Kitada M, Ogura F, Wakao S, Kohama M, Kira J-i, et al. Combined transplantation of bone marrow stromal cell-derived neural progenitor cells with a collagen sponge and basic fibroblast growth factor releasing microspheres enhances recovery after cerebral ischemia in rats. Tissue Eng Part A 2011;17 (15—16):1993—2004. Available from: https://doi.org/10.1089/ten.tea.2010.0585.

[46] Shrestha B, Coykendall K, Li Y, Moon A, Priyadarshani P, Yao L. Repair of injured spinal cord using biomaterial scaffolds and stem cells Stem Cell Res Ther 2014;BioMed Central Ltd. Available from: https://doi.org/10.1186/scrt480.

[47] Miranda M, Morici JF, Zanoni MB, Bekinschtein P. Brain-derived neurotrophic factor: a key molecule for memory in the healthy and the pathological brain Front Cell Neurosci 2019;13:1—25August. Available from: https://doi.org/10.3389/fncel.2019.00363.

[48] Limongi T, Rocchi A, Cesca F, Tan H, Miele E, Giugni A, et al. Delivery of brain-derived neurotrophic factor by 3D biocompatible polymeric scaffolds for neural tissue engineering and neuronal regeneration. Mol Neurobiol 2018;55(12):8788—98. Available from: https://doi.org/10.1007/s12035-018-1022-z.

[49] Liu C, Wang C, Zhao Q, Li X, Xu F, Yao X, et al. Incorporation and release of dual growth factors for nerve tissue engineering using nanofibrous bicomponent scaffolds. Biomed Mater (Bristol) 2018;13(4). Available from: https://doi.org/10.1088/1748-605X/aab693.

[50] Koutsopoulos S. Self-assembling peptide nanofiber hydrogels in tissue engineering and regenerative medicine: progress, design guidelines, and applications. J Biomed Mater Res - Part A 2016;104(4):1002—16. Available from: https://doi.org/10.1002/jbm.a.35638.

[51] Francis NL, Bennett NK, Halikere A, Pang ZP, Moghe PV. Self-assembling peptide nanofiber scaffolds for 3-D reprogramming and transplantation of human pluripotent stem cell-derived neurons. ACS Biomater Sci Eng 2016;2(6):1030—8. Available from: https://doi.org/10.1021/acsbiomaterials.6b00156.

[52] Marchini A, Raspa A, Pugliese R, Malek MAE, Pastori V, Lecchi M, et al. Multifunctionalized hydrogels foster HNSC maturation in 3D cultures and neural regeneration in spinal cord injuries. Proc Natl Acad Sci USA 2019;116(15):7483—92. Available from: https://doi.org/10.1073/pnas.1818392116.

[53] Soni S, Ruhela RK, Medhi B. Nanomedicine in central nervous system (CNS) disorders: a present and future prospective. Adv Pharm Bull 2016;6(3):319—35. Available from: https://doi.org/10.15171/apb.2016.044.

[54] Wang X, Huang X, Yang Z, Gallego-Perez D, Ma J, Zhao X, et al. Targeted delivery of tumor suppressor microRNA-1 by transferrin- conjugated lipopolyplex nanoparticles to patient-derived glioblastoma stem cells. Curr Pharm Biotechnol 2014;15(9):839—46. Available from: https://doi.org/10.2174/1389201015666141031105234.

[55] Dixit S, Novak T, Miller K, Zhu Y, Kenney ME, Broome AM. Transferrin receptor-targeted theranostic gold nanoparticles for photosensitizer delivery in brain tumors. Nanoscale 2015;7(5):1782—90. Available from: https://doi.org/10.1039/c4nr04853a.

[56] Liu Z, Jiang M, Kang T, Miao D, Gu G, Song Q, et al. Lactoferrin-modified PEG-Co-PCL nanoparticles for enhanced brain delivery of NAP peptide following intranasal administration. Biomaterials 2013;34 (15):3870—81. Available from: https://doi.org/10.1016/j.biomaterials.2013.02.003.

[57] Takenaga M, Ishihara T, Ohta Y, Tokura Y, Hamaguchi A, Igarashi R, et al. Nano PGE1 promoted the recovery from spinal cord injury-induced motor dysfunction through its accumulation and sustained release. J Controlled Release 2010;148(2):249—54. Available from: https://doi.org/10.1016/j.jconrel.2010.08.003.

[58] Gambaryan PY, Kondrasheva IG, Severin ES, Guseva AA, Kamensky AA. Increasing the efficiency of parkinson's disease treatment using a poly(lactic-co-glycolic acid) (PLGA) based L-DOPA delivery system. Exp Neurobiol 2014;23(3):246. Available from: https://doi.org/10.5607/en.2014.23.3.246.

[59] Kurakhmaeva KB, Djindjikhashvili IA, Petrov VE, Balabanyan VU, Voronina TA, Trofimov SS, et al. Brain targeting of nerve growth factor using poly(butyl cyanoacrylate) nanoparticles. J Drug Target 2009;17 (8):564—74. Available from: https://doi.org/10.1080/10611860903112842.

[60] Caruso Bavisotto C, Scalia F, Gammazza AM, Carlisi D, Bucchieri F, de Macario EC, et al. Extracellular vesicle-mediated cell—cell communication in the nervous system: focus on neurological diseases. Int J Mol Sci 2019;20(2):434. Available from: https://doi.org/10.3390/ijms20020434.

[61] Xin H, Li Y, Cui Y, Yang JJ, Zhang ZG, Chopp M. Systemic administration of exosomes released from mesenchymal stromal cells promote functional recovery and neurovascular plasticity after stroke in rats. J Cereb

Blood Flow Metab: Off J Int Soc Cereb Blood Flow Metabolism 2013;33(11):1711−15. Available from: https://doi.org/10.1038/jcbfm.2013.152.

[62] Lee M, Ban JJ, Kim KY, Jeon GS, Im W, Sung JJ, et al. Adipose-derived stem cell exosomes alleviate pathology of amyotrophic lateral sclerosis in vitro. Biochem Biophys Res Commun 2016;479(3):434−9. Available from: https://doi.org/10.1016/j.bbrc.2016.09.069.

[63] Yang L, Zhai Y, Hao Y, Zhu Z, Cheng G. The regulatory functionality of exosomes derived from HUMSCs in 3D culture for alzheimer's disease therapy. Small 2020;16(3):1906273. Available from: https://doi.org/10.1002/smll.201906273.

[64] Morel L, Regan M, Higashimori H, Ng SK, Esau C, Vidensky S, et al. Neuronal exosomal mirna-dependent translational regulation of astroglial glutamate transporter Glt1. J Biol Chem 2013;288(10):7105−16. Available from: https://doi.org/10.1074/jbc.M112.410944.

[65] Guo S, Perets N, Betzer O, Ben-Shaul S, Sheinin A, Michaelevski I, et al. Intranasal delivery of mesenchymal stem cell derived exosomes loaded with phosphatase and tensin homolog SiRNA repairs complete spinal cord injury. ACS Nano 2019;13(9):10015−28. Available from: https://doi.org/10.1021/acsnano.9b01892.

[66] Tian T, Zhang HX, He CP, Fan S, Zhu YL, Qi C, et al. Surface functionalized exosomes as targeted drug delivery vehicles for cerebral ischemia therapy. Biomaterials 2018;150:137−49. Available from: https://doi.org/10.1016/j.biomaterials.2017.10.012.

[67] Perets N, Betzer O, Shapira R, Brenstein S, Angel A, Sadan T, et al. Golden exosomes selectively target brain pathologies in neurodegenerative and neurodevelopmental disorders. Nano Lett 2019;19(6):3422−31. Available from: https://doi.org/10.1021/acs.nanolett.8b04148.

[68] Skardal A, Atala A. Biomaterials for integration with 3-D bioprinting. Ann Biomed Eng 2015;43(3):730−46. Available from: https://doi.org/10.1007/s10439-014-1207-1.

[69] Gopinathan J, Noh I. Recent trends in bioinks for 3D printing. Biomater Res 2018;22(1):1−15. Available from: https://doi.org/10.1186/s40824-018-0122-1.

[70] Zhuang P, Sun AX, An J, Chua CK, Chew SY. 3D neural tissue models: from spheroids to bioprinting. Biomaterials 2018;154:113−33. Available from: https://doi.org/10.1016/j.biomaterials.2017.10.002.

[71] Lozano R, Stevens L, Thompson BC, Gilmore KJ, Gorkin R, Stewart EM, et al. 3D printing of layered brain-like structures using peptide modified gellan gum substrates. Biomaterials 2015;67:264−73. Available from: https://doi.org/10.1016/j.biomaterials.2015.07.022.

[72] Lee S-J, Zhu W, Nowicki M, Lee G, Heo DN, Kim J, et al. 3D printing nano conductive multi-walled carbon nanotube scaffolds for nerve regeneration. J Neural Eng 2018;15:016018. Available from: https://doi.org/10.1088/1741-2552/aa95a5.

[73] Zhang Q, Phuong DN, Shi S, Burrell JC, Cullen DK, Le AD. 3D bio-printed scaffold-free nerve constructs with human gingiva-derived mesenchymal stem cells promote rat facial nerve regeneration. Sci Rep 2018;8 (1):1−11. Available from: https://doi.org/10.1038/s41598-018-24888-w.

[74] Sun Y, Yang C, Zhu X, Wang JJ, Liu XY, Yang XP, et al. 3D printing collagen/chitosan scaffold ameliorated axon regeneration and neurological recovery after spinal cord injury. J Biomed Mater Res - Part A 2019;107 (9):1898−908. Available from: https://doi.org/10.1002/jbm.a.36675.

[75] Vijayavenkataraman S, Thaharah S, Zhang S, Lu WF, Fuh JYH. Electrohydrodynamic jet 3D-printed PCL/PAA conductive scaffolds with tunable biodegradability as nerve guide conduits (NGCs) for peripheral nerve injury repair. Mater Des 2019;162:171−84. Available from: https://doi.org/10.1016/j.matdes.2018.11.044.

Respiratory Research

Lung disease and repair — Is regeneration the answer?

S.S. Pradeep Kumar and A. Maya Nandkumar

Division of Microbial Technology, Sree Chitra Tirunal Institute for Medical Sciences & Technology Thiruvananthapuram, India

14.1 Embryonic development

In humans the respiratory system does not carry out its chief physiological function of gas exchange until after birth. But the respiratory system starts forming precociously early by 28 days of intra uterine life. Lung development has three phases: (1) specification of lung progenitors, (2) Branching morphogenesis, (3) Sacculation & alveolarization. For [1] specification-initiation, lung endoderm is developed from ventral side of the fore gut endoderm and in this phase NK2 homeobox1 is expressed. Then a endoderm patterning in the homeobox1 begins with expression of Sox 2 and signals stage [2]. The cells expressing Sox 9 and Id2 are multipotent and generate alveolar Type I & II cells and lead to development of functional alveolar gas exchange units. With the development entering phase 3, Tumor protein P63, Keratin (krt) and podoplanin (Pdpn) markers for basal cells responsible for development of trachea and bronchi gets activated. Sox 2 is also responsible for the development of pulmonary cells which secrete neuroendocrines, responsible for post-natal immune response. Development of the airway epithelium is guided by Notch signaling which control the ciliated and non-ciliated cell development along with the secretary cells of the airways [3–5].

The lung has an elegant architecture carefully designed for optimal performance of its functions. It starts with the external nares leading to the trachea which branches into the right and left bronchi. Each bronchi divides into bronchioles which end in saccules called alveoli. The trachea, the bronchi are made up of epithelium consisting of ciliated cells, goblet cells, club cells undifferentiated goblet cells. It also contains the neuroendocrine cells located at the branching junctions in the bronchioles. The alveoli are made up of the Type I and Type II alveolar epithelial cells vascular endothelial cells, fibroblast and dendritic cells, etc.

The alveoli are the functional units of the lung where gas exchange takes place between the blood in the pulmonary circulation and inspired air in the lungs. In man lung development begins in the intrauterine life itself. Birth in humans occurs when lung development is at the saccular stage which arises from the foregut endoderm and consists of the differentiated alveolar type I epithelial cells (AECI) and alveolar type II epithelial cells(AECII), developing from common bipotent progenitor cells. Mechanical forces, Fibroblast growth factor are the major regulators of this differentiation. In this course of development AECII gain and retain stemness to propel the re-epithelialization if necessary on injury to the alveolar Type I pneumocytes.

The other cell types, are the endothelial cells forming from the lung endoderm, fibroblast form lung mesoderm. In rat lung fibroblasts, retinoic acid was essential for septation and elastin production [6]. The other cell types having a role in saccular development are the myofibroblasts which are formed by downstream signaling of the sonic hedgehog pathway and mobilization of Gli-1. Platelet derived growth factor A (PDGF-A) produced by the alveolar epithelial cells along with its receptor PDGFR- on the mesenchyme ensure ECM synthesis and maturation of the alveoli.

14.2 Stem cells in the lung

As our understanding the cellular architecture in the lung increased it became clear that lung indeed had various stem cell populations which could make regenerative therapy a reality. To meet the regional functionalities of the lung and conducting airways the epithelia within each domain are composed of distinct cell types. By virtue of the very function of respiration the lungs are continuously exposed to environmental insults causing injury warranting the need for re-epithelialization. A variety of stem/progenitor cells with functional specificity are responsible for both repair and general tissue homeostasis. As detailed in Fig. 14.1 the conducting airways consist of progenitor cells such as basal cells in the proximal airway, alveolar type II cells, club cells within neuroepithelial bodies (NEBs) or Bronchoalveolar duct junctions (BADJ), un-identified cells in submucosal glands (SMGs) etc. Several unique niches for specific progenitor cells are also being discovered. A subset of basal, secretory, and mucous cells in the SMGs of the proximal airways, variant club cells in the bronchioles, bronchoalveolar stem cells (BASCs) in bronchoalveolar-duct junctions (BADJ), and a subset of AEC II cells in alveolar space have been suggested as region-specific epithelial stem/progenitor cell populations in the adult lung of both mice and humans. The Trp63 + basal cells controlled by the Notch signaling pathway are responsible for the differentiation into secretory and multi-ciliated epithelial cells [7]. A rare population of progenitor cells called Lineage negative epithelial progenitor cells (LNEP) migrate from airways to the inflamed interstitium to effect repair [8,9].

Prof. Edward E. Morrisey, PhD, a professor of Cell and Developmental Biology, published in *Nature* about a population of stem cells resident in the human and mouse lung which get activated only on injury to regenerate the damaged tissue — an alveolar epithelial progenitor (AEP) which is embedded in Alveolar type II pneumocytes. They exhibit a conserved cell surface protein called TM4SF1 and using this marker they isolated these cells and succeeded in developing three-dimensional (3D) lung organoids [10] and could

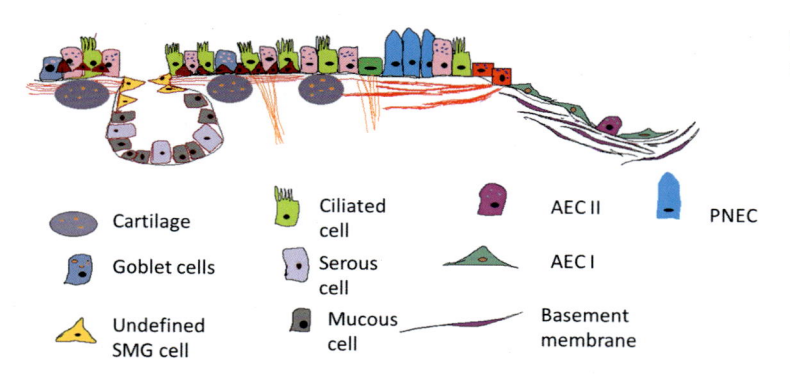

FIGURE 14.1 **FIGURE 14.1** Potential stem cells in respiratory tract.

Cartilage

Goblet cells

Undefined SMG cell

Ciliated cell

Serous cell

Mucous cell

AEC II

AEC I

Basement membrane

PNEC

promote lung regeneration. Another study showed that Distal airway stem cells expressing Trp63 and Keratin 5(Krt5) undergo proliferative expansion in response to influenza — H1N1 virus infection and damage in mice and develop into nascent alveoli at sites of interstitial lung inflammation [11] (Fig. 14.1).

In the distal airways Bronchoalveolar stem cells (BASC) repopulate the terminal bronchiolar and alveolar epithelia on injury and the Wnt signaling pathway plays a key role with GATA binding protein 6 (Gata6) a transcription factor in embryonic lung development. These discoveries show that these developmentally conserved pathways are also operational in regeneration [1]. With the discovery of induced pluripotent stem cells (iPSC) another class of stem cells were made available to study possibilities of lung regeneration. Recent publications show that iPSCs can be differentiated into lung epithelial cell lineages like AT2 cells which produce surfactant protein C and secretoglobulin Family 1 A secretory cells [1,12] indicating that regenerative therapy would become a reality in the near future.

14.3 Epithelial—mesenchymal interactions

About seventy years ago Dr Dorothy Rudnick demonstrated the significance of epithelial—mesenchymal interactions in lung development when she removed the mesenchyme from chick lung and demonstrated arrested lung development [13]. Lung epithelial cells secrete several growth factors and cytokines which are critical in development and maturation: PDGF which provides signals to myofibroblast and aids alveolar septation, which would later be responsible for alveolar wall thinning for development of the gas exchange surfaces in the alveoli [14] Any aberrant crosstalk between the epithelial and mesenchymal cells would lead to serious abnormalities in development. In fact airway remodeling is an important aspect of pathogenesis of many lung diseases.

In culture airway epithelial cells rapidly lose their properties and undergo cell death. But there are various studies including from our own group which has shown that presence of alveolar fibroblast help not only in retaining epithelia specific properties like surfactant protein B & C synthesis but also prolongs the duration of culture [15]. Mesenchymal cells play a critical role in regulating repair after injury. The proximal airways have the basal stem cells which are able to differentiate into secretory, ciliated and neuroendocrine cells. The other cell types of the proximal airway epithelium also have cellular plasticity and act as progenitor cells and contribute to alveolar homeostasis. These factors point to the fact that impaired epithelial—mesenchymal interactions and imbalance would result in alveolar disease specifically fibrosis. The role of retinoic acid in surfactant production by AECII cells [16] and the significance of cell—cell interactions and role of fibroblasts has been well elucidated using three dimensional *in vitro* models in alveolar -like architecture. In these studies co-culture with fibroblasts helped in maintaining alveolar specific functions and in prevention of senescence for the duration of culture in primary rat AECII cells *in vitro* [15,16]. The studies in the developing mouse lung show extensive remodeling by the mesenchymal cells in close association with alveolar epithelial cells depositing collagen and elastin network for alveolar maturation [17].

14.4 Role of mechanical forces in lung architecture

By the 24 week of gestation the distal airways dilate and start forming the air sacs which are responsible for formation of the gas exchange regions of the lung populated by the Type I pneumocytes. During this period its the alveolar progenitor cells which differentiate to form the type I pneumocytes. Glucocorticoids, Retinoic acid, parathyroid hormone, fibroblast growth factor, insulin like growth factor 1, etc. play a central role in lung maturation. At the same time it was observed that fetal breathing movements and inhalation of amniotic fluid resulted in mechanical forces which aid lung development and in fact regulate their differentiation into type I or type II cells.

The influence of biomechanical forces in lung development were difficult to study *in vivo, in vitro* experiments have amply proved this point and mouse studies have shown the dynamic ECM remodeling taking place in the developing lung [18]. The work of Jiao Li et al. has demonstrated a sophisticated development program controlled by mechanical forces and growth factor mediated signaling resulting in actin remodeling and apically enriched myosin strengthening to be pivotal factors in lung distal architecture development. These factors also are therefore critical in lung regeneration [19].

14.5 Lung regeneration in disease

The end stage lung diseases like Interstitial pulmonary fibrosis (IPF), Chronic obstructive pulmonary disease (COPD), Cystic fibrosis, etc. are the third leading cause of death worldwide. Clinically COPD is characterized by progressive loss of lung function accompanied by airflow obstruction. This develops from chronic inflammation, along with bronchitis and small airway disease. According to WHO the only lifesaving therapy is lung transplant. This is severely affected by lack of donor lungs and problems associated with lung transplantation. The option being regeneration of the damaged lung. The key here is the understanding of the disease development mechanism for a possible cure through regenerative technologies.

Hoffman et al. [20] showed that in the diseased lung structural fibroblasts differentiated into myofibroblast which secreted amounts of ECM proteins — collagen III, IV and V, elastin and lysyl oxidase the enzyme which plays a central role in cross linking structural fibers of the lung [21]. The other observation was a gradual decrease in population of AEC2. On the other hand a healthy adult lung is mostly quiescent but recent studies show that they have a remarkable capacity to regenerate after injury [22]. A reduction in AEC2 cells is thus a clear indication of loss of endogenous capacity for self-renewal. Pulmonary capillary endothelial cells (PCEC) are another group of cells exhibiting remarkable influence on lung cell proliferation and regeneration following injury. MMP14 was upregulated in these cells which promotes ACE2 and BASC cell proliferation through the release of ectodomain of HB-EGF the ligand for EGFR which in turn stimulates ACE2 multiplication [23]. Proliferation, differentiation and cellular reorganization in the adult lung is part of the repair mechanism and involves multiple signaling pathways such as Wnt/ ß catenin, notch, retinoic acid, fibroblast growth factor (FGF), etc. These events recapitulate the embryonic development to a limited extent [24,25]. Recent studies using single cell RNA sequencing in the adult lung showed conservation of injury responsive hypoxia/notch/ keratin response in lung epithelial stem cells [26]. When chronic diseases like COPD is considered the microenvironment is proinflammatory.

14.6 In vivo — Animal models

Animal models are even now considered as the gold standard for studying diseases, growth, regeneration, effect of drugs, etc. As far as lung is concerned mouse is the best model to study lung regeneration, drug efficacy, pharmacodynamic and toxicological profiles, etc. The mouse model developed for studying influenza. To overcome the short comings in the mouse model Bissig et al developed human liver chimeric mice using genome engineering and editing to create short palindromic sequence repeats (CRISPR/Cas9), creating and manipulating human lung chimeric mice for studies in lung regeneration after chronic injury and showed that they would be good models to use in pre-clinical lung regenerative study trials [27,28].

14.7 In vitro models

In vivo models although considered as gold standard in models to study lung physiology disease and repair cannot fully recapitulate in absolute terms the development and disease progression in human lung.

It's here that there is a need for *in vitro* models so that they can complement the studies using animal models. *In vitro* models can be created in multiple ways to address the different questions posed in such research. Scientist could design their own *in vitro* models or purchase it. Traditional monoculture models are the basic system, widely used and can be used for identifying molecules in the signal transduction pathways. The cell lines representing airway epithelium A549, 6HBE14o-(16HBE), BEAS2-B, Calu-3 [29] are commonly used and help in deciphering molecular mechanisms such as oxidative stress proinflammatory cytokine modulations, etc.

But such systems have limited applications and do not reflect the true nature of lung physiology. So there exists a need to represent lung physiology for a more detailed understanding of alveolar metabolism on drugs, pollutants, etc. Its in such a scenario the Air-liquid interphase (ALI) culture was tried. When such systems are used polarization of epithelial cells is visible and a cellular mixture more appropriate to the alveolar epithelium inclusive of basal cells, goblet cells neuroendocrine cells, alveolar type II cells is possible. The results obtained in such complex systems have the factor of cell—cell interactions playing a role.

EpiAirway (from Mat Tek) is one such model. This is a 3D model of human airway which can be used to analyse human-specific biological responses like muco-ciliary differentiation etc. It exhibits human specific tissue and cellular morphology with keratin 5 + basal cells, mucus producing goblet cells, functional tight junctions, beating cilia and fibroblasts for inflammatory and fibrotic research.

The ALI system evolved into the three dimensional (3D) models where cell—cell interactions and three dimensional architecture have a greater role. This evolution occurred with the realization that *in vivo* cells do not exist in isolation, but all the responses that are seen are a well-orchestrated multi-cellular interactive response given by tissues, organs and organ systems. Such 3D systems were developed using a combination of cell types — the alveolar epithelial cells, fibroblast, endothelial cells, immune cells, etc. Thus it was possible to get a better insight into the role of airway micro environment and cell—cell interactions in the lung. Examples of such 3D systems include organotypic cultures, multi-cellular spheroids, organoids and microfabricated models [30,31]. Organoids and spheroids both homotypic and heterotypic are where cells of the same lineage or multicellular lineage are embedded in Extracellular matrix (ECM) like matrigel, collagen, hyaluronic acid, alginate, etc. or hydrogels. These systems have proved useful in answering many of the complex pulmonary toxicological responses to drugs and pollutants. The model developed by Marrazo et al. [32] exhibited alveolar epithelial baso-lateral polarity, zona occludens and a continuous laminin layer ensuring conditions for trans-epithelial migration studies. Such in vitro systems being cost effective and adaptable they can be used to study cell—cell interactions, inhalation toxicity, etc.

14.8 Lung-on-a chip (LOC) model

The development of the lung-on-a chip models were a natural progression from 3D culture systems where both the 3D architecture and sheer stress associated with cyclic breathing activity is incorporated. The cell types seeded are ALEII, Fibroblast and human pulmonary vascular endothelial cells. To such models immune cells can be added if required.

Organ-on-a-chip is a microfluidic cell culture model which replicates organ specific microarchitecture and pathophysiology *in vitro*. Initially soft lithography was used for this microfluidics printing but now 3D printing is used casting molds in polydimethyl siloxane (PDMS).

Shrestra et al. [33] used an improved technique to achieve such molds. Using their method they fabricated a simple lung-on-a chip model which simulate *in vivo* airway air—liquid interphase model with calu-3 cells. Their validation showed the model replicated the 3D culture specific morphology, maintained barrier integrity, secreted mucus and expressed P-glycoprotein on the cell surface. The testing of the system was done using cigarette smoke extract and budesonite.

The many reasons for developing *in vitro* models is to simulate *in vivo* responses in vitro and Nikolic et al. showed that LOC can recapitulate pulmonary responses on challenge with drugs and toxins and that these responses were comparable to responses in animal models [30,31].

Microfabricated lung models which are not based on microfluids were also developed addressing responses to fungal pathogens [34]. Thus although they provide valuable molecular and cellular data on screening of chemicals the challenge lies in delineating what constitutes relevant exposure scenarios, due to lack of universal standardization norms.

14.9 Three dimensional printing of the lung

On 2nd May 2019 Jordan Miller and his team at Rice university Brown School of Engineering reported their success in three dimensional printing of the lung where they printed a lung structurally and functionally mimicking the human air sac with the ability to breathe.

Three D bioprinting as this technology is popularly known is the deposition of layer by layer of biomaterials and cells to build a complex organ or tissue. The cells are suspended in a complex hydrogel which supports the cells and gives it structural and spatial anchorage so that the complex organ may be modeled. So printing body parts may definitely be the next step in organ transplant. Bones cornea and skin are the front runners and complex organs and tissues follow. But one of the biggest bottle necks in this technology is our inability to build the complex network of vasculature and neuronal cells.

14.10 Future perspectives latest 4D printing

The latest development in the concept of regenerative technologies is 4D printing, currently a part of project in MIT Self-assembly lab. This is a process through which 3D printed objects responds to external stimuli like temperature, light, or others and transforms itself into another. So here the fourth dimension is transformation over time. Marcinhet Panhuis at The ARC Centre of excellence in Electro materials science in Australia is the leader and they have created a valve made of four hydrogels which confer the properties of toughness, softness, flexibility and strength. The valve is autonomous and opens to water but closes when it detects hot water.

Acknowledgments

Author expresses gratitude towards The Director SCTIMST and The Head, BMT wing for extending the facilities of the Institute for the writing of this manuscript.

Conflict of interest

There is no conflict of interest.

References

[1] Xuan-Ye C, Si-Yu X. Chronic lung disease, lung regeneration and future therapeutic strategies. Chronic Dis Transl Med 2018;4(2):103—8.

[2] Weiss DJ. Concise review: current status of stem cells and regenerative medicine in lung biology and diseases. Stem Cell 2014;32:16—25 [PMC free article] [PubMed] [Google Scholar].

[3] Tsao PN, Vasconcelos M, Izvolsky KI, Qian J, Lu J, Cardoso WV. Notch signaling controls the balance of ciliated and secretory cell fates in developing airways. Development 2009;136:2297—307 [PMC free article] [PubMed] [Google Scholar].

[4] Hong KU, Reynolds SD, Giangreco A, Hurley CM, Stripp BR. Clara cell secretory protein-expressing cells of the airway neuroepithelial body microenvironment include a label-retaining subset and are critical for epithelial renewal after progenitor cell depletion. Am J Respir Cell Mol Biol 2001;24:671—81 [PubMed] [Google Scholar].

[5] Giangreco A, Reynolds SD, Stripp BR. Terminal bronchioles harbor a unique airway stem cell population that localizes to the bronchoalveolar duct junction. Am J Pathol 2002;161:173—82 [PMC free article] [PubMed] [Google Scholar].

[6] McGowan SE, Harvey CS, Jackson SK. Retinoids, retinoic acid receptors and cytoplasmic retinoid binding proteins in perinatal rat lung fibroblast. Am J Phyiol Lung Cell Mol Physiol 1995;269(4Pt1):463—72 L.

[7] Branchfield K, Li R, Lungova V, Verheyden JM, McCulley D, Sun X. A three- dimensional study of alveologenesis in mouse lung. Dev Biol 2016;409:429—41. Available from: https://doi.org/10.1016/j.ydbio.2015.11.017.

[8] Willem M, Miosge N, Halfter W, Smyth N, Jannetti I, Burghart E, et al. Specific ablation of the nidogen-binding site in the laminin gamma1 chain interferes with kidney and lung development. Development 2002;129:2711—22.

[9] Rockich BE, Hrycaj SM, Shih HP, Nagy MS, Ferguson MAH, Kopp JL, et al. Sox9 plays multiple roles in the lung epithelium during branching morphogenesis. Proc Natl Acad Sci USA 2013;110:E4456—64. Available from: https://doi.org/10.1073/pnas.1311847110.

[10] Zacharias WJ, Frank DB, Zepp JA, Morley MP, Alkhaleel FA, Kong J, et al. Regeneration of the lung alveolus by an evolutionarily conserved epithelial progenitor. Nature 2018;. Available from: https://doi.org/10.1038/nature25786.

[11] Zuo W, Zhang T, Wu DZ. P63(+)Krt5(+) distal airway stem cells are essential for lung regeneration. Nature 2015;517:616—20 [PubMed] [Google Scholar].

[12] Guseh JS, Bores SA, Stanger BZ. Notch signaling promotes airway mucous metaplasia and inhibits alveolar development. Development 2009;136:1751—9 [PMC free article] [PubMed] [Google Scholar].

[13] Demaya F, Minoo P, Plopper CG, Schuger L, Shannon J, Torday JS. Mesenchymal epithelial interactions in lung development and repair: are modelling and remodeling the same process?

[14] Gouveia L, Betsholtz C, Andrea J. PDGF-A signaling is required for secondary alveolar septation and controls epithelial proliferation in the developing lung. Development 2018;145(7). Available from: https://doi.org/10.1242/dev.161976 pii: dev161976.

[15] Nandkumar MA, Yamato M, Kushida A, Konno C, Hirose M, Kikuchi A, et al. Two-dimensional cell sheet manipulation of heterotypically co-cultured lung cells utilizing temperature-responsive culture dishes results in long-term maintenance of differentiated epithelial functions. Biomaterials 2002;23(4):1121—30. Available from: https://doi.org/10.1016/S0142-9612(01)00225-3.

[16] Nandkumar MA, Ashna U, Thomas LV, Nair PD. Pulmonary surfactant expression analysis—role of cell—cell interactions and 3-D tissue-like architecture. Cell Biol Int 2015;39(3):272—82. Available from: https://doi.org/10.1002/cbin.10389.

[17] Luo Y, Li N, Chen H, et al. Spatial and temporal changes in extracellular elastin and laminin distribution during lung alveolar development. Sci Rep 2018;8:8334. Available from: https://doi.org/10.1038/s41598-018-26673-1.

[18] Hogan BLM. Integrating mechanical force into lung development. Dev Cell 2018;44(3):271−5.

[19] Li J, Wang Z, Chu Q, Jiang K, Li J, Tang N. The strength of mechanical forces determines the differentiation of alveolar epithelial cells. Dev Cell 2018;44(3):297−312.

[20] Paxson JA, Gruntman A, Parkin CD, Mazan MR, Davis A, Ingenito EP, et al. Age-dependent decline in mouse lung regeneration with loss of lung fibroblast clonogenicity and increased myofibroblastic differentiation. PLoS One 2011;6:e23232.

[21] Beers MF, Morrisey EE. The three R's of lung health and disease: repair, remodeling, and regeneration. J Clin Invest 2011;2065−73. Available from: https://doi.org/10.1172/JCI45961.

[22] Warburton D, Tefft D, Mailleux A, et al. Do lung remodeling repair, and regeneration recapitulate respiratory ontology? Am J Respir Crit Care Med 2001;164:S59−62.

[23] Mou H, Zhao R, Sherwood R. Generation of multipotent lung and airway progenitors from mouse ESCs and patient-specific cystic fibrosis iPSCs. Cell Stem Cell 2012;10:385−97 [PMC free article] [PubMed] [Google Scholar].

[24] Xi Y. Models for studying lung cell interactions and regeneration.

[25] Rafii S, Cao Z, Lis R, Siempos II, Chavez D, Shido K, et al. Platelet-derived SDF-1 primes the pulmonary capillary vascular niche to drive lung alveolar regeneration. Nat Cell Biol 2015;17:123−36.

[26] Rodríguez-Castillo JA, Pérez DB, Ntokou A, et al. Understanding alveolarization to induce lung regeneration. Respir Res 2018;19:148. Available from: https://doi.org/10.1186/s12931-018-0837-5.

[27] Barzi M, Pankowicz FP, Zorman B. A novel humanized mouse lacking murine P450 oxidoreductase for studying human drug metabolism. Nat Commun 2017;8:984 [PMC free article] [PubMed] [Google Scholar].

[28] Bissig-Choisat B, Kettlun-Leyton C, Legras XD. Novel patient-derived xenograft and cell line models for therapeutic testing of pediatric liver cancer. J Hepatol 2016;65:325−33 [PMC free article] [PubMed] [Google Scholar].

[29] Zhu Y, Chidekel A, Shaffer TH. Cultured human airway epithelial cells (Calu-3): a model of human respiratory function, structure, and inflammatory responses. Crit Care Res 2010;. Available from: https://doi.org/10.1155/2010/394578 394578. Pract.

[30] Nikolic M, Sustersic T, Filipovic N. In vitro models and on-chip systems: biomaterial interaction studies with tissues generated using lung epithelial and liver metabolic cell lines. Front Bioeng Biotechnol 2018;6:120. Available from: https://doi.org/10.3389/fbioe.2018.00120.

[31] Kreft ME, Jerman UD, Lasic E, et al. The characterization of the human cell line Calu-3 under different culture conditions and its use as an optimized in vitro model to investigate bronchial epithelial function. Eur J Pharm Sci 2015;69:1−9.

[32] Marrazzo P, Maccari S, Taddei A, et al. 3D reconstruction of the human airway mucosa in vitro as an experimental model to study NTHi infections. PLoS One 2016;11:e0153985 Crossref, Medline, Google Scholar.

[33] Shrestra J, Ghadiri M, Warkianin, et al. A rapid prototype lung-on-a chip model using 3D printed molds. In Organ-on-a-chip; 2020. https://doi.org/10.1016/j.ooc.2020.100001

[34] Grigoryan B, Paulsen SJ, Corbett DC, Sazer DW, Fortin CL, Zaita AJ, et al. Multivascular networks and functional intravascular topologies within biocompatible hydrogels. Science 2019;. Available from: https://doi.org/10.1126/science.aav9750.

Further reading

Cozens AL, Yezzi MJ, Kunzelmann K, et al. CFTR expression and chloride secretion in polarized immortal human bronchial epithelial cells. Am J Respir Cell Mol Biol 1994;10:38−47.

Huang SX, Islam MN, O'Neill J. Efficient generation of lung and airway epithelial cells from human pluripotent stem cells. Nat Biotechnol 2014;32:84−91 [PMC free article] [PubMed] [Google Scholar].

Vaughan AE, Brumwell AN, Xi Y. Lineage-negative progenitors mobilize to regenerate lung epithelium after major injury. Nature 2015;517:621−5 [PMC free article] [PubMed] [Google Scholar].

Key Enabling Technologies for Regenerative Medicine Future Outlook and Conclusions

3D printing in regenerative medicine

Aynur Unal[1] and Nidhi Arora[2]

[1]Digital Monzoukuri, Palo Alto, CA, United States [2]SS Infotech, Indore, MP, India

15.1 Introduction

Three-dimensional (3D) Printing, is also known as "Additive Manufacturing". The concept and technology for 3-D printing has been around for nearly 20 years and has grown into a billion dollar industry. Many items such as food, toys, watches, and television parts get fabricated by this printing. The shift from 3D printing of inanimate objects to living tissue [1] successfully is the feat that makes organ creation a possibility for the near future. Scientists are trying to take this technological know-how into the world of biology and medicine [2]. 3D printing of organs is now on the radar too [1,2]. What they have learnt so far is that the fabrication of non-living objects is easier than living body parts. 3D Printing is creating many innovative breakthroughs in medicine, pharmacology, manufacturing of objects. Last but not least, in regenerative medicine and drug testing, it stems from the knowledge gained in manufacturing composite materials, following layer by layer of deposition of the right materials according to given dimensions. In Regenerative Medicine, after many years of trials and errors in the labs, we have reduced the problem of having the right bio ink to do the printing. Bio Inks will be the focus of attention in the next decade. So far we have seen development in skin, and cardiology related components, the most recent contribution being designing tiny kidneys mimicking the functionalities of the real organs.

Having said that, the progress is still in growing phase and much more work needs to be done in both; the systems of the printers and the materials to be printed. It has been predicted that 3D Printing, in general, will grow at an exponential rate. Bioprinting is a fascinating subject and gives a lot of hope to humans who are in need of an organ replacement. Standardized Clinical Grade 3D printers and Bio compatible Bio inks will enable wider use of 3D Printing in real life situations [3]. Even if technology is still in its infancy, the attempts in both startups and academia are growing in this field.

Every day, there is a new result and new applications ranging from medical operational process related aids to real mimicking of transplantable organs obtained by using the cells of the patients, and hence eliminating the risk of rejection of the organ transplants. Imitating the functionalities of Skin, Kidney, prosthetic ovary, Bionic hand and female womb are only a few examples. In this paper we are taking up examples of Kidney and Skin. The number of startups developing new medicinal and pharmacological products is reviewed at the end. The problem of organ failure and transplantation needs to be addressed to avert a healthcare crisis, and in this speedy new and improved technique, we may have a solution which needs to be tested in real life cases and outside of the labs. The Future of Surgery will be different as a result of 3D printing and lab grown organs for sure. Exacting organ working means fewer chances of failure. 3-D Bioprinting is truly the most radical change to hit the medical horizon in the past few decades.

15.1.1 3D printers

3D printing is going to be relevant in diverse fields like engineering and medicine. ISO/ASTM 52900 classification standard, which was created in 2015, aims to standardize all terminology and classify each of the different types of 3D printers [4,5]. The technology is classified on the basis of different processes and materials available to be printed with. Different types of 3D printers are shown in detail in Table 15.1.

15.1.2 Bioprinting

In Bioprinting, cells, biomaterials and bioactive molecules are combined and assembled to fabricate bio-engineered structures for study of regenerative medicine [3,6]. The combination of these compounds is known as bioink [6]. In comparison with conventional additive manufacturing using cell free scaffolds, bioprinting includes living cells during the process of deposition, which requires special measures due to their sensitivity to temperature and mechanical tension as high pressure and shear stress, as well as pH and biocompatibility of surrounding materials. The 3D construct to be deposited is first designed using CAD software to obtain a 3D model. This model is then sliced into layers and each layer is decomposed in a succession of patterns to fill the full construct. Finally, the ready-to-deposit design is converted into a machine code (e.g., G-code, often used as a universal code for printers), which is read by the printer as a succession of instructions including various parameters such as pressure, feed rate and coordinates.

In the field of bioprinting, there are four main strategies- extrusion-based, drop based, laser based and stereolithography bioprinting [2−6]. Each technique has its own requirements and characteristics translated into specific advantages and limitations. Moreover, each technique accounts for a different range of resolution, manufacturing time and design limitations [6−8]. In addition to the intrinsic technique properties, bioprinting can be carried out as the fabrication of one or multiple materials simultaneously.

"Structures are printed with hydrogel, and the material is then solidified either physically or chemically such that the structures can be combined to create 3D shapes. Microextrusion printers allow for a wider selection of biomaterials since more viscous materials can be printed. Another advantage is that these printers can deposit very high cell densities. Although cell viability may be lower than that obtained with inkjet printers,

TABLE 15.1 Different types of 3D printing technology [4,5].

S. No	Name	3D printing technology	Materials	Dimensional accuracy	Common applications	Strengths	Weaknesses	Used for bioprinting
1.	Material extrusion	Fused deposition modeling, also called Fused Filament Fabrication (FFF)	Thermoplastic filament (PLA, ABS, PET, TPU)	± 0.5% (lower limit ± 0.5 mm)	Electrical housings; form and fit testings; jigs and fixtures; investment casting patterns	Best surface finish; full color and multi-material available, faster than SLA/DLP	Brittle, not sustainable for mechanical parts; higher cost than SLA/DLP for visual purposes	Blood vessel Cartilage and bone Muscle Liver-on-a-chip
2.	Vat polymerization	Stereolithography (SLA), Direct Light Processing (DLP)	Photopolymer resin (standard, castable, transparent, high temperature)	± 0.5% (lower limit ± 0.15 mm)	Injection mold-like polymer prototypes; jewelry (investment casting); dental applications; hearing aids	Smooth surface finish; fine feature details	Brittle, not suitable for mechanical parts	Vascular networks Skin Organ-on-a-chip
3.	Powder bed fusion	Selective Laser Sintering (SLS)	Thermoplastic powder (Nylon 6, Nylon 11, Nylon 12)	± 0.3% (lower limit ± 0.3 mm)	Functional parts; complex ducting (hollow designs); low run part production	Functional parts, good mechanical properties; complex geometries	Longer lead times; higher cost than FFF for functional applications	
4.	Material jetting	Material Jetting (MJ), Drop on Demand (DOD)	Photopolymer resin (standard, castable, transparent, high temperature)	± 0.1 mm	Full color product prototypes; injection mold-like prototypes; low run injection molds; medical models	Best surface finish; full color and multimaterial available	Brittle, not suitable for mechanical parts; higher cost than SLA/DLP for visual purposes	Skin Cartilage Bone Blood vessel

(Continued)

TABLE 15.1 (Continued)

S. No	Name	3D printing technology	Materials	Dimensional accuracy	Common applications	Strengths	Weaknesses	Used for bioprinting
5.	Binder jetting	Binder Jetting	Sand or metal powder: stainless/ bronze, full color sand, Silicia (sand casting)	± 0.2 mm (metal) or ± 0.3 mm (sand)	Functional metal parts; full color models; sand casting	Low-cost; Large build volumes; functional metal parts	Mechanical properties not as good as metal powder bed fusion	
6.	Sand binder jetting	Binder Jetting	Sand powder	± 0.3 mm (sand)				
7.	Metal binder jetting		metal powder	± 0.2 mm (metal)				
8.	Metal powder bed fusion	Direct Metal Laser Sintering (DMLS); Selective Laser Melting (SLM); Electron Beam Melting (EBM)	Metal powder: aluminum, stainless steel, titanium	± 0.1 mm	Functional metal parts (aerospace and automotive); Medical; Dental	Strongest, functional parts; complex geometries	Small build sizes; highest price point of all technologies	

it is in the range of 40−86%, depending on the size of nozzle and pressure of extrusion used" [7].

Laser assisted bioprinting is another type of printing system which is based on the principles of laser induced forward transfer. This involves the use of a pulsed laser beam, a focusing system and a "ribbon" that has a donor transport support, a layer of biological material, and a receiving substrate facing the ribbon [6−8,10]. Focused laser pulses are used to generate a high-pressure bubble that propels cell containing materials toward the collector substrate. Since laser bioprinting does not use nozzles, there are no cell clogging issues. Another advantage is the ability to print with high cell densities without affecting cell viability [11]. The main disadvantages however are the reduced overall flow rate as a result of the high resolution and also the possibility of metallic residues in the final construct [11,12].

In addition to laser-assisted bioprinting, other light-based 3D bioprinting techniques include digital light processing (DLP) and two photon polymerization (TPP)-based 3D bioprinting. DLP uses a digital micro mirror device to project a patterned mask of ultraviolet (UV)/visible range light onto a polymer solution, which in turn results in photo polymerization of the polymer in contact [13−15]. DLP can achieve high resolution with rapid printing speed regardless of the layer's complexity and area. In this method of 3D bioprinting, the dynamics of the polymerization can be regulated by modulating the power of the light source, the printing rate, and the type and concentrations of the photo initiators used. TPP, on the other hand, utilizes a focused near-infrared femtosecond laser of wavelength 800 nm to induce polymerization of the monomer solution [15]. TPP can provide a very high resolution beyond the light diffraction limit since two photon absorption only happens in the center region of the laser focal spot where the energy is above the threshold to trigger two photon absorption [4,15].

Fig. 15.1 shows the components of inkjet, microextrusion, and laser-assisted bioprinters.

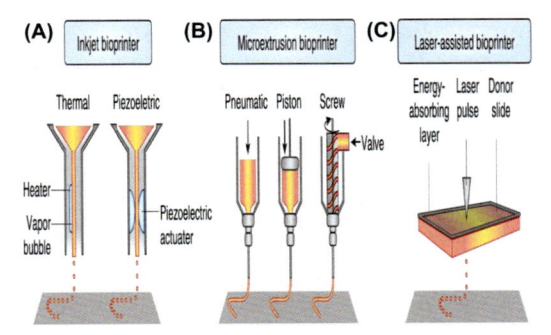

FIGURE 15.1 The types of bioprinter (A) inkjet (B) microextrusion (C) laser assisted. (A) In thermal inkjet printers, the print head is electrically heated to produce air-pressure pulses that force droplets from the nozzle, while acoustic printers use pulses formed by piezoelectric or ultrasound pressure. (B) Micro-extrusion printers use pneumatic or mechanical dispensing systems to extrude continuous beads of material and/or cells. (C) Laser-assisted printers use lasers focused on an absorbing substrate to generate pressures that propel cell-containing materials onto a collector substrate. *This figure was adapted from the original article of Murphy SV, Atala A. 3D bioprinting of tissues and organs. Nat Biotechnol 2014;32:773−85, https://doi.org/10.1038/nbt.2958. (Copyright 2014 Nature America, Inc.).*

15.1.3 The 3D bioprinting process

The layer-by-layer basic fabrication of 3D printing is as follows: [8,9]

1. Pre-Processing— Generation of a computerized design of the structure—blue print
2. Processing
   a. A layer of hydrogel (extracellular matrix) is printed or preset in a petri dish or container, which will function as the foundation base for the printed tissue.
   b. Bioink—usually made out of tissue spheroids or cultured cells. This bioink is loaded into the printer heads and disposed on the layer of hydrogel.
   c. Dispensing bioink or hydrogel repeatedly until many layers are formed
   d. As layers are built, the deposited bioink will fuse together forming 3D structure containing cells and hydrogel.
3. Maturation—Printed structure is then placed in an incubator and left to mature for 24 hours—1 week (depending on the materials used)

15.1.4 Bioprinting: diminishing the organ transplant shortage

Medical field has become the hotbed of innovation so far as creating organs to facilitate the transplants is concerned. Advanced and more capable Bioprinting processes with newer materials and novel fields of applications have achieved this. The biggest plus is the steady lowering costs of the operations. Scientists are raring to go and make Bioprinted replacement human organs. However, even as the possibilities are exciting, we are still in early stages of development and much work needs to be done.

Here in this paper, we present a review of 3D kidney and skin as case studies in following sections.

15.1.5 Bioprinted kidneys: what the future upholds

While it seems that bioprinting kidneys is still very much in its infancy, there has been a significant increase in the development and considerable progress made utilizing this technology. In this article, we've selected the three most promising institutes and companies around the globe and compiled their latest advancements.

Regenerative medicine replaces or regenerates human cells, tissues, and/or organs to restore or establish normal function [7,8,16,17]. In kidneys, this includes perfusion, filtration, secretion, and maintenance of homeostasis, with the ultimate aim of improving long-term patient QoL.

Kidney research initially aimed to achieve this by targeting and improving dialysis.

Unfortunately, the current limitations including the size of the structures mean that their lifespan is limited to days rather than years, which makes any estimation of appropriate scale model success speculative.

Wyss Institute team has advanced bioprinting [18] to the point of being able to fabricate a functional subunit of a kidney. Inside our kidneys, there are more than 1 million nephrons, each of which contains structures called proximal tubules. Proximal tubules reabsorb and transport nutrients back into the bloodstream after renal filtration.

The nephron filters the blood, processes nutrients and passes out waste from the blood. Blood is filtered by the capillary network (glomerulus). The filtrate is collected in the Bowman's capsule and passes through a series of renal tubules (proximal tube, loop of Henle and distal tube). These absorb water, minerals and glucose. Filtered fluids leave the nephron by the collecting duct and enter the renal pelvis.

Bioprinter proximal tubule contains living human cells and mimics several biological functions of human nephrons. This current work builds on their groundbreaking research which enables bioprinting of human tissues with multiple cell types and embedded vasculature at scale. The functional 3D renal architecture is the team's most recent bioprinting advance. A fugitive ink which is liquefied and removed later is printed on the top of an extracellular matrix [9]. It models the convoluted pathway of the proximal tubule. Another layer of extracellular matrix is deposited on the top of the printed feature.

The fugitive ink is removed. What remain is an open channel modeling a kidney's proximal tubule, with a perfusable inlet and outlet at each end. Living human kidney cells are pumped into the tubule, where they adhere to the channel's lining and form a tightly woven uniform layer of cells. As the tissue matures in 9–10 days, these cells form confluent epithelium that functions like the human proximal tubules inside our bodies. Cells in the bioprinted tubule recapitulate the morphology and function of native kidney cells. This is a significant advance from traditional 2D culture. Cross-section images of the cells in the tubule show polarization like cells have in vivo. Damage to proximal tubule constructs can be induced, allowing researchers to study effects of drugs or other factors. Next the team will integrate vasculature near the Bioprinted proximal tubules. Beyond pharmaceutical applications the teams future plans are to scale up their model to enable future clinical applications like implant or organ assistive device [9,18].

Dutch researchers have engineered 3D cellular constructs called organoids for kidney research using human induced pluripotent stem cells (iPSCs) and embryonic stem cells (ESCs). The organoids were successfully vascularized once implanted in vivo [19,20].

Now, Ton Rabelink and his team at the Leiden University Medical Center in the Netherlands have developed a kidney organoid which, once implanted in vivo, becomes vascularized and starts to become more mature [20]. The researchers were able to generate kidney organoids from ESCs and iPSCs obtained by reprogramming of somatic cells to the pluripotent state. Both cell types were able to develop into kidney organoids. They identified many structures in these kidney organoids, including the glomerulus surrounded by the Bowman's capsule, the renal tubules and the collecting duct. They also observed more specific features, including the presence of podocytes in the glomerulus (cells involved in the retention of plasma proteins from going into the urine), interstitial cells, which act as the kidney scaffold, endothelial cells (that form blood vessels) and pericytes, which are cells localized around the vessels. Upon implantation of the organoid under the renal capsule of mice, the researchers observed glomerular vascularization by the mouse endothelial cells, without needing further stimulation. Also, organoid maturation and organization was more advanced upon transplantation in vivo, compared with prolonged culture in vitro. This study presents the development of a more mature kidney organoid that is more comparable to adult kidneys. This research will be useful in many applications, such as drug screening, disease modeling and studying kidney regeneration.

Perhaps one of the biggest names in bioprinting, the award-winning Lewis Lab at Harvard University has been consistently advancing in kidney 3D printing [21]. The laboratory's name comes from Professor Jennifer A. Lewis, who leads the research group. According to its website, the laboratory works with 3D printing of soft functional materials for microvascular architectures for cell culture and tissue engineering, among other applications. The continuing efforts of the research team have led them to publish a new paper in March 2019, entitled "Renal reabsorption in 3D vascularized proximal tubule models". Their main development is the fabrication of a micro blood vessel that runs side by side with a proximal tubule. This enables the recreation of "active reabsorption of solutes via tubular—vascular exchange". In layman's terms, this means that the nutrient exchange between blood vessels and nephron structures could be reproduced in 3D printed tissues. This new advancement has immediate applications for drug testing and no doubt represents a step further in 3D printing a fully functional kidney.

The Wake Forest Institute for Regenerative Medicine [22] aims to develop laboratory-grown tissues and organs for various applications. Located in Winston-Salem, North Carolina, their scientists and physicians are responsible for the first successful implant of lab-created organs in humans. Led by Dr. Antony Atala, the institute works in the development of more than 30 different tissues and organs. In past studies, institute scientists have successfully grown kidney cells that were later implanted in animals. These cells eventually formed kidney structures and were even able to partially function in a "mini kidney". On the bioprinting side, the Wake Forest Institute initially used a modified desktop inkjet printer to produce 3D structures with grown cells. Then, in 2016, they successfully 3D printed living tissue structures using a specialized printer, designed by the institute's own scientists. With this new equipment, Dr. Atala and his team were able to develop a material that would hold cells together during and after the printing process. The next step was to 3D print a structured model of a kidney. Although only a macrostructure, or "prototype" as they called it, it could pave the way for developing a fully-functional kidney in the future. Additionally, the institute researchers have been working with yet another approach for treating chronic kidney diseases. In a 2019 paper, Dr. Atala and his colleges showed that fluid stem cells injected into a diseased kidney can lead to improved kidney function.

Organovo [23] is a publicly-traded medical company launched in 2007. Located in San Diego, California, the company is dedicated to developing functional human tissues using 3D printing techniques. Their main goal is to use lab-grown tissues for accelerating preclinical drug testing, but there's also a handful of regenerative medicine research, too. Towards this end, Organovo has developed its own bioprinter, the NovoGen MMX. Organovo's bioprinter is an extrusion-based printer. It includes two print heads, allowing the simultaneous use of both human cells and support structures (like scaffolds and hydrogel). The printer was recognized as one of the "Best Invention of 2010" by Time Magazine [19]. In 2015, Organovo announced the fabrication of (proximal tubular) kidney structures. The tissue was said to be only cell-based, with no scaffolds, showing real connections and architecture. In that same year, the company partnered with the University of Queensland's Institute for Molecular Bioscience, which in turn had shown the means to create a complete "mini kidney". In June 2019, the company announced the development of a process for automated production kidney organoids, potentially enabling the treatment of real patients with chronic renal disease. Organovo is now leading research collaboration along with Murdoch Children's Research Institute, the Royal Children's Hospital in Melbourne, and Ton Rabelink at Universiteit Leiden in the Netherlands.

15.1.6 3D printing skin: where are we now?

Skin is first to cover our body used to defense us from different injuries, wounds and burns.

It has different layers mainly the epidermis, dermis, and Subcutaneous tissue called hypodermis and subcutis. An (outermost) external layer called the epidermis protects from the external environment. Second thicker and deeper layer is the dermis. Subcutaneous tissue is to attach the skin to underlying bone and muscle as well as supplying it with blood vessels and nerves [5,24].

Detail structure of human skin has been depicted in Fig. 15.2.

"Recapturing the native skin physiology is still work in progress. Three-dimensional (3D) bioprinting involves layer-by-layer deposition of cells along with scaffolding

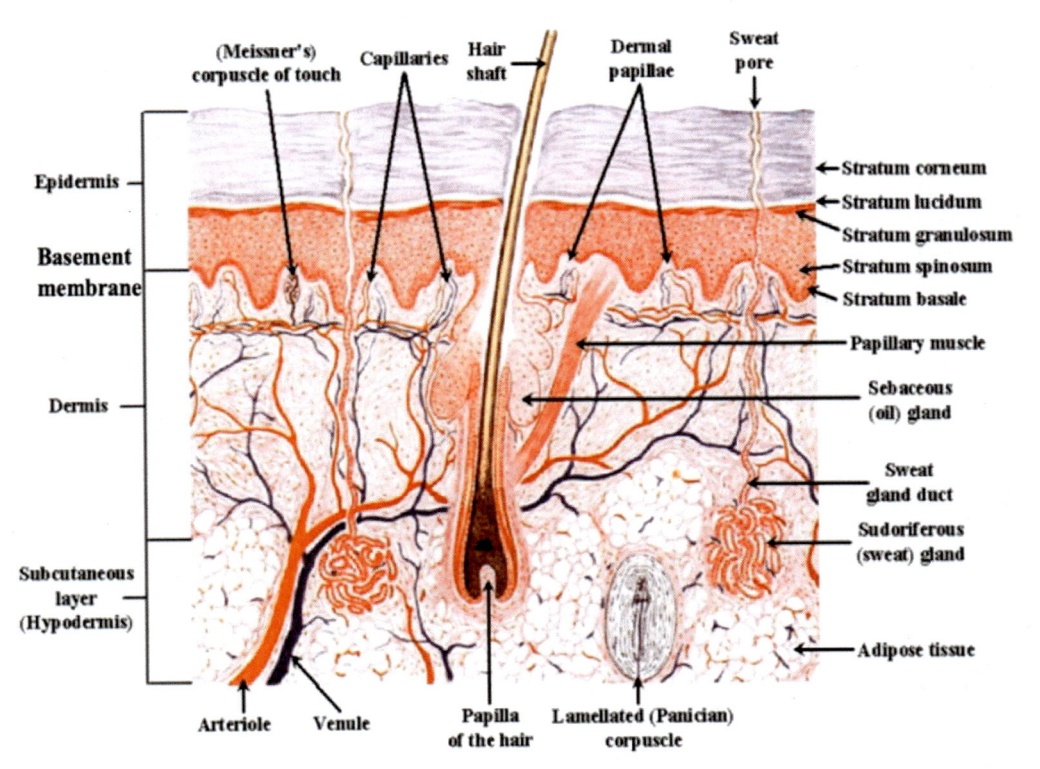

FIGURE 15.2 Structure of human skin depicting the different layers and appendages. *This figure was adapted from the original article of Varkey M, Visscher DO, van Zuijlen PPM, et al. Skin bioprinting: the future of burn wound reconstruction? Burn Trauma 2019;7, 4. https://doi.org/10.1186/s41038-019-0142-7.*

materials over the injured areas in wound treatment. Skin bioprinting can be done either in situ or in vitro. Both these approaches are similar except for the site of printing and tissue maturation. There are technological and regulatory hurdles in our efforts in clinical translation of bioprinted skin for burn reconstruction. Bioprinting may enable us to do more accurate placement of cell types and precise and reproducible fabrication of constructs to replace the injured or damaged sites. Overall, 3D bioprinting is a very transformative technology, and its use for wound reconstruction will lead to a paradigm shift in patient outcomes. In this review, we aim to introduce bio printing, the different stages involved, in vitro and in vivo skin bioprinting, and the various clinical and regulatory challenges in adoption of this technology" [7,24−26].

"3D Bioprinting can create accelerated wound healing and better scar quality as well as functional outcomes." [5,7,24].

Nano cellulose fibers can be used to create smart textiles by their layer by layer deposition to have bio-decomposable medical devices. The recent work on skin makes us think to somehow include the Nano cellulose fibers into skin care as smart textiles and in burn care as well which will occupy using the next few years to come as new bio decomposable smart textiles.

15.1.7 Considerations for bio printing skin

As explained in [7], the skin is a complex organ with a well-defined structure consisting of multiple layers and appendages and is made of several cell types as shown in Fig. 15.2. Therefore, to bioprint such a structure requires multiple cell types and biomaterials. The most superficial layer of the skin, the epidermis, is mainly composed of keratinocytes with varying degrees of differentiation and intertwined melanocytes near the lower layer of the epidermis. The epidermis is relatively thin (0.1−0.2 mm in depth) and attached to the underlying dermis via a highly specialized basement membrane [27]. Due to the relatively thin epidermis, laser-assisted bioprinting technology may be used to explore epidermal bioprinting [28]. Utilizing this technology, one may be able to recapitulate the epidermal morphology by printing consecutive layers of keratinocytes and melanocytes. The bioprinting technology could potentially be used to produce uniform pigmentation in patients [29]. The basement membrane is a thin, fibrous tissue composed of two layers, the basal lamina and the reticular connective tissue, which are connected with collagen type VII anchoring fibrils and fibrillin micro-fibrils [30]. The structure of the basement membrane becomes more complex deeper in the skin, where the tissue becomes several nanometers thick with many ECM components including collagen type IV, laminin, and various integrins and proteoglycans [31]. Bioprinting such a complex layer is a challenging and complex task, and therefore many researchers tend to rely on tissue self-assembly after printing [31,32].

The dermal layer can be found directly underneath the basement membrane in the skin and is composed of fibroblasts embedded in a complex ECM [33]. This layer also contains many different structures including all skin appendages, blood vessels, and nerves, which serve the epidermis. The reticular or deep dermis contains many ECM components including collagen and elastin; these elastic and reticular fibers give the skin its high elasticity and strength. In addition, the organization of these fibers also creates Langer's lines [34].

Therefore, this structure may be very important for the mechanical stability of bioprinted skin. Because this layer is thicker than the overlying epidermis, extrusion-based technology may be a good option as it can combine multiple cell types and biomaterials. The use of bioprinting will enable incorporation of other cell types in the dermis including hair follicles and sweat and sebaceous glands. This will enable regeneration of the skin tissue with structure and cellular composition resembling native tissue. In addition, bioprinting will enable control of the microarchitecture of the dermal tissue components, which may have a role in the formation of scar during the wound repair and healing process following injury [35]. Tailoring the microenvironment to facilitate tissue regeneration over repair may have some benefits in terms of better functional outcomes during the scar remodeling process [35]. The hypodermis lies directly below the dermis and consists mainly of adipose tissue that provides heat insulation, energy storage, protective padding [36], and a sliding system [37,38]. This last function has only recently become important in burn surgery because restoring the burned hypodermis with autologous fat injection has shown a remarkable improvement in scar pliability [38–40].

Scientists from the Universidad Carlos III de Madrid (UC3M), Hospital General Universitario Gregorio Marañón, CIEMAT (Center for Energy, Environmental and Technological Research), in collaboration with BioDan Group, developed prototype of a 3D bioprinter that can print functional human skin [41]. The bioprinter is able to recreate the natural structure of the skin with different layers. External layer to protect from external environment, second dermis and the last layer is to give elasticity and mechanical strength to the 3D printed skin, producing collagen. Bioprinted skin presents all characteristics of human skin, both at the molecular and macromolecular level [40]. Using 3 D Bioprinters we can print all three types of skin cells. We can do sound skin transplants and can match the skin depth and shape as per the need.

The Wake Forest Institute for Regenerative Medicine developed a bioprinter which is capable of printing all three types of skins. Printer can print skin directly on model patients. Patients' own skin cells can be used in printing in order to decrease the chances of rejection of grafts. It can help to heal very deep wounds [42].

The University of Toronto's has also developed handheld Bioprinter which is small in size and light weight. It also doesn't require washing and incubation. It prints human skin in form of strips. It is easy to operate therefore requires minimal operator training. They have created bioink using mixture of cells and proteins mainly protein hyaluronate. This ink is helpful in strengthening of cells and quick healing. They are sure that this printer will heal deep skin injuries in few minutes when it will be used on actual patients.

Researchers at the Singapore Institute of Manufacturing Technology are working on skin coloration. They are working on controlling melanin in the bioink in order to get accurate and exact skin color as of patients. They have got good results and success in it.

Organovo, One bioprinting company working with L'Oréal company are also working on other application of bioprinting. L'Oréal is a makeup company who test their makeup products on animals before launching it in market from a safety point of view. Now they are planning to test makeup products on 3D printed skins. Likewise drug companies are also testing their drugs on printed organs in order to avoid animal testers [43].

Researchers at the University Hospital of Dresden Technical University are working on development of printer which can print in space. They got success in creating a very viscous bioink that can be printed in zero gravity. This has gained/acquired the interest of NASA also as this could be very useful for astronauts going to Mars. It will allow them to heal unforeseen injuries with limited resources at other planet also.

A significant barrier to integration of 3d printed skin with host cell is the absence of a functioning vascular system in the skin grafts. Rensselaer Polytechnic Institute has developed a way to 3D print living skin, complete with blood vessels [44]. They have fabricated an implantable multilayered vascularized bioengineered skin graft using 3D bioprinting. The key elements include human endothelial cells, human pericyte cells, animal collagen and other structural cells typically found in a skin graft. They have grafted it onto a special type of mouse and got success. 3d printed skin was able to communicate and connect with the mouse's own vessels. Their work is helpful for people with diabetic or pressure ulcers but they are still working on it so that it can be used for burn patients.

Researchers at IIT Delhi, India have also developed the 3d bio-printed human skin. It is composed of two main layers, the inner dermis and outer epidermis, which are separated by an undulated morphology to provide structural stability. It also protects the two layers by not allowing cells to cross the junction. The epidermal layer consists of keratinocytes and melanocytes while the dermis is made of fibroblasts. Bio-printed skin is able to retain its original dimension without any substantial shrinkage for up to three weeks. Work was done in collaboration with ITC Ltd [43].

Now researchers are talking about using the stem cells for mass-produce cells compatible with all of the possible patients [44].

15.1.8 Reconstructive surgery for burn treatment

Surgical procedures in reconstructive surgery in burn treatment consists of [8,45,46].

1. burn wound excision
2. skin grafts
3. skin substitutes
4. functional and aesthetic are improved by surgical intervention after the stabilization and resuscitation of the patients. Most of the patients survive burn injuries.

Surgical wound closure and reconstructive surgery are typically performed to improve the functional and esthetic outcomes of burn wounds.

Primary closure of burn wounds involves direct wound closure following excision of the devitalized tissue. It is usually performed in small to moderate sized burn scars and takes into account Langer's lines of skin tension for an optimal esthetic outcome [8]. Recently, primary closure has also been performed in larger burn wounds in combination with skin-stretching devices [9−12].

When primary closure of a burn wound is not an option, additional surgery is required. A combination of excision and grafting is the preferred approach for the treatment of deeper dermal burns. The main goal of early excision is to remove devitalized tissue and prepare the wound for skin grafting; layers of burned tissue are excised until a viable wound bed is reached for grafting [3]. Early excision has been shown to be cost effective and reduce mortality and the length of hospital stay [13,14].

15.1.9 3D printing as a disruptive technology in burn care

3D Printing is a disruptive technology in Burn Care as explained by Atala et al. [24].

Conventional tissue-engineered skin substitutes are made by seeding cells on biodegradable scaffolds and allowed to mature, following which they are used for transplantation or in vitro testing. These skin substitutes have several limitations, they contain at most only two cell types, and since they are based on post-natal wound healing physiology, they do not stimulate regeneration of vasculature, nerves, sweat and sebaceous glands, hair follicles, and pigmentation. All these structures are essential to restore the complete anatomy and physiology of native skin; hence, there is an immense need to develop next generation tissue-engineered skin substitutes. Recent work from WAKE FOREST group demonstrates that bioprinting could be successfully used to close large full-thickness wounds [47]. Further, they have also shown that bioprinting could be very effectively used to precisely fabricate both soft and hard tissues with complex structures in an automated manner [48]. Bioprinting could revolutionize the field of burn care by replacing current off-the-shelf cellular or a cellular skin products and providing highly automated process of fabricating complex skin constructs to enhance functional outcome of burns.

15.1.10 Skin bioprinting—in situ and in vitro

To date, several studies have investigated skin bioprinting as a novel approach to reconstruct functional skin tissue [40,47,49−64]. Some of the advantages of fabrication of skin constructs using bioprinting compared to other conventional tissue engineering strategies are the automation and standardization for clinical application and precision in deposition of cells. Although conventional tissue engineering strategies (i.e., culturing cells on a scaffold and maturation in a bioreactor) might currently achieve similar results to bioprinting, there are still many aspects that require improvements in the production process of the skin, including the long production times to obtain large surfaces required to cover the entire burn wounds [64]. There are two different approaches to skin bioprinting: (1) in situ bioprinting and (2) in vitro bioprinting. Both these approaches are similar except for the site of printing and tissue maturation. In situ bioprinting involves direct printing of pre-cultured cells onto the site of injury for wound closure allowing for skin maturation at the wound site. The use of in situ bioprinting for burn wound reconstruction provides several advantages, including precise deposition of cells on the wound, elimination of the need for expensive and time- consuming in vitro differentiation, and the need for multiple surgeries [65]. In the case of in vitro bioprinting, printing is done in vitro and the bioprinted skin is allowed to mature in a bioreactor, after which it is transplanted to the wound site. Our group is working on developing approaches for in situ bioprinting [66]. An inkjet-based bio- printing system was developed to print primary human keratinocytes and fibroblasts on dorsal full-thickness ($3 \, cm \times 2.5 \, cm$) wounds in athymic nude mice. First, fibro- blasts (1.0×10^5 cells/cm^2) incorporated into fibrinogen/ collagen hydrogels were printed on the wounds, followed by a layer of keratinocytes (1.0×10^7 cells/cm^2) above the fibroblast layer [66]. Complete re-epithelialization was achieved in these relatively large wounds after 8 weeks. This bioprinting system involves the use of a novel cartridge-based delivery system for deposition of cells at the site of injury. A laser scanner scans the wound and creates a map of the missing skin, and fibroblasts and keratinocytes are printed directly on to this area. These cells then form the dermis and epidermis, respectively. This was further validated in a pig wound model, wherein larger wounds (10 cm × 10 cm) were treated by printing a layer of fibroblasts followed by keratinocytes (10 million cells each) [66]. Wound healing and complete re-epithelialization were observed by 8 weeks. This pivotal work shows the potential of using in situ bioprinting approaches for wound healing and skin regeneration. Clinical studies are currently in progress with this in situ bioprinting system.

In another study, amniotic fluid-derived stem cells (AFSCs) were bioprinted directly onto full-thickness dorsal skin wounds (2 cm × 2 cm) of nu/nu mice using a pressure-driven, computer-controlled bioprinting device [47]. AFSCs and bone marrow-derived mesenchymal stem cells were suspended in fibrin-collagen gel, mixed with thrombin solution (a crosslinking agent), and then printed onto the wound site. Two layers of fibrin-collagen gel and thrombin were printed on the wounds. Bioprinting enabled effective wound closure and re-epithelialization likely through a growth factor-mediated mechanism by the stem cells. These studies indicate the potential of using in situ bio- printing for treatment of large wounds and burns.

There are a few reports of in vitro skin printing from other groups. Laser-assisted bioprinting was used to print fibroblasts and keratinocytes embedded in collagen and fabricate simple skin equivalent structures [53]. The cells were shown to adhere together through the formation of gap junctions. In a similar study, fibroblasts and keratinocytes were printed in vitro on Matriderm® stabilizing matrix [52]. These skin constructs were subsequently tested in vivo, using a dorsal skin fold chamber model in nude mice. On full-thickness wounds, a multilayer epidermis with stratum corneum was observed in the explanted tissue after 11 days. Also, at this time, some blood vessels were found to be arising from the wound bed. In another report, dermal/epidermal-like distinctive layers were printed using an extrusion printer with primary adult human dermal fibroblasts and epidermal keratinocytes in a 3D collagen hydrogel. Epidermal and dermal structures were observed in these constructs; however, they did not show establishment of intercellular junctions [67]. More recently, Cubo et al. printed a human plasma-derived skin construct with fibro- blasts and keratinocytes [64]. The printed skin was analyzed in vitro and in vivo in an immunodeficient mouse model. The printed skin had a structure similar to native skin with identifiable stratum basale, stratum granulosum, and stratum corneum suggesting a functional epidermal layer and neovascular network formation [64]. In order to regenerate fully functional skin using bioprinting, other structures such as skin appendages (e.g., hair follicles, sweat glands, melanocytes, endothelial cells, and sebaceous glands) should be co-printed in the skin. Some recent studies have evaluated printing of melanocytes [51] and sweat glands [68,69] with varying results. Min and colleagues [51] co-printed melanocytes and keratinocytes on top of a dermal layer and showed terminal differentiation of keratinocytes and freckle-like pigmentations without the use of UV light or chemical stimuli. Huang and colleagues [69] bioprinted sweat glands using epidermal progenitor cells in a composite hydrogel based on gelatin and sodium alginate. They showed that the bioprinted 3D extracellular matrix (ECM) resulted in functional restoration of sweat glands in burned mice.

15.1.11 Stages of skin bioprinting

The process of skin bioprinting can be divided into three stages:

1. skin preprinting,
2. bioprinting, and
3. skin maturation.

- **Pre-printing** involves isolation of cells from the skin biopsy, expansion of cells, differentiation of cells, and preparation of the bioink, which is made of cells and biomaterial support materials. In the case of healthy skin, primary cells could be isolated, expanded, and used; however, in the case of injured skin, stem cells may need to be differentiated into epidermal and mesenchymal cells. Stem cells can be obtained from different sources including adipose, mesenchymal, perinatal, and induced pluripotent stem cells.
- **Bioprinting**: the print files that contain accurate surface information of complex 3D geometries are converted to the STereoLithog- raphy (STL) file format with coordinates for the print head path [70,71]. These files contain accurate surface information required to reconstruct the complex 3D model and can be designed using CAD-CAM graphic user interfaces or created from clinical images with input from magnetic resonance imaging (MRI) and computed tomography (CT) imaging [46,72,73]. The paths for the print heads are created by slicing the STL model into layers and creating bioprinter toolpaths that trace out the perimeter and interior features of each slice. The thickness of each of these slices determines the resolution of the printer and is usually in the $100-500$ μm range. Resolution is specific to the printer used; the smaller the resolution the better the quality but longer the print time. The bioprinter reads the STL files and layer-by-layer deposits the bioink to build the 3D tissue or organ from the series of 2D slices. High-quality image acquisition is essential for high-fidelity bioprinting. Clinical images can provide information regarding the in vivo cell distribution, and image processing tools can be used to determine anatomically realistic skin geometry.
- **Maturation stage:** This is especially critical in case of in vitro bioprinting, and immediately following printing, the skin constructs are fragile and need to be matured in a bioreactor for a few days prior to use for transplantation. When the skin is in situ bioprinted, maturation occurs on the body at the site of injury.

15.1.12 Bioink

The essential element for bioprinting: bioinks form the delivery medium that encapsulates the cells, minimize cell injury during the printing process, and provide a supportive microenvironment for maturation of the bioprinted skin. The choice of bioink is a critical aspect of bioprinting essential for the different cells to be deposited in specific patterns of the CAD models and is chosen with the desired biomechanical characteristics in mind. An appropriate choice of bioink is essential to provide the chemical and physical cues that facilitate necessary cell-ECM interactions; bioink properties not only affect cell growth, proliferation, and differentiation but also structure and function of the bioprinted skin.

It is essential that the chosen bio ink be biocompatible and cell supportive and facilitate functional differentiation of the cells into the skin [73]. Typically, the bioinks could physically serve as cell-laden hydrogels or sacrificial support materials that are removed immediately after printing or as mechanical sup- port materials that provide specific mechanical characteristics to the tissue. Bioinks can be fully natural materials such as collagen, fibrin, HA, and alginate, which could be used in the form of hydrogels for the cells or synthetic materials such as PCL, polylactide (PLA), polyglycolide (PGA), poly(lactic-co-glycolic acid) (PLGA), and poly-ethylene glycol (PEG) polymers or hybrid biomaterials that contain a combination of natural and synthetic materials, which could provide mechanical support [74]. Other bioinks that are typically used also include agarose-, silk-, cellulose-, and GelMA-based bioinks. Materials such as Pluronic F-127 could be used as sacrificial support materials that keep the cells together while printing and could be simply washed away following printing of the tissue construct [48].

15.1.13 Desired features of bio-ink

Printability of the bio ink indicates the ease with which it could be printed with good resolution and its ability to maintain its structure for post-printing skin maturation. Stability of the bioink formulation should be stable enough to provide architectural stability to the skin construct.

"Shape fidelity and printing resolution are important considerations when assessing the printability of the bioink" [75,76]. Other important bioink properties to consider include gelation kinetics, rheological characteristics, and material properties. Ideally, the viscosity of the bioink should be such that it is not only supportive for cell growth and differentiation but also suitable for printing, but in reality viscosities appropriate for bioprinting may not be supportive of cell viability. So, to achieve good printability and at the same time to ensure high cell viability, the printing conditions and bioink consistency need to be optimized. The biomechanical and structural characteristics of the skin are also important considerations for choice of bioink. As we advance in our ability to bioprint and potentially attempt to bioprint composite tissue that may contain a mix of soft and hard tissue such as the skin, skeletal muscle, and bone, we will need to develop some sort of standard or universal bioink that could support different tissue types without compromising functionality. Another important factor that should be considered is how quickly the material will degrade in the body; the cells should be able to degrade the scaffold at a rate that will match their ECM production and remodeling activity.

15.1.14 Technological challenges

Technological challenges confronted by Atala and his group are discussed below:
Numerous technological limitations at the different stages of the bioprinting process need to be overcome to ease and improve clinical translation of bioprinting technology [39].

Very large numbers of cells are required for printing transplant-ready skin; to bioprint skin with physiologically- equivalent cell numbers, billions of cells will be needed.

Current cell expansion technologies facilitate cell expansion in the range of millions, so innovative cell expansion technologies need to be developed [76]. Further, development of superior bioinks that allow for reproducible bioprinting of the skin with appropriate biomechanical properties is critical for clinical translation of the technology.

For composite tissue that contains different tissue types, the printing resolution will need to be improved to duplicate the intricate inner microarchitecture. The ability to print microscale features is necessary for optimal cellular function. Better control over the microarchitecture will enable fabrication of the skin capable of recapitulating the native form and function. Increasing the printing speed is another challenge; current approaches that facilitate higher printing speed such as extrusion bioprinting can compromise the integrity of cells and cause significant loss in their viability. CAD-CAM can also be used to predict the feasibility of the fabrication process by simulating relevant physical models using both classical formula calculations and finite element methods. Currently, the most widely used physical model for bioprinting is laminar multi-phase flow; although it is an oversimplified model and ignores issues related to inclusion of cells, the simulations are useful for checking and optimizing the feasibility of specific designs.

Building a functional vasculature is one of the most fundamental challenges in tissue engineering. The ability to 3D bioprint vasculature will enable fabrication of a preformed microvascular network that can better anastomose to the host circulation and achieve functional perfusion within the tissue-engineered skin construct [77,78]. The use of sacrificial inks to create 3D interconnecting networks, which can be removed after printing the entire construct, leaving hollow channels for the per- fusion of endothelial cells and formation of blood vessel network is a promising approach. Miller et al. have shown how 3D extrusion printing and cast molding could be combined to create a 3D-interconnected perfusable vasculature [79]. However, this molding technique is limited to the construction of simple block tissue architectures [79]. Recently, a bioprinting approach that enables the simultaneous printing of the vasculature structure and the surrounding cells for heterogeneous cell-laden tissue constructs has been reported by the research group of Prof. Lewis [80]. They have developed a method that involves the use of Pluronic F-127 as a fugitive bioink, which can be printed and dissolved under mild conditions, enabling printing of heterogeneous cell-laden tissue constructs with interconnecting vasculature networks [80].

There have also been attempts to bioprint the vascular network directly; Zhang et al. recently reported about direct bioprinting of vessel-like cellular microfluidic channels with hydrogels, such as alginate and chitosan, using a coaxial nozzle [81]. In very recently reported work from Prof. Lewis' lab, they have demonstrated bioprinting of 3D cell-laden, vascularized tissues that exceed 1 cm in thick- ness and can be perfused on chip for greater than 6 weeks [82]. They integrated parenchyma, stroma, and endothelium into a single thick tissue by co-printing multiple inks composed of human mesenchymal stem cells and human neonatal dermal fibroblasts within a customized fibrin-gelatin matrix alongside embedded vasculature, which was subsequently lined with human umbilical vein endothelial cells. This may open newer avenues for printing of pre-vascularized skin tissue.

To print vascularized skin models with complexity and resolution matching in vivo structures, print resolution needs to be improved and printing time reduced. The ability to bioprint hierarchical vascular networks while building complex tissues and the ability to recapitulate vascular flow in vitro [83] are critical for fabrication of transplantable organs.

Native skin has different cell types, each of them require different nutritional and metabolic support. Development of a standard or universal growth media for cells will be beneficial for growth and maturation of composite tissue constructs prior to transplantation. The cells also are in dynamic reciprocity with their microenvironment, which includes the ECM in which they are embedded in. The cells secrete proteins, proteases, and other metabolites onto the ECM, which facilitate dynamic homeostatic phase of tissue remodeling. Inclusion of native ECM in the bioink will ensure the presence of natural ligands and thus facilitate a suitable growth environment for the cells [76].

Also, the development of novel bioreactors to facilitate dynamic culture would facilitate physiologic like environment for the maturation of tissues that incorporate printed vasculatures [76].

In the coming decades, more accurate analytical and computational approaches to effectively study the proliferation and maturation of the bioprinted tissue need to be developed [76]. There has been a lot of effort to model bioprinted tissue with the corresponding printing parameters. For extrusion printing, relationships between dispensing pressure, printing time, and nozzle diameter have been tested and modeled [37]. In inkjet printers, cell settling that occurs during printing and is known to cause clogging of the nozzles has been modeled by both analytical and finite element methods [77–79]. For laser printing, the effects of laser energy, substrate film thickness, and hydrogel viscosity on cell viability [80] as well as droplet size [13,79], cell differentiation [81], and cell proliferation [81] have been studied. Researchers have also done post-printing modeling of cellular dynamics [82,83], fusion [83], deformation, and stiffness [84].

15.1.15 Clinical, regulatory and ethical requirements

Efficient and cost-effective advanced manufacturing techniques need to be developed and optimized to facilitate the use of bioprinted skin for clinical burn reconstruction. Bioprinted human physiologically relevant skin for burn reconstruction [48] should include different cell types. Active monitoring of cell yields and maintenance of quality parameters such as purity, potency, and viability for the different cell types during production is critical for clinical translation of bioprinted skin [73]. Also, since the bioinks contain ECM scaffold components, the quality of the scaffolds and potential for causing contamination and disease transmission will need to be checked along with real-time monitoring. Non-invasive release testing procedures will need to be established before the delivery of the bio printed tissues to the patient [84]. Also, to successfully translate organ bio printing to the clinic, robust automated protocols and procedures need to be established.

To ensure effective use of bio printed skin for burn reconstruction standards [48] for quality assurance of bioinks, bio printers and bio printed products are essential. A comprehensive regulatory framework involving quality control standards for every step of the process design of the model, selection of bio inks, bio printing process, validation of the printing, post-printing maturation, and product quality assessment prior to transplantation is essential. The Food and Drug Administration (FDA) recently issued a guidance document on "Technical Considerations for Additive Manufactured Devices" for production of medical devices [85]. All criteria applicable to engineered tissue will apply to bioprinted skin [39].

Tissue-engineered skin is typically considered as a "combination product" and Combination products include pharmaceuticals, medical devices, biologics, and their use involves the application of surgical procedures. New surgical procedures are not regulated by the FDA but by the Department of Health and Human Services and can be used on an "as needed" basis at the discretion of the concerned surgeon. However, surgically implantable engineered tissues, depending on their composition, are regulated by the FDA either as devices or biologics and need to be tested in clinical trials before a surgeon is allowed to use them. Currently, products that use stem cells or are derived from stem cells are treated by the FDA as somatic cellular therapies and are regulated as "biologics" under Section 351 of the Public Health Act [39]. As cellular therapies, they are also subject to FDA guidelines for the manufacture of human cells, tissues, and cellular and tissue based products found in part 1271 of the same act. Part 1271 establishes the requirements for donor eligibility procedures not found in the current Good Manufacturing Practices (GMP) guidelines of parts 210 and 211 [39]. These guidelines regulate the way stem cells are isolated, handled, and labeled. Also, engineered tissues typically used in research do not require FDA approval during animal and in vitro testing if they are not intended for use on humans. However, Title 21 of the Federal Code of Regulations defines certain restrictions with regard to shipping and disposal of these products.

Some of the novel features of 3-DP are associated with ethical questions or considerations. For instance, the ability of 3-DP to augment structures and functions of the human body suggests potential exploitation of this feature for human enhancement (e.g., proactively replacing bones with 3-DP alternative materials for function and performance). There is excitement and hope surrounding 3-DP, which may impact patient perceptions and expectations. This must be weighed against the uncertainty regarding safety and efficacy, and the ethics of offering experimental treatments.

Another concern is the shift toward a decentralized manufacturing process.

Current safety regulations rely on centralized manufacturing processes and may not be sufficient if manufacturing occurs at point-of-care. While some believe 3-DP may democratize access to personalized medicine, others believe complex 3-DP products—such as replacement organs, for example may only be accessible by those with substantial resources. This may depend on the funding and reimbursement structure, and the type of product or application.

15.1.16 Start-ups in bio printing

Scientists are racing to make replacement human organs with 3D printers. But while the technology's possibilities are exciting, already there are fears we could be 'playing God'.

An interesting reading related to the active startups is repeated here from [86] for easy access:

Cellink founded in 2016 by Erik Gatenholm and Héctor Martinez in Gothenburg, Sweden. It is a biotechnology start-up that designs bioinks and bioprinters for culturing different cell types to enable applications like patient-derived implants [86]. Cellink was the first company to provide a standardized bio-ink product for sale over the internet [86]. The company has ongoing collaborations with organizations including MedImmune, MIT and Takara Bio, and its printers are used for research at Harvard University, Merck, Novartis, the U.S. Army, Toyota, Johnson & Johnson and more. A stated goal of the company is to help address the existing global shortage of organs suitable for human transplantation. Scientists and commercial companies around the world are working on the startup project.

San Diego-based Organovo [23] has been around since 2007 and has had some success in printing parts of lung, kidney and heart muscle. In 2015, it announced a partnership with the cosmetics behemoth L'Oréal on a plan to supply 3D-printed skin. Ultimately, the goal is to eradicate the need for animal trials.

L'Oréal is committing massive resources to bioprinting. L'oreal is also working with the France-based startup Poietis [87]. The aim this time was to produce synthetic hair follicles. This, it turns out, is devilishly complex: there are more than 15 different cell types in each follicle and there is a cyclical process of fiber production that needs to be stimulated in vitro. The key is the bioprinter that Poietis has developed: most machines push bioink through a nozzle; theirs uses a laser that deposits cells one by one, at a rate of 10,000 drops per second, without damaging the cells. "The way it works is actually quite simple and is similar to inkjet printing," Fabien Guillemot, the CEO and chief scientific officer of Poietis, explained in the video announcing the collaboration. "It prints 3D structures, in this case, biological tissues, by successively layering microdrops of cells on a surface."

Poietis calls its innovation 4D bioprinting. "The fourth dimension is time," [87] "Because laser-assisted bioprinting technology can print the cells basically one at a time, it enables researchers to guide the interaction between the cells and their environment".

In the short to medium term, L'Oréal hopes that its sunscreens and age-defying serums will work more effectively because it can now endlessly test products on a material that reacts exactly like human skin. Perhaps your hair will look more lustrous after using its shampoo, but it's obvious that the impact of such technology could reach far beyond the cosmetics aisle of the supermarket.

If skin can be printed in a lab, then it's not a stretch to imagine it being used to treat severe burns. At present, skin grafts, which can lead to bleeding and infection and typically involve a long recovery time, are the most common forms of treatment for such burns.

Meanwhile, the developments in synthetic hair follicles appear to open the way for commercial products that reduce hair loss or even implants. "Obviously, our objective for the future is to be able to test innovative molecules using systems of follicles created in vitro," says José Cotovio, of L'Oréal's research and innovation department, "but also to increase our understanding of the key processes behind phenomena such as hair ageing, hair loss and hair growth." [87].

A much bigger hurdle that 3D bioprinting needs to overcome, they believe, are the costs. Although it is tempting to hope that the ability to make artificial organs will solve the problem of waiting lists, that is unlikely to be the case. "This is an extremely expensive technology that, if it is realized, only a few will be able to afford," warns Vermeulen. "There is a risk that the health inequalities and postcode lottery that currently exist will also make it unavailable for most people."

"Ideally, you'd like to think that exporting a relatively cheap bioprinter to a country or region with a non-optimal healthcare structure would enable people to have access to the therapies such a machine could provide. In reality, these printers can only work within an existing healthcare infrastructure that has the capacity to make use of it."

Cost has certainly been a prohibitive barrier in these early days of 3D bioprinting. The best machines, such as EnvisionTEC's 3D Bioplotter and RegenHU's 3DDiscovery, are priced in excess of £150,000 and, as a result, are usually only found in labs at universities. However, here, too, Cellink [88,89] is keen to shake things up a little. Although it started out supplying bioink, the company soon moved into hardware. On the table next to us in Gatenholm's office is "Bob", his pet name for the Inkredible + 3D bioprinter [88] that Cellink developed and which he carts around trade shows. The Inkredible + is an attractive machine: a little smaller than a hotel room minibar, it is clean and white and has blue LEDs. But what really catches the eye is the price. Cellink makes three 3D bioprinters, which cost from just £7600 to £29,900. The savings, Gatenholm explains, come in part from using cost-effective 3D printer components instead of super-expensive motor rail systems. Also, again in the spirit of the razors and blades business model, Cellink knows that the more people own 3D bioprinters, the more bioink it will sell.

Gatenholm is proud that his company is driving down the costs of 3D bioprinting. While Cellink's clients include MIT, Harvard and University College London, the company is also making the new technology available to hobbyists. Gatenholm doesn't know how these people will use their machines and inks — perhaps for printing tissues to test drugs or taking cells from a cancerous tumor and using multiple versions to work out how best to treat it — but that is what makes the new technology so exciting.

Researchers at Prellis Biologics aimed to resolve organ crisis using the 3D printing technique, but a faster version of it. The SFO-based company, in 2017, claimed to have reached unprecedented speeds, while printing quality human tissue, along with its active capillaries. This innovation could pave the way for the large-scale manufacture of tissues in the future.

Regarding this, co-founder of Prellis Biologics, Melanie Matheu, said, "A major goal in tissue engineering is to create viable human organs, but nobody could print tissue with the speed and resolution needed to form viable capillaries. At Prellis, we've now developed that technology, paving the way for important medical advances and, ultimately, functional organ replacements".

15.1.17 Challenges of 3D printing

A few years ago, with the introduction of techniques like 3D printing (i.e., organs such as the skin, bladder and blood vessels successfully printed via a 3D printer and transplanted into a recipient), it was expected that a large burden could be taken off acquiring these live human tissues for transplantation. But it was not that simple.

3D printing and organ transplant has known to be time consuming. The technique is a long-drawn process, where, during printing, the tissues require a constant supply of oxygen and other nutrients, a lack of which will cause the death of cells in 30 minutes or so [89].

This concern was overcome with vascularization/supplying blood through capillaries to the living tissues, as part of the process. Traditionally, thin sheets of biological material are printed, using this method, is constantly nourished.

However, yet another issue concerned scientists around the world—the slow rate of 3D bioprinting. In the time that the entire procedure was completed, which could take weeks, the living components involved would long be dead.

Cost has certainly been a prohibitive barrier in these early days of 3D bioprinting. The best machines, such as EnvisionTEC's 3D Bioplotter and RegenHU's 3DDiscovery, are priced in excess of £150,000 and, as a result, are usually only found in labs at universities.

15.1.18 Improved technique of 3D printing viable human organs

As described by the Prellis team in their white paper, there are four parameters to be considered while 3D printing viable human tissues: resolution, speed, complexity, and biocompatibility [90–92].

High resolution was a critical requirement [92] for designing intricate capillaries and microcellular elements. After much deliberation, the study used a resolution of 0.5 μm (10 times smaller than regular bio-printers) for the capillaries.

Speed was of vital importance due to the cells' various differentiation stages [90–92], and low survival time. A light-based polymerization technique called 3D Multi-Photon Holography was used to engineer quick three-dimensional printed structures. An entire block of tissue was formed, by this method, in less than 12 hours.

Todd Huffman, CEO of 3Scan, a digital tissue imaging and data analysis company, explained, "Vasculature is a key feature of complex tissues and is essential for engineering tissue with therapeutic value. Prellis' advancement represents a key milestone in the quest to engineer organs."

Complexity was another feature that was a requirement in research. Combined with the speed and resolution, the investigators were able to achieve a complex vasculature and scaffolding that, in the end, allowed the human tissue to survive. They did this with the help of 3D control of structural polymerization, and its capacity to print on other parts of the previously-printed material.

Biocompatibility or creating tissue that is compatible with living cells, was, and has always been, of "paramount importance."

15.2 Conclusions

This paper reviews recent advances in 3D bioprinting strategies, Bioink, organ printing in kidney and skin. It also gives brief review about startup industries in 3D organ printing. 3D bio printing being a transformative technology can be seen as a paradigm shift in organ transplantation and regenerative medicine. It will facilitate wider use and reduce the cost in organ transplant. This development can be used in medicine and pharmacology such as: organ replacement, Drug discovery, Toxicology testing, cosmetic and beauty industry.

Skin 3D printing is useful for better and more consistent diabetic, ulcer and burn patients' outcomes. Bio printing in skin reconstruction in burn treatment is yet to be achieved but seems to be a promising approach. Nano cellulose fibers can be used in such constructs effectively. It will enable accurate placement of all the different native skin cell types, precise and reproducible fabrication of constructs to replace injured or wounded skin. It will be helpful in wound reconstruction, faster wound closure, which is critical in the case of extensive burn injuries. Faster wound closure will reduce risk of potential infections and hence will result in faster healing, reduced scarring, and better cosmetic outcomes. This will also contribute to a reduction in the number of surgeries required and the length of stay in the hospital for patients.

Nano cellulose fibers can be used into skin care as well new bio decomposable smart textiles.

Further advances in terms of development of standardized clinical grade 3D bio printers and biocompatible bio inks will enable wider use of this technology in the clinic.

Organ printing can extend human lifespan by replacing the organ breaks with printed organs.

There are some ethical concerns. These range from fears over the quality and the efficacy of artificial organs and implants to the accusation that bioprinting will allow humans to "play God".

References

[1] Abouna GM. Organ shortage crisis: problems and possible solutions. Transpl Proc 2008;40(1):34−8.

[2] Zhang B, Luo Y, Ma L, Gao L, Li Y, Xue Q, et al. 3D bioprinting: an emerging technology full of opportunities and challenges. Bio-Des Manuf 2018;1(1):2−13.

[3] An Overview of Clinical Applications of 3-D Printing and Bioprinting. Ottawa: CADTH; 2019 Mar. (CADTH Issues in Emerging Health Technologies; Issue 175).

[4] Aimar A, Palermo A, Innocent B. The Role of 3D Printing in Medical Applications: A State of the Art, Hindawi J Healthc Eng 2019;10Article ID 5340616. Available from: https://doi.org/10.1155/2019/5340616.

[5] 3D Printing Technology Guide, All 10 Types of 3D Printing Technology, https://all3dp.com/1/types-of-3d-printers-3d-printing-technology/

[6] Zhang B, Gao L, Ma L, et al. 3D bioprinting: a novel avenue for manufacturing tissues and organs. Engineering 2019;5:777−94. Available from: https://doi.org/10.1016/j.eng.2019.03.009.

[7] Murphy SV, Atala A. 3D bioprinting of tissues and organs. Nat Biotechnol 2014;32:773−85. Available from: https://doi.org/10.1038/nbt.2958.

[8] Khademhosseini A, Camci-Unal G. 3D Bioprinting in Regenerative Engineering: Principles and Applications. CRC Press; 2018. 16.2.

[9] Urciuolo F, Casale C, Imparato G, Netti PA. Bioengineered skin substitutes: the role of extracellular matrix and vascularization in the healing of deep wounds. J Clin Med 2019;8:2083.

[10] Guillemot F, Souquet A, Catros S, Guillotin B. Laser-assisted cell printing: principle, physical parameters versus cell fate and perspectives in tissue engineering. Nanomed (Lond) 2010;5:507−15.

[11] Koch L, Gruene M, Unger C, Chichkov B. Laser assisted cell printing. Curr Pharm Biotechnol 2013;14:91−7.

[12] Hribar KC, Soman P, Warner J, Chung P, Chen S. Light-assisted direct-write of 3D functional biomaterials. Lab Chip 2014;14:268−75.

[13] Gruene M, Deiwick A, Koch L, Schlie S, Unger C, Hofmann N, et al. Laser printing of stem cells for biofabrication of scaffold-free autologous grafts. Tissue Eng Part C Methods 2011;17:79−87.

[14] Ferris CJ, Gilmore KG, Wallace GG. In het Panhuis M. Biofabrication: an overview of the approaches used for printing of living cells. Appl Microbiol Biotechnol 2013;97:4243−58.

[15] Zhu W, Ma X, Gou M, Mei D, Zhang K, Chen S. 3D printing of functional biomaterials for tissue engineering. Curr Opin Biotechnol 2016;40:103−12.

[16] Beheshtizadeh N, Lotfibakhshaiesh N, Pazhouhnia Z, et al. A review of 3D bio-printing for bone and skin tissue engineering: a commercial approach. J Mater Sci 2020;55:3729−49. Available from: https://doi.org/10.1007/s10853-019-04259-0.

[17] 3D Printed Kidney: The Latest Advancements, https://all3dp.com/2/3d-printed-kidney-the-latest-advancements/

[18] Homan K, Kolesky D, Skylar-Scott M, et al. Bioprinting of 3D Convoluted Renal Proximal Tubules on Perfusable Chips. Sci Rep 2016;6:34845. Available from: https://doi.org/10.1038/srep34845.

[19] van den Berg CW, Ritsma L, Avramut MC, Wiersma LE, et al. Renal subcapsular transplantation of PSC-derived kidney organoids induces neo-vasculogenesis and significant glomerular and tubular maturation in vivo. Stem Cell Rep 2018;10(3):751—65. Available from: https://doi.org/10.1016/j.stemcr.2018.01.041 ISSN: 2213—6711.

[20] One step closer for kidney tissue engineering, Biofabrication Research update, 2018 https://physicsworld.com/a/one-step-closer-for-kidney-tissue-engineering/.

[21] Building Toward a Kidney, Jennifer Lewis's quest to advance 3-D organ printing, 2017, https://harvardmagazine.com/2017/01/building-toward-a-kidney.

[22] George SK, Abolbashari M, Kim T-H, Zhang C, Allickson J, Jackson JD, et al. In Kap Ko, Anthony Atala, and James J. Yoo. Tissue Eng Part A. Nov 2019. 1493—503. https://doi.org/10.1089/ten.tea.2018.0371.

[23] Organovo, https://organovo.com/technology-platform/.

[24] Varkey M, Visscher DO, van Zuijlen PPM, et al. Skin bioprinting: the future of burn wound reconstruction? Burn Trauma 2019;7:4. Available from: https://doi.org/10.1186/s41038-019-0142-7.

[25] He P, Zhao J, Zhang J, et al. Bioprinting of skin constructs for wound healing. Burn Trauma 2018;6:5. Available from: https://doi.org/10.1186/s41038-017-0104-x.

[26] 3D printing skin: the most promising projects, https://all3dp.com/2/3d-printing-skin-the-most-promising-projects/.

[27] MacNeil S. Progress and opportunities for tissue-engineered skin. Nature 2007;445:874—80.

[28] Koch L, Kuhn S, Sorg H, Gruene M, Schlie S, Gaebel R, et al. Laser printing of skin cells and human stem cells. Tissue Eng Part C Methods 2010;16:847—54.

[29] Ng WL, Qi JTZ, Yeong WY, Naing MW. Proof-of-concept: 3D bioprinting of pigmented human skin constructs. Biofabrication 2018;10:025005.

[30] Paulsson M. Basement membrane proteins: structure, assembly, and cellular interactions. Crit Rev Biochem Mol Biol 1992;27:93—127.

[31] Jakab K, Norotte C, Marga F, Murphy K, Vunjak-Novakovic G, Forgacs G. Tissue engineering by self-assembly and bio-printing of living cells. Biofabrication 2010;2:022001.

[32] Climov M, Medeiros E, Farkash EA, Qiao J, Rousseau CF, Dong S, et al. Bioengineered self-assembled skin as an alternative to skin grafts. Plast Reconstr Surg Glob Open 2016;4:e731.

[33] Supp DM, Boyce ST. Engineered skin substitutes: practices and potentials. Clin Dermatol 2005;23:403—12.

[34] Langer K. On the anatomy and physiology of skin. Br J Plast Surg 1978;31:3—8.

[35] Varkey M, Ding J, Tredget EE, Wound Healing, Research G. The effect of keratinocytes on the biomechanical characteristics and pore microstructure of tissue engineered skin using deep dermal fibroblasts. Biomaterials 2014;35:9591—8.

[36] Markman B, Barton FE. Jr. Anatomy of the subcutaneous tissue of the trunk and lower extremity. Plast Reconstr Surg 1987;80:248—54.

[37] Lancerotto L, Stecco C, Macchi V, Porzionato A, Stecco A, De Caro R. Layers of the abdominal wall: anatomical investigation of subcutaneous tissue and superficial fascia. Surg Radiol Anat 2011;33:835—42.

[38] Jaspers ME, Brouwer KM, van Trier AJ, Groot ML, Middelkoop E, van Zuijlen PP. Effectiveness of autologous fat grafting in adherent scars: results obtained by a comprehensive scar evaluation protocol. Plast Reconstr Surg 2017;139:212—19.

[39] Varkey M, Atala A. Organ bioprinting: a closer look at ethics and policies. Wake F J Law Policy 2015;5:275—98.

[40] Pourchet L, Thepot A, Albouy M, Courtial E-J, Boher A, Blum L, et al. Human Skin 3D Bioprinting Using Scaffold-Free Approach. Adv Healthc Mater 2016;6. Available from: https://doi.org/10.1002/adhm.201601101.

[41] https://www.sciencedaily.com/releases/2019/11/191101111556.htm.

[42] Nourian Dehkordi A, Mirahmadi Babaheydari F, Chehelgerdi M, et al. Skin tissue engineering: wound healing based on stem-cell-based therapeutic strategies. Stem Cell Res Ther 2019;10:111. Available from: https://doi.org/10.1186/s13287-019-1212-2.

[43] http://bsp.iitd.ac.in/index.php/2019/08/31/research-of-the-month-3d-bio-printing-of-human-skin/.

[44] Baltazar T, Merola J, Catarino C, Xie CB, Kirkiles-Smith NC, Lee V, et al. 3D bioprinting of a vascularized and perfusable skin graft using human keratinocytes. Tissue Eng Part A 2019. Available from: http://doi.org/10.1089/ten.tea.2019.0201.

[45] Barret JP. Burns reconstruction. BMJ (Clin Res Ed) 2004;329(7460):274—6. Available from: https://doi.org/10.1136/bmj.329.7460.274.

[46] Varkey M, Atala A. Current challenges and future perspectives of bioprinting. In: Khademhosseini A, Camci-Unal G, editors. 3D bioprinting in regenerative engineering: principles and applications. Taylor & Francis; 2018.

[47] Skardal A, Mack D, Kapetanovic E, Atala A, Jackson JD, Yoo J, et al. Bioprinted amniotic fluid-derived stem cells accelerate healing of large skin wounds. Stem Cell Transl Med 2012;1:792—802.

[48] Kang HW, Lee SJ, Ko IK, Kengla C, Yoo JJ, Atala A. A 3D bioprinting system to produce human-scale tissue constructs with structural integrity. Nat Biotechnol 2016;34:312—19.

[49] Schiele NR, Chrisey DB, Corr DT. Gelatin-based laser direct-write technique for the precise spatial patterning of cells. Tissue Eng Part C Methods 2011;17:289—98.

[50] Rimann M, Bono E, Annaheim H, Bleisch M, Graf-Hausner U. Standardized 3D bioprinting of soft tissue models with human primary cells. J Lab Autom 2016;21:496—509.

[51] Min D, Lee W, Bae IH, Lee TR, Croce P, Yoo SS. Bioprinting of biomimetic skin containing melanocytes. Exp Dermatol 2017;27:453—9.

[52] Michael S, Sorg H, Peck CT, Koch L, Deiwick A, Chichkov B, et al. Tissue engineered skin substitutes created by laser-assisted bioprinting form skin-like structures in the dorsal skin fold chamber in mice. PLoS One 2013;8:e57741.

[53] Koch L, Deiwick A, Schlie S, Michael S, Gruene M, Coger V, et al. Skin tissue generation by laser cell printing. Biotechnol Bioeng 2012;109:1855—63.

[54] Enoch S, Shaaban H, Dunn KW. Informed consent should be obtained from patients to use products (skin substitutes) and dressings containing biological material. J Med Ethics 2005;31:2—6.

[55] Auger FA, Lacroix D, Germain L. Skin substitutes and wound healing. Skin Pharmacol Physiol 2009;22:94—102.

[56] Marga F, Jakab K, Khatiwala C, Shepherd B, Dorfman S, Hubbard B, et al. Toward engineering functional organ modules by additive manufacturing. Biofabrication 2012;4:022001.

[57] Visscher DO, Farre-Guasch E, Helder MN, Gibbs S, Forouzanfar T, van Zuijlen PP, et al. Advances in bioprinting technologies for craniofacial reconstruction. Trends Biotechnol 2016;34:700—10.

[58] Chang R, Nam J, Sun W. Effects of dispensing pressure and nozzle diameter on cell survival from solid freeform fabrication-based direct cell writing. Tissue Eng Part A 2008;14:41—8.

[59] Lim KS, Levato R, Costa PF, Castilho MD, Alcala-Orozco CR, van Dorenmalen KMA, et al. Bio-resin for high resolution lithography-based biofabrication of complex cell-laden constructs. Biofabrication 2018;10 034101.

[60] Lee JM, Yeong WY. Design and printing strategies in 3D bioprinting of cell- hydrogels: a review. Adv Healthc Mater 2016;5:2856—65.

[61] Gopinathan J, Noh I. Recent trends in bioinks for 3D printing. Biomater Res 2018;22:11.

[62] Lee W, Debasitis JC, Lee VK, Lee JH, Fischer K, Edminster K, et al. Multi- layered culture of human skin fibroblasts and keratinocytes through three- dimensional freeform fabrication. Biomaterials 2009;30:1587—95.

[63] Hou X, Liu S, Wang M, Wiraja C, Huang W, Chan P, et al. Layer-by-layer 3D constructs of fibroblasts in hydrogel for examining transdermal penetration capability of nanoparticles. SLAS Technol 2017;22:447—53.

[64] Cubo N, Garcia M, Del Canizo JF, Velasco D, Jorcano JL. 3D bioprinting of functional human skin: production and in vivo analysis. Biofabrication 2016;9:015006.

[65] Ozbolat IT. Bioprinting scale-up tissue and organ constructs for transplantation. Trends Biotechnol 2015;33:395—400.

[66] Binder K.W. In situ bioprinting of the skin: Wake Forest University Graduate School of Arts and Sciences; 2011.

[67] Zhang LGFJ, Leong K. 3D bioprinting and nanotechnology in tissue engineering and regenerative medicine. Academic Press; 2015.

[68] Liu N, Huang S, Yao B, Xie J, Wu X, Fu X. 3D bioprinting matrices with controlled pore structure and release function guide in vitro self- organization of sweat gland. Sci Rep 2016;6:34410.

[69] Huang S, Yao B, Xie J, Fu X. 3D bioprinted extracellular matrix mimics facilitate directed differentiation of epithelial progenitors for sweat gland regeneration. Acta Biomater 2016;32:170—7.

[70] Mondy WL, Cameron D, Timmermans JP, De Clerck N, Sasov A, Casteleyn C, et al. Computer-aided design of microvasculature systems for use in vascular scaffold production. Biofabrication 2009;1:035002.

[71] Mironov V, Visconti RP, Kasyanov V, Forgacs G, Drake CJ, Markwald RR. Organ printing: tissue spheroids as building blocks. Biomaterials 2009;30:2164—74.

[72] Keriquel V, Guillemot F, Arnault I, Guillotin B, Miraux S, Amedee J, et al. In vivo bioprinting for computer- and robotic-assisted medical intervention: preliminary study in mice. Biofabrication 2010;2:014101.

[73] Arai K, Iwanaga S, Toda H, Genci C, Nishiyama Y, Nakamura M. Three-dimensional inkjet biofabrication based on designed images. Biofabrication 2011;3:034113.

[74] Arslan-Yildiz A, El Assal R, Chen P, Guven S, Inci F, Demirci U. Towards artificial tissue models: past, present, and future of 3D bioprinting. Biofabrication 2016;8:014103.

[75] Kirchmajer DM, Gorkin R, Panhuis Min H. An overview of the suitability of hydrogel-forming polymers for extrusion-based 3D-printing. J Mater Chem B 2015;3:4105—17.

[76] Parak A, Pradeep P, du Toit LC, Kumar P, Choonara YE, Pillay V. Functionalizing bioinks for 3D bioprinting applications. Drug Discov Today 2018;S1359—6446(18):30243—5.

[77] Auger FA, Gibot L, Lacroix D. The pivotal role of vascularization in tissue engineering. Annu Rev Biomed Eng 2013;15:177—200.

[78] Laschke MW, Vollmar B, Menger MD. Inosculation: connecting the lifesustaining pipelines. Tissue Eng Part B Rev 2009;15:455—65.

[79] Miller JS, Stevens KR, Yang MT, Baker BM, Nguyen DH, Cohen DM, et al. Rapid casting of patterned vascular networks for perfusable engineered three-dimensional tissues. Nat Mater 2012;11:768—74.

[80] Kolesky DB, Truby RL, Gladman AS, Busbee TA, Homan KA, Lewis JA. 3D bioprinting of vascularized, heterogeneous cell-laden tissue constructs. Adv Mater 2014;26:3124—30.

[81] Yu Y, Zhang Y, Martin JA, Ozbolat IT. Evaluation of cell viability and functionality in vessel-like bioprintable cell-laden tubular channels. J Biomech Eng 2013;135:91011.

[82] Kolesky DB, Homan KA, Skylar-Scott MA, Lewis JA. Three-dimensional bioprinting of thick vascularized tissues. Proc Natl Acad Sci USA 2016;113:3179—84.

[83] Paulsen SJ, Miller JS. Tissue vascularization through 3D printing: will technology bring us flow? Dev Dyn 2015;244:629—40.

[84] Hunsberger J, Harrysson O, Shirwaiker R, Starly B, Wysk R, Cohen P, et al. Manufacturing road map for tissue engineering and regenerative medicine technologies. Stem Cell Transl Med 2015;4:130—5.

[85] Printing FsRiD. In: Document FG, editor. Technical considerations for additive manufactured devices. Rockville: FDA; 2017.

[86] Could 3D printing solve the organ transplant shortage? 2017, The Guardian, https://www.theguardian.com/technology/2017/jul/30/will-3d-printing-solve-the-organ-transplant-shortage [accessed 05.07.18].

[87] Engineering small vessels and micro-vascularisation for 3D models and regenerative medicine, https://poietis.com/events/engineering-small-vessels-and-micro-vascularisation-for-3d-models-and-regenerative-medicine/.

[88] Bioprinter Preps for ISS to 3D Print Beating Heart Tissue, https://cellink.com/bioprinter-preps-iss-3d-print-beating-heart-tissue/?tab = See%20our%20printers.

[89] https://www.evolving-science.com/bioengineering/3d-human-organs-00719.

[90] New tech allows much faster 3D-printing of human tissue with capillaries, 2018, New Atlas, https://newatlas.com/3d-printed-human-organs-capillaries/55173/, [accessed 09.18].

[91] Matheu M, et al. Human organ and tissue engineering: advances and challenges in addressing the medical crisis of the 21st century, 2018.

[92] Prellis biologics achieves unprecedented speed and resolution in 3d printing of human tissue with capillaries, 2018, Synbiobeta, https://synbiobeta.com/prellis-biologics-speed-and-resolution-in-3d-printing-of-human-tissue/ [accessed 10.07.18].

16

Role of umbilical cord stem cells in tissue engineering

Merlin Rajesh Lal[1] *and Oormila Kovilam*[2]

[1]LifeCell International (Pvt) Ltd, Chennai, India [2]St Joseph's Hospital, Indiana, United States

Stem cells are inevitable element in restoring tissue function and in cell-based tissue engineering strategies. Stem cells aid in tissue repair by replenishing the pro-healing cells and mediators owing to their self-renewal capacity. An active stem cell begets either diverse stem cell population or functional cell types by the process known as differentiation, accelerating the regenerative responses. The unique regeneration capability of stem cells elicits novel therapeutic avenues for injuries and chronic diseases such as diabetes, arthritis, heart diseases, and cancer. Stem cells are broadly classified as embryonic stem cells and adult stem cells. Embryonic stem cells (ESCs) derived from the inner cell mass of the early stage human embryo which can generate all cell types of the body (pluripotent). However, the undifferentiated ESCs can cause teratoma formation in vivo, coupled with the ethical issues associated to the source from it is derived restricts the use of ESCs for therapeutic applications.

Recently, the differentiated adult cells were reprogrammed by scientists to attain stem cells like pluripotency by genetic modifications. These cells are called as induced pluripotent stem cells (iPSCs). This innovation opened avenues to upgrade the understanding of how to specifically differentiate the stem cells to specific lineages. However, the challenges of using iPSCs for cell-based therapies warrants further investigation on the teratoma formation. The cells to form undifferentiated tissue mass is more in iPSCs than that of ESCs [1−4]. On the other hand, the adult stem cells or somatic stem cells are present in body after development with the function of growth, healing and replenishing the damaged cells. They are multipotent stem cells and are limited to differentiate into minimal lineages. The proliferation rate of ESCs are superior to the adult stem cells, but require xenogenic substrate for cell attachment, and high level of expertize and is expensive. Conversely, the multipotent adult stem cells easily adhere to plastic culture plates, are stable and pose less ethical concerns [5−8]. Moreover, it is still a challenge to achieve a uniform population of pluripotent stem cells in culture plates.

Mesenchymal stem cells (MSCs), neural stem cells, hematopoietic stem cells, and skin stem cells are adult stem cells present in human body. MSCs in bone marrow niche can differentiate to cartilage, bone and fat cells. Nerve cells, astroctes and oligodendrocytes are generated by neural cells, while the cells of blood tissues such as red blood cells (RBCs), white blood cells (WBCs) and platelets are generated and replenished from a unique population of hemaopoietic stem cells (HSCs). The keratinocytes are formed from skin stem cells. Adult stem cells can be derived from bone marrow, adipose tissue, liver and skin which may be autologous or allogeneic origin and its ability to differentiate to specific lineages differs with respect to the source of origin. On the other hand, the umbilical cords represent a rich reservoir of mesenchymal stem cells and haemaopoietic stem cells. The major focus of this chapter is to discuss the pros and cons of using the stem cells derived from umbilical cord and their application in tissue engineering and regenerative medicine.

16.1 Umbilical cord and MSCs

The umbilical cord consists of two umbilical arteries and an umbilical vein in the stroma that is filled with white mucopolysaccharides matrix called Wharton's jelly. The umbilical cord is covered with amniotic membrane which is the extension of placenta. The Wharton's jelly is a rich source of multipotent mesenchymal stem cells that can be easily harvested and expanded in vitro. Even though the numbers are limited, the umbilical cord blood forms an ideal source of MSCs that can be expanded for therapeutic applications. Since umbilical cords (UC) are clinically discarded tissue, there are no serious ethical concerns regarding the utilization of UC. The umbilical cord derived mesenchymal stem cells (UCMSC) that express MSC markers including CD29, CD44, CD73 CD90, CD105, CD144 and CD166 the presence of transcription factors octamer (Oct)-4, Nanog, and Sox-2; however, the hematopoietic markers including CD34, CD26, CD31, CD45 and HLA-DR are negative [9,10]. The immunological markers involved in tissue rejection are negative in UCMSC and exhibit superior proliferation rate than the bone marrow derived MSCs. Hence, UCMSCs have been considered as a viable allogenic source. The CD38⁻ cell population in human are considered to be a subtype of CD34 cells belonging to the hematopoietic stem cells (HSCs) while the CD38[+] cell population belong to the hematopoietic progenitor cells (HPCs) with few or no HSCs [11]. Conversely, there are increasing reports that umbilical cord blood possesses primitive HSCs that do not exhibit hematopoietic activity or phenotype matching the classical HSCs. Such cells (CD34⁻FltLin⁻) upon intrabone delivery get engrafted like a long-term repopulating HSCs (LT-HSCs); however, the engraftment was minimal when administered intravenously [12]. The International Society for Cellular Therapy states stringent clauses to distinguish between multipotent mesenchymal stem cells with other stem cells [13].

16.2 Transplantation biology of UCMSCs: biomaterials, differentiation and regeneration

Tissue engineering and regenerative medicine aims to regenerate, repair or replace the lost tissue or/and its function for which mesenchymal stem cells offer viable options. The two transplantation strategies include systemic delivery and deliveries at sight of damaged tissue (site-directed) are being widely practiced. In systemic administration the MSCs are administered into the systemic circulation, they migrate and get homed at injury site of the tissue such as brain, myocardium, spinal cord, or other organs and aid in repair of tissues. The possible mechanism of homing MSCs to the injured site is hypothesized to be similar to leukocyte recruitment across the endothelium of the blood vessels; however, the mechanism remains unclear. Generally, the MSCs express receptors for chemokines, integrins and adhesion molecules that aids in migration via vascular cell adhesion molecules and p-selectins [14−16]. Patients who received umbilical cord blood transplantation are being benefited by getting alleviated from life-threatening disease conditions coupled with the advantage of less possibility of graft versus host disease (GVHD). The effect of bone loss due to ovariectomy was studied on nude mice. After ovariectomy the mice were either administered with UCMSCs or human dermal fibroblast or UCMSCs conditioned medium. The animals which received UCMSCs did not exhibit any loss in bone mineral density compared to control. Moreover, the animals which received the UCMSC conditioned medium after ovariectomy had less bone loss than the control group, even though it is less than that of animals which received UCMSCs. Cell trafficking experiments did not show any of UCMSCs after 48 hours in the bones owing to the action bone mineral density recovery to be effected by paracrine signaling [17].

To ensure the delivery of MSCs to the site of injury, biomaterial scaffolds are being used. Biomaterial scaffolds provide mechanical cues and support the anchorage dependent cells facilitating the tissue engineering strategies. The use of biomaterials as scaffold for tissue repair applications has been reported in the 1960s [18]. Currently, there are number of scaffolds reported with tissue specific applications which are made of natural or synthetic biopolymers or combination of both. The natural polymers such as fibrin, collagen, alginate, chitosan, agarose, hyaluronic acid and synthetic polymers and synthetic polymers such as poly ethylene glycol (PEG), poly lactic acid (PLA), poly-glycolic acid (PGA), and their copolymer poly-lactic-*co*-glycolic acid (PLGA), poly ethylene glycol (PEG), polycaprolactone (PCL) are being used as such or in combinations. The hydrogel scaffolds are also widely used in tissue engineering which can be natural, synthetic or hybrid. In addition, the scaffolds have been developed from extracellular matrix (ECM) components which in turn form the intercellular connections or designed for eluting growth factors which eventually aids in tissue regeneration. MSCs along with diverse scaffolds have been studied extensively for its ability to generate functional tissues.

A previous study reported that UCMSCs when seeded to PCL/collagen nanoscaffolds exhibited prominent cell attachment, proliferation and differentiated to chondrogenic lineage as evidenced by the production of glycosaminoglycans (GAG) and hyaluronic acid production [19]. The differentiation of UCMSCs to adipocytes is minimal than bone marrow MSCs (BMSCs). UCMSCs express myocardial genes such as GATA-4, cTnT, α-actin and Cx-43 when cultured on myocardial induction medium. UCMSCs have also been studied for wound healing applications. Human UCMSCs were administered intravenously to adult Wistar rats with burn wounds improved the healing and neovascularization was reported in the rats which received UCMSCs than the control rats [20]. A dermal equivalent was developed with UCMSC incorporated fibrin-based scaffold and transplanted subcutaneously on immunocompetent mouse model with full thickness wounds. The wound healing was significantly increased without the formation of scar tissue than the control group which received tulle grease [21]. Similar observations were reported while using double layered collagen-fibrin membranes seeded with UCMSCs and implanted on wound repair. Moreover, the expression of genes responsible for neovascularisation, re-epithelialization, migration and fibroblast proliferation were upregulated and HGF, VEGF and other angiogenic growth factors were increased in circulation due to the presence of UCMSCs, suggesting their regenerative potential [22].

In addition, the UCMSC possess higher potential to differentiate to chondrogenic and osteogenic lineages when compared with that of the BMSCs and adipose derived MSCs. However, the in vivo studies on rat with scaffolds seeded with UCMSCs were inferior to BMSCs which could be due to the differences in the experimental condition, where Zhang e al., implanted cell seeded scaffolds subcutaneously to rat models in an attempt to study the chondrogenic potential of UCMSCs [23,24]. The proliferation rate of UCMSCs, human fetal bone marrow (hfMSCs) and human adult adipose tissue (hATMSCs) were studied using monolayer culture method and on three-dimensional polycaprolactone-tricalcium-phosphate scaffolds. The study revealed that UCMSCs and hfMSCs were predominated in expressing Oct-4 and Nanog, the embryonic pluripotency markers [24]. However, postimplantation of UCMSCs with polycaprolactone-tricalcium-phosphate scaffolds failed to mineralize as compared to hfMSCs. Conversely, the source to derive hfMSCs will be very limited and pose ethical concern.

Autologous chondrocytes implantation (ACI) is a well-known tissue engineering strategy to treat moderate size articular cartilage defect. An arthroscopic operation is performed to collect healthy cartilage biopsies from nonload bearing areas. The collected cartilage biopsies are used to isolate the chondrocytes and in vitro expansion. The expanded chondrocytes are implanted into the defect site and covered with a membrane to generate functional cartilage. Use of MSCs along with chondrocytes avoid the in vitro culture expansion of chondrocytes and the associated risks and can avoid the surgical interventions [25,26]. It was initially thought that the regeneration is affected by the differentiation of MSCs to chondrocytes at the site of implantation. However, the cell racking experiments revealed that the MSCs remained in the site of cartilage defect for first week and were not detected during the following weeks. This experiment revealed that MSCs would have acted initially and contributed to cartilage formation by prior to their differentiation suggesting that initial activation is required for cartilage regeneration which is independent of prolonged MSCs signaling. Matrix-associated ACI (MACI) is the next generation ACI where the cells are delivered on bi-layered collagen scaffolds. Several reports portray the regeneration of cartilage while long term follow up for up to five years showed further improvement. In another study, the polylactic acid (PLA) scaffolds seeded with UCMSCs were seeded and were implanted on rabbit models. After 12 weeks, the animals which received UCMSCs produced hyaline-like cartilage which was similar to the group of animal which were implanted with chondrocytes [27].

Parkinson's disease has been introduced in animal models using parkinsonian toxins 6-hydroxydopamine (6-OHDA), 1-methyl-4-phenyl-1, 2, 3, 6-tetrahydropyridine (MPTP) or rotenone and transplanted with UCMSCs. The administration of UCMSCs improved the behavioral recovery and decreased rotations by 50% in animals, however the release of dopamine by the UCMSCs warrants detailed investigation [28]. Also, the UCMSCs have been studied for wound healing applications. In another study, the human UCMSCs were administered intravenously to adult Wistar rats with burn wounds.

Improved the healing and neovascularization was reported in the rats which received UCMSCs compared to the control rats. In addition, the administration of 16 million UCMSCs intravenously was followed by intrathecal and intravenous implantation improved the spinal cord entrapment on a 62 year old paraplegic women improving the spinal function [29,30]. Another study reported very mild side effect of fever and respiratory tract infection on four patients (out of 41) with Crohn's disease after receiving the UCMSCs which could be due to retention of UCMSCs in the pulmonary capillaries which warrants further detailed investigation [31]. Intravenous administration of UCMSCs to patients with multiple sclerosis did not show any adverse side effects but improved the symptoms. Similar results were reported with patients with relapsing skin disease involving eczematous, xerosis and pruritus lesions who received UCMSCs [32,33].

Exosomes are class of extracellular vesicles secreted by various cells which contain genetic materials and proteins. The genetic materials include miRNA, premiRNA, and other noncoding RNA that are released to the target cells extending the cell to cell communication. The MSCs secreted exosomes are new alternative to cell free therapy for regeneration of tissues. The exosomes of MSCs are smaller than the cells, nonviable unlike the cells, nonimmunogenic and has no risk of tumor formation. Recently there are many reports on the therapeutic effects of UCMSCs exosomes on liver fibrosis. Liver injury on mouse model was created by CCl_4 and 250 µg of UCMSCs exosomes were administered in the liver. Three weeks postadministration reduced hepatic inflammations, collagen deposition and reduction in fibrous capsule surface were observed. Also, considerable reduction in the phosphorylation of Smad2, TGF-β1 and collagen types I suggesting that UCMSc-exosomes aids in healing CCL4 induced liver fibrosis by inhibiting the epithelial-mesenchymal transition [34].

16.3 Concerns and perspective

The major concerns on stem cells transplantation is the formation of undifferentiated cell mass or teratoma formation owing to the superior proliferation capacity. Wang et al., argued that the reason for this would be cross contamination with other cell lines like HT1080, HeLa, or other tumor cell lines within the system or used for transplantation. Since there is no obvious evidence showing source of MSCs crosses contamination with other cell lines this argument could not be ruled out. To study the ability of administered UCMSCs to inhibit tumor, leukemic tumor was induced in mouse model by injecting myelogenous leukemia cell line (K562 cells) and after tumor formation was confirmed transplanted with UCMSCs. The animal groups that received UCMSCs showed significant inhibition of K562 cells induced tumor [35].

Studies on UCMSCs exhibit the property of immunosuppression in vitro and in vivo by the secretion of soluble growth factors such as PGE2, galectin-1 and HLA-G5 which require the activation by proinflammatory cytokine IFN-γ [36]. The MSCs migrate to the injury site and contribute to the tissue repair by secreting cytokines, activation of circulator and resident stem cells and thus aiding in tissue repair and regeneration, referred as trophic effect. Scientists are inclined towards exploring these trophic roles of MSCs which are distinct from differentiation since the grafted MSC can modulate activity of surrounding cells by paracrine signaling [37]. Concurrent with this, there are many in vitro studies that showed improved tissue regeneration upon using MSC conditioned medium. The analysis of MSC conditioned medium revealed the presence of secreted growth factors, chemokines, cytokines and growth factors [28,38,39] which supports the trophic effect of UCMSCs by secreting soluble factors to modulate tissue repair. Previous studies on the paracrine factors secreted on the conditioned medium with UCMSCs showed VEGF-independent pathway of angiogenic responses [40]. The recent development and understanding on the UCMSCs revealed that the therapeutic effect is the function of cell to cell contact and by paracrine signaling; not by the engraftment and differentiation [41]. In the event of these, Arnold I Caplan who coined the word mesenchymal stem cells in 1991, later in 2010 proposed to change the name be referred as medicinal signaling cells [42–44]. Our knowledge on the signaling and regulatory pathways towards committed differentiation is still incomplete and there are tremendous hurdles in using stem cells to produce engineered tissues. However, much work remains to be done in the laboratory and the clinic to understand how to use these cells for cell-based regeneration.

References

[1] Jia F, et al. A nonviral minicircle vector for deriving human iPS cells. Nat Methods 2010;7(3):197–9. Available from: https://doi.org/10.1038/nmeth.1426 Nature Publishing Group.
[2] Sun N, et al. Feeder-free derivation of induced pluripotent stem cells from adult human adipose stem cells. Proc Natl Acad Sci 2009;106(37):15720–5. Available from: https://doi.org/10.1073/pnas.0908450106.
[3] Yu J, et al. Induced pluripotent stem cell lines derived from human somatic cells. Science 2007;318 (5858):1917–20. Available from: https://doi.org/10.1126/science.1151526.
[4] Zheng Z, et al. CDKN2B upregulation prevents teratoma formation in multipotent fibromodulin-reprogrammed cells. J Clin Invest 2019;129(8):3236–51. Available from: https://doi.org/10.1172/JCI125015.
[5] Friedenstein AJ, et al. Heterotopic of bone marrow. Analysis of precursor cells for osteogenic and hematopoietic tissues. Transplantation 1968;6(2):230–47. Available from: https://doi.org/10.1097/00007890-196803000-00009.
[6] McKee C, Chaudhry GR. Advances and challenges in stem cell culture. Colloids Surf B: Biointerfaces 2017;159:62–77. Available from: https://doi.org/10.1016/j.colsurfb.2017.07.051.
[7] Baksh D, Yao R, Tuan RS. Comparison of proliferative and multilineage differentiation potential of human mesenchymal stem cells derived from umbilical cord and bone marrow. Stem Cell 2007;25(2):1384–92. Available from: https://doi.org/10.1634/stemcells.2006-0709.
[8] Pittenger MF. Multilineage potential of adult human mesenchymal stem cells. Science 1999;284(5411):143–7. Available from: https://doi.org/10.1126/science.284.5411.143.
[9] Hordyjewska A, Popiołek Ł, Horecka A. Characteristics of hematopoietic stem cells of umbilical cord blood. Cytotechnology 2015;67(3):387–96. Available from: https://doi.org/10.1007/s10616-014-9796-y.
[10] Arpornmaeklong P, et al. Phenotypic characterization, osteoblastic differentiation, and bone regeneration capacity of human embryonic stem cell-derived mesenchymal stem cells. Stem Cell Dev 2009;18(7):955–68. Available from: https://doi.org/10.1089/scd.2008.0310.
[11] Park S-K, Won J-H. Usefulness of umbilical cord blood cells in era of hematopoiesis research. Int J Stem Cell 2009;2(2):90–6. Available from: https://doi.org/10.15283/ijsc.2009.2.2.90.
[12] Ratajczak J, et al. Hematopoietic differentiation of umbilical cord blood-derived very small embryonic/epiblast-like stem cells. Leukemia 2011;25(8):1278–85. Available from: https://doi.org/10.1038/leu.2011.73.
[13] Horwitz EM, et al. Clarification of the nomenclature for MSC: the International Society for Cellular Therapy position statement. Cytotherapy 2005;7(5):393–5. Available from: https://doi.org/10.1080/14653240500319234.
[14] Chen J, et al. Therapeutic benefit of intracerebral transplantation of bone marrow stromal cells after cerebral ischemia in rats. J Neurol Sci 2001;189(1–2):49–57. Available from: https://doi.org/10.1016/S0022-510X(01)00557-3.

[15] Qu C, et al. Treatment of traumatic brain injury in mice with marrow stromal cells. Brain Res 2008;1208:234−9. Available from: https://doi.org/10.1016/j.brainres.2008.02.042.

[16] Qu C, et al. Treatment of traumatic brain injury in mice with bone marrow stromal cell-impregnated collagen scaffolds. J Neurosurg 2009;111(4):658−65. Available from: https://doi.org/10.3171/2009.4.JNS081681.

[17] An JH, et al. Transplantation of human umbilical cord blood-derived mesenchymal stem cells or their conditioned medium prevents bone loss in ovariectomized nude mice. Tissue Eng Part A 2013;19(5−6):685−96. Available from: https://doi.org/10.1089/ten.tea.2012.0047.

[18] Chesterman PJ, Smith AU. Homotransplantation of articular cartilage and isolated chondrocytes: an experimental study in rabbits J Bone Jt Surg Br 1968;50-B(1):184−97Available at. Available from: http://www.bjj.boneandjoint.org.uk/content/50-B/1/184.

[19] Fong C-Y, et al. Human umbilical cord Wharton's jelly stem cells undergo enhanced chondrogenic differentiation when grown on nanofibrous scaffolds and in a sequential two-stage culture medium environment. Stem Cell Rev Rep 2012;8(1):195−209. Available from: https://doi.org/10.1007/s12015-011-9289-8.

[20] Liu L, et al. Human umbilical cord mesenchymal stem cells transplantation promotes cutaneous wound healing of severe burned rats. Plos One 2014;9(2):e88348. Available from: https://doi.org/10.1371/journal.pone.0088348.

[21] Montanucci P, et al. Human umbilical cord Wharton jelly-derived adult mesenchymal stem cells, in biohybrid scaffolds, for experimental skin regeneration. Stem Cell Int 2017;2017:1−13. Available from: https://doi.org/10.1155/2017/1472642.

[22] Lui L, et al. Human umbilical cord mesenchymal stem cellstransplantation promotes cutaneous wound healing of severe burned rats. PLoS One 2014;9(2):e88348. Available from: https://doi.org/10.1371/journal.pone.0088348.

[23] El Omar R, et al. Umbilical cord mesenchymal stem cells: the new gold standard for mesenchymal stem cell-based therapies? Tissue Eng − Part B: Rev 2014;20(5):523−44. Available from: https://doi.org/10.1089/ten.teb.2013.0664.

[24] Zhang Z-Y, et al. Superior osteogenic capacity for bone tissue engineering of fetal compared with perinatal and adult mesenchymal stem cells. Stem Cell 2009;27(1):126−37. Available from: https://doi.org/10.1634/stemcells.2008-0456.

[25] Ochs BG, et al. Remodeling of articular cartilage and subchondral bone after bone grafting and matrix-associated autologous chondrocyte implantation for osteochondritis dissecans of the knee. Am J Sports Med 2011;39(4):764−73. Available from: https://doi.org/10.1177/0363546510388896.

[26] Prasadam I, et al. Mixed cell therapy of bone marrow-derived mesenchymal stem cells and articular cartilage chondrocytes ameliorates osteoarthritis development. Lab Invest 2018;98(1):106−16. Available from: https://doi.org/10.1038/labinvest.2017.117.

[27] Revell CM, Athanasiou KA. Success rates and immunologic responses of autogenic, allogenic, and xenogenic treatments to repair articular cartilage defects. Tissue Eng Part B: Rev 2009;15(1):1−15. Available from: https://doi.org/10.1089/ten.teb.2008.0189.

[28] Boroujeni M, Gardaneh M. Umbilical cord: an unlimited source of cells differentiable towards dopaminergic neurons. Neural Regener Res 2017;12(7):1186. Available from: https://doi.org/10.4103/1673-5374.211201.

[29] Rahyussalim AJ, et al. Improvement of renal function after human umbilical cord mesenchymal stem cell treatment on chronic renal failure and thoracic spinal cord entrapment : a case report. J Med Case Rep 2020;11(1):1−12. Available from: https://doi.org/10.1186/s13256-017-1489-7.

[30] Bochon B, et al. Mesenchymal stem cells—potential applications in kidney diseases. Int J Mol Sci 2019;20 (10):2462. Available from: https://doi.org/10.3390/ijms20102462.

[31] Zhang J, et al. Umbilical cord mesenchymal stem cell treatment for Crohn's disease: a randomized controlled clinical trial. Gut Liver 2018;12(1):73−8. Available from: https://doi.org/10.5009/gnl17035.

[32] Chez M, et al. Safety and observations from a placebo-controlled, crossover study to assess use of autologous umbilical cord blood stem cells to improve symptoms in children with autism. Stem Cells Transl Med 2018;7 (4):333−41. Available from: https://doi.org/10.1002/sctm.17-0042.

[33] Kim H-S, et al. Clinical trial of human umbilical cord blood-derived stem cells for the treatment of moderate-to-severe atopic dermatitis: phase I/IIa studies. Stem Cells 2017;35(1):248−55. Available from: https://doi.org/10.1002/stem.2401.

[34] Yin F, Wang W-Y, Jiang W-H. Human umbilical cord mesenchymal stem cells ameliorate liver fibrosis in vitro and in vivo: from biological characteristics to therapeutic mechanisms. World J Stem Cell 2019;11 (8):548−64. Available from: https://doi.org/10.4252/wjsc.v11.i8.548.

[35] Wang Y, et al. Safety of mesenchymal stem cells for clinical application. Stem Cell Int 2012;2012:1−4. Available from: https://doi.org/10.1155/2012/652034.

[36] Nagamura-Inoue T. Umbilical cord-derived mesenchymal stem cells: their advantages and potential clinical utility. World J Stem Cell 2014;6(2):195. Available from: https://doi.org/10.4252/wjsc.v6.i2.195.

[37] Caplan AI, Dennis JE. Mesenchymal stem cells as trophic mediators. J Cell Biochem 2006;98(5):1076−84. Available from: https://doi.org/10.1002/jcb.20886.

[38] Dexheimer V, Frank S, Richter W. Proliferation as a requirement for in vitro chondrogenesis of human mesenchymal stem cells. Stem Cell Dev 2012;21(12):2160−9. Available from: https://doi.org/10.1089/scd.2011.0670.

[39] Hammar E, et al. Role of the Rho-ROCK (Rho-associated kinase) signaling pathway in the regulation of pancreatic β-cell function. Endocrinology 2009;150(5):2072−9. Available from: https://doi.org/10.1210/en.2008-1135.

[40] Kuchroo P, et al. Paracrine factors secreted by umbilical cord-derived mesenchymal stem cells induce angiogenesis in vitro by a VEGF-independent pathway. Stem Cell Dev 2015;24(4):437−50. Available from: https://doi.org/10.1089/scd.2014.0184.

[41] Fitzsimmons REB, et al. Mesenchymal stromal/stem cells in regenerative medicine and tissue engineering. Stem Cell Int 2018;2018:1−16. Available from: https://doi.org/10.1155/2018/8031718.

[42] Caplan AI. Mesenchymal stem cells. J Orthopaedic Res 1991;9(5):641−50. Available from: https://doi.org/10.1002/jor.1100090504.

[43] Caplan AI. What's in a name? Tissue Eng Part A 2010;16(8):2415−17. Available from: https://doi.org/10.1089/ten.tea.2010.0216.

[44] Caplan AI. Mesenchymal stem cells: time to change the name!. Stem Cells Transl Med 2017;6(6):1445−51. Available from: https://doi.org/10.1002/sctm.17-0051.

Index

Printed in the United States
By Bookmasters